Stress – From Molecules to Behavior

Edited by
Hermona Soreq, Alon Friedman,
and Daniela Kaufer

Stress – From Molecules to Behavior

A Comprehensive Analysis of the Neurobiology
of Stress Responses

Edited by
Hermona Soreq, Alon Friedman,
and Daniela Kaufer

WILEY-
BLACKWELL

WILEY-VCH Verlag GmbH & Co. KGaA

The Editors

Prof. Dr. Hermona Soreq
The Hebrew University of Jerusalem
Department of Biological Chemistry
Safra Campus – Givat Ram
Jerusalem 91904
Israel

Prof. Dr. Alon Friedman
Ben Gurion University of the Negev
Departments of Physiology and Neurosurgery
Zlotowksi Centre for Neuroscience
Beer Sheva 84105
Israel

Prof. Dr. Daniela Kaufer
University of California, Berkeley
Department of Integrative Biology
3060 Valley Life Sciences Bldg. 3140
Berkeley, CA 94720-3140
USA

Library of Congress Card No.: applied for

British Library Cataloguing-in-Publication Data
A catalogue record for this book is available from the British Library.

Bibliographic information published by the Deutsche Nationalbibliothek
The Deutsche Nationalbibliothek lists this publication in the Deutsche Nationalbibliografie; detailed bibliographic data are available on the Internet at http://dnb.d-nb.de

© 2010 WILEY-VCH Verlag GmbH & Co. KGaA, Weinheim

Printed in the Federal Republic of Germany
Printed on acid-free paper

Cover Design   Adam Design, Weinheim
Typesetting   SNP Best-set Typesetter Ltd., Hong Kong
Printing   betz-druck GmbH, Darmstadt
Bookbinding   Litges & Dopf Buchbinderei GmbH, Heppenheim

ISBN: 978-3-527-32374-6

Contents

Stress – From Molecules to Behavior. Edited by Hermona Soreq, Alon Friedman, and Daniela Kaufer
Copyright © 2010 WILEY-VCH Verlag GmbH & Co. KGaA, Weinheim
ISBN: 978-3-527-32374-6

Preface

This book is entitled "Stress: from Molecules to Behavior", which immediately highlights the complexity of the topic and the difficulty of assembling a decent representation of its various features and perspectives. To refine what we tried to cover in this volume, we added a subtitle statement attempting to present a comprehensive analysis of the neurobiology of stress responses, which in retrospect was even more presumptuous. One year later, we obviously realize that we aimed at the impossible; however, thanks to the efforts of those friends and colleagues who agreed to contribute chapters to this volume, we also learned that the assembled collection of chapters does provide a multi-leveled view of stress, substantially adding to our knowledge on this subject, while emphasizing to the readers the immense and intricate scope of the topic of stress.

Trying the impossible, in turn, brings to mind the 17th century poet Margaret Cavendish, who implemented conceptualization of mind through mathematics and is often called "the first poetess of science". One of her famous poems portrays the brain as a circle and states, as a metaphor, that *The Circle of the Brain Cannot be Squared*. Mirrored in poetic language, showing how mathematics in the 17th century influenced people's view of reality, Cavendish also explains in this succinct and powerful poem that man's attempt to take control over irrational nature will never cease; which is compatible with our ambitious effort.

The Circle of the Brain Cannot be Squared

A Circle round divided in four parts
Hath been great Study 'mongst the men of Arts;
Since Archimed's or Euclid's time, each Brain
Hath on a Line been stretched, yet all in Vain;
And every Thought hath been a Figure set,
Doubts Cyphers were, Hopes as Triangles met;
There was Division and Subtraction made,
And Lines drawn out, and Points exactly laid,
But none hath yet by Demonstration found
The way, by which to Square a Circle round:
For while the Brain is round, no Square will be,

Stress – From Molecules to Behavior. Edited by Hermona Soreq, Alon Friedman, and Daniela Kaufer
Copyright © 2010 WILEY-VCH Verlag GmbH & Co. KGaA, Weinheim
ISBN: 978-3-527-32374-6

While Thoughts divide, no Figures will agree.
And others did upon the same account,
Doubling the Cube to a great Number mount;
But some the Triangles did cut so small,
Till into equal Atoms they did fall:
For such is Man's curiosity and mind,
To seek for that, which is hardest to find.

The different chapters comprising this book are loosely grouped into more general stress-related topics. We open with the topic of *Systems*, encompassing studies on the role of stress in evolution by Lilach Hadany, on catecholamines and stress by Esther L. Sabban and on stress and the cholinergic system by Mariella De Biasi. These are clearly not the only systems that are relevant yet, when reading these chapters one gets a glimpse of the systems which might contribute to stress reactions and their significance. We then proceed with the topic of *Cells and circuits*, under which we collected diverse subjects such as the effects of stress on the function of hippocampal cells by Marian Joëls and Henk Karst, the "burning" issue of stress and adult neurogenesis in the mammalian central nervous system by Elizabeth D. Kirby and Daniela Kaufer and the state-of-the-art neurogenetic search for individual differences in reactivity to social stress in the laboratory and its mediation by common genetic polymorphisms by Richard P. Ebstein and co-workers.

Next, we moved on to *Cognition and behavior*, a highly complex topic by its own merit in which recent studies show causal links to specific molecules and cognitive functions. Ronald de Kloet and colleagues open this section with a coverage of corticosteroid hormones in stress and anxiety, as it is reflected in the role of particular receptor variants and environmental inputs. Thomas Blank and Joachim Spiess proceed with corticotropin-releasing factor (CRF) and CRF-related peptides, which together present a linkage between stress and anxiety. Sonia J. Lupien and co-authors then deal with the intriguing subject of stress, emotion and memory. Marta Weinstock closes this topic with the recently much-discussed contribution of early life stress to anxiety disorders.

The immune system and inflammatory responses are emerging as central components of stress reactions. In *Immune responses*, these are covered by Michael Shapira, who describes stress effects on immunity in vertebrates and invertebrates; by Michal Schwartz and co-workers, who present the concept of immunity to self as a mechanism to maintain resistance to mental stress and argue that boosting immunity can serve as a complement to psychological therapy; and by Inbal Goshen and Raz Yirmiya, whose chapter on brain interleukin-1 (IL-1) describes this pro-inflammatory cytokine as a mediator of stress-induced alterations in the activation of the hippocampal–pituitary–adrenal (HPA) axis, as well as in memory functioning and neural plasticity.

Last, but not least, we turn to studies of stress-related diseases in animal models and human patients. Hagit Cohen leads this section with a comprehensive coverage of post-traumatic stress disorder (PTSD) in animal models, then Alon

Friedman and Lev Pavlovsky proceed with the cholinergic model for PTSD as it progresses from acute stress and as it extends from a single neuron to functional networks and altered behavior. Stress-related vulnerability to other diseases is also represented, although the presented examples should be viewed as a sample of the very broad and diverse field rather than as in-depth coverage. The link between stress and neurodegeneration is covered by Amit Berson and co-workers, and a glimpse into the relevant clinical evidence and therapeutic implications of the inter-relationship between stress and neurotransmitter systems is provided by Hadar Shalev and Jonathan Cohen. The pertinent topic of the metabolic aspects of neuro-endocrine allostatic responses as it implicates on lifestyle-related diseases is also presented from the clinician's view point by Ronan M.G. Berg and Bente Klarlund Pedersen. The final chapter by Michal Horowitz and Esther Shohami, on environmental stress, bears a positive message by covering the added values of heat acclimation-mediated neuro-protection from traumatic brain injury.

Together, these chapters reflect a dynamic, evolving field of research of tremendous dimensions and spheres and possessing clinical and social implications that are of interest to many. We wish to thank all of the contributors to this volume, accept the blame for any errors or exclusion of important topics and promise our readers that there is much more to stress than is presented in this book.

Beer Sheva, Berkeley, Jerusalem
October 2009

Alon Friedman
Daniela Kaufer
Hermona Soreq

List of Contributors

Ronan M.G. Berg
University of Copenhagen
Centre of Inflammation and
Metabolism
Department of Infectious Diseases
and
University Hospital Rigshospitalet
Copenhagen Muscle Research
Centre
Blegdamsvej 9
2100 Copenhagen
Denmark

Amit Berson
The Hebrew University of
Jerusalem
Department of Biological Chemistry
Safra Campus – Givat Ram
Jerusalem 91904
Israel

Thomas Blank
University of Hawaii
Biosciences Building
John A. Burns School of Medicine
651 Ilalo Street
Honolulu, HI 96813
USA

Hagit Cohen
Ben-Gurion University of the Negev
Beer-Sheva Mental Health Center
Anxiety and Stress Research Unit
2 HaTzadik M'Yerushalyim
Beer-Sheva 84170
Israel

Jonathan Cohen
Ben-Gurion University of the Negev
Department of Physiology
Faculty of Health Sciences
P.O. Box 653
Beer Sheva 84105
Israel

Mariella De Biasi
Baylor College of Medicine
Department of Neuroscience
One Baylor Plaza
Houston, TX 77030
USA

E. Ronald de Kloet
Leiden University
Division of Medical Pharmacology
LACDR/LUMC
Gorlaeus Laboratories
P.O. Box 9502
2300 RA Leiden
The Netherlands

Stress – From Molecules to Behavior. Edited by Hermona Soreq, Alon Friedman, and Daniela Kaufer
Copyright © 2010 WILEY-VCH Verlag GmbH & Co. KGaA, Weinheim
ISBN: 978-3-527-32374-6

Roel H. DeRijk
Leiden University
Division of Medical Pharmacology
LACDR/LUMC
Gorlaeus Laboratories
P.O. Box 9502
2300 RA Leiden
The Netherlands

Richard P. Ebstein
The Hebrew University of
Jerusalem
Department of Psychology
Jerusalem 91905
Israel
and
S. Herzog Memorial Hospital
Givat Shaul
Jerusalem 91905
Israel

Alon Friedman
Ben-Gurion University of the Negev
Departments of Physiology and
Neurosurgery
Zlotowski Center for Neuroscience
Beer-Sheva 84105
Israel

Inbal Goshen
The Hebrew University of
Jerusalem
Department of Psychology
Mount Scopus
Jerusalem 91905
Israel

Inga Gritsenko
S. Herzog Memorial Hospital
Givat Shaul
Jerusalem 91905
Israel

Lilach Hadany
Tel Aviv University
Department of Plant Sciences
Faculty of Life Sciences
Tel Aviv 69978
Israel

Mor Hanan
The Hebrew University of
Jerusalem
Department of Biological
Chemistry
Safra Campus – Givat Ram
Jerusalem 91904
Israel

Michal Horowitz
The Hebrew University of
Jerusalem
Faculty of Dental Medicine
Laboratory of Environmental
Physiology
Jerusalem 91120
Israel

Salomon Israel
The Hebrew University of
Jerusalem
Department of Psychology
Mount Scopus
Jerusalem 91905
Israel

Marian Joëls
University Medical Center Utrecht
Department of Neuroscience and
Pharmacology
Universiteitsweg 100
3584 CG Utrecht
The Netherlands

Marsha Kaitz
The Hebrew University of
Jerusalem
Department of Psychology
Jerusalem 91905
Israel

Henk Karst
University Medical Center Utrecht
Department of Neuroscience and
Pharmacology
Universiteitsweg 100
3584 CG Utrecht
The Netherlands

Daniela Kaufer
University of California, Berkeley
Department of Integrative Biology
3060 Valley Life Sciences Bldg. 3140
Berkeley, CA 94720
USA

Elizabeth D. Kirby
University of California, Berkeley
Department of Integrative Biology
Helen Wills Neuroscience Institute
3060 Valley Life Sciences Bldg. 3140
Berkeley, CA 94720
USA

Efthimia Kitraki
University of Athens
School of Dentistry
Athens, 11527
Greece

Nitzan Kozlovsky
Ben-Gurion University of the Negev
Beer-Sheva Mental Health Center
Anxiety and Stress Research Unit
2 HaTzadik M'Yerushalyim
Beer-Sheva 84170
Israel

Elad Lerer
The Hebrew University of
Jerusalem
Human Genetics
Jerusalem 91905
Israel

Gil M. Lewitus
The Weizmann Institute of
Science
Department of Neurobiology
Rehovot 76100
Israel

Sonia J. Lupien
Fernand Seguin Mental Health
Research Centre
Louis H. Lafontaine Hospital
Centre for Studies on Human
Stress
7401 Hochelaga
Montreal, Quebec H1N 3M5
Canada

Françoise S. Maheu
CHU Sainte-Justine
Centre de Recherche
3175 Côte-Sainte-Catherine
Montreal, Quebec H3T 1C5
Canada

David Mankuta
Hadassah Medical Organization
Department of Labor and Delivery
Jerusalem 91905
Israel

Marie-France Marin
Fernand Seguin Mental Health
Research Centre
Louis H. Lafontaine Hospital
Centre for Studies on Human
Stress
7401 Hochelaga
Montreal, Quebec H1N 3M5
Canada

Lev Pavlovsky
Beilinson Campus
Department of Dermatology
Rabin Medical Center
Petah Tikva 49414
Israel

Bente Klarlund Pedersen
University of Copenhagen
Centre of Inflammation and
Metabolism
Department of Infectious Diseases
and
University Hospital Rigshospitalet
Copenhagen Muscle Research
Centre
Blegdamsvej 9
2100 Copenhagen
Denmark

Gal Richter-Levin
University of Haifa
The Institute for the Study of
Affective Neuroscience (ISAN)
Department of Psychology &
Department of Neurobiology and
Ethology
Haifa 31905
Israel

Esther L. Sabban
New York Medical College
Department of Biochemistry and
Molecular Biology
Basic Sciences Building
Valhalla, NY 10595
USA

Tania Elaine Schramek
Fernand Seguin Mental Health
Research Centre
Louis H. Lafontaine Hospital
Centre for Studies on Human
Stress
7401 Hochelaga
Montreal, Quebec H1N 3M5
Canada

Michal Schwartz
The Weizmann Institute of
Science
Department of Neurobiology
Rehovot 76100
Israel

Osnat Schwartz-Stav
Schneider's Children's Medical
Centre of Israel
Feinberg Child Study Centre
14 Kaplan St.
Petah Tikva 49202
Israel

Hadar Shalev
Soroka University Medical Center
Trauma Clinic
Psychiatric Department
Beer Sheva 84105
Israel

Idan Shalev
The Hebrew University of
Jerusalem
Department of Neurobiology
Mount Scopus
Jerusalem 91905
Israel

Michael Shapira
University of California, Berkeley
Department of Integrative Biology
3060 Valley Life Sciences Bldg. 3140
Berkeley, CA 94720
USA

Esther Shohami
The Hebrew University of
Jerusalem
Department of Pharmacology
School of Pharmacy
Jerusalem 91120
Israel

Hermona Soreq
The Hebrew University of
Jerusalem
Department of Biological Chemistry
Safra Campus – Givat Ram
Jerusalem 91904
Israel

Joachim Spiess
University of Hawaii
John A. Burns School of Medicine
Biosciences Building
651 Ilalo Street
Honolulu, HI 96813
USA

Florina Uzefovsky
The Hebrew University of
Jerusalem
Department of Psychology
Mount Scopus
Jerusalem 91905
Israel

Marta Weinstock
The Hebrew University of
Jerusalem
Department of Pharmacology
Ein Kerem
Jerusalem 91120
Israel

Raz Yirmiya
The Hebrew University of
Jerusalem
Department of Psychology
Mount Scopus
Jerusalem 91905
Israel

Joseph Zohar
Tel Aviv University
The Chaim Sheba Medical Center
Division of Psychiatry
Tel Hashomer 52621
Israel

Part I
Systems

1

On the Role of Stress in Evolution

Lilach Hadany

1.1
Introduction

Most of the work on stress considers its mechanisms and effects during the life-time of the stressed individual. In this chapter, we concentrate on the possible effects of stress on genes and populations. In particular, we consider the effects of the information contained in the stress response, especially in chronic stress: namely, that the stressed individual is maladapted to its current environment.

In addition to the physiological responses induced by stress, it can also have genomic responses. One type of response which is of particular interest is an increase in genetic variation, especially the mixing of different genotypes through recombination, sex and outcrossing. Here, we consider the evolution of such a genomic response and its possible implications for the long-term success of the population and for the evolution of complex traits.

When considering the evolution of a genomic response to stress, we can take one of two approaches: we can either consider what would be the "best" response at the level of the population (i.e., the response that would, on average, maximize the average fitness of the population), or consider the fate of a selectively neutral modifier allele [1] inducing the genetic response–would such an allele increase in frequency within the population, due to the forces of natural selection?

1.2
Stress Through the Gene's Eye: the Evolution of Stress-Induced Genetic Mixing

Let us consider the point of view of a gene that regulates genetic mixing–for example, recombining with a different genotype–in response to stress. This gene affects its own probability of moving to a different genetic background in the next generation. When would it be advantageous (at the level of the gene) to move to a different background? The answer depends on the quality of the current genetic background. If the current background is maladapted to the current environment

Stress – From Molecules to Behavior. Edited by Hermona Soreq, Alon Friedman, and Daniela Kaufer
Copyright © 2010 WILEY-VCH Verlag GmbH & Co. KGaA, Weinheim
ISBN: 978-3-527-32374-6

(e.g., includes multiple deleterious alleles), there is a much greater advantage in "taking the risk" of moving to a different, unknown, background. But how can the gene "know" the quality of the whole genome? One crucial source of information can be stress responses, which relay information about the well-being of the whole organism down to the molecular level. An individual carrying an unfit genome is more likely to be stressed, and the stress responses it experiences can affect the gene regulating genetic mixing. As a result, an increase in mixing is more likely to occur in the presence of stress.

An increase in genetic mixing can occur through various mechanisms, acting at different levels. Each of these mechanisms carries its own costs and benefits. Below we specifically discuss four of these mechanisms – recombination, sex, outcrossing and dispersal.

1.2.1
Stress-Induced Recombination

1.2.1.1 Classic Models of the Evolution of Recombination
The evolution of recombination has been the subject of scientific debate for over 70 years, see [2–7] for reviews. One major problem is that uniform recombination not only generates new advantageous combinations, but also breaks down existing good ones that were generated by selection [8, 9].

Models concentrating on population-level effects show that recombination might be advantageous only under limited conditions [10, 11]. Specifically, recombination can only be advantageous when associations between different loci in the genome result in decreased variation within the population (a situation termed negative linkage disequilibrium). Such associations can be generated by drift [12], synergistic epistasis [13, 14], or environmental changes [15–17]. In these cases recombination reduces linkage disequilibrium and leads to increased average fitness, resulting in a long-term advantage for the population as a whole.

The same question was also studied using modifier models, concentrating on the short-term dynamics of an allele affecting the rate of recombination. These models found that, in the absence of deleterious mutations or environmental changes, a recombination modifier tends to increase from rarity only if it *reduces* the recombination rate between selected loci [18], a result known as "the reduction principle". However, negative epistasis between deleterious mutations [19, 20], drift [21], or adaptation [22, 23], including inter-species interactions [24, 25], can explain the evolution of recombination modifiers under some circumstances. Nevertheless, none of these models fully accounts for the wide abundance of sex and recombination among higher eukaryotes.

1.2.1.2 The Evolution of Stress-Induced Recombination
When introducing the possibility of stress-induced recombination, radically different results are obtained. Let us first consider the modifier approach: a modifier regulating the level of recombination in a haploid organism according to the state

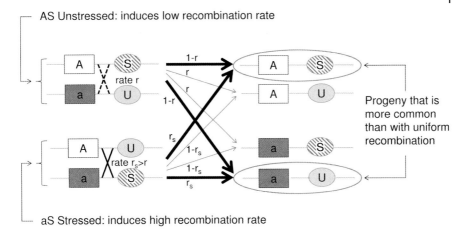

Figure 1.1 The "abandon ship" principle: intuition for fitness-associated mixing. An illustration of the short-term advantage of stress-induced recombination in a two-locus haploid model. The alleles: A—high fitness, a—low fitness, S—stress induced recombination, U—uniform recombination. The recombination products of the two double heterozygotes are shown, where the likelier-than-random changes are shown in bold arrows and the less-likely than random ones in gray arrows. We can see that if the allele S (for stress-induced recombination) induces a higher recombination rate than random when it is linked to the low fitness allele a and/or a lower recombination rate than random when it is linked to the high fitness allele A, then an association is generated between A and S, resulting in a short-term evolutionary advantage for S. A similar argument holds for mechanisms of genetic mixing other than recombination. (Adapted with permission from [2]).

of the individual—with stressed individuals recombining at a higher rate—can evolve under a very wide parameter range from any population where the rate of recombination is uniform and independent of stress [28]. This result was obtained for an analytical model with two loci (see Figure 1.1 for intuition), generalized using stochastic models with long genomes and finite populations and further confirmed using a QLE approach [26]. It means that increased recombination as a result of stress is expected to be a common pattern in the world. Empirical evidence strongly supports this prediction (see below).

What would be the effects of such a pattern of recombination on the population as a whole? Let us consider the effect of recombination in general and of stress-induced recombination in particular on the distribution of different allele combinations. In the absence of epistatic interactions and drift, uniform recombination does not alter the distribution of allele combinations from the one expected had alleles at different loci been entirely independent of each other. This situation is called linkage equilibrium, and the ineffectiveness of uniform recombination in that situation led many researchers to argue that recombination is advantageous only when the population is at linkage disequilibrium due to other forces acting on it [27]. However, this result changes when we consider recombination that is stress-induced. Such plasticity of the recombination response can allow it to generate good genetic combinations between alleles at other loci more than to break them down, thus generating "extreme" genotypes more often than random (a situation called positive linkage disequilibrium) and increasing variation in fitness within the population. Variation in fitness is the substrate on which selection can work—resulting in increased average fitness in the population. As a result, even under the conditions which are least favorable for the evolution of uniform recombination (i.e., an infinite population, no epistasis, linkage equilibrium), stress-induced recombination can still result in an increased average fitness [28].

1.2.1.3 Evidence for Stress-Induced Recombination

Evidence for stress-induced recombination has been found in a wide range of organisms and stressors. This includes elevated levels of recombination in response to DNA damage [29–33], starvation [34–41], temperature stress [42–46] and behavioral competition [47]. Further supportive evidence comes from studies which found an association between low fitness and recombination within the same population [48, 49]. See Refs. [5, 50] for reviews.

1.2.2
Stress and Sex

For recombination to be effective in the sense of mixing genetic material from different individuals, sexual reproduction needs to take place. However, sexual reproduction is one of the greatest puzzles in evolutionary theory: in addition to the difficulties associated with explaining recombination itself (mentioned above), sex is a very *costly* mechanism. First, it requires the generation of male offspring, thus resulting in the "twofold cost of males": An asexual mutant female would have only female offspring, identical to herself, producing twice as many copies of its genes in comparison with a similar sexual female. This advantage would be doubled in the next generation and—everything else being equal—would lead to a fast takeover of the sexual population by asexual mutants [51]. Furthermore, a sexual organism has to invest time and energy in courtship and mating and risks transmission of pathogens and conflicts between the sexes. Altogether, the wide abundance of sexual reproduction in the world strongly suggests that sexual organisms have significant advantages over asexual ones in both the short term and the long term, but these advantages are not yet fully understood.

We can study the evolution of sexual reproduction by considering the case of facultative sexuals—organisms that can reproduce either sexually or asexually—and the evolution of a gene that alters the investment in sex in such a system. When can a gene that results in increased investment in costly sexual reproduction spread in the population? What would be the long-term effects for the survival of such a population when it has to compete with asexual populations? This is a harder question than the question of recombination alone, as the cost of sex favors decreased investment in sexual reproduction.

If we consider facultative sexuals that have the same tendency to reproduce sexually at all times, then the evolution of sex is rather hard to explain. Indeed, classic models require rather extreme parameters in order to account for the evolution of sex [3]. Such extreme conditions do exist in certain systems, but do not seem to explain the ubiquity of sexual reproduction. As in the case of stress-induced recombination, the solution may lie in the fact that facultative sexuals, including viruses, protists, fungi and plants, are more likely to have sex when their own condition is poor—when they are stressed [31, 36, 37, 52–62]; see [63] for a review.

When we take stress into account, the picture is rather different. Reproducing sexually more often when the individual is stressed allows the allele regulating sexual reproduction to "abandon ship" and move to a different genetic background. But at what cost? Interestingly, even in the face of high cost, an allele promoting sexual reproduction under stress can evolve [64]. An intuition for that can be gained by considering the fate of this allele if it stays in a maladapted genome. It is almost surely doomed, together with its "neighbor genes", unless a mutation or a change of the environment happens first. Thus, from the allele's viewpoint, it is worth "paying" a significant cost to break away.

At first sight, it might seem that regardless of its association with stress the costs involved in sexual reproduction should be harmful for the population as a whole. However, taking stress into account alters the population-level effects as well. If the stressed individuals—the ones that are, on average, less fit in the current environment—are the ones most likely to reproduce sexually *and pay the cost of sex*, then they are less likely to transmit their genes to the next generation. As a result, the maladapted alleles they carry are more likely to be eliminated. In such a case the cost of sex becomes a component of natural selection and can actually be beneficial in the long term [64].

1.2.3
Stress and Outcrossing

It might seem that obligatory sexual organisms are much more limited in their ability to regulate genetic mixing. However, any mechanism that affects the probability of mating with a genetically dissimilar individual is in fact a mechanism that regulates the likelihood of genetic mixing. This includes many mate-preference mechanisms that are sensitive to familial relatedness (e.g., inbreeding avoidance) or physical similarity (e.g., opposites attract) and indirect mechanisms that increase the probability of meeting unrelated individuals—and thus also the probability of mating with them. In other words, all sorts of outcrossing are potential mechanisms for regulating genetic mixing.

The evolution of outcrossing is a riddle resembling the evolution of sex: in populations of self-compatible hermaphrodites, alleles for self-fertilization have an inherent advantage over alleles for outcrossing, as they help the spread of alleles that are identical by descent [65–67]. However, the effect of inbreeding is not identical to that of asexual reproduction—selfed offspring, in contrast with asexual ones, are not genetically identical to their parents. In particular, they might suffer from inbreeding depression: reduced fitness due to homozygotization of deleterious mutations.

However, the level of inbreeding depression is unlikely to be the same for all the individuals within the population. The selfed offspring of individuals carrying many deleterious mutations are expected to experience stronger inbreeding depression in comparison with the offspring of individuals carrying fewer deleterious mutations. As a result, we would expect stressed individuals to have a higher tendency for outcrossing. The effects of such plasticity on the entire population are less obvious: parallel to fitness-associated recombination [28] it could increase the average fitness of the population by generating positive linkage disequilibrium. Additionally, the costs of outcrossing would become part of natural selection, further benefitting the population as a whole. In contrast, outcrossing of individuals that carry more deleterious mutations would also result in increased heterozygosity in the population. This would lead to decreased selection on recessive deleterious alleles, resulting in a short-term disadvantage but also in increased potential for adaptation when the environment changes.

1.2.4
Stress and Dispersal

Dispersal combines several different effects. First, the dispersing individual increases its probability to mate with individuals that are unrelated to it. As a result, a gene regulating the probability of dispersal regulates its chance to change its genetic background. Second, dispersal increases the probability of the disperser to change its biotic and abiotic environment, including intra-species competition. Last, dispersal often carries costs of its own (e.g., the probability to die on the way, not to find a new territory, to lose time and energy) in addition to the inherent cost of outcrossing mentioned above.

The evolution of dispersal has classically been explained by inbreeding avoidance [68–70], kin competition [71, 72], habitat variability [73–75], or competition for resources [76, 77], including potential mates [78].

In most cases, the potential benefits for the dispersing individual are not the same for all individuals. As in previous examples, less-fit individuals have more to gain from changing their environment, the genes of their offspring, or both. In many species, there is evidence that less-fit individuals indeed disperse further than fitter ones [79–87]. We propose (L. Hadany *et al.*, in prep.) that fitness-associated dispersal could have evolved due to the advantages of stress-induced mixing, even in the absence of direct competition or environmental heterogeneity. We further suggest that the advantage of stress-induced dispersal may increase in a heterogeneous environment, where dispersal has a chance to change both the environment of the dispersing individual and the genetic background of the modifier allele affecting dispersal.

Dispersal has many ecological implications and is a particularly important factor for the survival of metapopulations. The effects of dispersal act in two opposite directions: on one hand, dispersal is necessary for colonization in empty patches [88] and can "rescue" sink populations from extinction [89–91]; on the other hand, dispersal can impede the stability of already colonized subpopulations [92–94]. The regulation of dispersal by stress could change the balance between these opposing effects and therefore may have important ecological implications that have not yet been explored.

1.3
The Effect of Stress-Induced Variation on the Evolvability of Complex Traits

The principle of stress-induced variation also offers a novel solution to the classic "peak shift" problem in population genetics [95]: For traits that are affected by multiple loci, there might exist allelic combinations that represent "local peaks" in the adaptive landscape (the plot of the fitness as a function of the underlying genotype, see Figure 1.2). When starting from such a combination, any single mutation would result in decreased fitness, but multiple mutations can lead to increased fitness. How can organisms evolve from one such co-adapted gene complex to a better one if the intermediate genotypes are less fit?

This question has been studied extensively, and most of the models that try to answer it assume that the population is divided into groups ("demes"), with a limited rate of migration between these groups. Under such conditions, rare

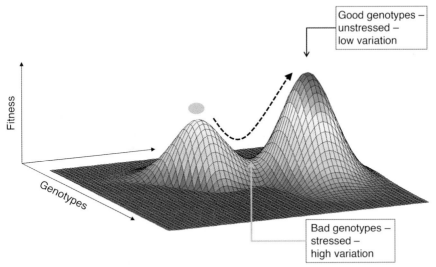

Figure 1.2 The problem of complex adaptation. The adaptive landscape is the plot of the fitness as a function of the genotype. The problem, first described by Wright [96] is how a species can evolve from one "adaptive peak" to a higher one, crossing a less-fit "valley". Increased variation under stress (when fitness is low) and/or decreased variation when not stressed (when fitness is high) can facilitate adaptation in such genetic systems.

advantageous combinations can appear and fix in one deme with the aid of random genetic drift and can later expand to other demes [95, 96]. However, these models typically require specific parameter settings, particularly a limited range of dispersal rates [97–100]. This has to do with the double-sided effect of genetic mixing on adaptive peak shifts: on one hand, mixing facilitates the appearance of rare advantageous combinations; on the other hand, recombination with different genotypes acts to break down these very combinations when rare, resulting in offspring within the "adaptive valley".

Stress-induced variation largely relieves that limitation. The effect has two aspects: more mixing when fitness is low increases the generation of new advantageous gene combinations by maladapted individuals in the "adaptive valley", whereas decreased mixing when fitness is high protects rare advantageous combinations when they appear. This basic principle works both for recombination [28] and for dispersal [101]. Stress-induced variation can thus facilitate the ability of a population to move between adaptive peaks, potentially allowing evolutionary breakthroughs and significant advantages in the very long term.

1.4
Stress-Induced Variation and Pathogen Evolution

Stress-induced variation has some ominous implications when it comes to the evolution of pathogens. One of these is its effect on the evolution of pathogens. Pathogens are constantly evolving to overcome their hosts' defenses, including medications. Parallel to other organisms, evidence suggests that the tendency of pathogens to produce genetic variation is affected by the stress they experience [102, 103]. Whenever man-made drugs become a major source of pathogen stress, our use of such drugs should be much more carefully considered. In particular, recent results suggest that massive use of drugs (including prophylactic and veterinarian use) can facilitate pathogen adaptation to these drugs not only through direct selection, but also through induction of increased generation of variation, resulting in a much more dramatic effect than previously thought [104]. This is particularly problematic when complex pathogen evolution (including multiple mutations, possibly with epistatic interactions between them) is required for drug resistance (V. Leontiev and L. Hadany, submitted).

1.5
Stress-Induced Mortality

This section discusses the most counter-intuitive effect of stress on evolution: the evolution of death. Normally, we do not consider early death as an evolved feature, but rather as an "accident", or an undesirable by-product of a trait with selective advantages in other contexts. A classic example is the increase in blood pressure associated with the "flight or fight" response – a response with obvious selective advantages in the short term, but disadvantages in the long term [105]. While some of the harmful effects of long-term exposure to stress can be easily explained in that way, others—such as the down-regulation of the immune response and the apoptosis of lymphocytes [106, 107]—are less so. Here, we suggest an alternative explanation for these harmful effects, without requiring a short-term benefit.

Let us start by assuming a structured population, where neighbors are likely to be more genetically related than randomly chosen individuals and where resources like space or food limit the number of individuals inhabiting a given area. Consider a gene that affects mortality in such a population. Under what circumstances would an allele whose only effect is to increase mortality (without any benefit to the individual) spread in the population? Stress can have a dramatic effect on the answer. An allele that occurs in a chronically stressed individual has very poor prospects: it is unlikely to be passed to the next generation, and in the rare case that it does, it would often still be linked to the deleterious genetic background that caused the stress to begin with. However, if the individual carrying this allele dies, then another individual—with a higher probability to carry the allele than a random individual—would take its place and would have better chances of passing its genes to the next generation. By that, other copies of the allele (possibly having a better genetic background) would have better chances of passing. Thus, if the condition of the first individual is poor enough, the increased mortality allele can do better by inducing the death of the individual carrying it than by helping it to survive. Indeed, such an allele can increase from rarity and fix in the population under the conditions described above [108].

We can regard the individual dying from stress under such circumstances as the ultimate altruist—it gives its life for the benefit of its neighbors and kin. This is advantageous only when the self-sacrificing individual has some information (in the form of stress) that it is not doing as well as others in this environment. This argument gives us some intuition about which kinds of stress are more likely to induce a harmful effect: social stress (e.g., losing a fight) is more indicative about the relative fitness of the individual than, for example, abiotic stress (e.g., heat). The former is thus more likely to result in evolved harmful effects.

What would be the effects of such an allele for stress-induced mortality at the level of the population as a whole? If such an allele takes over the population, it would lead to more effective natural selection. Deleterious mutations would be eliminated more efficiently (as their carriers would more often tend to die prematurely) and beneficial alleles would have a greater relative advantage (as their carriers do not tend to die prematurely). Altogether it would result in an increase in the average fitness of the population in the long term.

Of course, we can think of less drastic ways to help others which carry a cost for the helping individual. One way would be to decrease the use of the common resource of potential mates: indeed, stress-induced reduction in fertility is a well-documented phenomenon [109–111], which could have evolved along very similar lines. The results described above [108] suggest that many other kinds of altruistic behaviors are also likely to evolve to increase in times of stress.

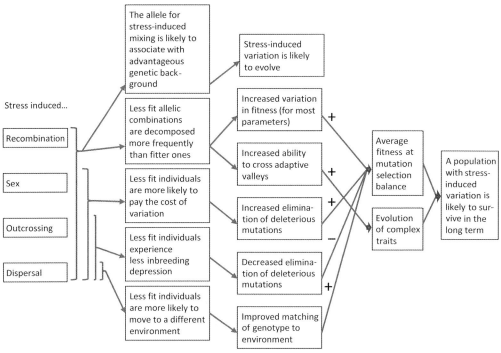

Figure 1.3 Key evolutionary implications of stress through four mechanisms of stress-induced variation: recombination, sex, outcrossing and dispersal.

Summary

In this chapter we reviewed some of the evolutionary implications of stress responses. We have seen that stress might have played an important role in the evolution of mechanisms of genetic mixing, shedding new light on the evolution of recombination, sex, outcrossing and dispersal. Stress-induced genetic variation can alter the evolution of species, both in the short term and in the long term – by allowing them to better adapt to their environments, especially when complex adaptation is required. Figure 1.3 summarizes some of the implications of stress-induced variation.

The evolutionary effects of stress can have significant ecological implications: from improved survival of metapopulations to more effective evolution of pathogens exposed to drugs. Evolutionary considerations might also suggest a new perspective on other possible responses to stress, including early death, reduced fertility, or altruistic behavior.

Acknowledgments

I wish to thank Tuvik Beker for many helpful comments on the manuscript. This work was supported in part by NSF grant 0639990.

References

1 Nei, M. (1967) Modification of linkage intensity by natural selection. *Genetics*, **57**, 625–642.

2 Hadany, L. and Comeron, J.M. (2008) Why are sex and recombination so common? *Ann. N. Y. Acad. Sci.*, **1133** (1), 26–43.

3 Michod, R.E. and Levin, B.R. (1988) *The Evolution of Sex*, Sinauer Press, Sunderland.

4 Kondrashov, A.S. (1993) Classification of hypotheses on the advantage of amphimixis. *J. Hered.*, **84** (5), 372–387.

5 Korol, A.B., Preygel, I.A. *et al.* (1994) *Recombination Variability and Evolution*, Chapman & Hall, London.

6 Barton, N.C. and B. (1998) Why sex and recombination? *Science*, **281**, 1986–1990.

7 Rice, W.R. (2002) Experimental tests of the adaptive significance of sexual recombination. *Nat Rev. Genet.*, **3** (4), 241–251.

8 Eshel, I. and Feldman, M.W. (1970) On the evolutionary effect of recombination. *Theor. Popul. Biol.*, **1** (1), 88–100.

9 Otto, S.P. and Lenormand, T. (2002) Resolving the paradox of sex and recombination. *Nat. Rev. Genet.*, **3** (4), 252–261.

10 Kimura, M. (1956) A model of a genetic system which leads to closer linkage by natural selection. *Evolution*, **10** (3), 278–287.

11 Lewontin, R.C. (1971) The effect of genetic linkage on the mean fitness of a population. *Proc. Natl. Acad. Sci. U.S.A.*, **68** (5), 984–986.

12 Muller, H.J. (1964) The relation of recombination to mutational advance. *Mutat. Res.*, **106**, 2–9.

13 Kimura, M. and Maruyama, T. (1966) Mutational load with epistatic gene interactions in fitness. *Genetics*, **54**, 1303–1312.

14 Kondrashov, A.S. (1988) Deleterious mutations and the evolution of sexual reproduction. *Nature*, **336** (6198), 435–440.

15 Fisher, R.A. (1930) *The Genetical Theory of Natural Selection*, Clarendon Press, Oxford.

16 Crow, J.F. and Kimura, M. (1965) Evolution in sexual and asexual populations. *Am. Nat.*, **99**, 439–450.

17 Peck, J.R. (1994) A ruby in the rubbish: beneficial mutations, deleterious mutations and the evolution of sex. *Genetics*, **137**, 597–606.

18 Liberman, U. and Feldman, M.W. (1986) A general reduction principle for genetic modifiers of recombination. *Theor. Popul. Biol.*, **30** (3), 341–371.

19 Feldman, M.W., Christiansen, F.B. *et al.* (1980) Evolution of recombination in a constant environment. *Proc. Natl. Acad. Sci. U.S.A.*, **77** (8), 4838–4841.

20 Barton, N.H. (1995) A general-model for the evolution of recombination. *Genet. Res.*, **65** (2), 123–144.

21 Otto, S.P. and Barton, N.H. (2001) Selection for recombination in small populations. *Evolution*, **55**, 1921–1931.

22 Charlesworth, B. (1976) Recombination modification in a fluctuating environment. *Genetics*, **83** (1), 181–195.

23 Otto, S.P. and Michalakis, Y. (1998) The evolution of recombination in changing environments. *Trends Ecol. Evol.*, **13** (4), 145–151.

24 Hamilton, W.D., Axelrod, R. *et al.* (1990) Sexual reproduction as an adaptation to resist parasites (a review). *Proc. Natl. Acad. Sci. U.S.A.*, **87**: 3566–3573.

25 Howard, R.S. and Lively, C.M. (1994) Parasitism, mutation accumulation and the maintenance of sex. *Nature*, **367** (6463), 554–557.

26 Agrawal, A.F., Hadany, L. *et al.* (2005)
The evolution of plastic recombination.
Genetics, **171** (2), 803–812.

27 Barton, N.H. and Charlesworth, B.
(1998) Why sex and recombination?
Science, **281** (5385), 1986–1990.

28 Hadany, L. and Beker, T. (2003)
On the evolutionary advantage of
fitness-associated recombination.
Genetics, **165** (4), 2167–2179.

29 Fabre, F. and Roman, H. (1977) Genetic
evidence for inducibility of recombination competence in yeast. *Proc. Natl.
Acad. Sci. U.S.A.*, **74** (4), 1667–1671.

30 Golub, E.I. and Low, K.B. (1983) Indirect
stimulation of genetic recombination.
Proc. Natl. Acad. Sci. U.S.A., **80** (5),
1401–1405.

31 Bernstein, C. (1987) Damage in DNA
of an infecting phage T4 shifts
reproduction from asexual to sexual
allowing rescue of its genes. *Genet. Res.*,
49 (3), 183–189.

32 Wojciechowski, M.F., Hoelzer, M.A.
et al. (1989) DNA repair and the
evolution of transformation in *Bacillus
subtilis*. II. Role of inducible repair.
Genetics, **121** (3), 411–422.

33 Kupiec, M. (2000) Damage-induced
recombination in the yeast
Saccharomyces cerevisiae. *Mutat. Res.*,
451 (1–2), 91–105.

34 Bergner, D.A. (1928) The relation
between larval nutrition and the
frequency of crossing over in the third
chromosome of *Drosophila Melanogaster*.
J. Exp. Zool., **50**, 107–163.

35 Neel, J.V. (1941) A relation between
larval nutrition and the frequency of
crossing over in the third chromosome
of *Drosophila melanogaster*. *Genetics*, **26**,
506–516.

36 Dubnau, D. (1991) Genetic competence
in *Bacillus subtilis*. *Microbiol. Rev.*, **55** (3),
395–424.

37 Redfield, R.J. (1993) Genes for breakfast:
the have-your-cake-and-eat-it-too of
bacterial transformation. *J. Hered.*, **84**
(5), 400–404.

38 Kon, N., Krawchuk, M.D. *et al.* (1997)
Transcription factor Mts1/Mts2
(Atf1/Pcr1, Gad7/Pcr1) activates the
M26 meiotic recombination hotspot in

*Schizosaccharomyces pombe. Proc. Natl.
Acad. Sci. U. S. A.*, **94** (25), 13765–13770.

39 Abdullah, M.F.F. and Borts, R.H. (2001)
Meiotic recombination frequencies are
affected by nutritional states in
Saccharomyces cerevisiae. *PNAS*, **98** (25),
14524–14529.

40 Davis, L. and Smith, G.R. (2001) Meiotic
recombination and chromosome
segregation in *Schizosaccharomyces
pombe*. *Proc. Natl. Acad. Sci. U.S.A.*,
98 (15), 8395–8402.

41 Jarmer, H., Berka, R. *et al.* (2002)
Transcriptome analysis documents
induced competence of *Bacillus subtilis*
during nitrogen limiting conditions.
FEMS Microbiol. Lett., **206** (2),
197–200.

42 Plough, H.H. (1917) The Effect of
Temperature on Linkage in the Second
Chromosome of *Drosophila*. *Proc. Natl.
Acad. Sci. U.S.A.*, **3** (9), 553–555.

43 Grell, R.F. (1978) A comparison of heat
and interchromosomal effects on
recombination and interference in
Drosophila melanogaster. *Genetics*, **89**,
65–77.

44 Tracey, M.L. and Dempsey, B. (1981)
Recombination rate variability in
Drosophila Melanogaster females
subjected to temperature stress.
J. Hered., **72**, 427–428.

45 Zhuchenko, A.A., Korol, A.B. *et al.*
(1986) The correlation between the
stability of the genotype and the change
in its recombination characteristics
under temperature influences. *Genetika*,
22, 966–974.

46 Parsons, P.A. (1988) Evolutionary rates:
effects of stress upon recombination.
Biol. J. Linn. Soc., **35**, 49–68.

47 Belyaev, D.K. and Borodin, P.M. (1982)
The influence of stress on variation and
its role in evolution. *Biol. Zentbl.*, **100**,
705–714.

48 Tucic, N., Ayala, F.J. *et al.* (1981)
Correlation between recombination
frequency and fitness in *Drosophila
melanogaster*. *Genet. Pol.*, **56**, 61–69.

49 Cvetkovic, D. and Tucic, N. (1986)
Female recombination rates and fitness
in *Drosophila melanogaster*. *Z. Zool. Syst.
Evolutionsforsch.*, **24** (3), 198–207.

50 Hoffman, A.A. and Parsons, P.A. (1991) *Evolutionary Genetics and Environmental Stress*, Oxford University Press, New York.

51 Maynard Smith, S.J. (1978) *The Evolution of Sex*, Cambridge University Press, Cambridge.

52 Bell, G. (1988) *Sex and Death in Protozoa*, Cambridge University Press, Cambridge.

53 Kassir, Y., Granot, D. *et al.* (1988) IME1, a positive regulator gene of meiosis in *S. cerevisiae*. *Cell*, **52** (6), 853–862.

54 Bernstein, C. and Johns, V. (1989) Sexual reproduction as a response to H_2O_2 damage in *Schizosaccharomyces pombe*. *J. Bacteriol.*, **171** (4), 1893–1897.

55 Harris, E.H. (1989) *The Chlamidomonas Sourcebook*, Academic Press, New York.

56 Kleiven, O.T., Larsson, P. *et al.* (1992) Sexual reproduction in daphnia-magna requires 3 stimuli. *Oikos*, **65** (2), 197–206.

57 Gemmill, A.W., Viney, M.E. *et al.* (1997) Host immune status determines sexuality in a parasitic nematode. *Evolution*, **51** (2), 393–401.

58 Mai, B. and Breeden, L. (2000) CLN1 and its repression by Xbp1 are important for efficient sporulation in budding yeast. *Mol. Cell. Biol.*, **20** (2), 478–487.

59 West, S.A., Gemmill, A.W. *et al.* (2001) Immune stress and facultative sex in a parasitic nematode. *J. Evol. Biol.*, **14** (2), 333–337.

60 Grishkan, I., Korol, A.B. *et al.* (2003) Ecological stress and sex evolution in soil microfungi. *Proc. Roy. Soc. Lond. B*, **270** (1510), 13–18.

61 Van Kleunen, M. and Fischer, M. (2003) Effects of four generations of density-dependent selection on life history traits and their plasticity in a clonally propagated plant. *J. Evol. Biol.*, **16** (3), 474–484.

62 Mitchell, S.E., Read, A.F. *et al.* (2004) The effect of a pathogen epidemic on the genetic structure and reproductive strategy of the crustacean *Daphnia magna*. *Ecol. Lett.*, **7** (9), 848–858.

63 Bell, G. (1982) *The Masterpiece of Nature: The Evolution and Genetics of Sexuality*, University of California Press, Berkeley.

64 Hadany, L. and Otto, S.P. (2007) The evolution of condition-dependent sex in the face of high costs. *Genetics*, **176** (3), 1713–1727.

65 Fisher, R.A. (1941) Average excess and average effect of a gene substitution. *Ann. Eugen.*, **11**, 53–63.

66 Williams, G.C. (1975) *Sex and Evolution*, Princeton University Press, Princeton, N.J.

67 Bateson, P. (1982) Preference for cousins in Japanese quail. *Nature*, **295** (5846), 236–237.

68 Bengtsson, B.O. (1978) Avoiding inbreeding – at what cost? *J. Theor. Biol.*, **73**, 439–444.

69 Pusey, A. and Wolf, M. (1996) Inbreeding avoidance in animals. *Trends Ecol. Evol.*, **11**, 201–206.

70 Roze, D. and Rousset, F. (2005) Inbreeding depression and the evolution of dispersal rates: a multilocus model. *Am. Nat.*, **166**, 708–721.

71 Hamilton, W.D. and May, R.M. (1977) Dispersal in stable habitats. *Nature*, **279**, 578–581.

72 Comins, H.N., Hamilton, W.D. *et al.* (1980) Evolutionary stable dispersal strategies. *J. Theor. Biol.*, **82**, 205–230.

73 McPeek, M.A. and Holt, R.D. (1992) The evolution of dispersal in spatially and temporally varying environments. *Am. Nat.*, **140** (6), 1010–1027.

74 Travis, J.M.J. and Dytham, C. (1998) The evolution of dispersal in a metapopulation: a spatially explicit, individual-based model. *Proc. Roy. Soc. Lond. B*, **265**, 17–23.

75 Clobert, J.J., Danchin, E. *et al.* (2001) *Dispersal*, Oxford University Press, Oxford.

76 Murray, B.B.G. (1967) Dispersal in vertebrates. *Ecology*, **48** (6), 975.

77 Lidicker, W.W.Z. (1975) The role of dispersal in the demography of small mammals, in *Mammalian Dispersal Patterns* (eds F.B. Golley, K. Petrusewicz and L. Ryszkowsky), Cambridge University Press, London, pp. 103–128.

78 Dobson, F.F.S. (1982) Competition for mates and predominant juvenile male dispersal in mammals. *Anim. Behav.*, **30** (4), 1183–1192.

79 Gaines, M.S. and McClenaghan, L.R. (1980) Dispersal in small mammals. *Annu. Rev. Ecol. Syst.*, **11** (1), 163–196.

80 Mech, L.L.D. (1987) Age, season, distance, direction, and social aspects of wolf dispersal from a Minnesota pack, in *Mammalian Dispersal Patterns* (eds B.D. Chepko Sade and T. Halpin), Chicago Press, Chicago, pp. 55–74.

81 Anderson, P.K. (1989) *Dispersal in Rodents: A Resident Fitness Hypothesis*, American Society of Mammalogists, Provo.

82 Hanski, I. (1991) Natal dispersal and social dominance in the common shrew *Sorex araneus*. *Oikos*, **62**, 48–58.

83 Gese, E.E.M., Ruff, R.L. *et al.* (1996) Social and nutritional factors influencing the dispersal of resident coyotes. *Anim. Behav.*, **52** (5), 1025–1043.

84 Alonso, J.C., Martin, E. *et al.* (1998) Proximate and ultimate causes of natal dispersal in the great bustard, *Otis tarda*. *Behav. Ecol.*, **9**, 243–252.

85 Altwegg, R., Ringsby, T.H. *et al.* (2000) Phenotypic correlates and consequences of dispersal in a metapopulation of house sparrows, *Passer domesticus*. *J. Anim. Ecol.*, **69**, 762–770.

86 Serrano, D.D., Tella, J.L. *et al.* (2001) Factors affecting breeding dispersal in the facultatively colonial lesser kestrel: individual experience vs. conspecific cues. *J. Anim. Ecol.*, **70** (4), 568–578.

87 Forero, M.G., Donazar, J.A. *et al.* (2002) Causes and fitness consequences of natal dispersal in a population of black kites. *Ecology*, **83**, 858–972.

88 Hanski, I. and Gilpin, M.E. (eds) (1997) *Metapopulation Biology: Ecology, Genetics, and Evolution*, Academic Press, San Diego.

89 Brown, J.H. and Kodric-Brown, A. (1977) Turnover rates in insular biogeography: effect of immigration on extinction. *Ecol. Lett.*, **58**, 445–449.

90 Fahrig, L. and Merriam, G. (1985) Habitat patch connectivity and population survival. *Ecol. Lett.*, **66**, 1762–1768.

91 Pulliam, R. (1988) Sources, sinks and population regulation. *Am. Nat.*, **132**, 652–661.

92 Holt, R.D. (1985) Population-dynamics in 2-patch environments – some anomalous consequences of an optimal habitat distribution. *Theor. Popul. Biol.*, **28** (2), 181–208.

93 Doak, D.F. (1995) Source–sink models and the problem of habitat degradation: general models and applications to the Yellowstone Grizzly. *Conserv. Biol.*, **9**, 1370–1379.

94 Amarasekare, P. (2004) The role of density-dependent dispersal in source-sink dynamics. *J. Theor. Biol.*, **226** (2), 159–168.

95 Wright (1931) Evolution in mendelian populations. *Genetics*, **16**, 97–159.

96 Wright, S.S.M. (1982) Character change, speciation, and the higher taxa. *Evolution*, **36**, 427.

97 Wade, M.J. and Goodnight, C.J. (1991) Wright's shifting balance theory: an experimental study. *Science*, **253** (5023), 1015–1018.

98 Moore, F.F.B.G. and Tonsor, S.J. (1994) A simulation of Wright's Shifting balance process: migration and the three phases. *Evolution* **48**, 69.

99 Coyne, J.A., Barton, N.H. *et al.* (1997) Perspective: a critique of Sewall Wright's shifting balance theory of evolution. *Evolution*, **51** (3), 643–671.

100 Peck, S.L., Ellner, S.P. *et al.* (2000) Varying migration and deme size and the feasibility of the shifting balance. *Evolution Int. J. Org. Evolution*, **54** (1), 324–327.

101 Hadany, L., Eshel, I. *et al.* (2004) No place like home: competition, dispersal and complex adaptation. *J. Evol. Biol.*, **17** (6), 1328–1336.

102 Foster, P.L. (2005) Stress responses and genetic variation in bacteria. *Mutat. Res.*, **569** (1–2), 3–11.

103 Venzke, S., Michel, N. *et al.* (2006) Expression of Nef downregulates CXCR4, the major coreceptor of human immunodeficiency virus, from the surfaces of target cells and thereby enhances resistance to superinfection. *J. Virol.*, **80** (22), 11141–11152.

104 Leontiev, V., Maury, W. *et al.* (2008) Drug-induced superinfection in HIV and the evolution of drug resistance. *Genet. Infect. Evol.*, **8**, 40–50.

105 McEwen, B. and Lasley, E.N. (2002) *The End of Stress As We Know It*, Joseph Henry Press.

106 Wyllie, A.H. (1980) Glucocorticoid-induced thymocyte apoptosis is associated with endogenous endonuclease activation. *Nature*, **284** (5756), 555–556.

107 Cohen, J.J. and Duke, R.C. (1984) Glucocorticoid activation of a calcium-dependent endonuclease in thymocyte nuclei leads to cell-death. *J. Immunol.*, **132** (1), 38–42.

108 Hadany, L., Beker, T. *et al.* (2006) Why is stress so deadly? An evolutionary perspective. *Proc. Biol. Sci.*, **273** (1588), 881–885.

109 Euker, J.S., Meites, J. *et al.* (1975) Effects of acute stress on serum Lh and prolactin in intact, castrate and dexamethasone-treated male rats. *Endocrinology*, **96** (1), 85–92.

110 Gray, G.D., Smith, E.R. *et al.* (1978) Neuroendocrine mechanisms mediating suppression of circulating testosterone levels associated with chronic stress in male rats. *Neuroendocrinology*, **25** (4), 247–256.

111 Rivier, C., Rivier, J. *et al.* (1986) Stress-induced inhibition of reproductive functions – role of endogenous cortico-tropin-releasing factor. *Science*, **231** (4738), 607–609.

2
Catecholamines and Stress

Esther L. Sabban

2.1
Rapid Stress-Triggered Changes in Catecholamines

The elevation of catecholamines (CAs) is one of the most rapid responses to stress. Stress triggers the release of CAs from the adrenal medulla, sympathetic nervous system and catecholaminergic neurons in the brain. The short-term elevations in CAs in the periphery and CNS are crucial to handling stress, overcoming the emergency and restoring homeostasis. The actions of epinephrine (Epi) and norepinephrine (NE) in the liver, adipocytes, skeletal muscles and the α- and β-cells of the pancreas directly affect metabolism toward the utilization of metabolic fuel. They increase fuel mobilization, triggering the breakdown of glycogen and mobilization of fats while suppressing the release of insulin and enhancing glucagon secretion. In the periphery, CAs also have marked effects on the cardiovascular system to raise blood pressure and increase cardiac output.

In response to stress, CAs in the brain stimulate certain limbic areas (e.g., hippocampus, amygdala, hypothalamus), whereas higher cognitive regions in the prefrontal cortex are inactivated. Norepinephrine in the brain is especially crucial in mediating the stress-induced increases in vigilance and alertness.

The regulation of CAs is highly stressor-specific. Adrenal medullary Epi release is mainly induced by hypoglycemia, glucopenia, immobilization (IMO), emotional stressors, etc. [1, 2]. Conversely, cold or pain (formalin) exposure does not activate Epi release very much but highly stimulates NE release from the sympathetic nerve terminals [1, 2].

Urinary CAs are also markedly elevated by stress. A significant rise in Epi and NE is observed in 24-h urine samples after exposure of rats to greater than 1 h of IMO stress[3]. When IMO is repeated several hours daily for a week, urinary Epi and NE are about double the levels attained with exposure to only a single IMO. Much of the excretion, especially of Epi, occurs during the IMO [3]. A comprehensive review of the structural and molecular genetic approaches to catecholaminergic systems in stress has recently appeared [87].

Stress – From Molecules to Behavior. Edited by Hermona Soreq, Alon Friedman, and Daniela Kaufer
Copyright © 2010 WILEY-VCH Verlag GmbH & Co. KGaA, Weinheim
ISBN: 978-3-527-32374-6

2.2
Catecholamines and Stress-Related Disorders

Although the response to stress is essential for life, when it is chronic, repeated or excessive, stress is associated with a variety of serious disorders. Stress is a major contributor to the development of cardiovascular disorders and psychiatric disorders, including PTSD, and affective disorders such as major depression [4–6].

Stress increases the body's susceptibility to infection, autoimmune diseases, chronic fatigue syndrome and cancer, and it increases the propensity of an individual to self-administer drugs of abuse. The CAs are intimately related to the pathophysiology of many stress related disorders [6, 7]. While a comprehensive discussion of the role of CAs in these disorders is beyond the scope of this work, some of the interesting findings related to the relationship between CAs and stress-related disorders are summarized below.

2.2.1
Cardiovascular Disease

Catecholamines are known to have a major impact on cardiovascular disease (reviewed in [6]). With acute myocardial infarction, there is a positive relationship between the size of the infarct and plasma NE levels. Catecholamines also accelerate the progression of myocardial cell damage. Protective effects of β-blockage on myocardial infarctions have been repeatedly shown.

Many studies have analyzed the relationship between sympathetic nerve activity, plasma NE and hypertension, as NE is a potent vasoconstrictor. Plasma NE levels, primarily from sympathetic nerve endings, but also from adrenal medulla, are markedly elevated with aging in humans and experimental animals, and they have been proposed to be involved with increased hypertension in the elderly [6]. Overall, it was concluded that sympathetic efferent nerves play an important role in the control of blood pressure in a number of situations. They appear to maintain hypertension, even when they are not the primary initiating cause. However, their role in the initiation or maintenance of essential hypertension is still controversial.

Congestive heart failure is also associated with activation of the SNS. While activation of the SNS may provide short-term hemodynamic support to the failing heart, over long periods of time it is deleterious. Preclinical and clinical studies suggest that the chronic sympathetic activation in congestive heart failure is a maladaptive response which accelerates the progressive worsening of the disease. Pharmacological approaches to block sympathetic activation are used to treat heart failure. These include inhibition of central sympathetic outflow, block of cardiac effects (β-adrenergic blockers) and inhibition of catecholamine biosynthesis.

One of the most striking effects of stress and CAs on the cardiovascular system is seen with Takotsubo cardiomyopathy, which is also called stress cardiomyopathy [8]. This increasingly recognized clinical syndrome is characterized by acute reversible apical ventricular dysfunction. It represents a form of Epi-mediated acute myocardial stunning. It is brought on by sudden emotional or physical stress and affects predominantly post-menopausal women.

2.2.2
Post-Traumatic Stress Disorder

A large body of data has provided compelling evidence for increased NE activity in traumatized humans with post-traumatic stress disorder (PTSD; reviewed by [9, 10]). PTSD is characterized neurochemically by reduced plasma cortisol, as well as by elevated CAs. Several studies have consistently reported increased urinary NE and/or Epi concentrations in patients with PTSD compared to controls with non-trauma-related anxiety [9, 11]. Increased NE levels in cerebral spinal fluid were also found to positively correlate with the severity of PTSD. It is hypothesized that stress-triggered elevations in NE may be related to the subsequent symptoms of re-experiences, intrusive memories and nightmares in individuals with PTSD. Interestingly, the non-specific β-adrenergic receptor blocker, propanolol, is effective in ameleriorating PTSD [12].

2.2.3
Depression

Maladaptation of the central NE system to stress is also implicated in depression (reviewed by [13, 14]). In this regard, some postmortem studies reveal increased levels of tyrosine hydroxylase (TH; the first and major rate-limiting enzyme in CA biosynthesis) and reduced levels of NE transporter in the locus coeruleus (LC) of subjects with major depression [15]. Chronic antidepressant treatment normalizes the stress-triggered elevation of TH immunoreactive protein in rat LC [16]. Moreover, depression-triggered bone loss is associated with a substantial increase in bone NE levels and can be blocked by propranolol, suggesting that the sympathetic nervous system (SNS) mediates the skeletal damage with stress-induced depression [17].

2.2.4
Immune Disorders

The SNS is crucial in regulation of the immune system. The primary pathway for neural regulation of the immune system is by the SNS and its main neurotransmitter NE, with no neuroanatomical evidence for a parasympathetic or vagal nerve supply to any immune organ. Activation of the SNS primarily inhibits the activity of cells associated with the innate immune systems, while it either enhances or inhibits the activity of cells associated with the acquired/adaptive immune system (reviewed by [18]). *In vitro* experiments with macrophages harvested from spleen and lymph nodes show that NE can dramatically inhibit the production and secretion of tumor necrosis factor α (TNFα) in response to endotoxin (lipopolysaccharide, LPS) [19]. Both splenic and plasma levels of TNFα are dramatically suppressed in stressed animals, and the inhibitory effects of stress are still present in adrenalectomized animals.

2.2.5
Pain

Stress can change the sensitivity to pain. Thus, Epi enhances bradykinin hyperalgesia and prolonged elevated plasma levels of Epi are proposed to sensitize bradykinin receptors with may contribute to chronic generalized pain syndromes [20].

2.3
Stress-Triggered Regulation of Catecholamine Biosynthetic Enzymes in Different Locations

2.3.1
Pathway of Catecholamine Biosynthesis

Up-regulation of the capacity to synthesize CAs in many catecholaminergic locations is a major adaptive, or maladaptive, response to stress. The pathway for the biosynthesis of CAs is shown in Figure 2.1. The biosynthetic pathway begins with tyrosine hydroxylase (TH, EC 1.14.16.2), the major rate-limiting enzyme in the biosynthesis of NE and other catecholamines. It catalyzes the hydroxylation of tyrosine to 3,4-dihydroxy-L-phenylalanine (L-Dopa). Tyrosine hydroxylase, given its role in the initial step in CA biosynthesis, is subject to intricate regulation (reviewed in [21–24]). This includes feedback inhibition, allosteric regulation and phosphorylation of the enzyme in the short term. TH has an N-terminal regulatory domain which contains four serine residues, which can be phosphorylated *in vitro*, *in situ* and *in vivo*. A range of protein kinases and phosphatases are able to modulate phosphorylation of these sites, some in a hierarchical fashion (reviewed by [25]). In the medium to long term, TH is regulated by enzyme stability, by changes in gene expression, due to increased transcription of TH and/or its essential cofactor, tetrahydrobiopterin (BH4), and in some cases by changes in mRNA stability and translational control.

GTP cyclohydrolase I (GTPCH) is the first and rate-limiting reaction in *de novo* biosynthesis of (6R)-5,6,7,8-tetrahydro-L-biopterin (BH4) the, essential cofactor for TH as well as for tryptophan hydroxylase, phenylalanine hydroxylase and the three isoforms of nitric oxide synthase.

L-Dopa is decarboxylated by aromatic acid decarboxylase (AADC), yielding dopamine. Dopamine β-hydroxylase (DBH, EC 1.14.17.1) catalyzes the formation of NE by hydroxylation of dopamine. NE, an important neurotransmitter/hormone

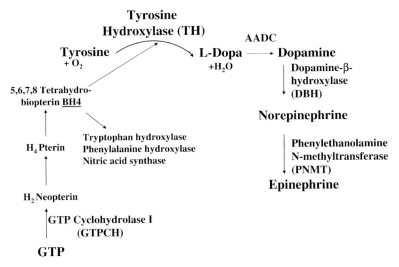

Figure 2.1 Pathway for catecholamine biosynthesis, including biosynthesis of tetrahydrobiopterin, essential cofactor for TH, as well as for tryptophan hydroxylase, phenylalanine hydroxylase and nitric oxide synthase.

in its own right, is also a precursor for Epi in cells which contain phenylethanolamine N-methyltransferase (PNMT).

In the periphery, NE is synthesized in the SNS and in the adrenal medulla. The sympathetic nerve endings are responsible for producing much of circulating NE, while the adrenal medulla synthesizes both NE and Epi and is the major source of circulating Epi. In the brain, most of the NE-neurons originate in the locus coeruleus (LC) and in the lateral tegmental areas of the brainstem.

Although stress increases expression of CA biosynthetic enzymes in various catecholaminergic locations, the mechanism differs, indicating selective regulation and a promise of selective pharmaceutical interventions.

2.3.2
Adrenomedullary Hormonal System

The adrenomedullary chromaffin cells receive input by mainly preganglionic inputs from cell bodies of intermediolateral columns of the spinal cord. They release the catecholamines (Epi and NE) directly into the bloodstream. The adrenal medulla is the major source of Epi, and adrenalectomy reduces the stress triggered rise in Epi by about 95%. In contrast, while the adrenal medulla also releases NE, the major source of circulating NE is from the sympathetic nervous system.

Repeated or chronic exposure to a variety of stressors increases activity of CA biosynthetic enzymes in the adrenal medulla (reviewed in [26–29]). Early studies showed that rise in TH activity induced by IMO or cold stress resulting from increases in the amount of enzyme protein and its synthetic rate [30–32]. Repeated or chronic exposure to stress also increases activity of DBH and to a lesser extent PNMT [32–34]. The elevation in adrenal TH and DBH measured *in vitro* corresponds well to the rate of CA synthesis measured *in vivo* in control and immobilized rats [35].

In the adrenal medulla, the activity of CA biosynthetic enzymes is associated with increased gene expression of CA biosynthetic enzymes (reviewed in [27, 29, 36]). Early work observed that, following several cold days at 4 °C, adrenal TH mRNA levels were about 3–4× their basal levels and TH protein and activity levels had nearly doubled [37–39]. Subsequently, it was shown that chronic or repeated exposure to many types of stressors affect gene expression of CA biosynthetic enzymes and leads to elevations of their mRNA levels. These include physical stressors such as IMO stress, exercise, hypoglycemia and a chronic mild stress model of depression. DBH and PNMT gene expression are also markedly induced by chronic or repeated stress [40, 41]. While the majority of these studies were performed in rats, an immobilization-induced increase in TH, DBH and PNMT mRNA levels in the adrenal medulla of mice is also reported.

In addition, to changes in CA biosynthetic enzymes, there is also up-regulation of biosynthesis of BH4, essential for TH activity. Gene expression of GTPCH, the rate-limiting enzyme for BH4 biosynthesis is markedly induced by stress [42]. The rat adrenal medulla expresses both the 1.2 and 3.6 Kb isoforms of GTPCH mRNA and the expression of both are increased by exposure to IMO stress. Furthermore, cold stress was found to lead to a doubling of biopterin content [43].

The dynamics of these stress-elicited changes in gene expression have been studied most thoroughly for immobilization stress (IMO). While an acute exposure to IMO is not sufficient to elicit elevation of activity or protein of the CA biosynthetic enzymes, surprisingly it triggered as high a maximal change in TH mRNA as observed with repeated IMO [44]. Compared to values in unstressed controls, TH mRNA levels increased by approximately 600%, DBH mRNA levels increased by 100%. While the increase in TH mRNA by a single IMO stress is large and rapid, it is transient (Figure 2.2). Within one day of cessation of the stress, TH mRNA levels returned to near basal levels and activation was not yet reflected in either increased immunoreactive protein levels or enzymatic activity. However, even with only a second exposure to the same stressor, there was "memory" of the first experience, such that the increase in TH mRNA was now much more prolonged and was sustained for longer periods of time after termination of the stress. It took nearly a week for TH mRNA levels to return to basal values. Only after repeated stress is there a significant rise in TH immunoreactive protein and enzymatic activity.

Transcriptional mechanisms were found to play an important role in the stress elicited elevation of TH and DBH gene expression in the adrenal medulla, although post-transcriptional mechanisms are likely also involved especially with acute stress [28]. A number of older studies reported that pretreatment of the animals with the actinomycin D, a non-specific inhibitor of transcription, prevented the stress triggered increase in TH and DBH activities. More recently stress was found to directly raise TH and DBH gene transcription. This was shown by run-on assays

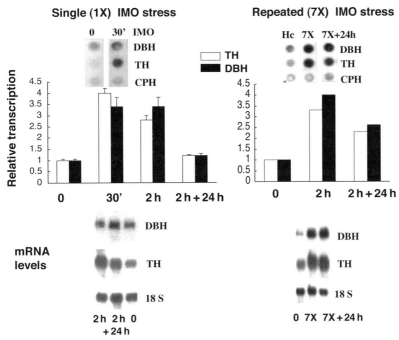

Figure 2.2 Effect of single and repeated IMO on the magnitude and persistence of elevation of TH and DBH transcription and mRNA levels. The changes in TH and DBH gene transcription and mRNA levels in rat adrenal medulla with single and repeated (2 h daily) IMO are shown. (Adapted from [46]).

of transcription, by hybridization with intron-specific probes and by transgenic animals in which the TH proximal promoter directed reporter activity [45–48]. Interestingly, these studies also found that the transcriptional response to acute IMO is large, but transient, while with repeated IMO, the elevation in transcription is sustained at these high levels. As shown in Figure 2.2, even a single exposure to IMO stress elevates TH and DBH transcription and mRNA levels. Nevertheless, one day after a single 2 h IMO, the levels had returned to near basal levels. However, exposure to repeated IMO stress leads to about a threefold increase in TH transcription and a marked increase in TH and DBH mRNA levels, which remain sustained one day later.

What are the factors and pathways initiating changes in catecholamine enzymes gene expression in the adrenal medulla? Experiments have been performed to determine whether the splanchnic nerve input, with release of acetylcholine and other neurotransmitters such as PACAP, or activation of the HPA axis are essential for the stress-triggered increase in catecholamine biosynthetic enzymes in the adrenal medulla.

It is very clear that activation of the HPA axis play a major role in the regulation of PNMT gene expression with stress. Hypophysectomy prevents the IMO-triggered induction of PNMT gene expression [41]. In the adrenal medulla of CRH-deficient mice, the induction of PNMT mRNA triggered by both single and repeated IMO is greatly attenuated compared to wild-type animals [49]. The regulation of PNMT by the HPA axis likely is mediated by glucocorticoids. Glucocorticoids are proposed to regulate PNMT in at least two ways: post-transcriptionally and transcriptionally [28, 50–52]. The PNMT promoter was shown to contain strong glucocorticoid response element(s) [50] and also a weak one [53]. The glucocorticoid receptor interacts with the PNMT promoter to control transcription, alone and also synergistically with Egr1 and AP-2 transcription factors [54]. The stress-triggered elevation of GTPCH gene expression in the adrenal medulla is also markedly influenced by the HPA axis [55].

The HPA axis is less important for the regulation by stress of TH or DBH gene expression in the adrenal medulla. In CRH knockout mice, IMO induced a similar increase in TH and DBH mRNA levels as seen in the wild-type animals [56]. Even in hypophysectomized rats IMO stress is still able to trigger an increase in TH mRNA levels [44]. However, splanchnic nerve section or the nicotinic antagonist chlorisondamine prevents the elevation of adrenal TH mRNA triggered by hypogly-cemia or cold stress [57–59]. Neither activation of the HPA axis nor splanchnic cholinergic innervation is essential for the induction of TH or PNMT mRNAs in response to a single IMO stress [36, 60].

TH mRNA levels are elevated in response to a single 2 h IMO, even in the adrenal medulla of hypophysectomized rats which have undergone splanchnic denervation [44]. This indicates that there are also non-pituitary, non-neuronal mechanisms involved with immobilization stress. However, the precise mechanism is still not clear.

An important question is how does the adrenal medullary response to acute stress differ from that of chronic exposure to stressful situations? When does the response becomes maladaptive? Recently we studied the dynamics of different transcriptional pathways associated with a brief or intermediate duration of a single stress or with repeated stress. The findings indicate that there is a dynamic interplay in converting a short-term to a long-term response [46, 61]. With even brief exposure to immobilization stress, there is increased, but transient, phosphorylation of CREB and induction of several transcription factors, including Egr1 and c-Fos [27, 62]. With repeated stress, the repertoire of transcription factor activation differs: prominent are the more sustained phosphorylation of CREB and the induction and extensive phosphorylation of Fra-2 [63, 64]. These changes may be critical in converting the acute beneficial effects of stress into prolonged detrimental consequences.

Microarray analysis has been helpful in distinguishing the clusters of genes differentially regulated by single and repeated stress. It has revealed distinct signaling pathways induced in the adrenal medulla following exposure to single and repeated IMO stress [65].

2.3.3
Sympathetic Nervous System

Originally the sympathoadrenal systems, comprising the SNS and the adrenal medulla, were considered as one system mediating the "fight or flight" response to stress [66]. Both respond to stress with increased release of catecholamines and with increased gene expression of catecholamine biosynthetic enzymes. However, it has become clear in recent years that these systems can be regulated separately [67, 68].

The SNS is the major source of norepinephrine (NE) and neuropeptide Y (NPY) in the periphery. It regulates multiple systems, including the cardiovascular, immune, gastrointestinal, respiratory, renal, endocrine and even bone. Recently the importance of sympathetic activation in regulation of adipose cells and obesity has been recognized.

Early studies established that exposure of rats to a variety of stressors triggers increased TH and DBH enzymatic activities in a number of sympathetic ganglia, including the superior cervical ganglia (SCG) and stellate ganglia (StG). This includes cold, IMO and swimming stress [30, 69–72]. Electric stimulation of pre-ganglionic nerves also increases TH and DBH activity in the SCG [73].

The rise in activity of NE biosynthetic enzymes in response to IMO or cold stress is associated with elevated TH immunoreactive protein and elevated levels of TH and DBH mRNAs in StG and SCG [74–76]. Exposure to even a single IMO increased TH mRNA levels two- to threefold in both SCG and StG. Peak levels, however, were reached 24–48 h after the stress exposure [75]. Repeated IMO further enhanced the expression of TH gene in sympathetic ganglia, eliciting an additional increase approximately twofold over the elevation attained with a single episode of this stressor. Under conditions of repeated IMO, increased TH immunoreactivity was observed in both StG and SCG [76]. Thus, increased CA biosynthetic capacity in the SNS likely is involved in elevation of NE in circulation and various target tissues leading to changes in the allosteric load and either adaptation or maladaptation to the stress.

The levels of TH activity in ganglia are also increased with aging. It is not known if this is the result of repeated exposure to stressors over the lifetime. [77]. However, the response of TH in the SCG to cold stress differs in young and old animals [78].

Stress not only elevates gene expression of NE biosynthetic enzymes in SCG, but also of NPY [76], which is co-localized in many of the NE synthesizing neurons of the SNS. NPY is released from sympathetic neurons in response to stress, particularly chronic ones (reviewed in [79]). NPY can enhance the vasoconstrictive effects of NE and plasma NPY is elevated in hypertensive men and women [80].

The response to stress may differ in different part of the SNS. Basal TH activity was measured in several different rat ganglia: the highest levels were found in the coelic ganglia and the next highest in the SCG. Selectivity was observed with swimming stress (15 min daily for 4 days) which increased TH activity primarily in stellate ganglia and celiac ganglia, but not in SCG [81]. An elegant study by Kiran and Ulus [82] revealed selectivity in the response of TH activity in different sympathetic ganglia to exposure to various stressors. In their study, rats were exposed to several stressors: (i) immobilization (IMO) stress (6 h daily for 4 days), (ii) glucopenia after 2-deoxy-D-glucose administration, (iii) hypercapnia (20% CO_2, 25% O_2, 55% N_2) and (iv) cold stress (64 h). In all cases the animals were killed one day after termination of the stress and TH activity was assayed in a variety of sympathetic ganglia and compared to untreated controls. IMO stress significantly increased TH activity mainly in the lumbar and sacral ganglia and slightly (about 30%) in the cervical ganglia, but no changes of TH activity were seen in the thoracic ganglia. In contrast, hypocapnia increased TH activity in celiac ganglia and primarily in thoracic and first two lumbar ganglia, but not in sacral ganglia. Prolonged exposure to cold had a significant effect in the celiac, L2, T9, T10 and cervical inferior ganglia, but not on other thoracic or lumbar ganglia. Glucopenia, which increases TH activity and gene expression in adrenal medulla and brain catecholaminergic locations, did not significantly alter TH activity in any of the sympathetic ganglia.

These studies showed selectivity of specific ganglia to particular stressors. It also revealed that stressors can increase TH activity selectively in a specific ganglia without triggering changes in all portions of the SNS. Selectivity in differential sympathetic drive to adipose tissue has also been observed after selective stressors, with various adipose tissues [83].

What is the mechanism for stress triggered activation of TH and DBH gene expression in sympathetic ganglia? Trans-synaptic nerve stimulation for regulation of NE biosynthetic enzyme gene expression in sympathetic ganglia system has been well demonstrated. Preganglionic nerve stimulation elicits increases TH and DBH activities and elevated TH protein and mRNA levels in rat SCG [73, 84]. Denervation of preganglionic inputs prevents the elevation of TH in response to reserpine. However, the contribution of trans-synaptic inputs to the stress-triggered changes are less certain.

The activation of the HPA axis by stress appears to be important in the regulation of the NE biosynthetic enzymes in the SNS. Injections of ACTH to rats is as effective as IMO stress in triggering elevation of TH and DBH mRNAs in SCG [76]. Moreover, ACTH still triggers increased TH and DBH gene expression in adrenalectomized rats in which basal ACTH was clamped with glucocorticoid pellets [85]. The results indicate that ACTH, at stress levels, can regulate TH and DBH gene expression independent of adrenal hormones. The specific receptor involved is still under investigation. However, we have also shown, for the first time, that sympathetic ganglia (SCG and stellate ganglia) express the melanocortin receptors (MCR), such as MC2R, the major ACTH receptor [86].

2.3.4
Noradrenergic Systems in the Brain

While stress activates various catecholaminergic nuclei in the brain with elevated expression of CA biosynthetic enzymes (reviewed in [87, 88]), the most information is known regarding the neurons originating in the locus coeruleus (LC). The

LC (A6 region) is comprised of about 3000 neurons (both sides) in the rat and about 24 000 neurons in humans. These neurons comprise a broad network of projections that extend throughout the neuroaxis and account for about 70% of all brain NE in primates [89]. The LC is implicated in controlling attention, arousal, vigilance, retrieval of emotional memories, state dependent cognitive functions and activity of the autonomic nervous system [90]. The LC is also implicated in opiate physical dependence and withdrawal.

Many types of stress were found to trigger elevated TH activity and protein levels in the LC. These include isolation, footshock, restraint or immobilization, chronic social stress, forced walking and chronic cold. In addition, several studies also examined changes in DBH activity or protein levels and found them elevated by stress (reviewed in [26, 27]). The stress-triggered increase in activity of NE-biosynthetic enzymes is accompanied, and in some cases preceded, by changes in their respective mRNAs [91–96].

Elevations in transcription and decreased turnover of the mRNAs are implicated in mediating the stress-triggered increase in TH and DBH gene expression in the LC [28, 48, 96]. Run-on assays of transcription revealed that both single and repeated immobilization stress triggers about a threefold increase in levels of transcription initiation for the TH and DBH genes [96]. Transcription of the GTPCH gene is also induced in the LC [96]. Footshock also increases TH transcription in LC, as determined by a different method (in situ hybridization with intron-specific probes) [97]. However, it appears that the elevation of TH transcription by stress in the LC is not nearly as high as occurs in the adrenal medulla [48].What is the mechanisms of stress triggered activation of gene expression in the LC? Stress leads to induction of a number of transcription factors. AP1 transcription factors such as c-Fos and Fra-2 and CREB are candidates for mediating this induction. Stress rapidly induces c-Fos in the LC; and we and others have shown that, while induction of c-Fos is characteristic of acute activation [88] in many brain areas, in the LC repeated stress continues to induce c-Fos expression. Induction of Fra-2 in the LC was observed in our studies with repeated stress [95]. Although the Egr1 is induced by stress in the adrenal medulla, induction of Egr1 in the brainstem by IMO stress does not overlap with TH expressing cells of the LC [98].

Phosphorylation of CREB (and the induction of CREB levels) is found with single (and especially with repeated) IMO stress and can activate TH transcription at the perfect consensus CRE motif. The role of cAMP pathway transcription factor CREB in regulation of gene expression in the LC received considerable attention in a series of elegant studies (reviewed in [99]). Multiple MAP kinases are also activated by repeated IMO, as shown by increased phosphorylation of p38, JNK1/2/3 and ERK1/2.

Thus, distinct alterations in transcriptional pathways following repeated, compared to single stress, may be involved in mediating long lasting neuronal remodeling and are implicated in the mechanisms by which acute beneficial responses to stress are converted into prolonged adaptive or maladaptive responses.

Summary

The catecholamines are among the first players recruited to help overcome the emergency threat to homeostasis. However, over-stimulated catecholaminergic systems are associated with many stress related pathologies. Stress triggers increased gene expression of catecholamine biosynthetic enzymes, which can have prolonged consequences. In many instances this is mediated by elevations in gene transcription. Importantly the molecular mechanism of activation of gene expression for catecholamine biosynthetic enzymes differs in the adrenal medulla, the sympathetic nervous system and the brain.

Acknowledgment

The support of NIH grant NS44218 is gratefully acknowledged.

References

1 Kvetnansky, R., Pacak, K., Sabban, E.L., Kopin, I.J. and Goldstein, D.S. (1998) Stressor specificity of peripheral catecholaminergic activation. *Adv. Pharmacol.*, **42**, 556–560.

2 Pacak, K., Palkovits, M., Yadid, G., Kvetnansky, R., Kopin, I.J. and Goldstein, D.S. (1998) Heterogeneous neurochemical responses to different stressors: a test of Selye's doctrine of nonspecificity. *Am. J. Physiol.*, **275**, R1247–R1255.

3 Kvetnansky, R. and Mikulaj, L. (1970) Adrenal and urinary catecholamines in rats during adaptation to repeated immobilization stress. *Endocrinology*, **87**, 738–743.

4 Chrousos, G.P. and Gold, P.W. (1992) The concepts of stress and stress system disorders. Overview of physical and behavioral homeostasis. *JAMA*, **267**, 1244–1252.

5 McEwen, B.S. (1998) Protective and damaging effects of stress mediators. *N. Engl. J. Med.*, **338**, 171–179.

6 Goldstein, D.S. (1995) Stress, Catecholamines and Cardiovascular Disease, Oxford University Press, Oxford.

7 Morilak, D.A., Barrera, G., Echevarria, D.J., Garcia, A.S., Hernandez, A., Ma, S. and Petre, C.O. (2005) Role of brain norepinephrine in the behavioral response to stress. *Prog. Neuropsychopharmacol. Biol. Psychiatry*, **29**, 1214–1224.

8 Lyon, A.R., Rees, P.S., Prasad, S., Poole-Wilson, P.A. and Harding, S.E. (2008) Stress (Takotsubo) cardiomyopathy – a novel pathophysiological hypothesis to explain catecholamine-induced acute myocardial stunning. *Nat. Clin. Pract. Cardiovasc. Med.*, **5**, 22–29.

9 Southwick, S.M., Bremner, J.D., Rasmusson, A., Morgan, C.A. 3rd, Arnsten, A. and Charney, D.S. (1999) Role of norepinephrine in the pathophysiology and treatment of posttraumatic stress disorder. *Biol. Psychiatry*, **46**, 1192–1204.

10 Southwick, S.M., Morgan, C.A. 3rd, Bremner, A.D., Grillon, C.G., Krystal, J.H., Nagy, L.M. and Charney, D.S. (1997) Noradrenergic alterations in posttraumatic stress disorder. *Ann. N.Y. Acad. Sci.*, **821**, 125–141.

11 Yehuda, R., Siever, L.J., Teicher, M.H., Levengood, R.A., Gerber, D.K., Schmeidler, J. and Yang, R.K. (1998) Plasma norepinephrine and 3-methoxy-4-hydroxyphenylglycol concentrations and severity of depression in combat posttraumatic stress disorder and major depressive disorder. *Biol. Psychiatry*, **44**, 56–63.

12 Orr, S.P., Milad, M.R., Metzger, L.J., Lasko, N.B., Gilbertson, M.W. and Pitman, R.K. (2006) Effects of beta blockade, PTSD diagnosis, and explicit threat on the extinction and retention of an aversively conditioned response. *Biol. Psychol.*, **73**, 262–271.

13 Anand, A. and Charney, D.S. (2000) Norepinephrine dysfunction in depression. *J. Clin. Psychiatry*, **61** (Suppl. 10), 16–24.

14 Leonard, B.E. (2001) Stress, norepinephrine and depression. *J. Psychiatry Neurosci.*, **26** (Suppl.) S11–S16.

15 Zhu, M.Y., Klimek, V., Dilley, G.E., Haycock, J.W., Stockmeier, C., Overholser, J.C., Meltzer, H.Y. and Ordway, G.A. (1999) Elevated levels of tyrosine hydroxylase in the locus coeruleus in major depression. *Biol. Psychiatry*, **46**, 1275–1286.

16 Melia, K.R., Nestler, E.J. and Duman, R.S. (1992) Chronic imipramine treatment normalizes levels of tyrosine hydroxylase in the locus coeruleus of chronically stressed rats. *Psychopharmacology (Berl.)*, **108**, 23–26.

17 Yirmiya, R., Goshen, I., Bajayo, A., Kreisel, T., Feldman, S., Tam, J., Trembovler, V., Csernus, V., Shohami, E. and Bab, I. (2006) Depression induces bone loss through stimulation of the sympathetic nervous system. *Proc. Natl. Acad. Sci. U. S. A.*, **103**, 16876–16881.

18 Nance, D.M. and Sanders, V.M. (2007) Autonomic innervation and regulation of the immune system (1987–2007). *Brain Behav. Immun.*, **21**, 736–745.

19 Ignatowski, T.A., Gallant, S. and Spengler, R.N. (1996) Temporal regulation by adrenergic receptor stimulation of macrophage (M phi)-derived tumor necrosis factor (TNF) production post-LPS challenge. *J. Neuroimmunol.*, **65**, 107–117.

20 Khasar, S.G., Green, P.G., Miao, F.J. and Levine, J.D. (2003) Vagal modulation of nociception is mediated by adrenomedullary epinephrine in the rat. *Eur. J. Neurosci.*, **17**, 909–915.

21 Kumer, S.C. and Vrana, K.E. (1996) Intricate regulation of tyrosine hydroxylase activity and gene expression. *J. Neurochem.*, **67**, 443–462.

22 Sabban, E.L. (1996) Synthesis of dopamine and its regulation, in *CNS Neurotransmitters and Neuromodulators: Dopamine* (ed. T. Stone), CNC Press, Boca Raton, Florida, pp. 1–20.

23 Flatmark, T. (2000) Catecholamine biosynthesis and physiological regulation in neuroendocrine cells. *Acta Physiol. Scand.*, **168**, 1–17.

24 Nagatsu, T. (1995) Tyrosine hydroxylase: human isoforms, structure and regulation in physiology and pathology. *Essays Biochem.*, **30**, 15–35.

25 Dunkley, P.R., Bobrovskaya, L., Graham, M.E., von Nagy-Felsobuki, E.I. and Dickson, P.W. (2004) Tyrosine hydroxylase phosphorylation: regulation and consequences. *J. Neurochem.*, **91**, 1025–1043.

26 Kvetnansky, R. and Sabban, E.L. (1993) Stress-induced changes in tyrosine hydroxylase and other catecholamine biosynthetic enzymes, in *Tyrosine Hydroxylase: From Dicscovery to Cloning* (eds M. Naoi and S.H. Parvez), VSP, Utrecht.

27 Sabban, E.L. and Kvetnansky, R. (2001) Stress-triggered activation of gene expression in catecholaminergic systems: dynamics of transcriptional events. *Trends Neurosci.*, **24**, 91–98.

28 Wong, D.L. and Tank, A.W. (2007) Stress-induced catecholaminergic function: transcriptional and post-transcriptional control. *Stress*, **10**, 121–130.

29 Kvetnansky, R. and McCarty, R. (2007) Adrenal medulla, in *Encyclopedia of Stress*, 2nd edn (ed. G. Fink), Academic Press, New York, pp. 52–59.

30 Thoenen, H. (1970) Induction of tyrosine hydroxylase in peripheral and central adrenergic neurones by cold-exposure of rats. *Nature*, **228**, 861–862.

31 Hoeldtke, R., Lloyd, T. and Kaufman, S. (1974) An immunochemical study of the induction of tyrosine hydroxylase in rat adrenal glands. *Biochem. Biophys. Res. Commun.*, **57**, 1045–1053.

32 Kvetnansky, R., Weise, V.K. and Kopin, I.J. (1970) Elevation of adrenal tyrosine hydroxylase and phenylethanolamine-N-methyl transferase by repeated immobilization of rats. *Endocrinology*, **87**, 744–749.

33 Kvetnansky, R., Gewirtz, G.P., Weise, V.K. and Kopin, I.J. (1970) Effect of hypophysectomy on immobilization-induced elevation of tyrosine hydroxylase and phenylethanolamine-N-methyl transferase in the rat adrenal. *Endocrinology*, **87**, 1323–1329.

34 Kvetnansky, R., Gewirtz, G.P., Weise, V.K. and Kopin, I.J. (1971) Enhanced synthesis of adrenal dopamine beta-hydroxylase induced by repeated immobilization in rats. *Mol. Pharmacol.*, **7**, 81–86.

35 Kvetnansky, R., Weise, V.K., Gewirtz, G.P. and Kopin, I.J. (1971) Synthesis of adrenal catecholamines in rats during and after immobilization stress. *Endocrinology*, **89**, 46–49.

36 Kvetnansky, R. and Sabban, E.L. (1998) Stress and molecular biology of neurotransmitter-related enzymes. *Ann. N. Y. Acad. Sci.*, **851**, 342–356.

37 Richard, F., Faucon-Biguet, N., Labatut, R., Rollet, D., Mallet, J. and Buda, M. (1988) Modulation of tyrosine hydroxylase gene expression in rat brain and adrenals by exposure to cold. *J. Neurosci. Res.*, **20**, 32–37.

38 Stachowiak, M., Sebbane, R., Stricker, E.M., Zigmond, M.J. and Kaplan, B.B. (1985) Effect of chronic cold exposure on tyrosine hydroxylase mRNA in rat adrenal gland. *Brain Res.*, **359**, 356–359.

39 Tank, A.W., Lewis, E.J., Chikaraishi, D.M. and Weiner, N. (1985) Elevation of RNA coding for tyrosine hydroxylase in rat adrenal gland by reserpine treatment and exposure to cold. *J. Neurochem.*, **45**, 1030–1033.

40 McMahon, A., Kvetnansky, R., Fukuhara, K., Weise, V.K., Kopin, I.J. and Sabban, E.L. (1992) Regulation of tyrosine hydroxylase and dopamine beta-hydro-xylase mRNA levels in rat adrenals by a single and repeated immobilization stress. *J. Neurochem.*, **58**, 2124–2130.

41 Viskupic, E., Kvetnansky, R., Sabban, E.L., Fukuhara, K., Weise, V.K., Kopin, I.J. and Schwartz, J.P. (1994) Increase in rat adrenal phenylethanolamine N-methyltransferase mRNA level caused by immobilization stress depends on intact pituitary-adrenocortical axis. *J. Neurochem.*, **63**, 808–814.

42 Serova, L.I., Nankova, B., Kvetnansky, R. and Sabban, E.L. (1997) Elevates GTP cyclohydrolase I mRNA levels in rat adrenals predominantly by hormonally mediated mechanisms. *Stress*, **1**, 135–144.

43 Baruchin, A., Weisberg, E.P., Miner, L.L., Ennis, D., Nisenbaum, L.K., Naylor, E., Stricker, E.M., Zigmond, M.J. and Kaplan, B.B. (1990) Effects of cold exposure on rat adrenal tyrosine hydroxylase: an analysis of RNA, protein, enzyme activity, and cofactor levels. *J. Neurochem.*, **54**, 1769–1775.

44 Nankova, B., Kvetnansky, R., McMahon, A., Viskupic, E., Hiremagalur, B., Frankle, G., Fukuhara, K., Kopin, I.J. and Sabban, E.L. (1994) Induction of tyrosine hydroxylase gene expression by a nonneuronal nonpituitary-mediated mechanism in immobilization stress. *Proc. Natl. Acad. Sci. U. S. A.*, **91**, 5937–5941.

45 Sun, B., Sterling, C.R. and Tank, A.W. (2003) Chronic nicotine treatment leads to sustained stimulation of tyrosine hydroxylase gene transcription rate in rat adrenal medulla. *J. Pharmacol. Exp. Ther.*, **304**, 575–588.

46 Nankova, B.B., Tank, A.W. and Sabban, E.L. (1999) Transient or sustained transcriptional activation of the genes encoding rat adrenomedullary catecholamine biosynthetic enzymes by different durations of immobilization stress. *Neuroscience*, **94**, 803–808.

47 Osterhout, C.A., Chikaraishi, D.M. and Tank, A.W. (1997) Induction of tyrosine hydroxylase protein and a transgene containing tyrosine hydroxylase 5' flanking sequences by stress in mouse adrenal gland. *J. Neurochem.*, **68**, 1071–1077.

48 Osterhout, C.A., Sterling, C.R., Chikaraishi, D.M. and Tank, A.W. (2005) Induction of tyrosine hydroxylase in the locus coeruleus of transgenic mice in response to stress or nicotine treatment: lack of activation of tyrosine hydroxylase promoter activity. *J. Neurochem.*, **94**, 731–741.

49 Kvetnansky, R., Kubovcakova, L., Tillinger, A., Micutkova, L., Krizanova, O. and Sabban, E.L. (2006) Gene expression of phenylethanolamine

N-methyltransferase in corticotropin-releasing hormone knockout mice during stress exposure. *Cell. Mol. Neurobiol.*, **26**, 735–754.

50 Tai, T.C., Claycomb, R., Her, S., Bloom, A.K. and Wong, D.L. (2002) Glucocorticoid responsiveness of the rat phenylethanolamine N-methyltransferase gene. *Mol. Pharmacol.*, **61**, 1385–1392.

51 Tai, T.C., Claycomb, R., Siddall, B.J., Bell, R.A., Kvetnansky, R. and Wong, D.L. (2007) Stress-induced changes in epinephrine expression in the adrenal medulla *in vivo*. *J. Neurochem.*, **101**, 1108–1118.

52 Wong, D.L., Lesage, A., Siddall, B. and Funder, J.W. (1992) Glucocorticoid regulation of phenylethanolamine N-methyltransferase *in vivo*. *FASEB J.*, **6**, 3310–3315.

53 Ross, M.E., Evinger, M.J., Hyman, S.E., Carroll, J.M., Mucke, L., Comb, M., Reis, D.J., Joh, T.H. and Goodman, H.M. (1990) Identification of a functional glucocorticoid response element in the phenylethanolamine N-methyltransferase promoter using fusion genes introduced into chromaffin cells in primary culture. *J. Neurosci.*, **10**, 520–530.

54 Wong, D.L., Siddall, B.J., Ebert, S.N., Bell, R.A. and Her, S. (1998) Phenylethanolamine N-methyltransferase gene expression: synergistic activation by Egr-1, AP-2 and the glucocorticoid receptor. *Brain Res. Mol. Brain Res.*, **61**, 154–161.

55 Serova, L., Nankova, B., Rivkin, M., Kvetnansky, R. and Sabban, E.L. (1997) Glucocorticoids elevate GTP cyclo-hydrolase I mRNA levels *in vivo* and in PC12 cells. *Brain Res. Mol. Brain Res.*, **48**, 251–258.

56 Kubovcakova, L., Tybitanclova, K., Sabban, E.L., Majzoub, J., Zorad, S., Vietor, I., Wagner, E.F., Krizanova, O. and Kvetnansky, R. (2004) Catechol-amine synthesizing enzymes and their modulation by immobilization stress in knockout mice. *Ann. N. Y. Acad. Sci.*, **1018**, 458–465.

57 Baruchin, A., Vollmer, R.R., Miner, L.L., Sell, S.L., Stricker, E.M. and Kaplan, B.B. (1993) Cold-induced increases in phenylethanolamine N-methyltransferase

(PNMT) mRNA are mediated by non-cholinergic mechanisms in the rat adrenal gland. *Neurochem. Res.*, **18**, 759–766.

58 Vietor, I., Rusnak, M., Viskupic, E., Blazicek, P., Sabban, E.L. and Kvetnansky, R. (1996) Glucoprivation by insulin leads to trans-synaptic increase in rat adrenal tyrosine hydroxylase mRNA levels. *Eur. J. Pharmacol.*, **313**, 119–127.

59 Stachowiak, M., Stricker, E.M., Zigmond, M.J. and Kaplan, B.B. (1988) A cholinergic antagonist blocks cold stress-induced alterations in rat adrenal tyrosine hydroxylase mRNA. *Brain Res.*, **427**, 193–195.

60 Kvetnansky, R., Nankova, B., Hiremagalur, B., Viskupic, E., Vietor, I., Rusnak, M., McMahon, A., Kopin, I.J. and Sabban, E.L. (1996) Induction of adrenal tyrosine hydroxylase mRNA by single immobilization stress occurs even after splanchnic transection and in the presence of cholinergic antagonists. *J. Neurochem.*, **66**, 138–146.

61 Sabban, E.L., Hebert, M.A., Liu, X., Nankova, B. and Serova, L. (2004) Differential effects of stress on gene transcription factors in catecholaminergic systems. *Ann. N. Y. Acad. Sci.*, **1032**, 130–140.

62 Papanikolaou, N.A. and Sabban, E.L. (1999) Sp1/Egr1 motif: a new candidate in the regulation of rat tyrosine hydroxylase gene transcription by immobilization stress. *J. Neurochem.*, **73**, 433–436.

63 Liu, X., Kvetnansky, R., Serova, L., Sollas, A. and Sabban, E.L. (2005) Increased susceptibility to transcriptional changes with novel stressor in adrenal medulla of rats exposed to prolonged cold stress. *Brain Res. Mol. Brain Res.*, **141**, 19–29.

64 Sabban, E.L., Liu, X., Serova, L., Gueorguiev, V. and Kvetnansky, R. (2006) Stress triggered changes in gene expression in adrenal medulla: transcriptional responses to acute and chronic stress. *Cell. Mol. Neurobiol.*, **26**, 843–854.

65 Liu, X., Serova, L., Kvetnansky, R. and Sabban, E.L. (2008) Identifying the stress transcriptome in the adrenal medulla following acute and repeated immo-

bilization. *Ann. N. Y. Acad. Sci.*, **1148**, 1–28.

66 Cannon, W. (1929) Organization of physiological homeostasis. *Physiol. Rev.*, **9**, 3900–3431.

67 Goldstein, D.S. (2003) Catecholamines and stress. *Endocr. Regul.*, **37**, 69–80.

68 Sabban, E.L., Nankova, B.B., Serova, L.I., Kvetnansky, R. and Liu, X. (2004) Molecular regulation of gene expression of catecholamine biosynthetic enzymes by stress: sympathetic ganglia versus adrenal medulla. *Ann. N. Y. Acad. Sci.*, **1018**, 370–377.

69 Molinoff, P.B., Brimijoin, S. and Axelrod, J. (1972) Induction of dopamine-hydroxylase and tyrosine hydroxylase in rat hearts and sympathetic ganglia. *J. Pharmacol. Exp. Ther.*, **182**, 116–129.

70 Otten, U., Paravicini, U., Oesch, F. and Thoenen, H. (1973) Time requirement for the single steps of trans-synaptic induction of tyrosine hydroxylase in the peripehral sympathetic nervous system. *Naunyn-Schmiedeberg's Arch. Pharmacol.*, **280**, 117–127.

71 Silberstein, S.D., Brimijoin, S., Molinoff, P.B. and Lemberger, L. (1972) Induction of dopamine-hydroxylase in rat superior cervical ganglia in organ culture. *J. Neurochem.*, **19**, 919–921.

72 Thoenen, H., Mueller, R.A. and Axelrod, J. (1969) Trans-synaptic induction of adrenal tyrosine hydroxylase. *J. Pharmacol. Exp. Ther.*, **169**, 249–254.

73 Zigmond, R.E. (1980) The long-term regulation of ganglionic tyrosine hydroxylase by preganglionic nerve activity. *Fed. Proc.*, **39**, 3003–3008.

74 Micutkova, L., Rychkova, N., Sabban, E.L., Krizanova, O. and Kvetnansky, R. (2003) Quantitation of changes in gene expression of norepinephrine biosynthetic enzymes in rat stellate ganglia induced by stress. *Neurochem. Int.*, **43**, 235–242.

75 Kvetnansky, R. (2004) Stressor specificity and effect of prior experience on catecholamine biosynthetic enzyme phenylethanolamine N-methyltransferase. *Ann. N. Y. Acad. Sci.*, **1032**, 117–129.

76 Nankova, B., Kvetnansky, R., Hiremagalur, B., Sabban, B., Rusnak, M. and Sabban, E.L. (1996) Immobilization stress elevates gene expression for catecholamine biosynthetic enzymes and some neuropeptides in rat sympathetic ganglia: effects of adrenocorticotropin and glucocorticoids. *Endocrinology*, **137**, 5597–5604.

77 Partanen, M., Waller, S.B., London, E.D. and Hervonen, A. (1985) Indices of neurotransmitter synthesis and release in aging sympathetic nervous system. *Neurobiol. Aging*, **6**, 227–232.

78 Andrews, T., Lincoln, J., Milner, P., Burnstock, G. and Cowen, T. (1993) Differential regulation of tyrosine hydroxylase protein and activity in rabbit sympathetic neurones after long-term cold exposure: altered responses in ageing. *Brain Res.*, **624**, 69–74.

79 Zukowska-Grojec, Z. (1995) Neuro-peptide, Y. A novel sympathetic stress hormone and more. *Ann. N. Y. Acad. Sci.*, **771**, 219–233.

80 Wocial, B., Ignatowska-Switalska, H., Pruszczyk, P., Jedrusik, P., Januszewicz, A., Lapinski, M., Januszewicz W. and Zukowska-Grojec, Z. (1995) Plasma neuropeptide Y and catecholamines in women and men with essential hypertension. *Blood Press.*, **4**, 143–147.

81 Ulus, I.H. and Wurtman, R.J. (1979) Selective response of rat peripheral sympathetic nervous system to various stimuli. *J. Physiol.*, **293**, 513–523.

82 Kiran, B.K. and Ulus, I.H. (1992) Selective response of rat peripheral sympathetic nervous system to various stress situations, in *Neuroendocrine and Molecular Approaches* (eds R. Kvetnansky, R. McCarty and J. Axelrod), Gordon and Breach Sci. Publ., New York, pp. 561–568.

83 Brito, N.A., Brito, M.N. and Bartness, T.J. (2008) Differential sympathetic drive to adipose tissues after food deprivation, cold exposure or glucoprivation. *Am. J. Physiol. Regul. Integr. Comp. Physiol.*, **294**, R1445–R1452.

84 Biguet, N.F., Rittenhouse, A.R., Mallet, J. and Zigmond, R.E. (1989) Preganglionic nerve stimulation increases mRNA levels for tyrosine hydroxylase in the rat superior cervical ganglion. *Neurosci. Lett.*, **104**, 189–194.

85 Serova, L.I., Gueorguiev, V., Cheng, S.Y. and Sabban, E.L. (2008) Adrenocorticotropic hormone elevates gene expression for catecholamine biosynthesis in rat superior cervical ganglia and locus coeruleus by an adrenal independent mechanism. *Neuroscience*, **153**, 1380–1389.

86 Nankova, B.B., Kvetnansky, R. and Sabban, E.L. (2003) Adrenocorticotropic hormone (MC-2) receptor mRNA is expressed in rat sympathetic ganglia and up-regulated by stress. *Neurosci. Lett.*, **344**, 149–152.

87 Kvetnansky, R., Sabban, E.L. and Palkovits, M. (2009) Catecholaminergic systems in stress: structural and molecular genetic approaches. *Physiol. Rev.*, **89**, 535–606.

88 Palkovits, M. (2002) Stress-related central neuronal regulatory circuits, in *Stress: Neural, Endocrine and Molecular Studies* (eds R. McCarty, G. Aguillera, E.L. Sabban and R. Kvetnansky), Taylor and Francis, London, pp. 1–11.

89 Foote, S.L., Bloom, F.E. and Aston-Jones, G. (1983) Nucleus locus ceruleus: new evidence of anatomical and physiological specificity. *Physiol. Rev.*, **63**, 844–914.

90 Valentino, R.J. and Van Bockstaele, E. (2008) Convergent regulation of locus coeruleus activity as an adaptive response to stress. *Eur. J. Pharmacol.*, **583**, 194–203.

91 Wang, P., Kitayama, I. and Nomura, J. (1998) Tyrosine hydroxylase gene expression in the locus coeruleus of depression-model rats and rats exposed to short-and long-term forced walking stress. *Life Sci.*, **62**, 2083–2092.

92 Rusnak, M., Zorad, S., Buckendahl, P., Sabban, E.L. and Kvetnansky, R. (1998) Tyrosine hydroxylase mRNA levels in locus ceruleus of rats during adaptation to long-term immobilization stress exposure. *Mol. Chem. Neuropathol.*, **33**, 249–258.

93 Watanabe, Y., McKittrick, C.R., Blanchard, D.C., Blanchard, R.J., McEwen, B.S. and Sakai, R.R. (1995) Effects of chronic social stress on tyrosine hydroxylase mRNA and protein levels. *Brain Res. Mol. Brain Res.*, **32**, 176–180.

94 Mamalaki, E., Kvetnansky, R., Brady, L.S., Gold, P.W. and Herkenham, M. (1992) Repeated immoiblization stress alters tyrosine hydroxylase, coricotropin-releasing hormone and corticosteroid receptor messenger ribonucleic acid levels in rat brain. *J. Neuroendocrinol.*, **4**, 689–699.

95 Hebert, M.A., Serova, L.I. and Sabban, E.L. (2005) Single and repeated immo-bilization stres differentially trigger induction and phosphorylation of several transcription factors and MAP kinases in the rat locus coeruleus. *J. Neurochem.*, **95**, 483–498.

96 Serova, L.I., Nankova, B.B., Feng, Z., Hong, J.S., Hutt, M. and Sabban, E.L. (1999) Heightened transcription for enzymes involved in norepinephrine biosynthesis in the rat locus coeruleus by immobilization stress. *Biol. Psychiatry*, **45**, 853–862.

97 Chang, M.S., Sved, A.F., Zigmond, M.J. and Austin, M.C. (2000) Increased transcription of the tyrosine hydroxylase gene in individual locus coeruleus neurons following footshock stress. *Neuroscience*, **101**, 131–139.

98 Hebert, M.A., Serova, L.I. and Sabban, E.L. (2005) Single and repeated immo-bilization stress differentially trigger induction and phosphorylation of several transcription factors and mitogen-activated protein kinases in the rat locus coeruleus. *J. Neurochem.*, **95**, 484–498.

99 Nestler, E.J., Alreja, M. and Aghajanian, G.K. (1999) Molecular control of locus coeruleus neurotransmission. *Biol. Psychiatry*, **46**, 1131–1139.

3
Stress and the Cholinergic System

Mariella De Biasi

3.1
Acetylcholine and Stress

Stress elicits a transient increase in ACh release [1–3]. This increase in ACh levels is an important contributor to the neuroendocrine, emotional and physiological response to stressors [4–6]. Cholinergic neurons are activated in rats and mice by psychological stressors such as sensory stimuli [7–10], social stress [4] and conditioned fear [11, 12]. Physical stressors such as cold temperature and restraint have also been shown to trigger brain ACh release [13–17]. In humans, alterations in cholinergic transmission can be produced by the stress associated with sleep deprivation [18]. The functional and systemic impact of changing brain ACh levels during stress is not completely understood, as the phenomenon can be viewed either as a promoter of the stress response or as part of a "coping" mechanism. Evidence for both interpretations is available. In fact, ACh can stimulate the hypothalamic–pituitary–adrenal (HPA) axis leading to the release of stress neurohormones and neuropeptides, including corticosterone (CORT), adrenocorticotropin (ACTH) and corticotrophin-releasing hormone (CRH) [19–21]. Acute intracerebral injection of the muscarinic agonist carbachol elevates plasma levels of corticosterone [4, 22]. Other studies, however, show that increased cholinergic levels are associated with reduced stress and anxiety. For example, systemic administration of the acetylcholinesterase (AChE) inhibitor, physostigmine decreases novelty-induced neophobia in rats [23], while lesion of the cholinergic neurons in the medial septum heightens the corticosterone response to acute stress [24]. These apparently conflicting findings likely reflect differences in cholinergic mechanisms that depend on the brain circuit considered, the presence of different cholinergic receptor subtypes and both the nature and duration of the stressor, as well as differences associated with the species, sex, or genetic background of the experimental animal.

Stress – From Molecules to Behavior. Edited by Hermona Soreq, Alon Friedman, and Daniela Kaufer
Copyright © 2010 WILEY-VCH Verlag GmbH & Co. KGaA, Weinheim
ISBN: 978-3-527-32374-6

3.1.1
Cholinergic Innervation of the Brain

The cholinergic innervation of the brain mainly arises from eight nuclei located either in the basal forebrain or in the brain stem. The basal forebrain cholinergic system consists of the medial septum nucleus, the vertical and the horizontal limbs of the diagonal band of Broca and the nucleus basalis of Meynert. The brain stem contains the pedunculopontine nucleus, the laterodorsal tegmental nucleus, the medial habenula and the parabigeminal nucleus [25]. Cholinergic neurons provide sparse, widespread innervation to the brain. The nuclei in the basal forebrain send projections to the neocortex, the limbic cortex and the amygdala, whereas the pontomesencephalic nuclei send cholinergic inputs to the thalamus, basal forebrain and brain stem [25–30]. The ability of the forebrain and brainstem cholinergic systems to influence large brain areas depends on their anatomical characteristics and the fact that some of the effects of acetylcholine may occur via volume transmission [31–33]. According to the principles of volume transmission, the ACh released from synapses or varicosities may diffuse some distance before activating cholinergic receptors at synapses or extrasynaptic sites [34]. An exception is represented by the cholinergic interneurons located in the striatum and olfactory tubercule which, unlike the previous groups of cholinergic cells, provide dense, local innervation [35].

3.1.2
Brain Cholinergic Receptors

Cholinergic systems use acetylcholine (ACh) as their neurotransmitter and act through two major transmembrane receptor subtypes: metabotropic muscarinic receptors [36, 37] and ionotropic nicotinic receptors [34]. Muscarinic acetylcholine receptors are G protein-coupled receptors that affect ionic channels and intracellular processes via a second messenger cascade. Nicotinic receptors are ligand-gated ion channels permeable to Na^+, K^+ and Ca^{2+} ions.

Muscarinic Receptors   There are five subtypes of muscarinic acetylcholine receptors (mAChR; M1 to M5) [38] which are all G protein-coupled receptors. In general, the M1, M3 and M5 receptors activate phospholipase C (PLC) via pertussis toxin-insensitive G proteins of the G_q family while the M_2 and M_4 receptors inhibit adenylyl cyclase via pertussis toxin-sensitive G proteins of the G_i/G_o family [39]. Muscarinic receptor stimulation affects signal transduction pathways that regulate several cellular functions, including cell growth and survival [40–42]. Receptor activation can also lead to the regulation of ion channel activity and trafficking, an effect that can be achieved not only through second messenger cascades but also by direct regulation of the channel by activated G proteins [43–46].

Nicotinic Receptors   Neuronal nAChRs are members of a protein superfamily that includes neuronal and muscle nAChRs and receptors for GABA, glycine and serotonin (5HT)$_3$. Eleven genes have been identified that encode neuronal nAChR subunits: eight α subunits (α2–α10) and three β subunits (β2–β4) [47–51]. The mammalian nervous system expresses transcripts for all subunits except α8, which is found only in avians [52]. Neuronal nAChRs are formed by the assembly of five subunits, each consisting of four membrane-spanning domains [47, 49]. The majority of functional neuronal nAChRs are composed of two α and three β subunits [53–57]. Heteromeric nAChR channels are comprised of α2, α3, α4, α5, α6, β2, β3 and β4 subunits. α7 can form either homopentamers or α/β hetero-pentamers [55, 58]. α9 and α10 are believed to assemble as α only heteropentamers [50, 51]. The α5 subunit participates in nAChR receptors with $\alpha_x\alpha_y\beta$ combinations [57, 59, 60] but cannot yield functional receptors when expressed alone or in combination with β subunits only [59]. Finally, the β3 subunit participates in $\alpha\beta_x\beta_y$ combinations [56] and, similar to α5, it cannot yield functional receptors unless co-expressed with another β subunit.

Although fast nicotinic transmission has been reported in few brain areas, nicotinic cholinergic mechanisms are presynaptic in the vast majority of the CNS [34]. Activation of presynaptic nAChRs increases the release of many different neurotransmitters, including dopamine, norepinephrine, serotonin, glutamate and GABA [34]. Enhanced neurotransmitter release is produced by a direct Ca^{2+} signal through nAChRs and an indirect Ca^{2+}-induced Ca^{2+} release from intracellular stores [61]. In addition to those two mechanisms, activation of nAChRs produces a depolarization that activates voltage-gated Ca^{2+} channels in the presynaptic terminal [62].

3.1.3
AChR Distribution in the CNS

Muscarinic Receptors   All five muscarinic receptor subtypes are expressed in the CNS [63]. M1 muscarinic receptors are located postsynaptically in the cortex, hippocampus, striatum and thalamus [63, 64]. M2 receptors are predominantly found in the hindbrain and thalamus, with lower levels of expression in cortex, hippocampus and striatum, where they have a presynaptic localization [37]. The density of M3 and M5 mAChRs is much lower than that of the M1 and the M2 subtypes. M3 receptors are expressed in cortex, hippocampus and striatum [64], whereas M5 receptors are located primarily in the substantia nigra pars compacta and ventral tegmental area. Low levels of M5 expression are also found in hippocampus, mamillary bodies, hypothalamus and lateral habenula [65]. Moderate expression levels of M4 mAChRs are found in all subdivisions of the cholinergic basal forebrain, striatal cholinergic interneurons and the brainstem [66]. In addition, m4 receptor mRNA can be detected in non-cholinergic cells such as hippocampal pyramidal neurons and dentate gyrus granule cells as well as neurons in the dorsal raphe and pontine reticular nuclei [66].

Nicotinic Receptors   The mRNA expression patterns of neuronal nAChR subunits in the CNS are diverse, yet sometimes overlapping [67, 68]. α4β2-containing nAChRs are the most abundant nAChR subtype in the brain, with their highest levels of expression in the thalamus, the isocortex, the dorsomedial hypothalamic nucleus and the ventral midbrain. α3-containing (α3*) nAChRs are highly expressed in pineal gland, medial habenula (MHb), dorsal nucleus of the vagus nerve and retinal ganglionic neurons. Moderate to low levels of expression are found in the anterior thalamus, nucleus of the solitary tract, area postrema, dopaminergic ventral midbrain and noradrenergic locus coeruleus, as well as the cortex and hippocampal or parahippocampal areas [67, 69–71]. α7* nAChRs are highly expressed in the cortex, hippocampus and subcortical limbic regions, with lower levels of expression in basal ganglia. nAChRs containing the α6 and β3 subunits are expressed in midbrain, striatum, superior colliculus and lateral geniculate nucleus and form two major receptor subtypes: α4α6β2β3 and α6β2β3 [71].

3.1.4
The Septohippocampal Pathway and Stress

The cholinergic component of the septohippocampal system is the main source of hippocampal acetylcholine. Besides influencing learning and memory processes, this cholinergic pathway is an important component of the neural circuits that control the physiological and behavioral response to stress. Hippocampal ACh levels increase in behaving animals both during learning paradigms and during stress [13, 72, 73]. De Groot and Nomikos [74] found that aversive stimuli and their memories lead to increased hippocampal ACh levels, a phenomenon supporting the notion that stress and aversive memories share a common neurochemical substrate in the hippocampus.

The negative effects of chronic stress on memory are well known [75] but the duration and the nature of the stressor are important factors to be considered. Similar to what happens during normal memory processes [76], the efflux of ACh induced in the hippocampus by acute stress probably contributes to the mechanisms that provide saliency to the stimulus [77–79]. During more prolonged stress, increased ACh levels in the hippocampus might initiate homeostatic mechanisms aimed at reducing anxiety levels [80]. Negative-feedback mechanisms involving circulating glucocorticoids trigger inhibition of the HPA axis by the hippocampus [81, 82]. Based on the fact that cholinergic blockade in the hippocampus enhances the HPA response to stress [83] increased hippocampal levels of ACh might also promote homeostatic responses that result in HPA inhibition. However, in the presence of severe, chronic stress, homeostasis is disrupted and cholinergic function is altered. As a consequence, learning and memory impairment is seen along with neurochemical changes. While acute stress does not impair learning in the passive avoidance test, chronic unpredictable stress does [84]. Chronic restraint stress produces memory deficits in a partially baited radial arm maze task [85]. The time-dependent nature of the effects of stress is apparent in several other behavioral tasks. For example, acute stress does not affect behavioral responses in the elevated plus maze [86] but chronic stress impairs performance in an attentional set-shifting test and increases anxiety in the elevated plus maze [87]. A large body of literature also documents the negative effects exerted by prolonged stress on learning and memory in humans [88–90].

3.1.5
Stress-Induced Molecular Adaptations in the Cholinergic System

The behavioral changes observed during stress reflect molecular adaptations that affect multiple neurotransmitter systems, including the cholinergic one [91]. The reduction of AChE staining in the medial septum [92] and the decrease in hippocampal AChE activity [93] are among the neurochemical changes produced by chronic stress or chronic administration of corticosterone. Interestingly, chronic stress increases the expression of the otherwise rare readthrough form of AChE (AChE-R; [94, 95]). When overexpressed, AChE-R produces anxiety-like behaviors in the mouse [96]. Some of the other stress-induced neurochemical adaptations observed depend on the duration of the stressor. For example, choline uptake increases during acute stress but decreases during chronic stress [72], the former effect possibly reflecting a homeostatic response and the latter reflecting cholinergic disruption.

Another characteristic phenomenon observed during stress is the change in sensitivity to cholinergic agonists. Ten days of restraint stress led to cholinergic hypersensitivity, reflected by the reduced ability of the anticholinergic drug, scopolamine, to induce amnesia in an otherwise normal passive avoidance test [97]. When mice were exposed to the same stress paradigm for 30 days, retention in the passive avoidance test was decreased and the amnesic properties of scopolamine were potentiated. Chronic restraint stress was also shown to produce sensitization to the depressant effect of the M1 mAChR agonist, oxotremorine, on locomotor behavior [98]. A totally different type of stress, chronic food restriction, enhanced the facilitatory effects of oxotremorine and reduced the impairing effects of atropine on memory consolidation [99]. The underlying mechanism of the increased sensitivity to cholinergic stimulation has been shown to be an increase in muscarinic binding sites which occurs without changes in ligand affinity [72, 100, 101]. How stress influences mAChR trafficking and turnover is yet to be established.

Some of the molecular changes produced by chronic stress are observed in stress-related disorders such as depression [102]. Patients with depressive disorders might express certain single nucleotide polymorphisms in the MCHR2 [103] and have been reported to be more sensitive to the behavioral and physiological effects of muscarinic cholinergic agonists [104–106]. Interestingly, animals with increased mAChR levels or muscarinic hypersensitivity show greater depression-like responses in behavioral tests [107, 108].

3.1.5.1 **The Nicotinic Cholinergic System and Stress**
The role of the nicotinic cholinergic system in stress and anxiety has been mostly studied using nicotine, a non-selective nAChR agonist. Those studies suggest that one of the functions of the nicotinic cholinergic system is the modulation of the HPA axis. Nicotine administration can elevate plasma ACTH and corticosterone in laboratory animals [109, 110], and smoking elevates circulating cortisol in humans [111, 112]. The effects of nicotine, however, depend on the dose

considered. The systemic injection of low doses of nicotine has anxiolytic effects, whereas higher nicotine doses have anxiogenic effects [113]. In addition, brain region-specific differences exist in the effects of the drug, which likely reflect the binding to nAChR subtypes with different pharmacological properties [114]. On the social interaction test, nicotine usually produces anxiogenic effects when injected directly in the septum and hippocampus but is anxiolityc when injected in the raphe [114, 115].

Like the muscarinic cholinergic system, the nicotinic system undergoes stress-induced plasticity. Chronic treatment with corticosterone resulting in hormone plasma levels comparable to those measured during stress reduces the binding sites for α-bungarotoxin, a toxin selective for α7* nAChRs, in hippocampus, striatum and hypothalamus [116]. Chronic immobilization decreases the number of [^{3}H] cytisine binding sites in the cerebral cortex of rats [117]. Such plasticity alters the effects of nicotine so that, while nicotine rises plasma corticosterone in unstressed rats, it has no effect on the levels of the hormone in stressed animals [118]. The effects are also seen at the behavioral level, as stress promotes sensitization to nicotine-induced ambulatory stimulation [119, 120].

Another notable phenomenon is the seeming ability of nicotine to counteract the behavioral and neurochemical effects of stress. Nicotine relieves the anxiety-like manifestations produced by overexpression of the stress-related AChE isoform, AChE-R [96] and "normalizes" the changes in striatal dopamine produced by chronic stress [121]. In addition, nicotine can prevent the impairment in short-term memory produced by chronic psychosocial stress [122]. These observation are particularly important because many smokers use cigarettes as a tool to attenuate stress and anxiety and maintain that smoking has a calming effect [123–125]. Therefore, the understanding of the mechanisms underlying the interaction between stress and nicotine holds the potential of providing new therapeutic avenues for smoking cessation.

3.2
Contribution of Genetically Engineered Mouse Models to the Understanding of the Role of Cholinergic Receptors in Stress

The creation of mutant mice carrying null mutations for cholinergic receptors has provided a new set of tools in the study of cholinergic function and its role in stress responses. The fact that both the behavioral and biochemical deficits resulting from stress exposure can be reversed by oxotremorine [85, 126] suggest an important role for M1 AChRs receptors in the stress response. So far, it has been shown that HPA hormone responses to restraint stress are decreased in M1 AChR knockout mice compared to their wild-type littermates [127]. Mice null for the M1 AChR also display reduced LTP in response to theta burst stimulation in the hippocampus and have impaired behavioral responses in tasks which probe non-matching-to-sample working memory and consolidation

[128]. No laboratory has examined the effects of stress on hippocampal function in the M1 AChR null mice, but given its effects on memory processes, the mutation is expected to affect stress-induced neuronal plasticity in this brain area. M2 AChR knockout mice are more responsive to cholinergic stimulation by oxotremorine and to the mild stress of a saline injection than their wild-type littermates, consistent with brain M2 being inhibitory autoreceptors. HPA hormone responses to restraint stress are also increased in the M2 AChR null mice [127]. The effects of the M2 null mutations highlight the well known sexual diergism in HPA responses, as male M2 AChR null mice display greater HPA responses to cholinergic stimulation by physostigmine and oxotremorine than do female mice [129]. Studies with M2 and M4 AChR null mice have shown that autoinhibition of AChE release is mostly mediated by M2 AChRs in the hippocampus and cerebral cortex and by the M4 AChRs in the striatum [130, 131], but no information is available on how the null mutations might influence the homeostatic response to stress.

Behavioral studies conducted on nAChR mutant mice confirm the notion that nAChRs participate in stress-related responses. Mice lacking the β4 [132], β3 [133] and α4 [134] nAChR subunits display reduced anxiety-related behaviors. Interestingly, these null mutations affect anxiety-like behaviors in a test-specific fashion, with no significant effects on the mildly anxiogenic OFA and LDB tests but a robust effect on the more stressful EPM test [132]. Furthermore, the α5 and β4 null mutations render mice less sensitive to some of the behavioral effects of nicotine [135, 136]. Because α5 and β4, together with α3, form a genomic cluster [137], have overlapping promoter regulatory elements [138] and assemble to form functional receptors [139], it is tempting to speculate that the effects observed so far on anxiety are derived from channels composed of β4, α3, α4 and α5 subunits, in various combinations. Given the diminished effects of nicotine on α3, α5 and β4 mutant mice [135, 136, 140], one might also speculate that α3, α5 and β4 nAChRs are important mediators of nicotine's effects on anxiety. The prominent role of β4* nAChRs in stress-related manifestations is supported by the finding that α7 and β2 nAChR null mice have normal anxiety-like responses [141, 142]. Furthermore, β4* nAChRs are highly expressed in the medial habenula, a brain region involved in stress, anxiety and depression [143–146].

Summary

The cholinergic system, through the activation of both muscarinic and nicotinic AChRs, is an important player in the physiological response to stress. When stress is prolonged and homeostasis is disrupted, the system becomes vulnerable and disease may occur. The availability of mouse models which carry null mutations for both muscarinic and nicotinic AChRs is helping to identify the mechanisms underlying the interaction between stress and the cholinergic system. This will ultimately provide insight into stress-related disease.

References

1 Imperato, A., Puglisi-Allegra, S., Casolini, P., Zocchi, A. and Angelucci, L. (1989) Stress-induced enhancement of dopamine and acetylcholine release in limbic structures: role of corticosterone. *Eur. J. Pharmacol.*, **165**, 337–338.

2 Imperato, A., Puglisi-Allegra, S., Casolini, P. and Angelucci, L. (1991) Changes in brain dopamine and acetylcholine release during and following stress are independent of the pituitary-adrenocortical axis. *Brain Res.*, **538**, 111–117.

3 Dazzi, L., Motzo, C., Imperato, A., Serra, M., Gessa, G.L. and Biggio, G. (1995) Modulation of basal and stress-induced release of acetylcholine and dopamine in rat brain by abecarnil and imidazenil, two anxioselective gamma-aminobutyric acidA receptor modulators. *J. Pharmacol. Exp. Ther.*, **273**, 241–247.

4 Bugajski, J. (1999) Social stress adapts signaling pathways involved in stimulation of the hypothalamic-pituitary-adrenal axis. *J. Physiol. Pharmacol.*, **50**, 367–379.

5 Janowsky, D. and Overstreet, D. (2000) The role of acetylcholine mechanisms in affective disorders, in Neuropsycho-pharmacology: The Fifth Generation of Progress, American College of Neuropsychopharmacology, pp. 3–14.

6 Brand, L., Groenewald, I., Stein, D.J., Wegener, G. and Harvey, B.H. (2008) Stress and re-stress increases conditioned taste aversion learning in rats: possible frontal cortical and hippocampal muscarinic receptor involvement. *Eur. J. Pharmacol.*, **586**, 205–211.

7 Arankowsky-Sandoval, G., Prospero-Garcia, O., Aguilar-Roblero, R. and Drucker-Colin, R. (1986) Cholinergic reduction of REM sleep duration is reverted by auditory stimulation. *Brain Res.*, **375**, 377–380.

8 Inglis, F.M. and Fibiger, H.C. (1995) Increases in hippocampal and frontal cortical acetylcholine release associated with presentation of sensory stimuli. *Neuroscience*, **66**, 81–86.

9 Degroot, A., Wade, M., Salhoff, C., Davis, R.J., Tzavara, E.T. and Nomikos, G.G. (2004) Exposure to an elevated platform increases plasma corticosterone and hippocampal acetylcholine in the rat: reversal by chlordiazepoxide. *Eur. J. Pharmacol.*, **493**, 103–109.

10 Smith, D.G., Davis, R.J., Gehlert, D.R. and Nomikos, G.G. (2006) Exposure to predator odor stress increases efflux of frontal cortex acetylcholine and monoamines in mice: comparisons with immobilization stress and reversal by chlordiazepoxide. *Brain Res.*, **1114**, 24–30.

11 Nail-Boucherie, K., Dourmap, N., Jaffard, R. and Costentin, J. (2000) Contextual fear conditioning is associated with an increase of acetyl-choline release in the hippocampus of rat. *Brain Res. Cogn. Brain Res.*, **9**, 193–197.

12 Calandreau, L., Trifilieff, P., Mons, N., Costes, L., Marien, M., Marighetto, A., Micheau, J., Jaffard, R. and Desmedt, A. (2006) Extracellular hippocampal acetylcholine level controls amygdala function and promotes adaptive conditioned emotional response. *J. Neurosci.*, **26**, 13556–13566.

13 Imperato, A., Puglisi-Allegra, S., Casolini, P. and Angelucci, L. (1991) Changes in brain dopamine and acetylcholine release during and following stress are independent of the pituitary-adrenocortical axis. *Brain Res.*, **538**, 111–117.

14 Mark, G.P., Rada, P.V. and Shors, T.J. (1996) Inescapable stress enhances extracellular acetylcholine in the rat hippocampus and prefrontal cortex but not the nucleus accumbens or amygdala. *Neuroscience*, **74**, 767–774.

15 Stillman, M.J., Shukitt-Hale, B., Coffey, B.P., Levy, A. and Lieberman, H.R. (1997) *In vivo* hippocampal acetylcholine release during exposure to acute stress. *Stress*, **1**, 191–200.

16 Mizoguchi, K., Yuzurihara, M., Ishige, A., Sasaki, H. and Tabira, T. (2001) Effect of chronic stress on cholinergic transmission in rat hippocampus. *Brain Res.*, **915**, 108–111.

17 Masuda, J., Mitsushima, D. and Kimura, F. (2004) Female rats living in small cages respond to restraint stress with both adrenocortical corticosterone release and acetylcholine release in the hippocampus. *Neurosci. Lett.*, **358**, 169–172.

18 Boonstra, T., Stins, J., Daffertshofer, A. and Beek, P. (2007) Effects of sleep deprivation on neural functioning: an integrative review. *Cell. Mol. Life Sci.*, **64**, 934–946.

19 Calogero, A.E., Bernardini, R., Gold, P.W. and Chrousos, G.P. (1988) Regulation of rat hypothalamic corticotropin-releasing hormone secretion *in vitro*: potential clinical implications. *Adv. Exp. Med. Biol.*, **245**, 167–181.

20 Calogero, A.E., Kamilaris, T.C., Johnson, E.O., Tartaglia, M.E. and Chrousos, G.P. (1990) Recovery of the rat hypothalamic–pituitary–adrenal axis after discontinuation of prolonged treatment with the synthetic glucocorticoid agonist dexamethasone. *Endocrinology*, **127**, 1574–1579.

21 Matta, S.G., Fu, Y., Valentine, J.D. and Sharp, B.M. (1990) Response of the hypothalamo–pituitary–adrenal axis to nicotine. *Psychoneuroendocrinology*, **23**, 103–113.

22 Bugajski, A.J., Zurowski, D., Thor, P. and Gadek-Michalska, A. (2007) Effect of subdiaphragmatic vagotomy and cholinergic agents in the hypothalamic-pituitary-adrenal axis activity. *J. Physiol. Pharmacol.*, **58**, 335–347.

23 Sienkiewicz-Jarosz, H., Członkowska, A.I., Siemiatkowski, M., Maciejak, P., Szyndler, J. and Planik, A. (2000) The effects of physostigmine and cholinergic receptor ligands on novelty-induced neophobia. *J. Neural. Transm.*, **107**, 1403–1412.

24 Helm, K.A., Ziegler, D.R. and Gallagher, M. (2004) Habituation to stress and dexamethasone suppression in rats with selective basal forebrain cholinergic lesions. *Hippocampus*, **14**, 628–635.

25 Mesulam, M.M., Mufson, E.J., Levey, A.I. and Wainer, B.H. (1983) Cholinergic innervation of cortex by the basal forebrain: cytochemistry and cortical connections of the septal area, diagonal band nuclei, nucleus basalis (Substantia innominata), and hypothalamus in the rhesus monkey. *J. Comp. Neurol.*, **214**, 170–197.

26 Woolf, N.J., Eckenstein, F. and Butcher, L.L. (1983) Cholinergic projections from the basal forebrain to the frontal cortex: a combined fluorescent tracer and immunohistochemical analysis in the rat. *Neurosci. Lett.*, **40**, 93–98.

27 Losier, B.J. and Semba, K. (1993) Dual projections of single cholinergic and aminergic brainstem neurons to the thalamus and basal forebrain in the rat. *Brain Res.*, **604**, 41–52.

28 Zhou, F.-M., Wilson, C. and Dani, J.A. (2003) Muscarinic and nicotinic cholinergic mechanisms in the mesostriatal dopamine systems. *Neuroscientist*, **9**, 23–36.

29 Butcher, L.L. and Woolf, N.J. (2004) Cholinergic neurons and networks revisited, in *The Rat Central Nervous System* (ed. G. Paxinos), Elsevier Academic, San Francisco, pp. 1257–1268.

30 Maskos, U. (2008) The cholinergic mesopontine tegmentum is a relatively neglected nicotinic master modulator of the dopaminergic system: relevance to drugs of abuse and pathology. *Br. J. Pharmacol.*, **153**, S438–S445.

31 Umbriaco, D., Watkins, K.C., Descarries, L., Cozzari, C. and Hartman, B.K. (1994) Ultrastructural and morphometric features of the acetylcholine innervation in adult rat parietal cortex: an electron microscopic study in serial sections. *J. Comp. Neurol.*, **348**, 351–373.

32 Agnati, L.F., Zoli, M., Strömberg, I. and Fuxe, K. (1995) Intercellular communication in the brain: wiring versus volume transmission. *Neuroscience*, **69**, 711–726.

33 Descarries, L., Gisiger, V. and Steriade, M. (1997) Diffuse transmission by acetylcholine in the CNS. *Prog. Neurobiol.*, **53**, 603–625.

34 Dani, J.A. and Bertrand, D. (2007) Nicotinic acetylcholine receptors and nicotinic cholinergic mechanisms of the central nervous system. *Annu. Rev. Pharmacol. Toxicol.*, **47**, 699–729.

35 Zhou, F.M., Wilson, C.J. and Dani, J.A. (2002) Cholinergic interneuron characteristics and nicotinic properties in the striatum. *J. Neurobiol.*, **53**, 590–605.

36 Bymaster, F.P., McKinzie, D.L., Felder, C.C. and Wess, J. (2003) Use of M1–M5 muscarinic receptor knockout mice as novel tools to delineate the physiological roles of the muscarinic cholinergic system. *Neurochem. Res.*, **28**, 437–442.

37 Wess, J., Duttaroy, A., Zhang, W., Gomeza, J., Cui, Y., Miyakawa, T., Bymaster, F.P., McKinzie, L., Felder, C.C., Lamping, K.G., Faraci, F.M., Deng, C. and Yamada, M. (2003) M 1-M 5 muscarinic receptor knockout mice as novel tools to study the physiological roles of the muscarinic cholinergic system. *Receptors Channels*, **9**, 279–290.

38 Nathanson, N.M. (2008) Synthesis, trafficking, and localization of muscarinic acetylcholine receptors. *Pharmacol. Ther.*, **119**, 33–43.

39 Lanzafame, A.A., Christopoulos, A. and Mitchelson, F. (2003) Cellular signaling mechanisms for muscarinic acetylcholine receptors. *Receptors Channels*, **9**, 241–260.

40 Marinissen, M.J. and Gutkind, J.S. (2001) G-protein-coupled receptors and signaling networks: emerging paradigms. *Trends Pharmacol. Sci.*, **22**, 368–376.

41 Tobin, A.B. and Budd, D.C. (2003) The anti-apoptotic response of the Gq/11-coupled muscarinic receptor family. *Biochem. Soc. Trans.*, **31**, 1182–1185.

42 Wessler, I., Kilbinger, H., Bittinger, F., Unger, R. and Kirkpatrick, C.J. (2003) The non-neuronal cholinergic system in humans: expression, function and pathophysiology. *Life Sci.*, **72**, 2055–2061.

43 Wickman, K. and Clapham, D.E. (1995) Ion channel regulation by G proteins. *Physiol. Rev.*, **75**, 865–885.

44 Herlitze, S., Garcia, D.E., Mackie, K., Hille, B., Scheuer, T. and Catterall, W.A. (1996) Modulation of Ca2+ channels [beta][gamma] G-protein py subunits. *Nature*, **380**, 258–262.

45 Cayouette, S., Lussier, M.P., Mathieu, E.-L., Bousquet, S.M. and Boulay, G. (2004) Exocytotic insertion of TRPC6 channel into the plasma membrane upon Gq protein-coupled receptor activation. *J. Biol. Chem.*, **279**, 7241–7246.

46 Singh, B.B., Lockwich, T.P., Bandyopadhyay, B.C., Liu, X., Bollimuntha, S., Brazer, S.-C., Combs, C., Das, S., Leenders, A.G.M., Sheng, Z.-H., Knepper, M.A., Ambudkar, S.V. and Ambudkar, I.S. (2004) VAMP2-dependent exocytosis regulates plasma membrane insertion of TRPC3 channels and contributes to agonist-stimulated Ca2+ influx. *Mol. Cell*, **15**, 635–646.

47 Sargent, P.B. (1993) The diversity of neuronal nicotinic acetylcholine receptors. *Annu. Rev. Neurosci.*, **16**, 403–443.

48 Elgoyen, A.B., Johnson, D.S., Boulter, J., Vetter, D.E. and Heinemann, S. (1994) α9: an acetylcholine receptor with novel pharmacological properties expressed in rat cochlear hair cells. *Cell*, **79**, 705–715.

49 McGehee, D.S. and Role, L.W. (1995) Physiological diversity of nicotinic acetylcholine receptors expressed by vertebrate neurons. *Annu. Rev. Physiol.*, **57**, 521–546.

50 Elgoyhen, A.B., Vetter, D.E., Katz, E., Rothlin, C.V., Heinemann, S.F. and Boulter, J. (2001) alpha10: a determinant of nicotinic cholinergic receptor function in mammalian vestibular and cochlear mechanosensory hair cells. *Proc. Natl. Acad. Sci. U. S. A.*, **98**, 3501–3506.

51 Sgard, F., Charpantier, E., Bertrand, S., Walker, N., Caput, D., Graham, D., Bertrand, D. and Besnard, F. (2002) A novel human nicotinic receptor subunit, alpha10, that confers functionality to the alpha9-subunit. *Mol. Pharmacol.*, **61**, 150–159.

52 Schoepfer, R., Conroy, W.G., Whiting, P., Gore, M. and Lindstrom, J. (1990) Brain alpha-bungarotoxin binding protein cDNAs and MAbs reveal subtypes of this branch of the ligand-gated ion channel gene superfamily. *Neuron*, **5**, 35–48.

53 Anand, R., Conroy, W.G., Schoepfer, R., Whiting, P. and Lindstrom, J. (1991) Neuronal nicotinic acetylcholine receptors expressed in Xenopus oocytes

have a pentameric quaternary structure. *J. Biol. Chem.*, **266**, 11192–11198.

54 Cooper, E., Couturier, S. and Ballivet, M. (1991) Pentameric structure and subunit stoichiometry of a neuronal nicotinic acetylcholine receptor. *Nature*, **350**, 235–238.

55 Seguela, P., Wadiche, J., Dineley-Miller, K., Dani, J.A. and Patrick, J.W. (1993) Molecular cloning, functional properties, and distribution of rat brain α7: a nicotinic cation channel highly permeable to calcium. *J. Neurosci.*, **13**, 596–604.

56 Boorman, J.P., Groot-Kormelink, P.J. and Sivilotti, L.G. (2000) Stoichiometry of human recombinant neuronal nicotinic receptors containing the b3 subunit expressed in Xenopus oocytes. *J. Physiol.*, **529** (Pt 3), 565–577.

57 Groot-Kormelink, P.J., Boorman, J.P. and Sivilotti, L.G. (2001) Formation of functional alpha3beta4alpha5 human neuronal nicotinic receptors in Xenopus oocytes: a reporter mutation approach. *Br. J. Pharmacol.*, **134**, 789–796.

58 Khiroug, S.S., Harkness, P.C., Lamb, P.W., Sudweeks, S.N., Khiroug, L., Millar, N.S. and Yakel, J.L. (2002) Rat nicotinic ACh receptor alpha7 and beta2 subunits co-assemble to form functional heteromeric nicotinic receptor channels. *J. Physiol.*, **540**, 425–434.

59 Ramirez-Latorre, J., Yu, C.R., Qu, X., Perin, F., Karlin, A. and Role, L. (1996) Functional contributions of alpha5 subunit to neuronal acetylcholine receptor channels. *Nature*, **380**, 347–351.

60 Gerzanich, V., Wang, F., Kuryatov, A. and Lindstrom, J. (1998) alpha 5 Subunit alters desensitization, pharmacology, Ca++ permeability and Ca++ modulation of human neuronal alpha 3 nicotinic receptors. *J. Pharmacol. Exp. Ther.*, **286**, 311–320.

61 Sharma, G. and Vijayaraghavan, S. (2003) Modulation of presynaptic store calcium induces release of glutamate and postsynaptic firing. *Neuron*, **38**, 929–939.

62 Tredway, T.L., Guo, J.-Z. and Chiappinelli, V.A. (1999) N-type voltage-dependent calcium channels mediate the nicotinic enhancement of

GABA release in chick brain. *J. Neurophysiol.*, **81**, 447–454.

63 Levey, A.I. (1993) Immunological localization of m1-m5 muscarinic acetylcholine receptors in peripheral tissues and brain. *Life Sci.*, **52**, 441–448.

64 Levey, A.I. (1996) Muscarinic acetylcholine receptor expression in memory circuits: implications for treatment of Alzheimer disease. *Proc. Natl Acad. Sci. USA*, **93**, 13541–13546.

65 Vilaro, M.T., Palacios, J.M. and Mengod, G. (1990) Localization of m5 muscarinic receptor mRNA in rat brain examined by in situ hybridization histochemistry. *Neurosci. Lett.*, **114**, 154–159.

66 Sugaya, K., Clamp, C., Bryan, D. and McKinney, M. (1997) mRNA for the m4 muscarinic receptor subtype is expressed in adult rat brain cholinergic neurons. *Mol. Brain Res.*, **50**, 305–313.

67 Gotti, C., Moretti, M., Gaimarri, A., Zanardi, A., Clementi, F. and Zoli, M. (2007) Heterogeneity and complexity of native brain nicotinic receptors. *Biochem. Pharmacol.*, **74**, 1102–1111.

68 De Biasi, M. and Salas, R. (2008) Influence of neuronal nicotinic receptors over nicotine addiction and withdrawal. *Exp. Biol. Med.*, **233**, 917–929.

69 Wada, E., Wada, K., Boulter, J., Deneris, E., Heinemann, S., Patrick, J. and Swanson, L.W. (1989) Distribution of alpha 2, alpha 3, alpha 4, and beta 2 neuronal nicotinic receptor subunit mRNAs in the central nervous system: a hybridization histochemical study in the rat. *J. Comp. Neurol.*, **284**, 314–335.

70 Zoli, M., Lena, C., Picciotto, M.R. and Changeux, J.P. (1998) Identification of four classes of brain nicotinic receptors using beta2 mutant mice. *J. Neurosci.*, **18**, 4461–4472.

71 Zoli, M., Moretti, M., Zanardi, A., McIntosh, J.M., Clementi, F. and Gotti, C. (2002) Identification of the nicotinic receptor subtypes expressed on dopaminergic terminals in the rat striatum. *J. Neurosci.*, **22**, 8785–8789.

72 Gilad, G.M., Mahon, B.D., Finkelstein, Y., Koffler, B. and Gilad, V.H. (1985) Stress-induced activation of the

hippocampal cholinergic system and the pituitary-adrenocortical axis. *Brain Res.*, **347**, 404–408.

73 Mizuno, T. and Kimura, F. (1997) Attenuated stress response of hippocampal acetylcholine release and adrenocortical secretion in aged rats. *Neurosci. Lett.*, **222**, 49–52.

74 Degroot, A. and Nomikos, G.G. (2005) Fluoxetine disrupts the integration of anxiety and aversive memories. *Neuropsychopharmacology*, **30**, 391–400.

75 Wolf, O.T. (2008) The influence of stress hormones on emotional memory: relevance for psychopathology. *Acta Psychol.*, **127**, 513–531.

76 Gold, P.E. (2003) Acetylcholine modulation of neural systems involved in learning and memory. *Neurobiol. Learn. Mem.*, **80**, 194–210.

77 Smeets, T., Otgaar, H., Candel, I. and Wolf, O.T. (2008) True or false? Memory is differentially affected by stress-induced cortisol elevations and sympathetic activity at consolidation and retrieval. *Psychoneuroendocrinology*, **33**, 1378–1386.

78 Shors, T.J. (2001) Acute stress rapidly and persistently enhances memory formation in the male rat. *Neurobiol. Learn. Mem.*, **75**, 10–29.

79 Duncko, R., Cornwell, B., Cui, L., Merikangas, K.R. and Grillon, C. (2007) Acute exposure to stress improves performance in trace eyeblink conditioning and spatial learning tasks in healthy men. *Learn. Mem.*, **14**, 329–335.

80 Smythe, J.W., Bhatnagar, S., Murphy, D., Timothy, C. and Costall, B. (1998) The effects of intrahippocampal scopolamine infusions on anxiety in rats as measured by the black-white box test. *Brain Res. Bull.*, **45**, 89–93.

81 Brown, E.S., Rush, A.J. and McEwen, B.S. (1999) Hippocampal remodeling and damage by corticosteroids: implications for mood disorders. *Neuropsychopharmacology*, **21**, 474–484.

82 McEwen, B.S. (2000) Protective and damaging effects of stress mediators: central role of the brain. *Prog. Brain Res.*, **122**, 25–34.

83 Bhatnagar, S., Costall, B. and Smythe, J.W. (1997) Hippocampal cholinergic blockade enhances hypothalamic-pituitary-adrenal responses to stress. *Brain Res.*, **766**, 244–248.

84 Das, A., Rai, D., Dikshit, M., Palit, G. and Nath, C. (2005) Nature of stress: differential effects on brain acetylcholinesterase activity and memory in rats. *Life Sci.*, **77**, 2299–2311.

85 Srikumar, B.N., Raju, T.R. and Shankaranarayana Rao, B.S. (2006) The involvement of cholinergic and noradrenergic systems in behavioral recovery following oxotremorine treatment to chronically stressed rats. *Neuroscience*, **143**, 679–688.

86 Daniels, W., de Klerk Uys, J., van Vuuren, P. and Stein, D. (2008) The development of behavioral and endocrine abnormalities in rats after repeated exposure to direct and indirect stress. *Neuropsychiatr. Dis. Treat.*, **4**, 451–464.

87 Bondi, C.O., Rodriguez, G., Gould, G.G., Frazer, A. and Morilak, D.A. (2008) Chronic unpredictable stress induces a cognitive deficit and anxiety-like behavior in rats that is prevented by chronic antidepressant drug treatment. *Neuropsychopharmacology*, **33**, 320–331.

88 McEwen, B.S. and Sapolsky, R.M. (1995) Stress and cognitive function. *Curr. Opin. Neurobiol.*, **5**, 205–216.

89 Sauro, M.D., Jorgensen, R.S. and Pedlow, C.T. (2003) Stress, glucocorticoids, and memory: a meta-analytic review. *Stress*, **6**, 235–245.

90 Shors, T.J. (2006) Stressful experience and learning across the lifespan. *Annu. Rev. Psychol.*, **57**, 55–85.

91 Jordan, S., Kramer, G.L., Zukas, P.K. and Petty, F. (1994) Previous stress increases *in vivo* biogenic amine response to swim stress. *Neurochem. Res.*, **19**, 1521–1525.

92 Tizabi, Y., Gilad, V.H. and Gilad, G.M. (1989) Effects of chronic stressors or corticosterone treatment on the septohippocampal cholinergic system of the rat. *Neurosci. Lett.*, **105**, 177–182.

93 Sunanda, S., Rao, B.S.S. and Raju, T.R. (2000) Restraint stress-induced alterations in the levels of biogenic

amines, amino acids, and AChE activity in the hippocampus. *Neurochem. Res.*, **25**, 1547–1552.

94 Soreq, H. and Seidman, S. (2001) Acetylcholinesterase – new roles for an old actor. *Nat. Rev. Neurosci.*, **2**, 294–302.

95 Meshorer, E., Erb, C., Gazit, R., Pavlovsky, L., Kaufer, D., Friedman, A., Glick, D., Ben-Arie, N. and Soreq, H. (2002) Alternative splicing and neuritic mRNA translocation under long-term neuronal hypersensitivity. *Science*, **295**, 508–512.

96 Salas, R., Main, A., Gangitano, D.A., Zimmerman, G., Ben-Ari, S., Soreq, H. and De Biasi, M. (2008) Nicotine relieves anxiogenic-like behavior in acetylcholinesterase-R overexpressing mice but not in wild-type mice. *Mol. Pharmacol.*, **74** (6), 1641–1648.

97 Zerbib, R. and Laborit, H. (1990) Chronic stress and memory: implication of the central cholinergic system. *Pharmacol. Biochem. Behav.*, **36**, 897–900.

98 Badiani, A., Castellano, C. and Oliverio, A. (1991) Effects of acute and chronic stress and of genotype on oxotremorine-induced locomotor depression of mice. *Behav. Neural Biol.*, **55**, 123–130.

99 Orsini, C., Castellano, C. and Cabib, S. (2001) Pharmacological evidence of muscarinic-cholinergic sensitization following chronic stress. *Psychopharmacology*, **155**, 144–147.

100 González, A.M. and Pazos, A. (1992) Modification of muscarinic acetylcholine receptors in the rat brain following chronic immobilization stress: an autoradiographic study. *Eur. J. Pharmacol.*, **223**, 25–31.

101 Myslivecek, J. and Kvetňanský, R. (2006) The effects of stress on muscarinic receptors. Heterologous receptor regulation: yes or no? *Auton. Autacoid Pharmacol.*, **26**, 235–251.

102 Janowsky, D., Davis, J., El-Yousef, M.K. and Sekerke, H.J. (1972) A Cholinergic-adrenergic hypothesis of mania and depression. *Lancet*, **300**, 632–635.

103 Wang, J.C., Hinrichs, A.L., Stock, H., Budde, J., Allen, R., Bertelsen, S., Kwon, J.M., Wu, W., Dick, D.M., Rice, J., Jones, K., Nurnberger, J.I. Jr., Tischfield, J., Porjesz, B., Edenberg, H.J., Hesselbrock, V., Crowe, R., Schuckit, M., Begleiter, H., Reich, T., Goate, A.M. and Bierut, L.J. (2004) Evidence of common and specific genetic effects: association of the muscarinic acetylcholine receptor M2 (CHRM2) gene with alcohol dependence and major depressive syndrome. *Hum. Mol. Genet.*, **13**, 1903–1911.

104 Meyerson, L.R., Wennogle, L.P., Abel, M.S., Coupet, J., Lippa, A.S., Rauh, C.E. and Beer, B. (1982) Human brain receptor alterations in suicide victims. *Pharmacol. Biochem. Behav.*, **17**, 159–163.

105 Dilsaver, S.C. (1986) Cholinergic mechanisms in depression. *Brain Res.*, **396**, 285–316.

106 Berger, M., Riemann, D., Hochli, D. and Speigel, R. (1989) The cholinergic rapid eye movement sleep induction test with RS 86. *Arch. Gen. Psychiatry*, **46**, 421–428.

107 Overstreet, D.H. (1993) The Flinders sensitive line rats: a genetic animal model of depression. *Neurosci. Biobehav. Rev.*, **17**,

108 Touma, C., Bunck, M., Glasl, L., Nussbaumer, M., Palme, R., Stein, H., Wolferstätter, M., Zeh, R., Zimbelmann, M., Holsboer, F. and Landgraf, R. (2008) Mice selected for high versus low stress reactivity: a new animal model for affective disorders. *Psychoneuroendocrinology*, **33**, 839–862.

109 Cam, G.R. and Bassett, J.R. (1983) The plasma levels of ACTH following exposure to stress or nicotine. *Arch. Int. Pharmacodyn. Ther.*, **264**, 154–167.

110 Balfour, D.J. (1989) Influence of nicotine on the release of monoamines in the brain. *Prog. Brain Res.*, **79**, 165–172.

111 Wilkins, J.N., Carlson, H.E., Van Vunakis, H., Hill, M.A., Gritz, E. and Jarvik, M.E. (1982) Nicotine from cigarette smoking increases circulating levels of cortisol, growth hormone, and prolactin in male chronic smokers. *Psychopharmacology (Berl.)*, **78**, 305–308.

112 Gossain, V., Sherma, N., Srivastava, L., Michelakis, A. and Rovner, D. (1986) Hormonal effects of smoking – II: effects on plasma cortisol, growth hormone, and prolactin. *Am. J. Med. Sci.*, **291**, 325–327.

113 File, S.E., Gonzalez, L.E. and Gallant, R. (1998) Role of the basolateral nucleus of the amygdala in the formation of a phobia. *Neuropsychopharmacology*, **19**, 397–405.

114 File, S.E., Cheeta, S. and Kenny, P.J. (2000) Neurobiological mechanisms by which nicotine mediates different types of anxiety. *Eur. J. Pharmacol.*, **393**, 231–236.

115 Ouagazzal, A.M., Kenny, P.J. and File, S.E. (1999) Modulation of behaviour on trials 1 and 2 in the elevated plus-maze test of anxiety after systemic and hippocampal administration of nicotine. *Psychopharmacology (Berl.)*, **144**, 54–60.

116 Stitzel, J.A., Farnham, D.A. and Collins, A.C. (1996) Chronic corticosterone treatment elicits dose-dependent changes in mouse brain alpha-bungarotoxin binding. *Neuroscience*, **72**, 791–799.

117 Takita, M. and Muramatsu, I. (1995) Alteration of brain nicotinic receptors induced by immobilization stress and nicotine in rats. *Brain Res.*, **681**, 190–192.

118 Balfour, D.J., Khullar, A.K. and Longden, A. (1975) Effects of nicotine on plasma corticosterone and brain amines in stressed and unstressed rats. *Pharmacol. Biochem. Behav.*, **3**, 179–184.

119 Kita, T., Okamotq, M., Kubq, K., Tanaka, T. and Nakashima, T. (1999) Enhancement of sensitization to nicotine-induced ambulatory stimulation by psychological stress in rats. *Prog. Neuropsychopharmacol. Biol. Psychiatry*, **23**, 893–903.

120 McCormick, C.M., Robarts, D., Gleason, E. and Kelsey, J.E. (2004) Stress during adolescence enhances locomotor sensitization to nicotine in adulthood in female, but not male, rats. *Horm. Behav.*, **46**, 458–466.

121 Salas, R. and De Biasi, M. (2008) Opposing actions of chronic stress and chronic nicotine on striatal function in mice. *Neurosci. Lett.*, **440**, 32–34.

122 Aleisa, A.M., Alzoubi, K.H., Gerges, N.Z. and Alkadhi, K.A. (2006) Nicotine blocks stress-induced impairment of spatial memory and long-term potentiation of the hippocampal CA1

region. *Int. J. Neuropsychopharmacol.*, **9**, 417–426.

123 Pomerleau, O.F. and Pomerleau, C.S. (1984) Neuroregulators and the reinforcement of smoking: towards a biobehavioral explanation. *Neurosci. Biobehav. Rev.*, **8**, 503–513.

124 Cohen, S. and Lichtenstein, E. (1990) Perceived stress, quitting smoking, and smoking relapse. *Health Psychol.*, **9**, 466–478.

125 Parrott, A.C. (1995) Stress modulation over the day in cigarette smokers. *Addiction*, **90**, 233–244.

126 File, S.E., Gonzalez, L.E. and Andrews, N. (1998) Endogenous acetylcholine in the dorsal hippocampus reduces anxiety through actions on nicotinic and muscarinic1 receptors. *Behav. Neurosci.*, **112**, 352–359.

127 Rhodes, M.E., Rubin, R.T., McKlveen, J.M., Karwoski, T.E., Fulton, B.A. and Czambel, R.K. (2008) Pituitary-adrenal responses to oxotremorine and acute stress in male and female M1 muscarinic receptor knockout mice: comparisons to M2 muscarinic receptor knockout mice. *J. Neuroendocrinol.*, **20**, 617–625.

128 Anagnostaras, S.G., Murphy, G.G., Hamilton, S.E., Mitchell, S.L., Rahnama, N.P., Nathanson, N.M. and Silva, A.J. (2003) Selective cognitive dysfunction in acetylcholine M1 muscarinic receptor mutant mice. *Nat. Neurosci.*, **6**, 51–58.

129 Rhodes, M.E., Billings T.E., Czambel R.K. and Rubin, R.T. (2005) Pituitary-adrenal responses to cholinergic stimulation and acute mild stress are differentially elevated in male and female M_2 muscarinic receptor knockout mice. *J. Neuroendocrinol.*, **17**, 817–826.

130 Zhang, W., Basile, A.S., Gomeza, J., Volpicelli, L.A., Levey, A.I. and Wess, J. (2002) Characterization of central inhibitory muscarinic autoreceptors by the use of muscarinic acetylcholine receptor knock-out mice. *J. Neurosci.*, **22**, 1709–1717.

131 Tzavara, E.T., Bymaster, F.P., Felder, C.C., Wade, M., Gomeza, J., Wess, J., McKinzie, D.L. and Nomikos, G.G.

(2003) Dysregulated hippocampal acetylcholine neurotransmission and impaired cognition in M2, M4 and M2//M4 muscarinic receptor knockout mice. *Mol. Psychiatry*, **8**, 673–679.

132 Salas, R., Pieri, F., Fung, B., Dani, J.A. and De Biasi, M. (2003) Altered anxiety-related responses in mutant mice lacking the β4 subunit of the nicotinic receptor. *J. Neurosci.*, **23**, 6255–6263.

133 Booker, T.K., Butt, C.M., Wehner, J.M., Heinemann, S.F. and Collins, A.C. (2007) Decreased anxiety-like behavior in beta3 nicotinic receptor subunit knockout mice. *Pharmacol. Biochem. Behav.*, **87**, 146–157.

134 Ross, S.A., Wong, J.Y., Clifford, J.J., Kinsella, A., Massalas, J.S., Horne, M.K., Scheffer, I.E., Kola, I., Waddington, J.L., Berkovic, S.F. and Drago, J. (2000) Phenotypic characterization of an alpha 4 neuronal nicotinic acetylcholine receptor subunit knock-out mouse. *J. Neurosci.*, **20**, 6431–6441.

135 Salas, R., Orr-Urtreger, A., Broide, R., Beaudet, A.L., Paylor, R. and De Biasi, M. (2003) The nicotinic acetylcholine receptor subunit alpha 5 mediates acute effects of nicotine *in vivo*. *Mol. Pharmacol.*, **63**, 1059–1066.

136 Salas, R., Cook, K.D., Bassetto, L. and De Biasi, M. (2004) The alpha3 and beta4 nicotinic acetylcholine receptor subunits are necessary for nicotine-induced seizures and hypolocomotion in mice. *Neuropharmacology*, **47**, 401–407.

137 Boulter, J., O'Shea-Greenfield, A., Duvoisin, R.M., Connolly, J.G., Wada, E., Jensen, A., Gardner, P.D., Ballivet, M., Deneris, E.S., McKinnon, D. *et al.* (1990) Alpha 3, alpha 5, and beta 4: three members of the rat neuronal nicotinic acetylcholine receptor-related gene family form a gene cluster. *J. Biol. Chem.*, **265**, 4472–4482.

138 McDonough, J., Francis, N., Miller, T. and Deneris, E.S. (2000) Regulation of transcription in the neuronal nicotinic receptor subunit gene cluster by a neuron-selective enhancer and ETS domain factors. *J. Biol. Chem.*, **275**, 28962–28970.

139 Wang, F., Gerzanich, V., Wells, G.B., Anand, R., Peng, X., Keyser, K. and Lindstrom, J. (1996) Assembly of human neuronal nicotinic receptor alpha5 subunits with alpha3, beta2, and beta4 subunits. *J. Biol. Chem.*, **271**, 17656–17665.

140 Salas, R., Pieri, F. and De Biasi, M. (2004) Decreased signs of nicotine withdrawal in mice null for the β4 nicotinic acetylcholine receptor subunit. *J. Neurosci.*, **24**, 10035–10039.

141 Maskos, U., Molles, B.E., Pons, S., Besson, M., Guiard, B.P., Guilloux, J.P., Evrard, A., Cazala, P., Cormier, A., Mameli-Engvall, M., Dufour, N., Cloez-Tayarani, I., Bemelmans, A.P., Mallet, J., Gardier, A.M., David, V., Faure, P., Granon, S. and Changeux, J.P. (2005) Nicotine reinforcement and cognition restored by targeted expression of nicotinic receptors. *Nature*, **436**, 103–107.

142 Salas, R., Main, A., Gangitano, D. and De Biasi, M. (2007) Decreased withdrawal symptoms but normal tolerance to nicotine in mice null for the [alpha]7 nicotinic acetylcholine receptor subunit. *Neuropharmacology*, **53**, 863–869.

143 Christoph, G.R., Leonzio, R.J. and Wilcox, K.S. (1986) Stimulation of the lateral habenula inhibits dopamine-containing neurons in the substantia nigra and ventral tegmental area of the rat. *J. Neurosci.*, **6**, 613–619.

144 Amat, J., Sparks, P.D., Matus-Amat, P., Griggs, J., Watkins, L.R. and Maier, S.F. (2001) The role of the habenular complex in the elevation of dorsal raphe nucleus serotonin and the changes in the behavioral responses produced by uncontrollable stress. *Brain Res.*, **917**, 118–126.

145 Heldt, S.A. and Ressler, K.J. (2006) Lesions of the habenula produce stress- and dopamine-dependent alterations in prepulse inhibition and locomotion. *Brain Res.*, **1073–1074**, 229–239.

146 Matsumoto, M. and Hikosaka, O. (2007) Lateral habenula as a source of negative reward signals in dopamine neurons. *Nature*, **447**, 1111–1115.

Part II
Cells and Circuits

Stress – From Molecules to Behavior. Edited by Hermona Soreq, Alon Friedman, and Daniela Kaufer
Copyright © 2010 WILEY-VCH Verlag GmbH & Co. KGaA, Weinheim
ISBN: 978-3-527-32374-6

4
Effects of Stress on the Function of Hippocampal Cells

Marian Joëls and Henk Karst

4.1
Introduction

Changes in the environment of an organism which potentially drive essential body functions away from their optimal range typically lead to a stress response, aimed at restoring a situation of homeostasis [1, 2]. As part of the stress response two systems are activated, that is, the sympatho-adrenomedullar system and the hypothalamic–pituitary–adrenal (HPA) axis (Figure 4.1). The former leads to enhanced levels of adrenaline in the circulation which indirectly causes enhanced release of noradrenaline from specific synapses in the brain. Activation of the HPA axis leads to increased release of corticosteroid hormones (cortisol in humans, corticosterone in most rodents) from the adrenal cortex. Corticosteroid hormones easily pass the blood–brain barrier and thus reach all cells in the brain. However, they are only effective in those cells that carry receptors for the hormone. Corticosteroid release is pulsatile, with inter-pulse intervals of 1–2 h [3]. The peak of the pulses varies throughout the day, with low amplitudes around the start of the inactive period and high amplitudes toward the onset of the active period of the day, resulting in an overall circadian pattern. Exposure to stressful events causes a premature and large surge of corticosteroids.

Two receptor types for corticosterone have been distinguished in the brain [1]: first, the mineralocorticoid receptor (MR). This receptor is expressed in discrete areas, such as in the principal neurons of the hippocampus. Corticosterone, cortisol and aldosterone bind with high affinity to the MR. Due to its high affinity, the MR is nearly always occupied, even at the troughs of the ultradian pulses and circadian rhythm [4]. The second receptor type is the glucocorticoid receptor (GR), which is widely expressed in brain and in neurons and glial cells alike. This receptor has a 10-fold lower affinity for corticosterone than the MR and displays an even lower affinity for aldosterone. The GR is only partly occupied at rest, but is activated at the peak of high-amplitude ultradian pulses as well as after stress. Upon activation, MRs and GRs dissociate from chaperone proteins and translocate to the nucleus. They bind either as dimers to response elements in the promoter

Stress – From Molecules to Behavior. Edited by Hermona Soreq, Alon Friedman, and Daniela Kaufer
Copyright © 2010 WILEY-VCH Verlag GmbH & Co. KGaA, Weinheim
ISBN: 978-3-527-32374-6

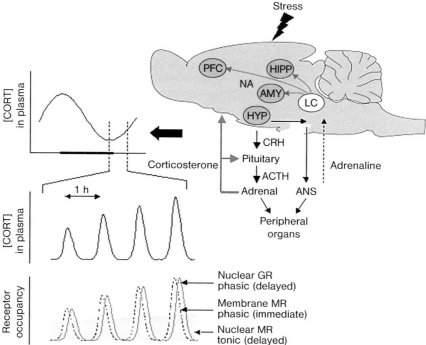

Figure 4.1 *Upper right*: Exposure to a stressful event is processed in limbic areas which project to the paraventricular nucleus in the hypothalamus (HYP). From there the autonomic nervous system (ANS) is activated, resulting in the release of adrenaline; this in turn can indirectly elevate the release of noradrenaline (NA) in brain, from projections originating in the locus coeruleus (LC). Stress also leads to the release of CRH, which results in pituitary secretion of ACTH. As a result corticosterone is released from the adrenal cortex. Corticosterone acts on many peripheral organs but also feeds back on the pituitary and hypothalamus, thus turning down the activity of the axis. Corticosterone furthermore reaches extrahypothalamic brain regions such as the prefrontal cortex (PFC), amygdala (AMY) and hippocampus (HIPP). *Left*: The graph in the middle illustrates the pulsatile release of corticosterone in time (overall duration of this fragment is approx. 4 h; 1 h interpulse intervals). The peak amplitude of the pulses varies throughout the day, with a large amplitude at the start of the active period and with low amplitudes at the end of that period [3]. Overall this results in a circadian release pattern as shown at the top; this graph reflects a typical 24 h pattern (fat black line is active period). The graph at the bottom summarizes the occupation of the corticosteroid receptors in hippocampal cells. Due to the very high affinity of the nuclear MR, this receptor is already extensively occupied during the circadian trough (gray background). Moreover, the high affinity also explains why receptors remain occupied during the interpulse intervals. Hence occupation is tonic, as are the delayed effects mediated via this receptor. Occupancy of the GR follows a more phasic pattern [4], which most likely follows the fluctuations in circulating hormone levels with a slight delay. The effects mediated via this receptor are slow in onset, due to the gene-dependent mechanism of action. In addition to these nuclear steroid receptors there is recent evidence for the existence of an MR located in the plasma membrane of limbic neurons. In view of the immediate effects seen after activation of this receptor, this receptor is postulated to follow the pulsatile pattern of circulating hormone levels not only with respect to its occupancy but also with regard to its functional outcome. (Adapted from [5, 6]).

region of specific genes – usually causing transcriptional activation – or interact as monomers with other transcription factors, which often results in repression of gene transcription [7, 8]. In addition to acting via these gene-mediated pathways, corticosteroid receptors can also change brain function in a non-genomic manner [5, 9]. This allows corticosteroid hormones to act over a very wide span of time, from minutes to many hours. The hippocampus is one of the main targets of corticosteroid hormones. Pyramidal neurons in the CA1 area express high amounts of MR and GR, as do granule cells in the dentate gyrus [1]. In the CA3 region, the expression of MR is much higher than of GR, which is exceptional in brain since nearly all cells have higher amounts of GR than MR. Studies over the past two decades have addressed the question how corticosterone changes the function of rodent hippocampal cells. Here, we summarize current views on non-genomic and genomic modulations by corticosterone of hippocampal ion currents and neurotransmitter responses, and the potential implication for the encoding of stressful information.

4.2
Non-Genomic Effects of Corticosterone

Although it was reported decades ago that corticosterone can alter hippocampal cell activity over a time-frame that is probably too fast for a gene-mediated pathway [10–12], non-genomic actions have only recently been resolved in any detail. It was found that in CA1 neurons corticosterone increases the frequency of miniature excitatory postsynaptic currents (EPSCs), which each represent the spontaneous release of one glutamate-containing vesicle [13]. Additional experiments indicated that the enhanced mEPSC frequency is due to an increased release probability of the vesicles; this agrees with earlier *in vivo* microdialysis experiments [14]. The corticosteroid effect required at least 10 nM corticosterone (slightly higher than the K_d of the GR), was visible within minutes and quickly reversed. The latter points to a non-genomic mechanism, which was confirmed by the observation that mEPSC frequency is still enhanced by corticosterone in the presence of a protein synthesis inhibitor. A conjugate of corticosterone and bovine serum albumine (CORT-BSA), which cannot pass the plasma membrane, was as effective as corticosterone, suggesting that corticosterone binds to a membrane receptor mediating these rapid effects.

In an attempt to resolve the nature of this receptor, studies first focused on the MR and GR, since in mammalian (as opposed to amphibian [15]) brain the existence of a specific membrane receptor is still not fully understood [16]. Despite the relatively high effective dose (10 nM), pharmacological experiments did not confirm involvement of the GR [13]. Similarly, in mice where the GR was deleted in a forebrain-specific manner corticosterone still increased the mESPC frequency. By contrast, aldosterone turned out to be even more effective than corticosterone in enhancing the mEPSC frequency, while the MR-antagonist spironolactone completely blocked the effect. Moreover, corticosterone was unable to change mEPSC frequency in forebrain-specific MR knockout mice. Collectively these data indicate that corticosterone increases the release probability of glutamate-containing vesicles in CA1 terminals, probably through a membrane located MR that displays a 10-fold lower affinity than the nuclear MR [13]. This puts the role of the MR in the stress response in an entirely new perspective, since it would provide the brain with the means to quickly respond to fluctuations in circulating hormone level, for example, shortly after stress [5].

Follow-up studies resolved the signaling pathway of rapid corticosteroid effects in more detail [17]. It turned out that corticosterone, via the MR, activates the presynaptic ERK1/2 pathway. In contrast to rapid effects in the hypothalamus [18, 19], this does not involve candidate retrograde messengers such as endocannabinoids or nitric oxide, so that most likely the receptor is presynaptically located. Interestingly, in addition to enhancing the mEPSC frequency, corticosterone was also found to decrease a specific K-conductance, that is, the transient IA [17]. Suppression of the IA depends on the MR, but in this case the receptor is postsynaptically located and (via a G protein) is coupled to the postsynaptic ERK1/2 pathway. Through these dual pathways, corticosterone is predicted to enhance cellular excitability shortly after stress: the hormone enhances the probability that incoming signals cause the release of glutamate-containing vesicles and postsynaptically increases the likelihood that this results in the generation of an action potential. This may contribute to the observation that the presence of corticosterone during high-frequency stimulation of the afferent pathway to the CA1 facilitates synaptic strengthening [20].

Some caution, though, is recommended, since stress results in exposure of hippocampal cells not only to corticosterone but also to noradrenaline and peptides like corticotropin releasing hormone (see also Chapter 8), which collectively and in interaction alter the firing properties of the cells. Moreover, the principal neurons of the CA3 region and dentate gyrus also express high amounts of MRs, but rapid effects of corticosterone on these cells have not been examined in detail. Nevertheless, such effects could be very important for the overall functionality of the hippocampus. Therefore, *in vivo* experiments in freely moving animals will have to demonstrate if and to what extent hippocampal excitability alters during the first 10–20 min after stress.

4.3
Genomic Effects of Corticosterone

Most effects of corticosterone take place via intracellularly located MRs and GRs which (upon binding of the hormone) translocate to the nucleus, thus starting gene-mediated effects which alter cell function with a delay of at least 1 h [7, 8]. Below, we review the slow gene-mediated effects on hippocampal ion conductances and on responses to prevalent neurotransmitters, such as glutamate, noradrenaline and serotonin.

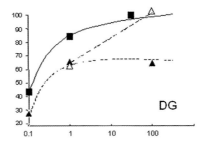

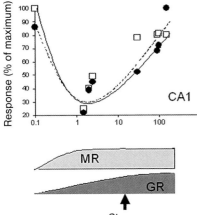

Figure 4.2 Dose–response relationships for cellular effects of corticosterone in the CA1 hippocampal area and dentate gyrus (DG). The diagrams show hormone responses expressed as a percentage of the maximal response. The concentration of corticosterone is a rough estimate of the local concentration, based on the solutions perfused on *in vitro* preparations or derived from the plasma concentration when fluctuations in hormone levels were accomplished *in vivo*. In the CA1 area, both the amplitude of depolarization-induced calcium currents (open squares) and the hyperpolarization caused by serotonin-1A receptor activation (filled circles) display a clear U shaped dose-dependency. The descending limb is linked to activation of MRs (depicted as the bottom), while the ascending limb is associated with gradual GR activation, as occurs after stress. DG granule neurons show a clear effect on the field potential (filled squares) and single cell response (filled triangle) caused by activation of glutamatergic AMPA receptors; this effect is linked to MR activation. Although these cells also abundantly express GRs high doses of corticosterone do not give additional changes in the signal, except when tested in chronically stressed rats (open triangles). (Reproduced with permission from [24]).

4.3.1
Ion Currents

Of all ion conductances tested so far in CA1 cells, the voltage-dependent calcium currents are most conspicuously affected by corticosterone. Activation of MR serves to reduce the calcium current amplitude, compared to the situation where both steroid receptors are unoccupied (e.g., after adrenalectomy) [21]. A pulse of corticosterone (30–100 nM) increases the calcium current amplitude, with a delay of several hours [21–23]. Overall, there is a U-shaped corticosteroid dose dependency for the calcium current amplitude (Figure 4.2) [24]. The enhanced amplitude of the calcium current was not only seen after *in vitro* application of a pulse of corticosterone but also after *in vivo* exposure of rats to a stressful situation [25]. Follow-up studies revealed that the L-type rather than N-type high-voltage-activated currents form a target for corticosterone [26]. The hormone appears to enhance the number of channels in the membrane rather than the single channel conductance.

The intracellular signaling pathway is still not fully resolved. The enhancement of the calcium current amplitude by corticosterone requires protein synthesis and involves binding of GR homodimers to the DNA [22, 23]. The poreforming α-subunits of the L-type calcium channel, though, do not appear to be a target for transcriptional regulation by GR [26]. However, with various techniques a consistent GR-dependent upregulation of the auxiliary β-4 subunit expression was observed. This subunit is involved in gating properties of the calcium channels but also plays a role in trafficking of channels from the intracellular compartment toward the plasma membrane [27]. The latter could provide a mechanism by which the number of effective channels increases after corticosterone exposure.

Not all hippocampal cells show an increased calcium current amplitude when corticosteroid levels rise. Interestingly, CA3 pyramidal cells (similarly to CA1 cells) have increased calcium current amplitude after elevations in corticosteroid level [28], despite the fact that these cells express GRs at a relatively low level. By contrast, dentate granule cells which abundantly express GRs in addition to MRs do not exhibit an enhanced calcium current amplitude in response to a pulse of corticosterone [29]. However, when animals have a history of chronic stress, a pulse of corticosterone does significantly enhance calcium currents. Apparently, local cell-specific and environment-dependent factors determine the overall outcome of GR-dependent pathways.

In the CA1 area, the enhanced calcium influx after GR activation was found to be coupled to a less efficient extrusion [30], in line with a suppression of the plasma membrane Ca(2+)-ATPase-1 [31, 32]. Overall this exposes cells to a higher calcium load upon depolarization. This may have consequences for calcium-dependent processes, such as the highly calcium-sensitive IK(Ca). Activation of this current slows down the firing frequency of CA1 pyramidal neurons during a period of depolarization [33]. The slow deactivation of the current is apparent from a lingering hyperpolarization (afterhyperpolarization, AHP) when the depolarizing pulse is terminated. In accordance with expectations, GR activation has been described to slowly attenuate the firing frequency and increase the AHP amplitude of CA1 pyramidal neurons [34, 35]. Effectively this means that, after a period of aroused activity, corticosterone gradually attenuates cell firing in the CA1 region, which is in line with the overall normalizing role of glucocorticoids at the end of the stress response. Again, this is subject to regional differentiation. For instance, in the basolateral amygdala GR activation causes increased high-voltage-activated current amplitude [36], yet this is not translated into a stronger firing frequency accommodation and AHP amplitude [37, 38]. It was argued that local differences in calcium channel subunit expression may contribute to the absence of attenuated firing during depolarization [38]. If these differences also occur *in vivo*, they could signify that arousal of amygdala activity in association with emotional situations is less effectively contained via genomic GR actions so that such situations are more effectively encoded.

Increased calcium load not only changes cell firing in limbic cells but also imposes a risk on limbic cells if they are depolarized for a prolonged period of time. This may happen during epileptic or ischemic insults. In that case cells could become exposed to a strong calcium load, facilitating delayed cell death. It was indeed observed that the progress and outcome of epilepsy or ischemia is negatively influenced by exposure to high corticosteroid concentrations [39–42].

Compared to high-voltage-activated calcium currents, voltage-dependent sodium and potassium currents are far less sensitive to stress or corticosterone application [43, 44]. Some effects have been described for the Ih, but the IA or delayed rectifying currents do not form a major target for gene-mediated steroid modulation.

4.3.2
Amino Acid Responses

Hippocampal cells receive major excitatory inputs mediated by glutamate. Within a restricted time-window (2–4 h) GR activation may enhance aspects of glutamate transmission [45]. However, most studies at the single cell or field potential level have reported either no effect of GR activation [35] or reduced transmission, particularly with repeated stimulation of afferent pathways and with rather high concentrations of the hormone [10, 11, 46]. This issue needs much more investigation, not only because of the conflicting results but also because many of the effects observed so far developed rather quickly for transcription and translation to be involved.

Also in other limbic regions, the effects of corticosteroids on glutamate transmission have been studied. In the dentate gyrus a pulse of corticosterone does not change AMPA receptor-mediated responses [47]. However, if animals also had experienced a period of chronic stress, the same pulse of corticosterone evoked a strong enhancement of AMPA receptor-mediated synaptic responses (Figure 4.2). In all cases NMDA receptor-mediated responses were unaffected. In the basolateral amygdala no apparent effects of corticosterone were reported on evoked or spontaneous AMPA and NMDA receptor-mediated responses in one study [37], but preliminary results suggest that mEPSC frequency may be lastingly altered secondary to a rapid enhancement [48].

While reports on GR-dependent modulation of glutamate transmission are quite sparse and equivocal, hormonal effects on hippocampal GABAergic transmission are even less investigated. There is some evidence that GR activation leads to reduction of GABA receptor-mediated responses in CA1 hippocampal cells [49], at least with repeated stimulation of Schaffer collateral afferents [46]; whether or not these effects are gene-mediated is not clear. In the dentate gyrus, acute stress within 30 min upregulates GABAa receptor delta subunit expression and increases the tonic inhibition mediated via delta subunit-containing GABAa receptors [50, 51]. In view of the timescale, this probably is not a gene-mediated effect; moreover, it appears to be caused by deoxycorticosterone metabolites rather than by corticosterone itself. The most explicit reports so far come from the basolateral amygdala, where corticosterone reduces GABAa receptor-mediated inhibitory postsynaptic potentials evoked by stimulation of the external capsule, in a delayed manner [37]. This effect is not caused by a change in GABA release but rather by a positive shift of the GABAa receptor reversal potential.

Overall, the slow gene-mediated effect of corticosterone on glutamate receptor-mediated synaptic transmission in limbic cells is presently unclear and seems to depend on the time that has elapsed after corticosteroid exposure, the corticosteroid concentration, the cellular context and the history of the animal. GABAergic responses, if affected, are generally reduced after a pulse of corticosterone, particularly in the basolateral amygdala. Interestingly, at the network level glucocorticoids and stress strongly affect long-term potentiation (LTP) [52], a form of synaptic strengthening that critically depends on AMPA and NMDA receptor-mediated transmission [53]. Based on *in vitro* and *in vivo* observations, an inverted U-shaped corticosteroid dose dependency was postulated with respect to LTP in the CA1 region [54]. Stress and GR activation were consistently found to suppress induction of LTP in the CA1 area in a delayed manner. In the dentate gyrus, the effect of stress and glucocorticoids on LTP is very sensitive to the nature, timing and severity of the stress, so that next to facilitation of long-term potentiation stress has also been described to either induce suppression or be ineffective (reviewed in [55]).

4.3.3
Aminergic Responses

Clear effects of corticosteroid hormones have been described with respect to responses to monoaminergic transmitters. One series of studies focused on responses to serotonin (5-HT) mediated by the 5-HT1A receptor which is prevalent in the hippocampal CA1 region [56]. This receptor is (through a G protein) coupled to an inwardly rectifying potassium conductance, so that upon binding of 5-HT to the receptor the membrane is hyperpolarized and cell firing is inhibited [57]. With low levels of corticosterone, 5-HT1A receptor-mediated responses were found to be small, compared to a situation where both the MR and GR are unoccupied (reviewed in [55]). Extensive GR activation, for example, after stress, slowly increases the 5-HT1A receptor-mediated response, yielding a U-shaped steroid dose dependency (Figure 4.2). A similar effect of corticosterone was found in the dentate gyrus, although here the effects were less marked.

The corticosteroid effect on 5-HT1A responsiveness depends on binding of GR homodimers to the DNA [23], although the nature of the critical responsive genes has remained elusive. Transcriptional regulation of the 5-HT1A receptor is not the main pathway underlying the enhanced 5-HT responsiveness (reviewed in [55]). Likewise, GRs do not target modulators of 5-HT1A receptor signaling, such as RGS-4, SGK, or N-CAMs [58], or more downstream molecules in the 5-HT1A receptor signaling pathway, for example, G proteins [59] or the inwardly rectifying K channel itself [60]. It is thought that GRs regulate a gene product that is involved in more general processes, like trafficking of receptor molecules or proteins involved in the translational machinery.

A second transmitter that is clearly affected by GR activation is noradrenaline. There is evidence that at the short term corticosterone (via a non-genomic pathway)

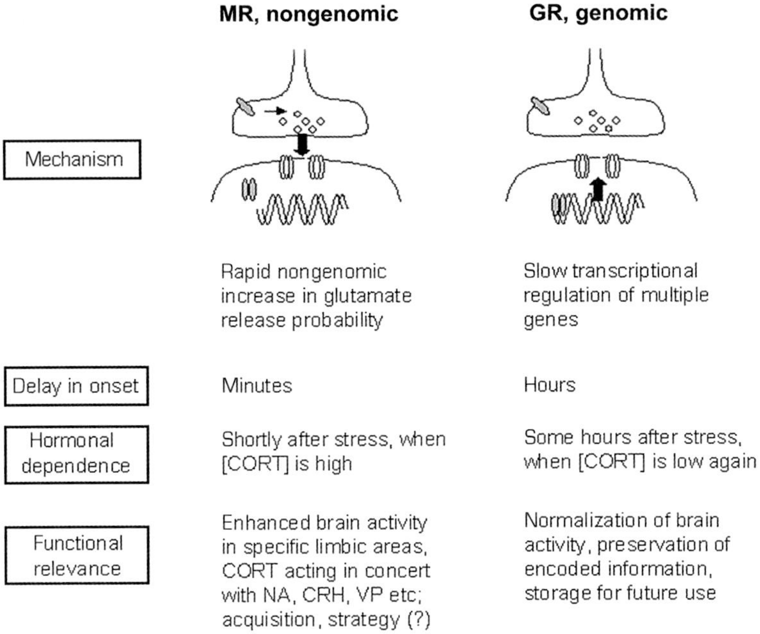

MR, nongenomic   **GR, genomic**

Mechanism	Rapid nongenomic increase in glutamate release probability	Slow transcriptional regulation of multiple genes
Delay in onset	Minutes	Hours
Hormonal dependence	Shortly after stress, when [CORT] is high	Some hours after stress, when [CORT] is low again
Functional relevance	Enhanced brain activity in specific limbic areas, CORT acting in concert with NA, CRH, VP etc; acquisition, strategy (?)	Normalization of brain activity, preservation of encoded information, storage for future use

Figure 4.3 Signaling pathways activated via MR and GR in the hippocampus after stress. When corticosteroid levels rise, for example, shortly after stress exposure, lower affinity membrane MRs are activated which – within minutes – enhance the release probability of glutamate containing vesicles (left). At the same time, GRs are translocated to the nucleus where they affect gene transcription primarily through transrepression. These effects become apparent some hours after the stress response, that is, by the time that corticosteroid levels are back to the pre-stress level and rapid non-genomic effects have subsided (right). These distinct MR- and GR-mediated actions operate in a bimodal stress system, in which patterns of neural and humoral mediators serve to organize stress reactions in the acute phase and in the management of the late recovery phase. Thus, the low affinity membrane MR synergizes with, for example, CRH and vasopressin in the organization of the neuroendocrine, sympathetic and behavioral fight, flight, fright reactions. Complementary to this, GR genomic actions dampen the initial stress reactions. Elipses depict receptor molecules: AMPA receptors (open) and corticosteroid receptors (gray).

promotes the effect of noradrenaline [61]. However, slow and presumably gene-mediated pathways cause a delayed suppression of noradrenergic effects. Thus, in the CA1 area activation of β-adrenoceptors diminishes IK(Ca) function, so that cell firing frequency is enhanced and the AHP amplitude decreased [62]; this enhances the cellular excitability.

Pretreatment of these cells with a GR agonist attenuated the effect of β-adrenoceptor activation, so that the enhanced excitability is gradually reversed by glucocorticoids [34].

4.3.4
Implications for Hippocampal Function

Although most of the cellular functions by corticosterone in the hippocampus are based on *in vitro* findings, it is possible to make predictions what these corticosteroid actions would mean for hippocampal function *in vivo*, after stress [6] (Figure 4.3).

One of the main new and unexpected insights over recent years is that corticosterone is able to enhance excitability shortly after stress, through non-genomic actions. *In vivo* microdialysis [14] support the electrophysiological indications [13] that shortly after stress corticosterone enhances the release of glutamate, while the level of GABA depends on the nature of the stressor [63].

The likelihood that the release of glutamate results in repetitive firing of CA1 pyramidal neurons is enhanced, since corticosterone suppresses the IA [13] These phenomena may contribute to the fact that corticosterone facilitates the induction of LTP in a rapid manner [20]. This is not unique for corticosterone. Thus, other stress-related hormones like noradrenaline (via β-adrenoceptors) [64, 65], corticotropin-releasing hormone [66] and vasopressin [67] all promote the induction of LTP.

To what extent these cellular observations in the CA1 area can be extrapolated to other hippocampal regions or the amygdala nuclei awaits further investigation. Studies so far indicate that rapid modulatory effects of stress in the dentate gyrus may be more complex. For instance, swim stress promotes the transition of short-term to long-term LTP through an MR-requiring process [51, 68], involving activation of MAPK [69]. However, swim stress also increases tonic GABAergic inhibition in this region, via metabolites of deoxycorticosterone [70]. Moreover, the involvement of input from the amygdala to the dentate gyrus seems to play an important and permissive role in the effects of stress [71–73]. Stressors with a clear emotional component (and hence amygdala activation) most likely lead to an overall enhanced activity of the dentate gyrus network.

When corticosterone is released after stress and reaches the hippocampus it not only changes cellular excitability at the short term but also initiate gene-mediated and GR-dependent pathways that some hours later affect hippocampal function. In the CA1 region, most of the delayed effects suppress overall excitability (Figure 4.3). For instance, the transfer of steady excitatory input is attenuated through an increased function of the IK(Ca) [34, 35]. Arousal promoted by β-adrenoceptors shortly after stress is gradually reversed via GR-dependent actions [34]. Inhibitory actions mediated by 5-HT1A receptors are enhanced via a process requiring DNA-binding of GR homodimers [23]. All of these effects serve to contain the initial response to stress, so that hippocampal activity is restored and normalized to the pre-stress level; it prevents the stress response from inappropriately overshooting. At the same time, information unrelated to the stressful experience and arriving at the same area some hours later encounters a heightened threshold for synaptic strengthening [52], so that the signal-to-noise ratio is increased and stress-related information is preserved.

Again, involvement of the amygdala–dentate pathway in a stressful event may adapt this picture. Preliminary observations support that stress and glucocorticoids do not suppress and normalize activity in the basolateral amygdala several hours after a pulse of corticosterone but, rather, induce a state of prolonged hyperexcitability [37, 38]. If so, this could signify that strongly emotional, stressful situations shift the balance between the CA1 versus amygdala–dentate networks toward the latter [74]. One could speculate that this contributes to the observation that in humans stress promotes memory [75] particularly for emotional versus neutral information [76, 77].

Clearly, to get a better understanding of the cellular and network processes contributing to the effects of stress on learning and memory, studies should not be confined to the CA1 area, but extended to the dentate gyrus, amygdalar nuclei and the prefrontal cortex.

The observations discussed in this chapter refer to brief periods of stress exposure. The responsiveness of hippocampal (and amygdalar) neurons to accidental elevations in corticosterone level, however, may change dramatically against a history of repetitive stress or early life trauma. Thus, extensive studies over the past decade have exemplified that neurons both in the CA1 and dentate regions can display stronger, weaker, or even entirely opposite responses to corticosterone after chronic unpredictable stress (reviewed in [55]). For instance, the calcium current amplitude of CA1 neurons is slowly enhanced after GR activation in naïve animals, but reduced after chronic stress [78]. The complexity of the dendritic tree of CA1 neurons is not affected by corticosterone in naïve rats, but is reduced when animals also have a history of chronic stress [79]. Similarly, corticosterone strongly impairs induction of LTP in adult animals which received a high amount of licking–grooming from their mothers during the first postnatal week; these animals show very effective LTP induction when corticosteroid levels are low [80]. By contrast, LTP in offspring from low licking–grooming mothers is poor when corticosteroid levels are low, but markedly enhanced after a pulse of corticosterone. It is expected that such changes in cellular function are translated into behavioral alterations. Indeed, offspring from low licking–grooming mothers show poor performance in moderately stressful learning situations but outperform the high licking–grooming offspring when learning occurs under very stressful conditions. This suggests that stress in early life programs organisms to optimally perform under comparable conditions later in life.

Clearly, there is still a great paucity in knowledge about corticosteroid actions on cell and network function during behaviorally relevant situations, in freely moving animals. Only with such information will it be possible to really link the here described effects of corticosteroids on cell and network function to behavioral observations.

Summary

Stress in rodents activates the sympathetic nervous system as well as the hypothalamic–pituitary–adrenal axis, resulting in enhanced exposure of brain cells to noradrenaline and corticosterone respectively. Electrophysiological studies over the past decades have revealed multiple actions of corticosteroid hormones on limbic neurons, particularly in the hippocampal CA1 area. The current view is that shortly after stress corticosterone increases the excitability of CA1 neurons via non-genomic pathways. In concert with other stress hormones corticosterone also rapidly facilitates synaptic strengthening, which may promote the initial stages of encoding of stressful information. As soon as corticosterone reaches limbic cells, however, it also starts a gene-mediated pathway which exerts gradual but lasting effects on the function of these cells. In the CA1 area, the overall result of these gene-dependent effects is a slow reversal of earlier raised excitability, which will prevent the stress response from overshooting. At the same time, information unrelated to the stressful experience will encounter a heightened threshold for synaptic strengthening, so that the signal-to-noise ratio is increased and stress-related information is preserved. Preliminary evidence suggests that in the amygdala corticosterone induces a long-lasting increase in excitability. This could signify that stressful events with a strong emotional component have a more lasting impact on limbic networks, leading to particularly effective encoding of such information.

References

1 De Kloet, E.R., Joëls, M. and Holsboer, F. (2005) Stress and the brain: from adaptation to disease. *Nat. Rev. Neurosci.*, **6**, 463–475.

2 McEwen, B.S. (2007) Physiology and neurobiology of stress and adaptation: central role of the brain. *Physiol. Rev.*, **87**, 873–904.

3 Young, E.A., Abelson, J. and Lightman, S.L. (2004) Cortisol pulsatility and its role in stress regulation and health. *Front. Neuroendocrinol.*, **25**, 69–76.

4 Conway-Campbell, B.L., McKenna, M.A., Wiles, C.C., Atkinson, H.C., de Kloet, E.R. and Lightman, S.L. (2007) Proteasome-dependent down-regulation of activated nuclear hippocampal glucocorticoid receptors determines dynamic responses to corticosterone. *Endocrinology*, **148**, 5470–5477.

5 Joëls, M., Karst, H., DeRijk, R. and de Kloet, E.R. (2008) The coming out of the brain mineralocorticoid receptor. *Trends Neurosci.*, **31**, 1–7.

6 Joëls, M., Pu, Z., Wiegert, O., Oitzl, M.S. and Krugers, H.J. (2006) Learning under stress: how does it work? *Trends Cogn. Sci.*, **10**, 152–158.

7 Pascual-Le Tallec, L. and Lombes, M. (2005) The mineralocorticoid receptor: a journey exploring its diversity and specificity of action. *Mol. Endocrinol.*, **19**, 2211–2221.

8 Zhou, J. and Cidlowski, J.A. (2005) The human glucocorticoid receptor: one gene, multiple proteins and diverse responses. *Steroids*, **70**, 407–417.

9 Tasker, J.G., Di, S. and Malcher-Lopes, R. (2006) Minireview: rapid glucocorticoid signaling via membrane-associated receptors. *Endocrinology*, **147**, 5549–5556.

10 Pfaff, D.W., Silva, M.T. and Weiss, J.M. (1971) Telemetered recording of hormone effects on hippocampal neurons. *Science*, **172**, 394–395.

11 Vidal, C., Jordan, W. and Zieglgans-berger, W. (1986) Corticosterone reduces excitability of hippocampal pyramidal cells *in vitro*. *Brain Res.*, **383**, 54–59.

12 Dubrovsky, B., Gijsbers, K., Filipini, D. and Birmingham, M.K. (1993) Effects of adrenocortical steroids on long-term potentiation in the limbic system: basic mechanisms and behavioral consequences. *Cell. Mol. Neurobiol.*, **13**, 399–414.

13 Karst, H., Berger, S., Turiault, M., Tronche, F., Schütz, G. and Joëls, M. (2005) Mineralocorticoid receptors are indispensable for nongenomic modulation of hippocampal glutamate transmission by corticosterone. *Proc. Natl. Acad. Sci. U. S. A.*, **102**, 19204–19207.

14 Venero, C. and Borrell, J. (1999) Rapid glucocorticoid effects on excitatory amino acid levels in the hippocampus: a microdialysis study in freely moving rats. *Eur. J. Neurosci.*, **11**, 2465–2473.

15 Orchinik, M., Murray, T.F. and Moore, F.L. (1991) A corticosteroid receptor in neuronal membranes. *Science*, **252**, 1848–1851.

16 Orchinik, M., Hastings, N., Witt, D. and McEwen, B.S. (1997) High-affinity binding of corticosterone to mammalian neuronal membranes: possible role of corticosteroid binding globulin. *J. Steroid. Biochem. Mol. Biol.*, **60**, 229–236.

17 Olijslagers, J.E., de Kloet, E.R., Elgersma, Y., van Woerden, G.M., Joëls, M. and Karst, H. (2008) Rapid changes in hippocampal CA1 pyramidal cell function via pre- as well as postsynaptic membrane mineralocorticoid receptors. *Eur. J. Neurosci.*, **27**, 2542–2550.

18 Di, S., Malcher-Lopes, R., Halmos, K.C. and Tasker, J.G. (2003) Nongenomic glucocorticoid inhibition via endo-cannabinoid release in the hypo-thalamus: a fast feedback mechanism. *J. Neurosci.*, **23**, 4850–4857.

19 Malcher-Lopes, R., Di, S., Marcheselli, V.S., Weng, F.J., Stuart, C.T., Bazan, N.G. and Tasker, J.G. (2006) Opposing crosstalk between leptin and gluco-corticoids rapidly modulates synaptic excitation via endocannabinoid release. *J. Neurosci.*, **26**, 6643–6650.

20 Wiegert, O., Joëls, M. and Krugers, H. (2006) Timing is essential for rapid effects of corticosterone on synaptic potentiation in the mouse hippocampus. *Learn. Mem.*, **13**, 110–113.

21 Karst, H., Wadman, W.J. and Joëls, M. (1994) Corticosteroid receptor-dependent modulation of calcium currents in rat hippocampal CA1 neurons. *Brain Res.*, **649**, 234–242.

22 Kerr, D.S., Campbell, L.W., Thibault, O. and Landfield, P.W. (1992) Hippocampal glucocorticoid receptor activation enhances voltage-dependent Ca conduc-tances: relevance to brain aging. *Proc. Natl. Acad. Sci. U.S.A.*, **89**, 8527–8531.

23 Karst, H., Karten, Y.J., Reichardt, H.M., de Kloet, E.R., Schütz, G. and Joëls, M. (2000) Corticosteroid actions in hippo-campus require DNA binding of glucocorticoid receptor homodimers. *Nat. Neurosci.*, **3**, 977–978.

24 Joëls, M. (2006) Corticosteroid effects in the brain: U-shape it. *Trends Pharmacol. Sci.*, **27**, 244–250.

25 Joëls, M., Velzing, E., Nair, S., Verkuyl, J.M. and Karst, H. (2003) Acute stress increases calcium current amplitude in rat hippocampus: temporal changes in physiology and gene expression. *Eur. J. Neurosci.*, **18**, 1315–1324.

26 Chameau, P., Qin, Y., Spijker, S., Smit, G. and Joëls, M. (2007) Glucocorticoids specifically enhance L-type calcium current amplitude and affect calcium channel subunit expression in the mouse hippocampus. *J. Neurophysiol.*, **97**, 5–14.

27 Hidalgo, P. and Neely, A. (2007) Multiplicity of protein interactions and functions of the voltage-gated calcium channel beta-subunit. *Cell Calcium*, **42**, 389–396.

28 Kole, M.H., Koolhaas, J.M., Luiten, P.G., Fuchs, E. and Ca2+, H. (2001) currents and the excitability of pyramidal neurons in the hippocampal CA3 subfield in rats depend on corticosterone and time of day. *Neurosci. Lett.*, **307**, 53–56.

29 Van Gemert, N. and Joëls, M. (2006) Effect of chronic stress and mifepristone treatment on voltage-dependent Ca2+ currents in rat hippocampal dentate gyrus. *J. Neuroendocrinol.*, **18**, 732–741.

30 Joëls, M., Werkman, T., Karst, H. and Wadman, W. (1998) Corticosteroids and calcium homeostasis: implication for

neuroprotection and neurodegeneration, in *New Frontiers in Stress Research, Modulation of Brain Function* (eds E.R. de Kloet, D. Ben Nathan, E. Grauer and R. Levy), Harwood Acad. Publ., pp. 95–104.

31 Bhargava, A., Meijer, O.C., Dallman, M.F. and Pearce, D. (2000) Plasma membrane calcium pump isoform 1 gene expression is repressed by corticosterone and stress in rat hippocampus. *J. Neurosci.*, **20**, 3129–3138.

32 Bhargava, A., Mathias, R.S., McCormick, J.A., Dallman, M.F. and Pearce, D. (2002) Glucocorticoids prolong Ca(2+) transients in hippocampal-derived H19-7 neurons by repressing the plasma membrane Ca(2+)-ATPase-1. *Mol. Endocrinol.*, **16**, 1629–1637.

33 Faber, E.S. and Sah, P. (2003) Calcium-activated potassium channels: multiple contributions to neuronal function. *Neuroscientist*, **9**, 181–194.

34 Joëls, M. and De Kloet, E.R. (1989) Effects of glucocorticoids and norepinephrine on the excitability in the hippocampus. *Science*, **245**, 1502–1505.

35 Kerr, D.S., Campbell, L.W., Hao, S.Y. and Landfield, P.W. (1989) Corticosteroid modulation of hippocampal potentials: increased effect with aging. *Science*, **245**, 1505–1509.

36 Karst, H., Nair, S., Velzing, E., Rumpff-van Essen, L., Slagter, E., Shinnick-Gallagher, P. and Joëls, M. (2002) Glucocorticoids alter calcium conductances and calcium channel subunit expression in basolateral amygdala neurons. *Eur. J. Neurosci.*, **16**, 1083108–1083109.

37 Duvarci, S. and Pare, D. (2007) Glucocorticoids enhance the excitability of principal basolateral amygdala neurons. *J. Neurosci.*, **27**, 4482–4491.

38 Liebmann, L., Karst, H., Sidiropoulou, K., van Gemert, N., Meijer, O.C., Poirazi, P. and Joëls, M. (2008) Differential effects of corticosterone on the sAHP in the basolateral amygdala and CA1 region: possible role of calcium channel subunits. *J. Neurophysiol.*, **99**, 958–968.

39 Stein-Behrens, B.A., Elliott, E.M., Miller, C.A., Schilling, J.W., Newcombe, R. and Sapolsky, R.M. (1992) Glucocorticoids exacerbate kainic acid-induced extracellular accumulation of excitatory amino acids in the rat hippocampus. *J. Neurochem.*, **58**, 1730–1735.

40 Smith-Swintosky, V.L., Pettigrew, L.C., Sapolsky, R.M., Phares, C., Craddock, S.D., Brooke, S.M. and Mattson, M.P. (1996) Metyrapone, an inhibitor of glucocorticoid production, reduces brain injury induced by focal and global ischemia and seizures. *J. Cereb. Blood Flow Metab.*, **16**, 585–598.

41 Karst, H., De Kloet, E.R. and Joëls, M. (1999) Episodic corticosterone treatment accelerates kindling epileptogenesis and triggers long-term changes in hippocampal CA1 cells, in the fully kindled state. *Eur. J. Neurosci.*, **11**, 889–898.

42 Krugers, H.J., Maslam, S., Korf, J. and Joëls, M. (2000) The corticosterone synthesis inhibitor metyrapone prevents Hypoxia/Ischemia-induced loss of synaptic function in the rat hippocampus. *Stroke*, **31**, 1162–1172.

43 Karst, H., Wadman, W.J. and Joëls, M. (1993) Long-term control by cortico-steroids of the inward rectifier in rat CA1 pyramidal neurons, *in vitro*. *Brain Res.*, **612**, 172–179.

44 Werkman, T.R., van der Linden, S. and Joëls, M. (1997) Corticosteroid effects on sodium and calcium currents in acutely dissociated rat CA1 hippocampal neurons. *Neuroscience*, **78**, 663–672.

45 Karst, H. and Joëls, M. (2005) Corticosterone slowly enhances miniature excitatory postsynaptic current amplitude in mice CA1 hippocampal cells. *J. Neurophysiol.*, **94**, 3479–3486.

46 Joëls, M. and de Kloet, E.R. (1993) Corticosteroid actions on amino acid-mediated transmission in rat CA1 hippocampal cells. *J. Neurosci.*, **13**, 4082–4090.

47 Karst, H. and Joëls, M. (2003) Effect of chronic stress on synaptic currents in rat hippocampal dentate gyrus neurons. *J. Neurophysiol.*, **89**, 625–633.

48 Karst, H., Berger, S., Schütz, G. and Joëls, M. (2008) Rapid effect of corticosterone in the amygdala. Abstract SfN Meeting.

49 Teschemacher, A., Zeise, M.L. and Zieglgänsberger, W. (1996)

Corticosterone-induced decrease of inhibitory postsynaptic potentials in rat hippocampal pyramidal neurons *in vitro* depends on cytosolic factors. *Neurosci. Lett.*, **215**, 83–86.

50 Stell, B.M., Brickley, S.G., Tang, C.Y., Farrant, M. and Mody, I. (2003) Neuroactive steroids reduce neuronal excitability by selectively enhancing tonic inhibition mediated by d subunit-containing GABAA receptors. *Proc. Natl. Acad. Sci. U. S. A.*, **100**, 14439–14444.

51 Maguire, J. and Mody, I. (2007) Neurosteroid synthesis-mediated regulation of GABA(A) receptors: relevance to the ovarian cycle and stress. *J. Neurosci.*, **27**, 2155–2162.

52 Kim, J.J. and Diamond, D.M. (2002) The stressed hippocampus, synaptic plasticity and lost memories. *Nat. Rev. Neurosci.*, **3**, 453–462.

53 Bannerman, D.M., Rawlins, J.N. and Good, M.A. (2006) The drugs don't work-or do they? Pharmacological and transgenic studies of the contribution of NMDA and GluR-A-containing AMPA receptors to hippocampal-dependent memory. *Psychopharmacology (Berl.)*, **188**, 552–566.

54 Diamond, D.M., Bennett, M.C., Fleshner, M. and Rose, G.M. (1992) Inverted-U relationship between the level of peripheral corticosterone and the magnitude of hippocampal primed burst potentiation. *Hippocampus*, **2**, 421–430.

55 Joëls, M., Karst, H., Krugers, H.J. and Lucassen, P.J. (2007) Chronic stress: implications for neuronal morphology, function and neurogenesis. *Front. Neuroendocrinol.*, **28**, 72–96.

56 Pucadyil, T.J., Kalipatnapu, S. and Chattopadhyay, A. (2005) The serotonin 1A receptor: a representative member of the serotonin receptor family. *Cell. Mol. Neurobiol.*, **25**, 553–580.

57 Andrade, R. (1998) Regulation of membrane excitability in the central nervous system by serotonin receptor subtypes. *Ann. N. Y. Acad. Sci.*, **861**, 190–203.

58 Van Gemert, N.G., Meijer, O.C., Morsink, M.C. and Joëls, M. (2006) Effect of brief corticosterone administration on SGK1 and RGS4 mRNA expression in rat hippocampus. *Stress*, **9**, 165–170.

59 Okuhara, D.Y., Beck, S.G. and Muma, N.A. (1997) Corticosterone alters G protein alpha-subunit levels in the rat hippocampus. *Brain Res.*, **745**, 144–151.

60 Muma, N.A. and Beck, S.G. (1999) Corticosteroids alter G protein inwardly rectifying potassium channels protein levels in hippocampal subfields. *Brain Res.*, **839**, 331–335.

61 Pu, Z., Krugers, H.J. and Joëls, M. (2007) Corticosterone time-dependently modulates beta-adrenergic effects on long-term potentiation in the hippo-campal dentate gyrus. *Learn. Mem.*, **14**, 359–367.

62 Harley, C.W. (2007) Norepinephrine and the dentate gyrus. *Prog. Brain Res.*, **163**, 299–318.

63 De Groote, L. and Linthorst, A.C. (2007) Exposure to novelty and forced swimming evoke stressor-dependent changes in extracellular GABA in the rat hippocampus. *Neuroscience*, **148**, 794–805.

64 Katsuki, H., Izumi, Y. and Zorumski, C.F. (1997) Noradrenergic regulation of synaptic plasticity in the hippocampal CA1 region. *J. Neurophysiol.*, **77**, 3013–3020.

65 Kemp, A. and Manahan-Vaughan, D. (2008) Beta-adrenoreceptors comprise a critical element in learning-facilitated long-term plasticity. *Cereb. Cortex*, **18**, 1326–1334.

66 Blank, T., Nijholt, I., Eckart, K. and Spiess, J. (2002) Priming of long-term potentiation in mouse hippocampus by corticotropin-releasing factor and acute stress: implications for hippocampus-dependent learning. *J. Neurosci.*, **22**, 3788–3794.

67 Chepkova, A.N., French, P., de Wied, D., Ontskul, A.H., Ramakers, G.M., Skrebitski, V.G., Gispen, W.H. and Urban, I.J. (1995) Long-lasting enhancement of synaptic excitability of CA1/subiculum neurons of the rat ventral hippocampus by vasopressin and vasopressin(4–8). *Brain Res.*, **701**, 255–266.

68 Korz, V. and Frey, J.U. (2003) Stress-related modulation of hippocampal long-term potentiation in rats:

Involvement of adrenal steroid receptors. *J. Neurosci.*, **19**, 7281–7287.

69 Ahmed, T., Frey, J.U. and Korz, V. (2006) Long-term effects of brief acute stress on cellular signaling and hippocampal LTP. *J. Neurosci.*, **15**, 3951–3958.

70 Reddy, D.S. and Rogawski, M.A. (2002) Stress-induced deoxycorticosterone-derived neurosteroids modulate GABAA receptor function and seizure susceptibility. *J. Neurosci.*, **22**, 3795–3805.

71 Akirav, I. and Richter-Levin, G. (1999) Biphasic modulation of hippocampal plasticity by behavioral stress and basolateral amygdala stimulation in the rat. *J. Neurosci.*, **19**, 10530–10535.

72 Akirav, I. and Richter-Levin, G. (2002) Mechanisms of amygdala modulation of hippocampal plasticity. *J. Neurosci.*, **22**, 9912–9921.

73 Kim, J.J., Lee, H.J., Han, J.S. and Packard, M.G. (2001) Amygdala is critical for stress-induced modulation of hippocampal long-term potentiation and learning. *J. Neurosci.*, **21**, 5222–5228.

74 Kavushansky, A., Vouimba, R.M., Cohen, H. and Richter-Levin, G. (2006) Activity and plasticity in the CA1, the dentate gyrus, and the amygdala following controllable vs. uncontrollable water stress. *Hippocampus*, **16**, 35–42.

75 Abercrombie, H.C., Kalin, N.H., Thurow, M.E., Rosenkranz, M.A. and Davidson, R.J. (2003) Cortisol variation in humans affects memory for emotionally laden and neutral information. *Behav. Neurosci.*, **117**, 505–516.

76 Buchanan, T.W. and Lovallo, W.R. (2001) Enhanced memory for emotional material following stress-level cortisol treatment in humans. *Psychoneuroendocrinology*, **26**, 307–317.

77 Kuhlmann, S. and Wolf, O.T. (2006) Arousal and cortisol interact in modulating memory consolidation in healthy young men. *Behav. Neurosci.*, **120**, 217–223.

78 Karst, H. and Joëls, M. (2007) Brief RU 38486 treatment normalizes the effects of chronic stress on calcium currents in rat hippocampal CA1 neurons. *Neuropsychopharmacology*, **32**, 1830–1839.

79 Alfarez, D.N., Karst, H., Velzing, E.H., Joëls, M. and Krugers, H.J. (2008) Opposite effects of glucocorticoid receptor activation on hippocampal CA1 dendritic complexity in chronically stressed and handled animals. *Hippocampus*, **18**, 20–28.

80 Champagne, D.L., Bagot, R.C., van Hasselt, F., Ramakers, G., Meaney, M.J., de Kloet, E.R., Joëls, M. and Krugers, H. (2008) Maternal care and hippocampal plasticity: evidence for experience-dependent structural plasticity, altered synaptic functioning, and differential responsiveness to glucocorticoids and stress. *J. Neurosci.*, **28**, 6037–6045.

5
Stress and Adult Neurogenesis in the Mammalian Central Nervous System

Elizabeth D. Kirby and Daniela Kaufer

5.1
Introduction

While the developing brain often receives the most credit for ability to change in response to the environment, the adult CNS is also characterized by a great deal of plasticity. The ability to create new neurons – known as neurogenesis – is one of the most intriguing aspects of adult CNS plasticity both because it reveals a unique cell population that continues to proliferate beyond development and because it is highly sensitive to environmental input. One of the most powerful and ecologically relevant factors known to affect adult neurogenesis is the experience of a stressor [1, 2]. Indeed, stress and stress hormones were the first exogenous factors shown to influence the production of new neurons in the adult mammalian brain [1, 3, 4]. This chapter reviews the nature of the stress effect on adult mammalian neurogenesis: the when, where, how and why of stress-related alteration in new neuron birth, survival and differentiation in the adult CNS. In brief, it appears that several aspects of adult mammalian neurogenesis are sensitive to environmental stress, primarily via exposure to the stress hormones, glucocorticoids. However, this effect depends on the neurogenic region, the timing and duration of the stressor and the phase of neurogenesis being examined.

5.2
Adult Neurogenesis: a Brief Primer

First demonstrated by Altman and colleagues [5, 6], adult mammalian neurogenesis occurs in the dentate gyrus subgranular zone (SGZ) and the subventricular zone (SVZ) of every mammalian species investigated to date, including rats [5–7], mice [8, 9], tree shrews [4], monkeys [10] and humans [11]. The process of neurogenesis consists of at least three overlapping phases: proliferation, survival and differentiation/migration. In the SGZ, cells are born (proliferate) at the rate of

Stress – From Molecules to Behavior. Edited by Hermona Soreq, Alon Friedman, and Daniela Kaufer
Copyright © 2010 WILEY-VCH Verlag GmbH & Co. KGaA, Weinheim
ISBN: 978-3-527-32374-6

approximately 9000 per day [12] and around half of these newly generated cells survive beyond one month [13], migrating a short distance out of the SGZ into the granule cell layer of the dentate gyrus. Of the surviving cells, typically 80–90% become neurons under standard conditions [12]. The remainder of the cells differentiate into glial cells, oligodendrocytes or epithelial cells. In the SVZ, by contrast, about 30 000 cells are born per day [14]. These new cells migrate along the rostral migratory stream to the olfactory bulb. Of the approximately 50% of these cells that survive [14, 15], 75–99% differentiate into olfactory neurons [14, 15]. Importantly, new neurons in both the dentate gyrus and the olfactory bulb seem to integrate functionally into existing networks [16, 17].

5.3
Measuring Neurogenesis: How to Find New Neurons

There are three broad categories of method that have been used to identify new cells in the adult CNS: exogenous DNA replication markers, endogenous cell cycle-associated proteins and retroviruses.

5.3.1
Using DNA Replication to Detect New Cells

The most common method of identifying new cells relies on the incorporation of labeled nucleotides during DNA replication. In early studies of neurogenesis, tritiated thymidine ($[^3H]dT$) was injected systemically and autoradiography was used to identify cells that had incorporated the tagged nucleotide into their DNA, presumably during S phase of cell division. More recently, $[^3H]dT$ has been replaced by 5′-bromodeoxyuridine (BrdU), a thymidine analog which can be detected using immunohistochemistry. BrdU is typically preferred because it requires less tissue-processing time than $[^3H]dT$ labeling and because it can be more easily combined with antibody labeling of other proteins. However, the underlying principle of marking new cells by detecting DNA division remains the same.

The timing of BrdU (or $[^3H]dT$) injection and the chase period before sacrifice both must be optimized to assess changes in proliferation and survival/differentiation. For effects on proliferation, BrdU is injected immediately after the manipulation, allowing for an assessment of the pool of currently dividing cells. The animals are then sacrificed sometime between 2 h and a few days later [1, 18]. To investigate effects on survival and differentiation, BrdU is typically administered before beginning an experimental manipulation and the animals are then sacrificed after the manipulation is complete [19, 20].

There are several potential pitfalls of using BrdU to label new cells. First, neurotoxicity of even relatively low doses of BrdU has been reported [21], suggesting that BrdU incorporation might alter cellular function. However, many other studies have used high, repeated doses of BrdU without any reported ill-effect. Second, it is possible that BrdU might be incorporated into cells during DNA repair, but the size of this repair-related contribution to the population of labeled cells is likely very small [22]. Third, the bioavailability of BrdU could change in response to an environmental manipulation through changes in blood flow or blood–brain barrier permeability. Such a change in bioavailability could either create false or mask real group differences in proliferative cell labeling. One way to control for this limitation of BrdU is through the use of endogenous markers of cell division, as described below.

5.3.2
Endogenous Markers of Cell Cycle

A number of proteins are expressed exclusively during the cell division cycle and can be used as markers to quantify the number of dividing cells. The most commonly used marker is Ki-67, a nuclear protein expressed throughout the cell cycle, which has been successfully combined with BrdU to assess proliferation levels in the adult rodent hippocampus [13, 23]. Proliferating cell nuclear antigen (PCNA), a DNA polymerase-associated protein, is also used to evaluate proliferating cell number [24]. While these markers and others like them have the advantage of being endogenous, they only allow for the assessment of proliferation and, within that, only the subset of cells that are within S phase at the very moment of sacrifice are marked.

5.3.3
Retroviral Labeling of New Cells

Newly born cells can also be detected through retroviral infection. Stereotaxically injected retrovirus preferentially infects cells undergoing division and, if the virus carries a reporter gene such as GFP, it can be used to visualize mitotically active cells at the time of infection. This approach has been most powerful when used to investigate the electrophysiological and morphological properties of newly born cells [16]. While staining for BrdU or other cell cycle markers requires processing that destroys neuronal function, retroviral labeling with direct visualization of a fluorescent reporter protein allows for the targeting of newly born cells in tissue slice preparations for electrophysiology. Unfortunately, retroviral infection is somewhat sporadic and unpredictable, making this method inappropriate for quantification of neurogenesis.

5.3.4
Determining Cell Fate

The most common method of demonstrating the fate of new cells is by combining staining for BrdU with staining for lineage-specific proteins. Markers such as doublecortin (DCX), neuron-specific class III β-tubulin (TUJ1), polysialylated neuronal cell adhesion molecule (PSA-NCAM) and turned on after division (TOAD-64) are commonly used as markers of immature neurons [18, 25–27].

To mark mature neurons, the most common proteins used are neuron-specific nuclear protein (NeuN), neuron-specific enolase (NSE) and microtubule-associated protein (MAP-2) [18, 28, 29].

5.4
Stress-Induced Alteration in Cell Proliferation

The effects of stress on cell proliferation can be divided into those due to acute stress and those due to repeated, chronic stress. This division is based on experimental evidence that stress has different effects on cell division depending on whether exposure is short-term or long-term (see Table 5.1).

5.4.1
Acute Stress

Acute stress has been shown to inhibit cell proliferation in the adult SGZ in a number of species, including rat [30, 31, 36], mouse [30], tree shrew [4] and marmoset monkeys [47] and using a number of different stressors, including brief restraint [30], exposure to fox odor [31] and subordination stress [4, 47]. This acute stress-induced suppression of proliferation appears to be driven by exposure to high levels of glucocorticoids (GCs) – while acute stress effects on SGZ proliferation are blocked by removing GCs through adrenalectomy [31], acute corticosterone injections decrease cell proliferation in the SGZ much as acute stress does [1, 35]. Moreover, we have shown that treating neural precursor cells with GCs *in vitro* leads to a dose-dependent reduction in proliferation [48]. By contrast, cell proliferation in the SVZ does not appear to be sensitive to stress [33, 34], suggesting that the differences in BrdU labeling found in the SGZ following acute stress/GC exposure reflect true proliferation differences, not just changes in systemic label bioavailability.

The acute stress-induced suppression of adult hippocampal proliferation rate is relatively short-lived, dissipating within 24 h in rodents [32, 36]. The reduction in new cell number due to this brief suppression of proliferation rate remains detectable for up to one week [31, 34]. By three weeks, however, the number of surviving BrdU positive cells is similar in stressed animals and controls [31], implying either a floor effect in number of surviving cells over the long term, or an enhanced resistance of cells born during acute stress exposure to later cell death.

It should be noted that a few studies have not found an acute stress effect on adult SGZ proliferation [38–41]. Some of these studies can be explained by the timing of the BrdU injection – by administering BrdU before the stressor, labeled cells were those dividing immediately before stress began. This timing makes the BrdU+ population a more appropriate measure of short-term survival than proliferation [39, 40]. It is also possible that control animals experienced some environmental stressor that the investigators were unaware of or that the stressor used was not severe enough (i.e., did not lead to a great enough rise in corticosterone) to affect proliferating cells.

Table 5.1 Proliferation under acute stress and chronic stress.

Species	Manipulation	Washout time after stress	New cell marker	Brain area	Effect on cell numbers compared to controls	Ref.
Acute stress						
Male tree shrew	1 h social stress	2 h	BrdU after stress	SGZ	Decrease	[4]
Male marmoset monkey	1 h social stress	2 h	BrdU after stress	SGZ	Decrease	[47]
Male rat	45 min restraint or 30 min uncontrollable shock	2 h	BrdU after stress	SGZ	Decrease	[30]
Male mouse	50 min restraint/ rotation	2 h	BrdU after stress	SGZ	Decrease	[30]
Male rat	2 h fox odor exposure	2 h, 1 week and 3 weeks	BrdU after stress	SGZ	Decrease at 2 h and 1 week; none at 3 weeks	[31]
Male rat	1 h cold immobilization (am) then 30 min cold swim (pm)	48 h	BrdU 24 h after second stress session	SGZ	None	[32]
Male rat	1 day inescapable shock with escape test on day 2	7 days	BrdU 7 days after stress; sacrifice 2 h later	SGZ	Decrease	[33]
Male rat	1 day inescapable shock with escape test on day 2	7 days	BrdU 7 days after stress; sacrifice 2 h later	SVZ	None	[33]
Male rat	1 h fox odor exposure	2 h and 1 week	BrdU after stress	SGZ	Decrease at both 2 h and 1 week	[34]
Male rat	Corticosterone injection	2 h	[3H]dT after cort injection	SGZ	Decrease	[35]
Male rat	Corticosterone injection once daily for 2 days	1 h	[3H]dT after second cort injection	SGZ	Decrease	[1]

Table 5.1 *Continued*

Species	Manipulation	Washout time after stress	New cell marker	Brain area	Effect on cell numbers compared to controls	Ref.
Male rat	ADX	5 weeks	[3H]dT after 1 week recovery from ADX	SGZ	Increase (only in NSE+, not NSE– cells)	[1]
Male rat	Unpredictable stress	24 h	BrdU after stress	SGZ	Decrease	[36]
Male rat	Unpredictable stress	24 h	Ki-67	SGZ	None	[36]
Male rat	1 h fox odor exposure	24 h	BrdU after 15 min stress	SGZ	Decrease	[37]
Female rat	1 h fox odor exposure	24 h	BrdU after 15 min stress	SGZ	None	[37]
Male rat	2 h or 6 h restraint	2 h	BrdU after stress	SGZ	None	[38]
Male rat	20 min exposure to TMT (fox urine odor)	2 h	BrdU (given before stress); Ki-67	SGZ	None	[39]
Male rat	20 min social dominance stress	2 h	BrdU (given before stress)	SGZ	None	[40]
Male rat	3 h restraint	0 h	Ki-67	SGZ	None	[41]
Chronic stress						
Male rat	6 days DEX injection	12 h or 28 days	BrdU after last DEX injection	SGZ	Decrease at 12 h; no difference at 28 days	[42]
Male rat	3 weeks variable stressors, twice daily	48 h	BrdU 24 h after stress; Ki-67	SGZ	Decrease	[32, 99]
Male rat	3 weeks variable stressors, twice daily	3 weeks	BrdU 24 h before sacrifice (3 weeks after stress); Ki-67	SGZ	Decrease	[32, 99]
Male rat	3 weeks restraint, 6 h/day	26 h	BrdU 24 h after last restraint session	SGZ	Decrease	[38]

Table 5.1 *Continued*

Species	Manipulation	Washout time after stress	New cell marker	Brain area	Effect on cell numbers compared to controls	Ref.
Male rat	18 days social defeat	24 h	BrdU after last defeat session	SGZ	Decrease	[19]
Male rat	14 days restraint, 6 h/day	24 h	Ki-67	SGZ	Decrease	[41]
Male rat	3 weeks variable stressors	24 h	BrdU after last stress session	SGZ	Decrease	[36]
Male tree shrew	7 days psychosocial conflict	24 h	BrdU after last stress exposure	SGZ	Decrease	[43]
Male mice	24 days variable stress	3 days	BrdU twice daily for 2 days after stress	SGZ	Decrease	[44]
Female rat	8 days foot shock	24 h	BrdU once daily for last 5 days of stress	SGZ	None	[45]
Male rat	19 days variable stress	6–8 h	BrdU 4–6 h after last stressor (2 h before sacrifice)	SGZ	None	[46]
Male mouse	1–3 weeks daily forced swim	7 days (for BrdU); 0 h (in vitro assay)	BrdU 5× every 3 h after stress; clonal colony forming assay	OB/ SVZ	Decrease with 3 weeks but not 1 week	[51]

The majority of work on acute stress and adult SGZ cell proliferation has used male subjects. However, basal rates of adult neurogenesis differ in male and female rodents [49], suggesting that neurogenic stress response may also vary by sex. Indeed, it has been reported that while acute fox odor exposure suppresses cell proliferation in the SGZ of male rats, it does not do so in females [37]. Given that administration of estradiol to female rats increases SGZ cell proliferation [49], high circulating estrogen levels in female animals may be neuroprotective for mitotic cells, preventing GC-induced suppression of cell division. More research is needed to probe this potential sex difference in proliferative response to stress and what mechanism(s) might underlie it.

5.4.2
Chronic Stress

As with single, acute stress exposure, repeated exposure to a variety of stressors (social defeat [19, 43], 2–3 weeks of daily restraint [38, 41] or 3 weeks of daily exposure to rotating stressors such as cold swim, shock, etc [32, 36]) has also been shown to suppress cell proliferation in the adult SGZ of many species, including rats [19, 38], mice [44] and tree shrews [43]. Like acute stress, the effect of chronic stress on dividing cells also appears to be driven by GC exposure [42]. Importantly, the effectiveness of chronic stress at reducing proliferation suggests that proliferating cells are not capable of a great degree of adaptation to high circulating glucocorticoids over time.

The effect of chronic stress is more enduring than that of acute stress. While the suppressive effects of acute stress and GC treatment seem to be alleviated within 24 h, 24 h following the end of chronic stress, the proliferative capacity of the SGZ is still diminished [32, 38, 41]. Indeed, chronic stress-induced suppression of SGZ proliferation may extend up to 3 weeks after the end of the stressor [32, 36]. Importantly, at these later time points following stress, circulating GCs have returned to normal levels [50], indicating that prolonged exposure to high levels of GCs leads to a change in the proliferating pool of cells that persists even when hormone levels have returned to baseline.

Research on sex differences in the effect of chronic stress on SGZ proliferation is more limited than with acute stress. One study has found that females are resistant to chronic stress effects of cell proliferation/survival [45]. However, this effect was dependent on the social environment and the BrdU injection regimen used in that study made disentanglement of proliferation and survival effects difficult. More research is needed to determine how chronic stress effects female SGZ cell proliferation.

Chronic stress may also suppress proliferation in the adult SVZ. Using BrdU labeling and *in vitro* colony forming assays, Hitoshi *et al.* [51] recently showed that the number of rapidly dividing progenitor cells in the adult mouse SVZ is decreased by three weeks, but not one week, of daily exposure to forced cold swim. Moreover, *in vitro*, they showed that corticosterone treatment can suppress proliferation of SVZ neural precursors, suggesting that the stress effect is driven by GCs. They also showed that the chronic-stress induced suppression of proliferation was still present after three weeks recovery, much as is the case with chronic stress effects in the SGZ. More research is needed to expand upon these initial findings, but it seems that the SVZ may be sensitive to GCs when exposure is prolonged.

5.4.3
Cell Cycle Arrest Versus Progenitor Death

There are two main possibilities for how stress, acute or chronic, might impact the number of newly born cells in the adult dentate gyrus: GCs could cause a slowing of the cell cycle of progenitors/stem cells or they could eliminate progenitor/stem cells by causing them to exit the cell cycle permanently. The rapid time course of onset of stress effects on proliferation and the speed of recovery of proliferative ability following acute stress both suggest that stress acts through a slowing of the cell cycle rather than by causing a reduction in the progenitor pool. In support of this hypothesis, one study using adult hippocampal neural progenitors *in vitro* showed that 24 h of treatment with a synthetic GC receptor (GR) agonist, dexamethasone (DEX), led not only to a decrease in new cell production, but also to an upregulation in p21, a cdk-inhibitor that is associated with cell cycle arrest in G1 [42]. Upregulation of p21 and accumulation of cells in G1 was also found when HT-22 cells, a neuronal mouse cell line, were treated with DEX [52]. Moreover, *in vivo*, chronic stress has been shown to cause upregulation of p27Kip1, another cdk-inhibitor associated with cell cycle arrest at the G1/S checkpoint [53]. Altogether, these studies suggest that GCs somehow block the transition from G1 to S phase.

5.5
Stress-Induced Alteration of New Cell Survival

Approximately half of newly born cells in the adult dentate gyrus die within the first few weeks after birth [12, 13]. This pruning of new cells, while dramatic, is also highly sensitive to environmental input [25, 54]. Though less research has been done on the effects of stress and GCs on new cell survival than on proliferation, stress-related suppression of new cell survival has been well demonstrated (see Table 5.2).

Daily stress in the form of social defeat [19] or variable stressors such as cold exposure, shock, overcrowding, etc. [46, 55] have both been shown to decrease the number of surviving cells from those born prior to stress exposure in the adult dentate gyrus (though see [32, 38]). As with stress effects on proliferation, stress-induced suppression of cell survival appears to be mediated by GC secretion; while adrenalectomy leads to an increase in cell survival rates that is eliminated by low level corticosterone replacement, injections of high levels of corticosterone suppresses new cell survival [20]. However, the sensitivity of newly born dentate cells to GCs appears to be somewhat temporally limited; in rats, administration of corticosterone after cells are 18 days old does not lead to changes in number of surviving BrdU positive cells [20]. More work is needed to determine whether these findings extend to females and/or cell survival in the SVZ/olfactory bulb.

5.6
Stress-Induced Alteration of Cell Fate Choice

After proliferation and survival, selection of neuronal fate is the final phase that characterizes neurogenesis. Under normal circumstances, about 70–90% of

Table 5.2 Cell survival.

Species	Manipulation	New cell marker	Brain area	Effect on cell numbers compared to controls	Ref.
Male rat	0–28 days corticosterone injection once daily	BrdU before first injection	DG	Decrease cells with cort on days 1–7, 1–18, 1–28; no difference with cort days 19–28	[20]
Male rat	18 days social defeat	BrdU twice daily for 2 days before stress	DG	Decrease	[19]
Male rat	19 days variable stressors (twice daily)	BrdU once daily for 4 days before stress	DG	Decrease	[46]
Male rat	3 weeks variable stress (twice daily)	BrdU before stress	DG	No change	[99]
Male rat	3 weeks restraint (6 h/day)	BrdU once daily for 3 days before stress	DG	No change	[38]

newly born, surviving cells in the dentate gyrus express neuronal fate markers [12, 14]. Whether stress exposure during differentiation can alter this cell fate choice is currently unclear (see Table 5.3). Heine *et al.* [32], for example, found that injecting BrdU before three weeks of daily restraint stress did not lead to differences in the percentage of labeled cells that express NeuN. Several others have similarly found no difference in neuronal marker co-labeling with BrdU [19, 46], including one study that used *in vitro* hippocampal precursors allowed to differentiate in the presence of glucocorticoids [56]. In contrast to these studies, however, one detailed study by Wong and Herbert [18] does demonstrate a consistent inhibitory effect of GCs on neuronal fate choice. By injecting male rats with high levels of corticosterone over different time periods after a BrdU injection, they found that the percentage of new, BrdU-labeled cells expressing neuronal fate markers (DCX, NeuN) was decreased by corticosterone treatment over any 7-day period during the first month. More recently, we have confirmed and extended these findings, showing both *in vivo* and *in vitro* that stress/GCs drive newly proliferative dentate cells away from neuronal fate toward an oligodendrocyte fate [48]. Though more work is needed to confirm these findings given the contrary past studies, they nevertheless suggest that neuronal fate choice can be influenced by GCs [20].

Table 5.3 Neuronal fate choice.

Species	Manipulation	New cell marker	Neuronal fate marker	Brain area	Effect on neuron numbers compared to controls	Ref.
Male rat	7–28 days corticosterone injection	BrdU given before corticosterone regimen	DCX	SGZ	Decrease in % BrdU/DCX+ with cort days 1–7 (but not days 1–28)	[18]
Male rat	9–28 days corticosterone injection	BrdU given before corticosterone regimen	NeuN	SGZ	Decrease in % BrdU/NeuN+ with cort days 19–27 (but not days 1–9 or days 10–18)	[18]
Male rat	ADX	BrdU given 24 h after ADX	DCX (day 7); NeuN (day 28)	SGZ	No change in % DCX/BrdU+ or % NeuN/BrdU+	[18]
Male rat	3 weeks variable stress (twice daily)	BrdU given before stress	NeuN	SGZ	No change in % NeuN/BrdU+	[32]
Male rat	18 days social defeat	BrdU given twice daily for 2 days before defeat	NeuN	SGZ	No change in %	[19]
Rat hippocampal progenitors	2 weeks GC treatment *in vitro*	N/A	NeuN	Hippocampus	No change in NeuN+	[56]
Male rat	19 days variable stress (twice daily)	BrdU given once daily for 4 days before stress	Calbindin D28k	DG	No change	[46]

5.7
Mechanism of Stress-Induced Changes in Adult Neurogenesis

Though it seems clear that the effect of stress on adult neurogenesis is mediated primarily by stress hormones (e.g., GCs), there are many competing potential molecular mechanisms for how the GCs might alter precursor division and new cell survival/differentiation. Overall, the extra- and intracellular cascades leading from increased GCs to inhibition of cell proliferation, survival and/or neuronal differentiation are not well characterized. It is not even certain whether GCs act directly on new cells, through neighboring cells in the niche, or both.

5.7.1
Direct Effects of Glucocorticoids on Adult Neurogenesis

When stress effects on adult hippocampal cell proliferation were first reported, work using labeling for GR combined with [^{3}H]dT autoradiography seemed to indicate that proliferating cells lacked receptors for GCs [57], meaning that GC effects on proliferation must occur through some other intermediary molecule/cell. More recently, however, evidence has emerged to suggest that a portion of recently proliferative cells do express GR [58]. Moreover, we show that GCs suppress proliferation *in vitro* in a GR-dependent manner, suggesting that the GR in neural precursors is not only present but also functionally active and relevant [42]. Nonetheless, other mechanisms might contribute, as well, to account for the full effect of stress on proliferation *in vivo* (see Section 5.7.2, below).

GCs may also act directly on newly born cells to influence survival and/or differentiation. In support of this mechanism, the percent of cells expressing GR *in vivo* is reported to increase to over 50% by 7 days after BrdU injection and to almost 100% by 4 weeks of age [58]. Therefore, during the time window in which newly born cells are sensitive to GC-induced suppression of survival and neuronal differentiation [18, 20], these cells have the molecular machinery necessary to respond directly to GCs. More research is necessary to determine if GCs do act directly on maturing cells in the adult hippocampus *in vivo* and, if so, what intracellular cascade(s) are involved.

5.7.2
Indirect Effects of Glucocorticoids on Adult Neurogenesis

Though some proliferating cells do express GR [58], work on contributing factors such as excitatory amino acids, serotonin and growth factors indicate a role for indirect mechanisms, as well.

5.7.2.1 Excitatory Amino Acids
In the adult hippocampus, levels of excitatory amino acids such a glutamate are known to rise prominently during stress [59], making EAAs one probable contributor to stress-induced suppression of adult hippocampal cell proliferation. Indeed, much as stress inhibits cell proliferation and ADX enhances it, administration of NMDA, a stimulator of glutamate receptors, leads to decreased BrdU labeling while blockade of the NMDA receptor leads to increased BrdU labeling [4, 28, 35]. More directly, a study by Cameron *et al.* [35] showed that systemic administration of an NMDA receptor antagonist, MK-801, before acute corticosterone injection blocked the normal corticosterone-associated suppression of [^{3}H]dT labeling. The

NMDA-dependency of stress effects on cell proliferation likely extends to chronic stress effects as well – one study in male mice found that daily injection of agmatine, an NMDA-R antagonist, blocked the effect of 24 days of variable stress on hippocampal cell proliferation [44]. These studies strongly suggest that stress effects on proliferating cells are mediated by increases in the EAA glutamate and its actions on the NMDA receptor. Because only a small portion of dividing cells in the SGZ show NMDA-R expression [60], NMDA-R activation probably affects cell division in the adult hippocampus via its effects on the mature neuronal population. In addition, little work has been done on whether NMDA-receptor dependency of stress effects on adult neurogenesis extends beyond proliferation to effects on cell survival and differentiation.

5.7.2.2 **Serotonin**
A number of studies have shown that adult hippocampal neurogenesis is influenced by serotonergic input. In normal adult rodents, chronic treatment with selective serotonin uptake inhibitors such as fluoxetine leads to upregulation of SGZ proliferation [61, 62] while serotonin depletion leads to suppression of cell division in the adult rodent hippocampus [63]. The effect of serotonin seems to be limited to division of early progenitors, with little effect on cell survival or differentiation [61, 62]. In addition to regulating basal levels of cell proliferation, serotonin may play a role in the effects of GCs on cell proliferation. Twenty days of chronic, variable stress has been shown to result in lower $5HT_{1A}$ mRNA levels in the DG [64]. Moreover, acute stress-induced reductions in SGZ and SVZ cell proliferation have been shown to be reversed or blocked by treatment with SSRIs [33, 51, 65].

5.7.2.3 **Growth Factors**
A number of growth factors are known to influence neurogenesis in the adult DG, including BDNF, FGF-2, EGF, IGF and TGF-α (for a review, see [14]). Research on manipulations that increase GC levels yet do not impair hippocampal neurogenesis have suggested that two of these factors may be involved in GC effects on neurogenesis: BDNF and FGF-2.

Voluntary exercise, enriched environment and cognitive tasks all have been shown to lead to elevations in circulating GCs [66–69] and yet, paradoxically, to increases in hippocampal neurogenesis, primarily through increased cell survival [8, 25, 70, 71]. Interestingly, these manipulations also all lead to elevations of hippocampal BDNF [72–74]. Given that BDNF has been shown to promote cell survival of adult rodent progenitor cells [56], it seems likely that BDNF may be acting to block the effects of high GCs caused by exercise, enrichment and/or learning. FGF-2 may also influence stress effects on adult neurogenesis. Experiments using controllable and uncontrollable shock have shown that controllability is associated with both an increase in hippocampal FGF-2 and a block of stress-induced suppression of cell proliferation, despite normal elevations in plasma corticosterone [75–77]. More research is needed to clarify the role of these and other growth factors in stress-related regulation of adult neurogenesis.

5.7.3
Intracellular Mechanisms of Glucocorticoids Effects

Regardless of whether GCs act directly on NPCs or through other intermediary cells, there is still the question of what intracellular cascades are the final mediators of the NPC response to stress. The Akt/PI3K and MAPK/ERK pathways are both good candidate intracellular mediators of GC effects because both of these pathways play key roles in regulating NPC proliferation and in signaling from several extracellular factors that can influence GC effects on NPC proliferation (FGF-2, BDNF, IGF, etc.) [78–82]. Activation of Akt appears to be essential for FGF-2-mediated stimulation of NPC proliferation [78], suggesting that this pathway is an essential regulator of NPC mitosis. The MAPK/ERK pathway also regulates NPC proliferation and has been implicated in control of NPC proliferation in response to a number of stimuli such as hypoxia [80] and seizures [79]. GR activation may interact with either or both of these cascades to influence NPC proliferation but more work is needed to determine how/whether this occurs.

5.7.4
An Overall Picture of Mechanism: Putting the Pieces Together

How do all the parts of the stress effect on neurogenesis fit together? Given the direct effects of GCs on *in vitro* neuronal precursors, it seems clear that GCs could have some direct effect on cell proliferation. However, *in vivo*, the indirect mechanisms probably contribute, as well. Therefore, it is likely that those precursors that do express GR respond directly to elevated GCs by initiating genetic programs that slow the cell cycle, upregulating cell cycle inhibitors such as p21 and p27Kip1, among others. Simultaneously, in precursors without GR, and perhaps also in some precursors with GR, elevated GCs may act through glutamatergic transmission, serotonergic transmission from the DRN and/or growth factors. The downstream cascades that lead from these extracellular signals to slowing of the cell cycle are not well characterized but could rely on changes in signaling cascades known to regulate stem cell division such as the Akt/PI3K and MAPK/ERK pathways. How stress effects new cell survival and differentiation is less well studied than the mechanism of proliferation effects. Given the higher percentage of cells expressing GR over time, though, it seems possible that this mechanism could rely more heavily on GR activation than proliferative effects do. However, BDNF may also play a large role, particularly in cell survival, as discussed above.

5.7.5
Function of Regulation of Adult Neurogenesis by Stress and Glucocorticoids

The question of whether changes in adult hippocampal neurogenesis due to stress have functional or behavioral implications is largely unanswered. Indeed, the exact function of adult hippocampal neurogenesis is itself somewhat unsettled. Given that the hippocampus is important for many aspects of learning and memory [83] and that adult neurogenesis is enhanced by hippocampus-dependent [25, 54] but not by hippocampus-independent tasks [84], some have hypothesized that, though small in absolute numbers, these newly born neurons have disproportionate influence on hippocampal function. In support of this hypothesis, electrophysiological studies have shown that newly born neurons are more excitable than mature granule neurons, having lower thresholds for LTP and greater response amplitudes than older neurons [85–87]. More direct evidence for the functional importance of adult hippocampal neurogenesis comes from studies where cell mitosis is reduced or eliminated experimentally. The DNA methylating agent, methylazomethanol acetate (MAM), inhibits mitosis and, when administered systemically, has been shown to impair memory in a trace conditioning task [88] and to block enrichment-induced improvement in object recognition memory [27]. Irradiation, a more localized method of inhibiting cell mitosis [89], has also been shown to cause hippocampal memory deficits [90]. A number of computational approaches have suggested essential roles for new neurons in hippocampal function despite their limited quantities, as well [91, 92]. In summary, while there are some contradictory findings, it seems at least possible that adult neurogenesis may contribute to hippocampal cognitive function in a behaviorally relevant manner.

Even if adult neurogenesis does play an important role in hippocampus-dependent cognition, whether stress-induced suppression of neurogenesis has any functional consequences is still unclear. Stress certainly does alter hippocampal function at the behavioral level, leading to impaired hippocampus-dependent learning and memory performance in a variety of tasks [93–95]. However, stress influences many aspects of hippocampal physiology and morphology other than adult neurogenesis (such as LTP [96] and CA3 dendritic branching [97]), making attribution of any single stress effect to neurogenic changes difficult. The change in the number of new neurons due to an acute episode of stress or GC exposure is transient (see cell proliferation and survival sections), suggesting that effects of acute stress on hippocampal function are likely minimal. Changes due to chronic stress, however, are longer-lasting and may have more relevance due to prolonged inhibition of proliferation combined with reduction of cell survival and inhibition of neuronal fate choice. Whether the scale of change observed is enough to have functional or behavioral relevance *in vivo* has not been well investigated. The finding that enrichment-induced gains in neurogenesis may be functionally relevant [27] is encouraging, but that finding has been questioned more recently [98]. Though there is some evidence pointing to a possible role of adult neurogenesis in normal hippocampal function, more research is needed to determine if stress-induced loss of new neurons contributes to any of the known effects of stress on hippocampal function.

Summary

While there is a great deal of evidence pointing to a role for stress in regulation of adult neurogenesis, many questions remain unanswered. Exposure to high levels of GCs whether through stress or exogenous administration, inhibits cell proliferation, survival and possibly neuronal differentiation. The inhibition of proliferation is by far the best studied phenomenon. It appears to rely on a slowing of the cell cycle, but the molecular mechanism is largely unclear. Excitatory amino acids, serotonin and growth factors all seem to play a role but which of these factors acts directly on proliferative cells and what intracellular cascades trigger the slowing of progression through the cell cycle is uncertain. Less is known about stress effects on cell survival and differentiation, but the greater presence of GR as cells mature suggests that direct actions of GCs could play more of a role here than they could in proliferation. Finally, the functional consequences of stress-associated regulation of adult neurogenesis are largely undetermined. This will perhaps become clearer, though, as more work is done to determine the functional meaning of adult neurogenesis.

Future work on stress and adult neurogenesis should focus on better utilizing *in vitro* models of adult neurogenesis to clarify the cellular mechanism of stress-related suppression of neurogenesis. While the many *in vivo* studies investigating adult hippocampal neurogenesis have provided solid evidence for the existence of a stress effect and have given some hints at mechanism, by using neuronal progenitors derived from the adult brain, more detailed and controlled investigation can be done on what molecular cascades are activated by GCs, how other compounds such as EAAs and serotonin contribute to GC effects and what intracellular cascades and genetic programs are activated by exposure to stress hormones. The results of such investigations would provide not only a better understanding of stress-response biology, but would also contribute to a greater understanding of the regulation of plasticity in the adult brain.

References

1 Cameron, H.A. and Gould, E. (1994) Adult neurogenesis is regulated by adrenal steroids in the dentate gyrus. *Neuroscience*, **61**, 203–209.

2 Mirescu, C. and Gould, E. (2006) Stress and adult neurogenesis. *Hippocampus*, **16**, 233–238.

3 Gould, E., Cameron, H.A., Daniels, D.C., Woolley, C.S. and McEwen, B.S. (1992) Adrenal hormones suppress cell division in the adult rat dentate gyrus. *J. Neurosci.*, **12**, 3642–3650.

4 Gould, E., McEwen, B.S., Tanapat, P., Galea, L.A. and Fuchs, E. (1997) Neurogenesis in the dentate gyrus of the adult tree shrew is regulated by psychosocial stress and NMDA receptor activation. *J. Neurosci.*, **17**, 2492–2498.

5 Altman, J. and Das, G.D. (1965) Autoradiographic and histological evidence of postnatal hippoampal neurogenesis in rats. *J. Comp. Neurol.*, **124**, 319–335.

6 Altman, J. (1969) Autoradiographic and histological studies of postnatal neurogenesis. IV. Cell proliferation and migration in the anterior forebrain, with special reference to presisting neurogenesis in the olfactory bulb. *J. Comp. Neurol.*, **137**, 433–457.

7 Kaplan, M.S. and Dell, D.H. (1977) Neurogenesis in the adult rat: electron microscopic analysis of light radiographs. *Science*, **197**, 1092–1094.

8 van Praag, H., Shubert, T., Zhao, C. and Gage, F.H. (2005) Exercise enhances

learning and hippocampal neurogenesis in aged mice. *J. Neurosci.*, **25**, 8680–8685.

9 Kempermann, G., Gast, D., Kronenberg, G., Yamaguchi, M. and Gage, F.H. (2003) Early determination and long-term persistence of adult-generated new neurons in the hippocampus of mice. *Development*, **130**, 391–399.

10 Gould, E., Reeves, A.J., Fallah, M., Tanapat, P., Gross, C.G. and Fuchs, E. (1999) Hippocampal neurogenesis in adult Old World pirmates. *Proc. Natl Acad. Sci. USA*, **96**, 5263–5267.

11 Eriksson, P.S. (1998) Neurogenesis in the adult human hippocampus. *Nat. Med.*, **4**, 1313–1317.

12 Cameron, H.A. and McKay, R.D.G. (2001) Adult neurogenesis produces a large pool of new granule cells in the dentate gyrus. *J. Comp. Neurol.*, **435**, 406–417.

13 Dayer, A.G., Ford, A.A., Cleaver, K.M., Yassaee, M. and Cameron, H.A. (2003) Short-term and long-term survival of new neurons in the rat dentate gyrus. *J. Comp. Neurol.*, **460**, 563–572.

14 Abrous, D.N., Koehl, M. and Le Moal, M. (2005) Adult neurogenesis: from precursors to network and physiology. *Physiol. Rev.*, **85**, 523–569.

15 Lois, C. and Alvarez-Buylla, A. (1994) Long-distance neuronal migration in the adult mammalian brain. *Science*, **264**, 1145–1148.

16 van Praag, H., Schinder, A.F., Christie, B.R., Toni, N., Palmer, T.D. and Gage, F.H. (2002) Functional neurogenesis in the adult hippocampus. *Nature*, **415**, 1030–1034.

17 Carlen, M., Cassidy, R.M., Brismar, H., Smith, G.A., Enquist, L.W. and Frisen, J. (2002) Functional integration of adult-born neurons. *Curr. Biol.*, **12**, 606–608.

18 Wong, E.Y. and Herbert, J. (2006) Raised circulating corticosterone inhibits neuronal differentiation of progenitor cells in the adult hippocampus. *Neuroscience*, **137**, 83–92.

19 Czeh, B., Welt, T., Fischer, A.K., Erhardt, A., Schmitt, W., Muller, M.B., Toschi, N.F.E. *et al.* (2002) Chronic psychosocial stress and concomitant repetitive transcranial magnetic stimulation: effects on stress hormone levels and adult hippocampal neurogenesis. *Biol. Psychiatry*, **52**, 1057–1065.

20 Wong, E.Y. and Herbert, J. (2004) The corticoid environment: a determining factor for neural progenitors' survival in the adult hippocampus. *Eur. J. Neurosci.*, **20**, 2491–2498.

21 Drapeau, E., Mayo, W., Aurousseau, C., Le Moal, M., Piazza, P. and Abrous, D.N. (2003) Spatial memory performances of aged rats in the water maze predict levels of hippocampal neurogenesis. *Proc. Natl Acad. Sci. USA*, **100**, 14385–14390.

22 Cooper-Kuhn, C.M. and Kuhn, H.G. (2002) Is it all DNA repair? Methodological considerations for detecting neurogenesis in the adult brain. *Brain Res. Dev. Brain Res.*, **134**, 13–21.

23 Kee, N., Sivalingam, S., Boonstra, R. and Wojtowicz, J.M. (2002) The utility of Ki-67 and BrdU as proliferative markers of adult neurogenesis. *J. Neurosci. Methods*, **115**, 97–105.

24 Cameron, H.A. and Gould, E. (1996) Distinct populations of cells in the adult dentate gyrus undergo mitosis or apoptosis in response to adrenalectomy. *J. Comp. Neurol.*, **369**, 56–63.

25 Gould, E., Beylin, A., Tanapat, P., Reeves, A. and Shors, T.J. (1999) Learning enhances adult neurogenesis in the hippocampal formation. *Nat. Neurosci.*, **2**, 260–265.

26 Montaron, M.F., Petry, K.G., Rodriguez, J.J., Marinelli, M., Aurousseau, C., Rougon, G., Le Moal, M. *et al.* (1999) Adrenalectomy increases neurogenesis but not PSA-NCAM expression in aged dentate gyrus. *Eur. J. Neurosci.*, **11**, 1479–1485.

27 Bruel-Jungerman, E., Laroche, S. and Rampon, C. (2005) New neurons in the dentate gyrus are involved in the expression of enhanced long-term memory following environmental enrichment. *Eur. J. Neurosci.*, **21**, 513–521.

28 Cameron, H.A., McEwen, B.S. and Gould, E. (1995) Regulation of adult neurogenesis by excitatory input and NMDA receptor activation in the dentate gyrus. *J. Neurosci.*, **15**, 4687–4692.

29 Gross, C.G. (2000) Neurogenesis in the adult brain: death of a dogma. *Nat. Rev. Neurosci.*, **1**, 67–73.

30 Koo, J.W. and Duman, R.S. (2008) Il-1Beta is an essantil mediator of the antineurogenic and anhedonic effects of stress. *Proc. Natl Acad. Sci. USA*, **105**, 751–765.

31 Tanapat, P., Hastings, N.B., Rydel, T.A., Galea, L.A. and Gould, E. (2001) Exposure to fox odor inhibits cell proliferation in the hippocampus of adult rats via an adrenal hormone-dependent mechanism. *J. Comp. Neurol.*, **437**, 496–504.

32 Heine, V.M., Maslam, S., Zareno, J., Joels, M. and Lucassen, P. (2004) Suppressed proliferation and apoptotic changes in the rat dentate gyrus after acute and chronic stress are reversible. *Eur. J. Neurosci.*, **19**, 131–144.

33 Malberg, J.E. and Duman, R.S. (2003) Cell proliferation in adult hippocampus is decreased by inescapable stress: reversal by fluoxetine treatment. *Neuropsychopharmacology*, **28**, 1562–1571.

34 Mirescu, C., Peters, J.D. and Gould, E. (2004) Early life experience alters response of adult neurogenesis to stress. *Nat. Neurosci.*, **7**, 841–846.

35 Cameron, H.A., Tanapat, P. and Gould, E. (1998) Adrenal steroids and N-methyl-D-aspartate receptor activation regulate neurogenesis in the dentate gyrus of adult rats through a common pathway. *Neuroscience*, **82**, 349–354.

36 Joels, M., Karst, H., Alfarez, D., Heine, V.M., Qin, Y., van Riel, E., Verkuyl, M. *et al.* (2004) Effects of chronic stress on structure and cell function in rat hippocampus and hypothalamus. *Stress*, **7**, 221–231.

37 Falconer, E.M. and Galea, L.A. (2003) Sex differences in cell proliferation, cell death and defensive behavior following acute predator odor stress in adult rats. *Brain Res.*, **975**, 22–36.

38 Pham, K., Nacher, J., Hof, P.R. and McEwen, B.S. (2003) Repeated restraint stress suppresses neurogenesis and induces biphasic PSA-NCAM expression in the adult rat dentate gyrus. *Eur. J. Neurosci.*, **17**, 879–886.

39 Thomas, R.M., Urban, J.H. and Peterson, D.A. (2006) Acute exposure to predator odor elicits a robust increase in corticosterone and a decrease in activity without altering proliferation in the adult rat hippocampus. *Exp. Neurol.*, **201**, 308–315.

40 Thomas, R.M., Hotsenpiller, G. and Peterson, D.A. (2007) Acute psychosocial stress reduces cell survival in adult hippocampal neurogenesis without altering proliferation. *J. Neurosci.*, **27**, 2734–2743.

41 Rosenbrock, H., Koros, E., Bloching, A., Podhorna, J. and Borsini, F. (2005) Effect of chronic intermittent restraint stress on hippocampal expression of marker proteins for synaptic plasticity and progenitor cell proliferation in rats. *Brain Res.*, **1040**, 55–63.

42 Kim, J.B., Ju, J.Y., Kim, J.H., Kim, T.-Y., Yang, B.-H., Lee, Y.-S. and Son, H. (2004) Dexamethasone inhibits proliferation of adult hippocampal neurogenesis *in vivo* and *in vitro*. *Brain Res.*, **1027**, 1–10.

43 Czeh, B., Michaelis, T., Watanabe, T., Frahm, J., de Biurrun, G., van Kampen, M., Bartolomucci, A. *et al.* (2001) Stress-induced changes in cerebral metabolites, hippocampal volume, and cell proliferation are prevented by antidepressant treatment with tianeptine. *Proc. Natl Acad. Sci. USA*, **98**, 12796–12801.

44 Li, Y.-F., Chen, H.-X., Liu, Y.-Z., Liu, Y.-Q. and Li, J. (2006) Agmatine increases proliferation of cultured hippocampal progenitor cells and hippocampal neurogenesis in chronically stressed mice. *Acta Pharmacol. Sin.*, **27**, 1395–1400.

45 Westenbroek, C., Der Boer, J.A., Veenhuis, M. and Ter Horst, G.J. (2004) Chronic stress and social housing differentially affect neurogenesis in male and female rats. *Brain Res. Bull.*, **64**, 303–308.

46 Lee, K.J., Kim, S.J., Kim, S.W., Choi, S.H., Shin, Y.C., Park, S.H., Moon, B.H. *et al.* (2006) Chronic mild stress decreases survival, but not proliferation, of new-born cells in adult rat hippocampus. *Exp. Mol. Med.*, **38**, 44–54.

47 Gould, E., Tanapat, P., McEwen, B.S., Flugge, G. and Fuchs, E. (1998) Proliferation of granule cell precursors in the dentate gyrus of adult monkeys is diminished by stress. *Proc. Natl Acad. Sci. USA*, **95**, 3168–3171.

48 Chetty, S., Mirescu, C., Bentley, G.E. and Kaufer, D. (2007) Dose-dependent effects of glucocorticoids on neural precursor cells in the adult dentate gyrus, *Neuroscience Meeting Planner, San Diego, CA: Society for Neuroscience, 2007*, Program No. 298.6.

49 Tanapat, P., Hastings, N.B., Reeves, A.J. and Gould, E. (1999) Estrogen stimulates a transient increase in the number of new neurons in the dentate gyrus of the adult female rat. *J Neurosci.*, **19**, 5792–5801.

50 Ostrander, M.M., Ulrich-Lai, Y.M., Choi, D.C., Richtand, N.M. and Herman, J.P. (2006) Hypoactivity of the hypothalamo-pituitary-adrenocortical axis during recovery from chronic variable stress. *Endocrinology*, **147**, 2008–2017.

51 Hitoshi, S., Maruta, N., Higashi, M., Kumar, A., Kato, N. and Ikenaka, K. (2007) Antidepressant drugs reverse the loss of adult neural stem cells following chronic stress. *J. Neurosci. Res.*, **85**, 3574–3585.

52 Crochemore, C., Michaelidis, T.M., Fischer, D., Loeffler, J.-P. and Almeida, O.F.X. (2002) Enhancement of p53 activity and inhibition of neural cell proliferation by glucocorticoid receptor activation. *FASEB J.*, **16**, 761–770.

53 Heine, V.M., Maslam, S., Joëls, M. and Lucassen, P.J. (2004) Increased P27KIP1 protein expression in the dentate gyrus of chronically stressed rats indicates G1 arrest involvement. *Neuroscience*, **129**, 593–601.

54 Ambrogini, P., Cuppini, R., Cuppini, C., Ciaroni, S., Cecchini, T., Ferri, P., Sartini, S. *et al.* (2000) Spatial learning affects immature granule cell survival in adult rat dentate gyrus. *Neurosci. Lett.*, **286**, 21–24.

55 Oomen, C.A., Mayer, J.L., De Kloet, E.R., Joels, M. and Lucassen, P. (2007) Brief treatment with the glucocorticoid receptor antagonist mifepristone normalizes the reduction in neurogenesis after chronic stress. *Eur. J. Neurosci.*, **26**, 3395–3401.

56 Palmer, T.D., Takahashi, J. and Gage, F.H. (1997) The adult rat hippocampus contains primordial neural stem cells. *Mol. Cell. Neurosci.*, **8**, 389–404.

57 Cameron, H.A., Woolley, C.S., McEwen, B.S. and Gould, E. (1993) Differentiation of newly born neurons and glia in the dentate gyrus of the adult rat. *Neuroscience*, **56**, 337–344.

58 Garcia, A., Steiner, B., Kronenberg, G., Bick-Sander, A. and Kempermann, G. (2004) Age-dependent expression of glucocorticoid – and mineralocorticoid – receptors on neural precursor cell populations in the adult murine hippocampus. *Aging Cell*, **3**, 363–371.

59 Gilad, G.M., Gilad, V.H., Wyatt, R.J. and Tizabi, Y. (1990) Region-specific stress-induced increase of glutamate uptake and release in rat forebrain. *Brain Res.*, **525**, 335–338.

60 Nacher, J., Varea, E., Blasco-Ibanez, J.M., Gomez-Climent, M.A., Castillo-Gomez, E., Crespo, C., Martinez-Guijarro, F.J. *et al.* (2007) N-methyl-D-aspartate receptor expression during adult neurogenesis in the rat dentate gyrus. *Neuroscience*, **144**, 855–864.

61 Encinas, J.M., Vaahtokari, A. and Enikolopov, G. (2006) Fluoxetine targets early progenitor cells in the adult brain. *Proc. Natl Acad. Sci. USA*, **103**, 8233–8238.

62 Malberg, J.E., Eisch, A.J., Nestler, E.J. and Duman, R.S. (2000) Chronic antidepressant treatment increases neurogenesis in adult rat hippocampus. *J. Neurosci.*, **20**, 9104–9110.

63 Brezun, J.M. and Daszuta, A. (1999) Depletion in serotonin decreases neurogenesis in the dentate gyrus and the subventricular zone of adult rats. *Neuroscience*, **89**, 999–1002.

64 Xu, Y., Ku, B., Cui, L., Li, X., Barish, P.A., Foster, T.C. and Ogle, W.O. (2007) Curcumin reverses impaired hippocampal neurogenesis and increases serotonin receptor 1A mRNA and brain-derived neurotrophic factor expression in chronically stressed rats. *Brain Res.*, **1162**, 9–18.

65 Huang, A.M. and Herbert, J. (2006) Stimulation of neurogenesis in the hippocampus of the adult rat by fluoxetine requires rhythmic change in corticosterone. *Biol. Psychiatry*, **59**, 619–624.

66 Benaroya-Milshtein, N., Hollander, N., Apter, A., Kukulansky, T., Raz, N., Wilf, A., Yaniv, I. *et al.* (2004) Environmental enrichment in mice decreases anxiety, attenuates stress responses and enhances natural killer cell activity. *Eur. J. Neurosci.*, **20**, 1341–1347.

67 Moncek, F., Duncko, R., Johansson, B.B. and Jezova, D. (2004) Effect of environmental enrichment on stress related systems in rats. *J. Endocrinol.*, **16**, 423–431.

68 Droste, S.K., Gesing, A., Ulbricht, S., Müller, M.B., Linthorst, A.C. and Reul, J.M. (2003) Effects of long-term voluntary exercise on the mouse hypothalamic-pituitary-adrenocortical axis. *Endocrinology*, **144**, 3012–3023.

69 Aguilar-Vallesa, A., Sáncheza, E., de Gortarib, P., Balderasc, I., Ramírez-Amayad, V., Bermúdez-Rattonic, F. and Joseph-Bravoa, P. (2005) Analysis of the stress response in rats trained in the water-maze: differential expression of corticotropin-releasing hormone, CRH-R1, glucocorticoid receptors and brain-derived neurotrophic factor in limbic regions. *Neuroendocrinology*, **82**, 306–319.

70 van Praag, H., Kempermann, G. and Gage, F.H. (1999) Running increases cell proliferation and neurogenesis in the adult mouse dentate gyrus. *Nat. Neurosci.*, **2**, 266–270.

71 Kempermann, G., Kuhn, H.G. and Gage, F.H. (1997) More hippocampal neurons in adult mice living in an enriched environment. *Nature*, **386**, 493–495.

72 Falkenberg, T., Mohammed, A.K., Henriksson, B., Persson, H., Winblad, B. and Lindefors, N. (1992) Increased expression of brain-derived neurotrophic factor mRNA in rat hippocampus is associated with improved spatial memory and enriched environment. *Neurosci. Lett.*, **138**, 153–156.

73 Russo-Neustadt, A.A., Beard, R.C., Huang, Y.M. and Cotman, C.W. (2000) Physical activity and antidepressant treatment potentiate the expression of specific brain-derived neurotrophic factor transcripts in the rat hippocampus. *Neuroscience*, **101**, 305–312.

74 Kesslak, J.P., So, V., Choi, J., Cotman, C.W. and Gomez-Pinilla, F. (1998) Learning upregulates brain-derived neurotrophic factor messenger ribonucleic acid: a mechanism to facilitate encoding and circuit maintenance? *Behav. Neurosci.*, **112**, 1012–1019.

75 Bland, S.T., Schmid, M.J., Greenwood, B.N., Watkins, L.R. and Maier, S.F. (2006) Behavioral control of the stressor modulates stress-induced changes in neurogenesis and fibroblast growth factor-2. *Neuroreport*, **17**, 593–597.

76 Shors, T.J., Mathew, J., Sisti, H.M., Edgecomb, C., Beckoff, S. and Dalla, C. (2007) Neurogenesis and helplessness are mediated by controllability in males but not in females. *Biol. Psychiatry*, **62**, 487–495.

77 Maier, S.F., Ryan, S.M., Barksdale, C.M. and Kalin, N.H. (1986) Stressor controllability and the pituitary-adrenal system. *Behav. Neurosci.*, **100**, 669–674.

78 Peltier, J., O'Neill, A.O. and Schaffer, D.V. (2007) PI3K/Akt and CREB regulate adult neural hippocampal progenitor proliferation and differentiation. *Dev. Neurobiol.*, **67**, 1358–1361.

79 Choi, Y.-S., Cho, H.-Y., Hoyt, K.R., Naegele, J.R. and Obrietan, K. (2008) IGF-1 receptor-mediated ERK/MAPK signaling couples status epilepticus to progenitor cell prolfieration in the subgranular layer of the dentate gyrus. *Glia*, **56**, 791–800.

80 Zhou, L., Villar, K.D., Dong, Z. and Miller, C.A. (2004) Neurogenesis response to hypoxia-induced cell death: map kinase signal transduction mechanisms. *Brain Res.*, **1021**, 8–19.

81 Patapoutian, A. and Reichardt, L.F. (2001) Trk receptors: mediators of neurotrophin action. *Curr. Opin. Neurobiol.*, **11**, 272–280.

82 Eswarakumar, V.P., Lax, I. and Schlessinger, J. (2005) Cellular signaling by fibroblast growth factor receptors.

Cytokine Growth Factor Rev., **16**, 139–149.

83 Eichenbaum, H. (2000) A cortical-hippocampal system for declarative memory. *Nat. Rev. Neurosci.*, **1**, 41–50.

84 Van der Borght, K., Havekes, R., Bos, T., Eggen, B.J. and Van der Zee, E.A. (2007) Exercise improves memory acquisition and retrieval in the Y-maze task: relationship with hippocampal neuro-genesis. *Behav. Neurosci.*, **121**, 324–334.

85 Schmidt-Heiber, C., Jonas, P. and Bischofberger, J. (2004) Enhanced synaptic plasticity in newly generated granule cells of the adult hippocampus. *Nature*, **429**, 184–187.

86 Ge, S., Yang, C.-H., Hsu, K.-S., Ming, G.-L. and Song, H. (2007) A critical period for enhanced synaptic plasticity in newly generated neurons of the adult brain. *Neuron*, **54**, 559–566.

87 Snyder, J.S., Kee, N. and Wojtomicz, J.M. (2001) Effects of adult neurogenesis on synaptic plasticity in the rat dentate gyrus. *J. Neurophysiol.*, **85**, 2423–2431.

88 Shors, T.J., Miesegaes, G., Beylin, A., Zhao, M., Rydel, T. and Gould, E. (2001) Neurogenesis in the adult is involved in the formation of trace memories. *Nature*, **410**, 372–376.

89 McGinn, M.J., Sun, D. and Colello, R.J. (2008) Utilizing x-irradiation to selectively eliminate neural stem/progenitor cells from neurogenic regions of the mammalian brain. *J. Neurosci. Methods*, **170**, 9–15.

90 Snyder, J.S., Hong, N.S., McDonald, R.J. and Wojtomicz, J.M. (2005) A role for adult neurogenesis in spatial long-term memory. *Neuroscience*, **130**, 843–852.

91 Chambers, R.A., Potenza, M.N., Hoffman, R.E. and Miranker, W. (2004) Simulated apoptosis/neurogenesis regulates learning and memory capabilities of adaptive neural networks. *Neuropsychopharmacology*, **29**, 747–758.

92 Aimone, J.B., Wiles, J. and Gage, F.H. (2006) Potential role for adult neurogenesis in the encoding of time in new memories. *Nat. Neurosci.*, **9**, 723–727.

93 Kitraki, E., Kremmyda, O., Youlatos, D., Alexis, M. and Kittas, C. (2004) Spatial performance and corticosteroid receptor status in the 21-day restraint stress paradigm. *Ann. N. Y. Acad. Sci.*, **1018**, 323–327.

94 Bodnoff, S.R., Humphreys, A.G., Lehman, J.C., Diamond, D.M., Rose, G.M. and Meaney, M.J. (1995) Enduring effects of chronic corticosterone treatment on spatial learning, synaptic plasticity, and hippocampal neuropathology in young and mid-aged rats. *J. Neurosci.*, **15**, 61–69.

95 Radecki, D.T., Brown, L.M., Martinez, J. and Teyler, T.J. (2005) BDNF protects against stress-induced impairments in spatial learning and memory and LTP. *Hippocampus*, **15**, 246–253.

96 Watanabe, Y., Gould, E. and McEwen, B.S. (1992) Stress induces atrophy of apical dendrites of hippocampal CA3 pyramidal neurons. *Brain Res.*, **588**, 341–345.

97 Pavlides, C., Nivon, L.G. and McEwen, B.S. (2002) Effects of chronic stress on hippocampal long-term potentiation. *Hippocampus*, **12**, 245–257.

98 Meshi, D., Drew, M.R., Saxe, M., Ansorge, M.S., David, D., Santarelli, L., Malapani, C. *et al.* (2006) Hippocampal neurogenesis is not required for behavioral effects of environmental enrichment. *Nat. Neurosci.*, **9**, 729–731.

99 Heine, V.M., Zareno, J., Maslam, S., Joels, M. and Lucassen, P. (2005) Chronic stress in the adult dentate gyrus reduces cell proliferation near the vascular and VEGF and Flk-1 protein expression. *Eur. J. Neurosci.*, **21**, 1304–1314.

6

Individual Differences in Reactivity to Social Stress in the Laboratory and Its Mediation by Common Genetic Polymorphisms

Idan Shalev, Elad Lerer, Salomon Israel, Florina Uzefovsky, Inga Gritsenko, David Mankuta, Marsha Kaitz, and Richard P. Ebstein

6.1
Stressors and HPA Axis Regulation: Naturalistic Studies

The hypothalamic–pituitary–adrenal (HPA) axis is a major pathway in regulating stress responses, and animal and human studies have shown that exaggerated response as well as under response of the HPA axis is associated with a spectrum of disease conditions [1, 2]. When the brain perceives stress, corticotrophin-releasing hormone (CRH) is released from the hypothalamus, which in turn stimulates the synthesis and release of adrenocorticotrophic hormone (ACTH) from the pituitary gland, leading to cortisol synthesis and release by the adrenal cortex. There is abundant evidence that brain and body adaptations to acute and chronic stress are critical for physical and mental health [2].

Both physical stressors and psychological stressors can elicit heightened responses from the HPA axis [3, 4]. Similar to physical stressors (including externally imposed stimuli such as electric shock, noise exposure, prolonged exercise), psychological stressors also modulate the HPA axis. The effect of stressors on HPA reactivity has been reviewed by Michaud *et al.* [5]. The majority of naturalistic studies involve occupational stressors (job strain, burnout, work-related noise exposure, unemployment, academic stressors) followed by social stressors (marital conflict, loneliness, interpersonal stress, caregiving), medical procedures and sports-related stressors. Across studies, the cortisol level of humans following naturalistic stressors has been shown to have a moderate size effect [6]. Indeed, the magnitude of the cortisol increase is generally much smaller than that evident in stressed rodents (where eight- to 10-fold elevations are common), or in response to a laboratory challenge such as the Trier social stress test (TSST; see below), where two- to fourfold increases have been reported.

Stress – From Molecules to Behavior. Edited by Hermona Soreq, Alon Friedman, and Daniela Kaufer
Copyright © 2010 WILEY-VCH Verlag GmbH & Co. KGaA, Weinheim
ISBN: 978-3-527-32374-6

6.2
Trier Social Stress Test

A robust strategy beyond naturalistic studies in understanding how social stress impacts on the HPA axis is the Trier social stress test (TSST) [7]. The TSST is a protocol that has been used in many studies to investigate the stress response in human adults. The primary stressor is social and is related to the anticipation of "performing" in front of strangers, "the unknown", the non-responsiveness of "interviewers" who are supposedly judging the subjects' performance and the fear of failure or embarrassment in front of others. In operational terms, the TSST consists of an anticipation period (10 min) and a test period (10 min) in which the subjects have to deliver a free speech and perform mental arithmetic in front of an audience with a camera and microphone situated between the interviewers clearly visible to the interviewee. Many studies demonstrate that this protocol induces considerable changes in the concentration of ACTH, cortisol (serum and saliva), growth hormone and prolactin, as well as significant increases in heart rate and blood pressure. As for salivary cortisol levels, the TSST reliably led to two- to fourfold elevations above baseline with similar peak cortisol concentrations (Figure 6.1). Changes in salivary cortisol in the TSST are characterized by marked individual differences. Among the numerous factors that account for the variance in TSST cortisol responses are gender [8–11], ethnicity [12], smoking [13, 14], stress-related disorders [15, 16], education level [17], suckling [18], birth weight and gestational age [19, 20], sports training [21], acute couple interactions [22], temperament [23–25] and diet during pregnancy [26]. These studies and many others show that the TSST can serve as an important tool for psychobiological research.

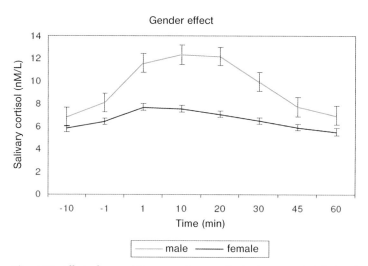

Figure 6.1 Effect of sex and time on salivary cortisol values sampled during the TSST.

6.3
Genes and Regulation of the HPA Axis

A crucial variable that impacts and interacts across all kinds of stressors to modulate HPA axis reactivity is an individual's particular genotype (see Table 6.1 for a summary). Indeed, there is considerable impact of heredity on basal free cortisol, and analyses of results across five comparable twin studies estimated heritability to be 62% for this measure [33]. Similarly, HPA axis responses to moderate

Table 6.1 Summary of common and not so common polymorphisms that modulate HPA axis reactivity during the TSST.

Gene	Polymorphism	Allele	Frequency	Number of Ss TSST	Ref.
Glucocorticoid receptor (GR)[a] NR3C1	BclI RFLP N363S ER22/23EK	G/C	0.63/0.37 0.08 0.04	112	[27, 28]
Mineralcorticoid receptor (MR) NR3C2	MRI180V	A/G Compared (MR180I = 83 to MR180V, n = 27)	G = 0.12 A = 0.88	110	[28–30]
Micro-opioid receptor (MOR)[b]	A118G	AA = 59 AG = 14 GG = 1; AA = 59 vs AG + GG = 15	79.7 18.9 1.3 79.7 20.2	74	[31]
GABRA6	SNP T1521C	CC = 15 CT = 29 TT = 12	26.7 51.7 21.4	56	[32]

a Glucocorticoid receptor (GR).
Common BclI restriction fragment length polymorphism (RFLP) in the GR gene presumably located in intron 2.
AAT-to-AGT point mutation in exon 2 causing an asparagines to serine amino acid change in codon 363.
Codon 23 change from AGG to AAG results in an amino acid change from arginine to lysine (ER22/23EK).
Because of small numbers of Ss carrying the rare alleles the following four GR genotypes were statistically compared: (i) BclI CC and N363S AA (wild type, n = 36; the term wild type was chosen to enhance the comprehensibility of the text; this decision was solely based on the fact that the BclI C allele and the N363S A allele showed the highest frequencies); (ii) BclI CC and N363S AG or GG (N363S AG or GG (363S carrier, n = 10); (iii) BclI CG and N363S AA (BclI G heterozygote, n = 38); and (iv) BclI GG and N363SAA(BclI G homozygote, n = 18).
b Mineralcorticoid receptor.
Exon 2 of the MR gene at position NR3C2 c.538A_G, changing an isoleucine to valine (codon 180, ATT to GTT, MRI180V; rs5522 A/G).

psychosocial stress elicited during the TSST [7] have also been shown to be moderately heritable [34]. Surprisingly, only a limited number of studies have focused on specific polymorphisms that contribute to HPA axis reactivity. Expectedly, the first studies of genes and the TSST have examined the glucocorticoid (*GR*) receptor (*NRC31*), an obvious candidate for modulating psychosocial stress response. In the first of a series of studies, Wust *et al.* [27] used a sample of 112 healthy to estimate the impact of three *GR* gene polymorphisms (BclI RFLP, N363S, ER22/23EK – see Table 6.1) on cortisol responses to the TSST. Compared with subjects with the wild-type *GR* genotype (*n* = 36), 363S allele carriers (*n* = 10) showed significantly increased salivary cortisol responses to stress, whereas the BclI genotype GG (*n* = 18) was associated with a diminished cortisol response. A subsequent investigation by this group observed significant sex-specific associations between *GR* gene polymorphisms and HPA axis responses to psychosocial stress [29]. These *GR* polymorphisms were also shown to confer differential sensitivity to glucocorticoids measured in subdermal blood vessels and peripheral leukocytes in 206 healthy individuals [35].

In addition to the *GR* gene, the mineralocorticoid (*NR3C2*) receptor gene [30] has also been examined in relation to the human stress responses in the TSST. Mineralocorticoid receptors (MR) mediate the action of aldosterone on sodium resorption in kidney tubular cells, but in brain they respond to the glucocorticoid cortisol in stress regulation and cognitive processes. A single SNP was examined that is located in exon 2 of the MR gene changing an isoleucine (I) to valine (V). Carriers of the MR180V allele showed higher saliva ($P < 0.01$), plasma cortisol ($P < 0.01$) and heart rate responses ($P < 0.05$) to the TSST than non-carriers (MR180I).

The HPA axis and the sympathetic adrenomedullary system are regulated by several neuronal pathways, including the inhibitory gamma-aminobutyric acid (GABA)ergic system [36]. This suggested to Uhart *et al.* that a gene coding for the GABA(A)alpha6 receptor would be worth examining using the TSST paradigm [32]. Fifty-six healthy subjects were genotyped to determine the influence of the T1521C single nucleotide polymorphism (SNP) in the GABA(A)alpha6 receptor subunit gene (*GABRA6*). Adrenocorticotropin (ACTH), cortisol, diastolic blood pressure and mean blood pressure responses to the TSST were significantly greater in subjects homozygous for the T allele or heterozygous compared to subjects homozygous for the C allele. Behavioral data was also collected employing the *revised NEO personality inventory* (NEO PI-R); subjects homozygous for the C allele scored significantly lower on the extraversion factor compared to subjects homozygous for the T allele or heterozygous. These results suggest that the T1521C polymorphism in the *GABRA6* gene is associated with specific personality characteristics as well as a marked attenuation in hormonal and blood pressure responses to psychological stress.

A polymorphism in the micro-opioid receptor gene (*MOR*; A118G) has been shown to increase beta-endorphin binding affinity, theoretically placing greater inhibitory tone on hypothalamic corticotropin-releasing hormone (CRH) neurons. This SNP results in an Asn40Asp substitution in the extracellular

N-terminal domain of *MOR* and presumably the loss of a glycosylation site [37]. Chong *et al.* [31] hypothesized that the minor allele (G) would predict cortisol responses to both pharmacological (naloxone) and psychological (stress) activation of the HPA axis. Healthy subjects completed a naloxone challenge and/or the TSSt. Among subjects expressing a G allele, there was a higher cortisol response to naloxone ($P = 0.046$), but a lower cortisol response to the TSST ($P = 0.044$).

6.4
Jerusalem Studies

6.4.1
Experimental Design

Our research team examined the role of several common polymorphisms that we hypothesized are likely to partially modulate HPA axis response to social stress under controlled laboratory conditions. Respondents were primarily college students at Israeli institutions of tertiary education, recruited by word of mouth and advertisements on campus notice boards for a study on human genetics and personality. Selection criteria ensured that all subjects were <35 years old, had no history of psychiatric or endocrine illness, were currently non-smokers, were not pregnant or had not given birth in the past year and were not using medication on a regular basis besides single-phase oral contraceptives. All together, 102 subjects participated in the experiment.

Salivary cortisol was sampled eight times during the 90-min session: after the 10-min introduction and before entering room B the first time (sample 1), just prior to role-playing (sample 2), after phase 1 and 2 of the TSST (sample 3) and after filling out questionnaires (samples 4–6 at 10-min intervals, samples 7–8 at 15-min intervals). A Salivette plug (Sarstedt, Germany) was used to collect saliva. Saliva samples were kept at room temperature throughout the session and were then immediately centrifuged at 4000 rpm at 24 °C for 10 min and then assayed using an electrochemiluminescence immunoassay with a specific cortisol kit for saliva samples (Roche Diagnostics, USA) in an Elecsys 2010 analyzer. The lower detection limit of the assay was 0.5 nM/L. Subjects' blood pressure and heart rate were measured using a digital automatic blood pressure monitor (Omron R7), in parallel with the collection of their saliva.

Statistical analyses of cortisol data used: (i) raw cortisol values at eight time-points; (ii) area under the curve with respect to ground (AUCg), reflecting change over time with respect to absolute zero; (iii) area under the curve with respect to increase (AUCi), emphasizing the changes over time with respect to basal levels; (iv) slope rise from basal to peak values, reflecting HPA reactivity to stress; and (vi) slope decline from the peak cortisol value across subsequent samples, reflecting HPA recovery of the stress response. All values were log-transformed to correct skewed distributions. The different formulas for the area under the curve were derived from the trapezoid formula, as described by Pruessner *et al.* [38]. Slopes were derived by using linear trendlines in an Excel spreadsheet. The importance of discerning reactivity and recovery has been discussed by Ramsay and Lewis [39], among others. Reactivity and recovery slopes were highly correlated ($r = -0.764$, $P < 0.001$ controlling for sex). Blood pressure and heart rate values were reduced to four measures from 10 min prior to the TSST to 10 min after (samples 1–4) to evaluate the fast sympathetic response [8]. We also compared the peak systolic rise in blood pressure and heart rate in the TSST. Peak level was defined as the highest level reached during the session.

6.4.2
Genes Associated with TSST Response

We examined the following polymorphic genes for association with TSST (see Tables 6.2 and 6.3 for a summary) that are essential to: (i) nerve growth and maintenance (*BDNF*); (ii) serotonergic (*MAOA, SLC6A4*); and (iii) dopaminergic neurotransmission (*COMT, DRD4, DRD5, DAT1*); (iv) social cognition and social stress (*AVPR1a*); (v) sexual dimorphism (*ESR, ESR2, AN*); and (vi) cytokine activity (*IL1RN*). As detailed below, these genes are all characterized by functional polymorphisms and have been widely investigated for their role in psychopathology as well as normal behavior. We hypothesized that some or all of these genes are good candidates to modulate response to social stress. Altogether of the 13 genes we examined, five genes showed evidence for association with the TSST response. The most significant results were obtained for brain-derived neurotrophic factor (BDNF) [47]. Notably, we observed a significant gender × genotype interaction mirroring the known gender effect on social stress responses [7, 29]. As shown in Figure 6.1, women subjects show reduced response to social stress compared to males. There was a significant within subject effect of time × sex interaction (GLM: $F = 11.52$, $P = 0.0002$, DF $= 3.13$, estimated effect size $\eta^2 = 0.11$).

6.5
Growth Factors

6.5.1
BDNF Gene

The *BDNF* gene, like other peptide growth factors, encodes a precursor peptide (proBDNF), which is proteolytically cleaved to form the mature protein [48]. Only one frequent, non-conservative polymorphism in the human *BDNF* gene (dbSNP number rs6265) has been identified, a single nucleotide polymorphism (SNP) at nucleotide 196 (G/A) producing an amino acid substitution (valine to methionine) at codon 66 (Val66Met). Several investigations using brain imagery and

Table 6.2 Polymorphisms examined in the Jerusalem study positively associated with salivary cortisol changes in the TSST.

Gene	Polymorphism	Genotype frequency	Gender effect	Number of Ss	Measure	P
BDNF	Val66Met	val/val = 68 val/met = 29	Yes	97	GLM: time × sex × gene (within Ss)	0.0003
					AUCi × BDNF × sex	0.006
AVPR1a promoter	RS3 short/long	Short/short = 21 short/long = 45 long/long = 25	Yes	91	GLM: time × sex × gene (within Ss)	0.003
					AUCi × RS3 × sex	0.007
AVPR1a promoter	RS1 short/long	Short/short = 25 short/long = 41 long/long = 26	Yes	92	GLM: time × sex × gene (within Ss)	0.007
					AUCi × RS1 × sex	0.041
SLC6A4 Intron 2	VNTR	10/10 = 8 10/12 = 39 12/12 = 44	Yes	91	GLM: time × sex × gene (within Ss)	0.014
MAOA promoter		3/3 = 14 3/4 = 18 4/4 = 12	F	44 F	GLM: time × MAO-A (within Ss)	0.0007
					AUCi × MAO-A	0.039
ESR1 estrogen receptor ERα promoter	VNTR	Short/short = 16 short/long = 21 long/long = 6	M	43 M	GLM: time × gene (within Ss)	0.039

transfection experiments have demonstrated the functional importance of the Val66Met polymorphism on BDNF expression levels [49, 50]. Furthermore, the Val66Met polymorphism has been associated with stress-related dysfunction, particularly depression in animal models [51, 52] and anxiety-related personality traits [53–55] and clinical depression in humans [56–58].

The role of *BDNF* in maintaining neuronal integrity across the life span, its relation to neuronal integrity, its suggested role in the etiology of some stress-related traits or disorders [58, 59] and the presence of a common *BDNF* gene

Table 6.3 Summary of common polymorphisms that modulate HPA axis reactivity during the TSST with no association in the Jerusalem study.

Gene	Polymorphism	Allele	Genotype frequency
SLC6A4 5-HTTLPR	Promoter region 44 bp insertion/deletion [40]	short/long	short/short = 25 short/long = 49 long/long = 21
Dopamine D4 receptor DRD4	Exon3 48 bp repeat [41]	no 7 repeat/ 7 repeat	No 7 repeat = 61 7 repeat = 32
DRD4	5′ UTR 120 bp tandem duplication [42]	short/long	short/short = 47 short/long = 26 long/long = 3
Dopamine D5 receptor DRD5	148 bp repeat upstream of the promoter region [43]	no 7 repeat/ 7 repeat	no 7 repeat = 29 7 repeat = 62
Dopamine transporter DAT1	3′ UTR VNTR [44]	9/10/11 repeat analyzed 9/10 and 10/10 groups	9/9 = 9 9/10 = 44 10/10 = 39 9/11 = 2 10/11 = 3
Catechol-o-methyltrans-ferase COMT	val158met rs4680 [45]	A(1947)G SNP	val/val = 32 val/met = 44 met/met = 18
Interleukin-1 receptor antagonist IL1RN	Intron 2 VNTR 86-bp repeat [46]	A1 A2	A1/A1 = 36 A1/A2 = 42 A2/A2 = 14
ESR2 Estrogen receptor ERβ	VNTR	short/long	short/short = 12 short/long = 33 long/long = 56
Androgen receptor AN	CAG repeat in exon 1 (X chr) short/long	short/long	M: short = 23 long = 24 F: short/short = 10 short/long = 26 long/long = 19

variant that modulates expression suggested to us that the Val66Met polymorphism would be worth investigating for its role in modulating salivary cortisol and autonomic responses in a group of non-clinical subjects participating in the TSST paradigm. Additionally, the sex bias observed in clinical depression [60], the sex differences in depression-related behaviors in *BDNF* conditional knockout mice [51], the effects of estrogen on brain *BDNF* [61, 62] and the higher reactivity of the HPA axis to social stress in men compared to women [7] prompted us to hypothesize that the *BDNF* Val66Met variant shows a differential effect on acute social stress responses in men and women.

As shown in Figure 6.2, tests of within-subjects effects showed a significant three-way interaction (GLM: time $\times$ *BDNF* $\times$ sex: $F = 6.15$ $P = 0.0002$, DF = 3.33). There was no significant effect of time $\times$ *BDNF* (GLM: $P = 0.595$, DF = 3.33). Notably, in male subjects, val/val homozygotes showed an overall higher response than val/met heterozygotes whereas in female subjects the opposite was observed. Sex and BDNF genotype explained 6% of the variance in individual differences in salivary cortisol ($\eta^2 = 0.06$). When men and women were analyzed separately there was a significant *BDNF* x time interaction ($F = 4.217$, $P = 0.007$) for male subjects and a similar, albeit opposite allele, trend in female subjects ($F = 2.179$, $P = 0.076$). For female subjects the effect of the *BDNF* genotype was significant when AUCi was used as a measure of cortisol rise ($F = 6.296$, $P = 0.016$; AUCi val/met = 133.6 $\pm$ 182.38 SD vs val/val = 20.65 $\pm$ 125.1 SD) and for male subjects ($F = 4.01$, $P = 0.051$; AUCi val/met = 114.1 $\pm$ 185.2 SD vs val/val = 343.1 $\pm$ 434.9 SD).

We also analyzed the slopes of the cortisol rise and decline in these subjects (dependent variable) grouped by sex and *BDNF* Val66Met polymorphism (fixed factors) using univariate GLM analyses. Similar results for sex and *BDNF* Val66Met were obtained as those observed for repeated measures and AUC analyses. Univariate analysis showed a significant Sex $\times$ *BDNF* Val66Met interaction during the reactivity stage (basal to peak) of the change in slope (univariate: $F = 7.24$, $P = 0.009$), as well as the recovery/regulatory (peak to endpoint after recovery) phase (univariate: $F = 5.65$, $P = 0.020$). Results using both measures shows the significant gender difference between women and men carrying the (val/val) genotype and no significant gender difference between women and men carrying the (val/met) genotype, emphasizing even further the significance gender by genotype interaction by using not only the eight time-points and total area under the curve analysis as reported above, but also by using the peak cortisol value as a set point for the two phases of the endocrine response (i.e., reactivity/ arousal versus recovery/regulatory).

Figure 6.3 shows systolic blood pressure changes in male and female subjects during the TSST. A test of the between subjects effect showed a significant interaction between time $\times$ sex in the repeated measures GLM analysis ($F = 5.17$, $P = 0.025$), with men's responses being higher then women's responses at all time points, and a significant time $\times$ *BDNF* $\times$ sex interaction (GLM: $F = 5.88$, $P = 0.017$). Overall, similar results were obtained for diastolic blood pressure and for heart rate.

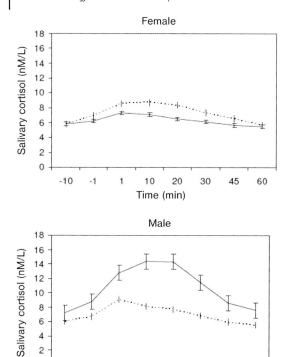

Figure 6.2 Modulation of salivary cortisol by BDNF val66met and sex during the TSST. The SNP (rs6265; A/G; val/met) was genotyped using high resolution melt (HRM) analysis (see [63] for a general description of the method). After PCR amplification, homoduplexes (val/val, met/met) and heteroduplexes (val/met) are formed. Each duplex has a characteristic melting temperature, and the sum of all transitions can be observed by melting curve analysis. The presence of heteroduplexes changes the melting curve shape of amplified heterozygotes compared with homozygotes. The changes are small but can be reliably detected with high-resolution melting analysis. HRM distinguished between val/val, val/met and met/met amplicons and the method was verified by comparison with results obtained by genotyping the same samples using the SNaPShot procedure (Applied Biosystems) and analyzing the reaction products on a DNA analyzer (ABI 310). PCR reactions were performed using 5 µl Thermo-Start Master Mix (Thermo Scientific), 2 µl primers (2.5 µM), 1 µl SYTO9 (dye) and 1 µl of water to total of 9 µl total volume and an additional 1 µl of genomic DNA. All PCR reactions and HRM analysis were performed on a Rotor-Gene 3000 (Corbett Life Science, Australia), using the following primers that produced a 142-bp amplicon: F5'GACATCATTGGCTGACACTT'3; R5'CTCCAAAGGCACTTGACTAC'3. PCR reaction conditions were as follows: activating enzyme step at 95.0°C for 15 min, 45 cycles of denaturation at 95.0°C for 5 s, reannealing at 58°C for 15 s and extension at 72°C for 10 s; at the end of each cycle fluorescence was measured. The reaction proceeded to a hold at 40°C for 2 min, a second hold at 65°C for 2 min and then the melt procedure ramped from 65°C to 95°C, rising by 0.1°C every 3 s, when fluorescence was acquired. In the present sample only two individuals were met/met homozygotes and they were not included in the analysis.

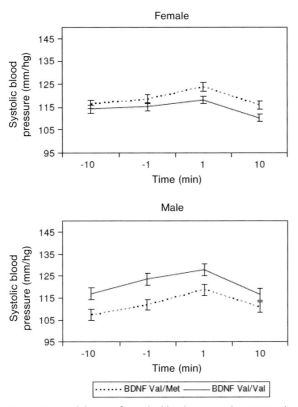

Figure 6.3 Modulation of systolic blood pressure by BDNF val66met and sex in the TSST.

6.6
Neuropeptides

6.6.1
Arginine Vasopressin 1a Receptor (AVPR1a)

Animal studies have shown that vasopressin has myriad functions in the CNS, including important behavioral actions modulating aggression, affiliation, learning and memory, as well as stress, anxiety and depression (reviewed in [64]. In the vole, a small rodent extensively studied as a model of social communication, the length of the promoter region repeat elements dramatically determine affiliative behaviors by regulating region-specific receptor expression in the brain [65] and we have shown that in humans similar repeat regions contribute to altruistic or prosocial behavior [66]. Similar to the vole longer, we have shown that the *AVPR1a* promoter repeat regions predict greater gene transcription in postmortem human hippocampal specimens [66]. The role of *AVPR1a* in modifying social interactions in humans and other mammals suggested that promoter region repeat length variations of this gene would also contribute to individual differences in response to social stress. Indeed, there was a significant three-way interaction (GLM: time $\times$ *AVPR1a* $\times$ sex: $P = 0.003$). The *AVPR1a* RS3 microsatellites were grouped by length into long and short repeats as we previously described [66].

Notably, this is the third observation by us that the configuration of the *AVPR1a* RS3 promoter repeat alleles into long and short categories makes sense biologically. We had previously shown that *AVPR1a* mRNA levels in postmortem hippocampal samples are highest in samples characterized by the long/long RS3 genotype [66]. We also showed that at the behavioral level allocation of funds in the Dictator game showed a dose-response effect and that short/short subjects gave the least amount of money whereas short/long gave intermediary sums and long/long individuals gave the most to the anonymous other[66]. We know observed a similar pattern in response to social stress. For male subjects short/ short individuals show the greatest rise in salivary cortisol, short/long subjects show intermediary rises and long/long genotypes are least responsive to social stress. Moreover, Meyer-Lindenberg *et al.* have shown for RS3, longer variants were associated with significantly stronger activation of amygdala in a function imaging study [67]. Altogether, it appears that similar to the vole [65] length of the *AVPR1a* microsatellite regions in humans also has functional consequences at the level of gene transcription, brain activation and social behavior.

6.7
Serotonin Transmission

6.7.1
Serotonin Transporter (SLC6A4)

The serotonin transporter is characterized by two repeat regions, a 44-base pair (bp) deletion/insertion (5-HTTLPR) in the promoter [68] and a 17-bp variable number of tandem repeats (correspond to 10- or 12-repeat units of the 17-bp VNTR) in the second intron (STin2 VNTR) [69] that are involved in the regulation of emotion and social cognition. Interestingly, highly significant evidence for association between the STin2.12 allele of the STin2 VNTR polymorphism ($P = 0.000\,14$) and schizophrenia was found from 12 population-based association studies consisting of 2177 cases and 2369 control subjects [70]. However, no association was observed between the 5-HTTLPR and schizophrenia.

In our study, there was an effect of gender and a gender $\times$ gene interaction (GLM: time $\times$ STin2 10/12 repeats $\times$ sex: $P = 0.014$).

6.7.2
Monoamine Oxidase A (MAOA)

MAOA codes for monoamine oxidase A, which is important in the metabolism of serotonin and dopamine. There is a functional repeat polymorphism in the promoter region that regulates transcription of this gene. Two alleles (3 and 4 repeats) are most common and constitute >97% of the observed alleles. Functional characterization in a luciferase assay demonstrated that the longer alleles (4 and 5) were more active than allele 3 [71]. In a seminal study by Caspi *et al.* [72] it was demonstrated that maltreated children with a genotype conferring high levels of *MAOA* expression (allele 4) were less likely to develop antisocial problems. An imaging study by Pezawas *et al.* [73] showed that the low expression variant (allele 3), associated with increased risk of violent behavior, predicted pronounced limbic volume reductions and hyper-responsive amygdala during emotional arousal, with diminished reactivity of regulatory prefrontal regions, compared with the high expression allele. In men, the low expression allele is also associated with changes in orbitofrontal volume, amygdala and hippocampus hyper-reactivity during aversive recall and impaired cingulate activation during cognitive inhibition.

Because the behavioral effects of *MAOA* variation have been consistently more penetrant in males in both animal [74] and human [75] studies, we also expected that some physiological differences including *MAOA* impact on cortisol rises in the TSST would be more pronounced in males than females. However to our surprise, whereas women who were 3/3 homozygote had very significant increases in salivary cortisol in the TSST protocol compared to women with 4/4 genotypes (GLM: time × MAO-A, $P = 0.0007$), in male subjects there was no significant effect of *MAOA* genotype.

6.8
Sex Steroids

A great deal of evidence shows that the prevalence, incidence and morbidity risk for depression are higher in females than in males and fluctuations in sex hormone levels are considered to be involved in the etiology [60]. It therefore seems a reasonable notion that variations in genes coding for common variants in sex steroids would modulate social stress responses. As detailed below, we examined two estrogen receptor genes and the androgen receptor gene for a possible role in regulation of cortisol responses in the TSST.

6.8.1
Estrogen Receptor Alpha (ESR1)

A thymine-adenine (TA)n dinucleotide repeat polymorphism in the 5′ promoter region of the ESR1 gene might affect ER activity (see [76]). In Caucasian populations, the frequency of repeat lengths exhibits a bimodal distribution peaking sharply at $(TA)_{14}$ followed by very low allele frequency between $(TA)_{16}$ and $(TA)_{19}$ and a shallower peak at $(TA)_{23}$. Most prior epidemiologic research has compared long versus short alleles, defined in various ways. Longer alleles are associated with lower ESR1 expression. Somewhat counter intuitively, we observed an association between the *ESR1* and salivary cortisol in the TSST only in male subjects (GLM: time × gene, *P* = 0.039) and not in female subjects.

Effects of *ESR1* polymorphisms in male subjects, however, have been reported. For example, associations between long *ESR1* (TA)$_n$ repeats have been observed with prostate cancer [76], coronary heart disease in men [77] (but also see [78]) and angiographic extent of coronary artery disease in men and women [79]. In our study, men with the long/long ESR1 genotype (reduced ESR1 expression) showed the lowest rise in TSST salivary cortisol. However, these results need to be tempered by the small number of long/long homozygote males (*n* = 6).

6.8.2
Genes Not Associated with TSST Response

Surprisingly, no association was observed between the extensively studied [68] serotonin transporter promoter-region (5-HTTLPR) and TSST, although as discussed above we did find an association with the intron 2 VNTR (summarized in Table 6.3). We expected that 5-HTTLPR, which is associated with depression (and traumatic life events [72]), personality traits of anxiety [80] and increased amygdala neuronal activity in response to fear-inducing stimuli [81], would have effected the cortisol response in the TSST.

We also failed to observe an association between the dopamine D4 receptor (DRD4; exon III VNTR or the 120-bp promoter region repeat [82]). The exon III VNTR has been associated with personality traits such as novelty seeking [83, 84] and attention deficit hyperactivity disorder[85]. Following meta-analysis, the DRD4 gene including the 120-bp tandem repeat has also been associated with schizophrenia [86]. A dopamine gene also of interest is the dopamine D5 receptor (*DRD5*) that is characterized by a highly polymorphic microsatellite dinucleotide repeat discovered by Sherrington *et al.* [43], located 18.5 kb from the first exon. This microsatellite predisposes (especially the most common 148-bp allele) to ADHD in some populations [87]. Two recent meta-analyses confirm that the DRD5 microsatellite confers a small but significant risk for ADHD [88, 89]. No association was observed between the 148-bp allele and cortisol response in the TSST. No association was observed with the dopamine transporter (DAT1) 3′UTR that has been associated in some studies with ADHD [90, 91] nor with an important metabolizing enzyme of dopamine, catechol-o-methyltransferase (COMT). The COMT Val158Met polymorphism has been associated with frontal lobe function and schizophrenia [92].

The X-linked (Xq11–12) androgen receptor (*AN*) gene is characterized by a CAG microsatellite in exon one that codes for variable-length glutamine repeats in the N-terminal domain of the AR protein repeat region. Elimination of the CAG tract in both human and rat *AN* gene resulted in elevated transcriptional activation activity, strongly suggesting that the presence of the polyglutamine tract is inhibitory to transactivation [93]. Longer CAG repeats decrease transcriptional activity The CAG repeat has been associated with prostate cancer [94], cognitive ability in older men [95] and coronary heart disease [96], among many disorders.

No association was observed between *AN* CAG repeat length and TSST in our experiments.

Cytokines modulate the biological substrate of depressive symptoms and regulate stress-responsive systems; and interleukin-1, its antagonist (*IL1RN*) and IL-1 receptors are all present in the brain [97]. Segman *et al.* [98] tested *IL1RN* as a candidate for involvement in ADHD and found significant evidence for biased transmission of the 4-repeat allele and non-transmission of the 2-repeat allele. However, a more recent investigation failed to replicate these findings [99]. These investigations suggested to us that the *IL1RN* polymorphism might be involved in regulation of social stress, but no association was observed in the current study.

Summary

Adequate adaptation to environmental stressors is crucial to successful functioning across a range of human situations [100]. However, inappropriate response to long-term social stress may result in physical and psychiatric pathology [2]. Among the factors that contribute to appropriate stress response are an individual's particular genetic background. Genes are hypothesized to either protect or sensitize the stress response generating resiliency in some people and leading to psychopathology in others. We have reviewed the role of common genetic polymorphisms, including those from our own investigations, in modulating a laboratory-based social stress test.

Quite naturally, the family of genes that were first examined for regulation of social stress were those coding for the glucocorticoid and mineralcortiocoid receptors, genes that are most proximate to regulation of the HPA axis [27, 29, 30, 34, 35, 101, 102] and represent primary targets for cortisol's initial actions. We have employed a parallel strategy and our first investigations examined common polymorphisms encoded by elements of neuronal transmission that also regulate HPA axis and modulate its function including neurotrophic factors (*BDNF*), arginine vasopressin (*AVPR1a*), dopaminergic elements (*DRD4*, *DAT1*, *DRD5*, *COMT*), serotonergic neurotransmission (*SLC6A4*, *MAOA*), sex steroids (*AN*, *ESR1*, *ESR2*) and cytokines (*IL1RN*). Serotonin [103, 104], vasopressin [105, 106], dopamine [107], sex steroids [105, 108], cytokines [97, 109] and neurotrophic factors [51, 52, 55] all have profound effects on the development, growth, maintenance and adaptation of the HPA axis across developmental time. It is therefore not surprising that a number of the genes we have examined, characterized by polymorphic and functional variants, are associated with social stress response in the TSST.

An important influence on HPA reactivity is the role of environment. There is increasing interest as to the mechanisms by which environmental challenges modulate gene effects on complex behavioral traits [110–112]. Epigenetic changes in DNA, mediated by such chemical modifications as methylation and acetylation of the DNA itself, may start as early as the fetal stage [113, 114] and provide a conduit for environmental signals to influence not only behavior but also physical wellbeing [115–117]. In an intriguing study [118], prenatal exposure to increased third trimester maternal depressed/anxious mood was associated with increased methylation of the glucocorticoid receptor gene (NR3C1) at a predicted NGFI-A binding site. Increased NR3C1 methylation at this site was also associated with increased salivary cortisol stress responses at three months, controlling for prenatal SRI exposure, postnatal age and pre- and postnatal maternal mood.

There have been a number of studies of modifiable environmental factors and HPA axis reactivity. For example, among unfit women, aging is associated with greater HPA axis reactivity to psychological stress but higher aerobic fitness among such women can attenuate these age-related changes, as indicated by a blunted cortisol response to psychological stress [119]. These findings suggest that exercise training may be an effective way of modifying some of the neuroendocrine changes associated with aging. However, we are not aware of any studies of controlled social stress experiments that examined both gene × environmental challenges on changes in salivary cortisol.

Importantly, our own investigations (summarized in Tables 6.2 and 6.3) and the other studies cited in Table 6.1, indicate that interpreting effects on HPA axis reactivity across non-clinical groups as well as in clinical subjects, needs to be informed by genetic information. Of course, other variables especially gender are also critical in interpreting such group and diagnostic differences. Not only do men and women differ in their responses to social stress but allelic differences between sexes can either enhance or diminish the role of gender effects. Future studies of social stress need to consider genetic polymorphisms that can markedly modify interpretation of HPA axis reactivity across group difference.

Acknowledgments

This research was supported by the Israel Science Foundation (grant No. 389/05) and partially by Phillip Morris USA and Phillip Morris International (R.P.E.).

References

1 Dickerson, S.S. and Kemeny, M.E. (2004) Acute stressors and cortisol responses: a theoretical integration and synthesis of laboratory research. *Psychol. Bull.*, **130**, 355–391.

2 McEwen, B.S. (2008) Central effects of stress hormones in health and disease: understanding the protective and damaging effects of stress and stress mediators. *Eur. J. Pharmacol.*, **583**, 174–185.

3 Ehlert, U. and Straub, R. (1998) Physiological and emotional response to psychological stressors in psychiatric and psychosomatic disorders. *Ann. N. Y. Acad. Sci.*, **851**, 477–486.

4 Levine, S. (2000) Influence of psychological variables on the activity of the hypothalamic-pituitary-adrenal axis. *Eur. J. Pharmacol.*, **405**, 149–160.

5 Michaud, K., Matheson, K., Kelly, O. and Anisman, H. (2008) Impact of stressors in a natural context on release of cortisol in healthy adult humans: a meta-analysis. *Stress*, **11**, 177–197.

6 Cohen, J. (1988) *Statistical Power Analysis for the Behavioral Sciences*, 2nd edn, Lawrence Erlbaum, Hillsdale, NJ.

7 Kirschbaum, C., Pirke, K.M. and Hellhammer, D.H. (1993) The "Trier Social Stress Test" – a tool for investigating psychobiological stress responses in a laboratory setting. *Neuropsychobiology*, **28**, 76–81.

8 Kirschbaum, C., Kudielka, B.M., Gaab, J., Schommer, N.C. and Hellhammer, D.H. (1999) Impact of gender, menstrual cycle phase, and oral contraceptives on the activity of the hypothalamus-pituitary-adrenal axis. *Psychosom. Med.*, **61**, 154–162.

9 Kudielka, B.M. and Kirschbaum, C. (2005) Sex differences in HPA axis responses to stress: a review. *Biol. Psychol.*, **69**, 113–132.

10 Kelly, M.M., Tyrka, A.R., Anderson, G.M., Price, L.H. and Carpenter, L.L. (2008) Sex differences in emotional and physiological responses to the Trier Social Stress Test. *J. Behav. Ther. Exp. Psychiatry*, **39**, 87–98.

11 Kirschbaum, C., Wust, S. and Hellhammer, D. (1992) Consistent sex differences in cortisol responses to psychological stress. *Psychosom. Med.*, **54**, 648–657.

12 Chong, R.Y., Uhart, M., McCaul, M.E., Johnson, E. and Wand, G.S. (2008) Whites have a more robust hypothalamic-pituitary-adrenal axis response to a psychological stressor than blacks. *Psychoneuroendocrinology*, **33**, 246–254.

13 Kirschbaum, C., Scherer, G. and Strasburger, C.J. (1994) Pituitary and adrenal hormone responses to pharmacological, physical, and psychological stimulation in habitual smokers and nonsmokers. *Clin. Investig.*, **72**, 804–810.

14 Back, S.E., Waldrop, A.E., Saladin, M.E., Yeatts, S.D., Simpson, A., McRae, A.L., Upadhyaya, H.P., Contini Sisson, R., Spratt, E.G., Allen, J., Kreek, M.J. and Brady, K.T. (2008) Effects of gender and cigarette smoking on reactivity to psychological and pharmacological stress provocation. *Psychoneuroendocrinology*, **33**, 560–568.

15 Heim, C., Ehlert, U. and Hellhammer, D.H. (2000) The potential role of hypocortisolism in the pathophysiology of stress-related bodily disorders. *Psychoneuroendocrinology*, **25**, 1–35.

16 Simeon, D., Knutelska, M., Yehuda, R., Putnam, F., Schmeidler J. and Smith, L.M. (2006) Hypothalamic-pituitary-adrenal axis function in dissociative disorders, posttraumatic stress disorder, and healthy volunteers. *Biol. Psychiatry*, **61**, 966–973.

17 Fiocco, A.J., Joober, R. and Lupien, S.J. (2007) Education modulates cortisol reactivity to the Trier Social Stress Test in middle-aged adults. *Psychoneuroendocrinology*, **32**, 1158–1163.

18 Heinrichs, M., Meinlschmidt, G., Neumann, I., Wagner, S., Kirschbaum, C., Ehlert, U. and Hellhammer, D.H. (2001) Effects of suckling on hypothalamic-pituitary-adrenal axis responses to psychosocial stress in postpartum lactating women. *J. Clin. Endocrinol. Metab.*, **86**, 4798–4804.

19 Kajantie, E., Feldt, K., Raikkonen, K., Phillips, D.I., Osmond, C., Heinonen, K., Pesonen, A.K., Andersson, S., Barker, D.J. and Eriksson, J.G. (2007) Body size at birth predicts hypothalamic-pituitary-adrenal axis response to psychosocial stress at age 60 to 70 years. *J. Clin. Endocrinol. Metab.*, **92**, 4094–4100.

20 Feldt, K., Raikkonen, K., Eriksson, J.G., Andersson, S., Osmond, C., Barker, D.J., Phillips, D.I. and Kajantie, E. (2007) Cardiovascular reactivity to psychological stressors in late adulthood is predicted by gestational age at birth. *J. Hum. Hypertens.*, **21**, 401–410.

21 Rimmele, U., Zellweger, B.C., Marti, B., Seiler, R., Mohiyeddini, C., Ehlert, U. and Heinrichs, M. (2007) Trained men show lower cortisol, heart rate and psychological responses to psychosocial stress compared with untrained men. *Psychoneuroendocrinology*, **32**, 627–635.

22 Ditzen, B., Neumann, I.D., Bodenmann, G., von Dawans, B., Turner, R.A., Ehlert, U. and Heinrichs, M. (2007) Effects of different kinds of couple interaction on cortisol and heart rate responses to stress in women. *Psychoneuroendocrinology*, **32**, 565–574.

23 Giesbrecht, T., Smeets, T., Merckelbach, H. and Jelicic, M. (2007) Depersonalization experiences in

undergraduates are related to heightened stress cortisol responses. *J. Nerv. Ment. Dis.*, **195**, 282–287.

24 Tyrka, A.R., Wier, L.M., Anderson, G.M., Wilkinson, C.W., Price, L.H. and Carpenter, L.L. (2007) Temperament and response to the Trier Social Stress Test. *Acta Psychiatr. Scand.*, **115**, 395–402.

25 Wirtz, P.H., Elsenbruch, S., Emini, L., Rudisuli, K., Groessbauer, S. and Ehlert, U. (2007) Perfectionism and the cortisol response to psychosocial stress in men. *Psychosom. Med.*, **69**, 249–255.

26 Reynolds, R.M., Godfrey, K.M., Barker, M., Osmond, C. and Phillips, D.I. (2007) Stress responsiveness in adult life: influence of mother's diet in late pregnancy. *J. Clin. Endocrinol. Metab.*, **92**, 2208–2210.

27 Wust, S., Van Rossum, E.F., Federenko, I.S., Koper, J.W., Kumsta, R. and Hellhammer, D.H. (2004) Common polymorphisms in the glucocorticoid receptor gene are associated with adrenocortical responses to psychosocial stress. *J. Clin. Endocrinol. Metab.*, **89**, 565–573.

28 Derijk, R.H. and de Kloet, E.R. (2008) Corticosteroid receptor polymorphisms: determinants of vulnerability and resilience. *Eur. J. Pharmacol.*, **583**, 303–311.

29 Kumsta, R., Entringer, S., Koper, J.W., van Rossum, E.F., Hellhammer, D.H. and Wust, S. (2007) Sex specific associations between common glucocorticoid receptor gene variants and hypothalamus-pituitary-adrenal axis responses to psychosocial stress. *Biol. Psychiatry*, **62**, 863–869.

30 DeRijk, R.H., Wust, S., Meijer, O.C., Zennaro, M.C., Federenko, I.S., Hellhammer, D.H., Giacchetti, G., Vreugdenhil, E., Zitman, F.G. and de Kloet, E.R. (2006) A common polymorphism in the mineralocorticoid receptor modulates stress responsiveness. *J. Clin. Endocrinol. Metab.*, **91**, 5083–5089.

31 Chong, R.Y., Oswald, L., Yang, X., Uhart, M., Lin, P.I. and Wand, G.S. (2006) The Micro-opioid receptor polymorphism A118G predicts cortisol

responses to naloxone and stress. *Neuropsychopharmacology*, **31**, 204–211.

32 Uhart, M., McCaul, M.E., Oswald, L.M., Choi, L. and Wand, G.S. (2004) GABRA6 gene polymorphism and an attenuated stress response. *Mol. Psychiatry*, **9**, 998–1006.

33 Bartels, M., Van den Berg, M., Sluyter, F., Boomsma, D.I. and de Geus, E.J. (2003) Heritability of cortisol levels: review and simultaneous analysis of twin studies. *Psychoneuroendocrinology*, **28**, 121–137.

34 Federenko, I.S., Nagamine, M., Hellhammer, D.H., Wadhwa, P.D. and Wust, S. (2004) The heritability of hypothalamus pituitary adrenal axis responses to psychosocial stress is context dependent. *J. Clin. Endocrinol. Metab.*, **89**, 6244–6250.

35 Kumsta, R., Entringer, S., Koper, J.W., van Rossum, E.F., Hellhammer, D.H. and Wust, S. (2008) Glucocorticoid receptor gene polymorphisms and glucocorticoid sensitivity of subdermal blood vessels and leukocytes. *Biol. Psychol.*, **79**, 179–184.

36 Herman, J.P., Mueller, N.K. and Figueiredo, H. (2004) Role of GABA and glutamate circuitry in hypothalamo-pituitary-adrenocortical stress integration. *Ann. N. Y. Acad. Sci.*, **1018**, 35–45.

37 Bergen, A.W., Kokoszka, J., Peterson, R., Long, J.C., Virkkunen, M., Linnoila, M. and Goldman, D. (1997) Mu opioid receptor gene variants: lack of association with alcohol dependence. *Mol. Psychiatry*, **2**, 490–494.

38 Pruessner, J.C., Kirschbaum, C., Meinlschmid, G. and Hellhammer, D.H. (2003) Two formulas for computation of the area under the curve represent measures of total hormone concentration versus time-dependent change. *Psychoneuroendocrinology*, **28**, 916–931.

39 Ramsay, D. and Lewis, M. (2003) Reactivity and regulation in cortisol and behavioral responses to stress. *Child Dev.*, **74**, 456–464.

40 Heils, A., Teufel, A., Petri, S., Stober, G., Riederer, P., Bengel, D. and Lesch, K.P.

(1996) Allelic variation of human serotonin transporter gene expression. *J. Neurochem.*, **66**, 2621–2624.

41 Van Tol, H.H., Wu, C.M., Guan, H.C., Ohara, K., Bunzow, J.R., Civelli, O., Kennedy, J., Seeman, P., Niznik, H.B. and Jovanovic, V. (1992) Multiple dopamine D4 receptor variants in the human population [see comments]. *Nature*, **358**, 149–152.

42 Seaman, M.I., Fisher, J.B., Chang, F. and Kidd, K.K. (1999) Tandem duplication polymorphism upstream of the dopamine D4 receptor gene (DRD4). *Am. J. Med. Genet.*, **88**, 705–709.

43 Sherrington, R., Mankoo, B., Attwood, J., Kalsi, G., Curtis, D., Buetow, K., Povey, S. and Gurling, H. (1993) Cloning of the human dopamine D5 receptor gene and identification of a highly polymorphic microsatellite for the DRD5 locus that shows tight linkage to the chromosome 4p reference marker RAF. *Genomics*, **18**, 423–425.

44 Vandenbergh, D.J., Persico, A.M., Hawkins, A.L., Griffin, C.A., Li, X., Jabs, E.W. and Uhl, G.R. (1992) Human dopamine transporter gene (DAT1) maps to chromosome 5p15.3 and displays a VNTR. *Genomics*, **14**, 1104–1106.

45 Lachman, H.M., Papolos, D.F., Saito, T., Yu, Y.M., Szumlanski, C.L. and Weinshilboum, R.M. (1996) Human catechol-O-methyltransferase pharmacogenetics: description of a functional polymorphism and its potential application to neuropsychiatric disorders. *Pharmacogenetics*, **6**, 243–250.

46 Tarlow, J.K., Blakemore, A.I., Lennard, A., Solari, R., Hughes, H.N., Steinkasserer, A. and Duff, G.W. (1993) Polymorphism in human IL-1 receptor antagonist gene intron 2 is caused by variable numbers of an 86-bp tandem repeat. *Hum. Genet.*, **91**, 403–404.

47 Shalev, I., Lerer, E., Salomon, I., Uzefovsky, F., Gritsenko, I., Mankuta, D., Kaitz, M. and Ebstein, R.P. (2008) BDNF Val66Met polymorphism is associated with HPA axis reactivity to psychological stress. *Psychoneuro-endocrinology*, **34**, 382–388.

48 Seidah, N.G., Benjannet, S., Pareek, S., Chretien, M. and Murphy, R.A. (1996) Cellular processing of the neurotrophin precursors of NT3 and BDNF by the mammalian proprotein convertases. *FEBS Lett.*, **379**, 247–250.

49 Bueller, J.A., Aftab, M., Sen, S., Gomez-Hassan, D., Burmeister, M. and Zubieta, J.K. (2006) BDNF Val66Met allele is associated with reduced hippocampal volume in healthy subjects. *Biol. Psychiatry*, **59**, 812–815.

50 Egan, M.F., Kojima, M., Callicott, J.H., Goldberg, T.E., Kolachana, B.S., Bertolino, A., Zaitsev, E., Gold, B., Goldman, D., Dean, M., Lu, B. and Weinberger, D.R. (2003) The BDNF val66met polymorphism affects activity-dependent secretion of BDNF and human memory and hippocampal function. *Cell*, **112**, 257–269.

51 Monteggia, L.M., Luikart, B., Barrot, M., Theobold, D., Malkovska, I., Nef, S., Parada, L.F. and Nestler, E.J. (2007) Brain-derived neurotrophic factor conditional knockouts show gender differences in depression-related behaviors. *Biol. Psychiatry*, **61**, 187–197.

52 Berton, O., McClung, C.A., Dileone, R.J., Krishnan, V., Renthal, W., Russo, S.J., Graham, D., Tsankova, N.M., Bolanos, C.A., Rios, M., Monteggia, L.M., Self, D.W. and Nestler, E.J. (2006) Essential role of BDNF in the mesolimbic dopamine pathway in social defeat stress. *Science*, **311**, 864–868.

53 Lang, U.E., Hellweg, R., Kalus, P., Bajbouj, M., Lenzen, K.P., Sander, T., Kunz, D. and Gallinat, J. (2005) Association of a functional BDNF polymorphism and anxiety-related personality traits. *Psychopharmacology (Berl.)*, **180**, 95–99.

54 Sen, S., Nesse, R.M., Stoltenberg, S.F., Li, S., Gleiberman, L., Chakravarti, A., Weder, A.B. and Burmeister, M. (2003) A BDNF coding variant is associated with the NEO personality inventory domain neuroticism, a risk factor for depression. *Neuropsychopharmacology*, **28**, 397–401.

55 Hunnerkopf, R., Strobel, A., Gutknecht, L., Brocke, B. and Lesch, K.P. (2007)

Interaction between BDNF Val66Met and dopamine transporter gene variation influences anxiety-related traits. *Neuropsychopharmacology*, **32**, 2552–2560.

56 Schule, C., Zill, P., Baghai, T.C., Eser, D., Zwanzger, P., Wenig, N., Rupprecht, R. and Bondy, B. (2006) Brain-derived neurotrophic factor Val66Met polymorphism and dexamethasone/CRH test results in depressed patients. *Psychoneuroendocrinology*, **31**, 1019–1025.

57 Taylor, W.D., Zuchner, S., McQuoid, D.R., Payne, M.E., Macfall, J.R., Steffens, D.C., Speer, M.C. and Krishnan, K.R. (2008) The Brain-Derived neurotrophic factor VAL66MET polymorphism and cerebral white matter hyperintensities in late-life depression. *Am. J. Geriatr. Psychiatry*, **16**, 263–271.

58 Gotlib, I.H., Joormann, J., Minor, K.L. and Hallmayer, J. (2007) HPA axis reactivity: a mechanism underlying the associations among 5-HTTLPR, stress, and depression. *Biol. Psychiatry*, **63**, 847–851.

59 Groves, J.O. (2007) Is it time to reassess the BDNF hypothesis of depression? *Mol. Psychiatry*, **12**, 1079–1088.

60 Grigoriadis, S. and Robinson, G.E. (2007) Gender issues in depression. *Ann. Clin. Psychiatry*, **19**, 247–255.

61 Meltser, I., Tahera, Y., Simpson, E., Hultcrantz, M., Charitidi, K., Gustafsson, J.K. and Canlon, B. (2008) Estrogen receptor beta protects against acoustic trauma in mice. *J. Clin. Invest.*, **118**, 1563–1570.

62 Sasahara, K., Shikimi, H., Haraguchi, S., Sakamoto, H., Honda, S., Harada, N. and Tsutsui, K. (2007) Mode of action and functional significance of estrogen-inducing dendritic growth, spinogenesis, and synaptogenesis in the developing Purkinje cell. *J. Neurosci.*, **27**, 7408–7417.

63 Liew, M., Pryor, R., Palais, R., Meadows, C., Erali, M., Lyon, E. and Wittwer, C. (2004) Genotyping of single-nucleotide polymorphisms by high-resolution melting of small amplicons. *Clin. Chem.*, **50**, 1156–1164.

64 Caldwell, H.K., Lee, H.J., Macbeth, A.H. and Young, W.S. 3rd (2008) Vasopressin: behavioral roles of an "original"

neuropeptide. *Prog. Neurobiol.*, **84**, 1–24.

65 Hammock, E.A. and Young, L.J. (2005) Microsatellite instability generates diversity in brain and sociobehavioral traits. *Science*, **308**, 1630–1634.

66 Knafo, A., Israel, S., Darvasi, A., Bachner-Melman, R., Uzefovsky, F., Cohen, L., Feldman, E., Lerer, E., Laiba, E., Raz, Y., Nemanov, L., Gritsenko, I., Dina, C., Agam, G., Dean, B., Bornstein, G. and Ebstein, R.P. (2008) Individual differences in allocation of funds in the dictator game associated with length of the arginine vasopressin 1a receptor RS3 promoter region and correlation between RS3 length and hippocampal mRNA. *Genes Brain Behav.*, **7**, 266–275.

67 Meyer-Lindenberg, A., Kolachana, B., Gold, B., Olsh, A., Nicodemus, K.K., Mattay, V., Dean, M. and Weinberger, D.R. (2008) Genetic variation in AVPR1A linked to autism predicts amygdala activation and personality traits in healthy humans. *Mol. Psychiatry*, doi: 10.1038/mp.2008.54.

68 Canli, T. and Lesch, K.P. (2007) Long story short: the serotonin transporter in emotion regulation and social cognition. *Nat. Neurosci.*, **10**, 1103–1109.

69 Hranilovic, D., Stefulj, J., Schwab, S., Borrmann-Hassenbach, M., Albus, M., Jernej, B. and Wildenauer, D. (2004) Serotonin transporter promoter and intron 2 polymorphisms: relationship between allelic variants and gene expression. *Biol. Psychiatry*, **55**, 1090–1094.

70 Fan, J.B. and Sklar, P. (2005) Meta-analysis reveals association between serotonin transporter gene STin2 VNTR polymorphism and schizophrenia. *Mol. Psychiatry*, **10**, 928–938, 891.

71 Deckert, J., Catalano, M., Syagailo, Y.V., Bosi, M., Okladnova, O., Di Bella, D., Nothen, M.M., Maffei, P., Franke, P., Fritze, J., Maier, W., Propping, P., Beckmann, H., Bellodi, L. and Lesch, K.P. (1999) Excess of high activity monoamine oxidase A gene promoter alleles in female patients with panic disorder. *Hum. Mol. Genet.*, **8**, 621–624.

72 Caspi, A., Sugden, K., Moffitt, T.E., Taylor, A., Craig, I.W., Harrington, H., McClay, J., Mill, J., Martin, J., Braithwaite, A. and Poulton, R. (2003) Influence of life stress on depression: moderation by a polymorphism in the 5-HTT gene. *Science*, **301**, 386–389.

73 Pezawas, L., Meyer-Lindenberg, A., Drabant, E.M., Verchinski, B.A., Munoz, K.E., Kolachana, B.S., Egan, M.F., Mattay, V.S., Hariri, A.R. and Weinberger, D.R. (2005) 5-HTTLPR polymorphism impacts human cingulate-amygdala interactions: a genetic susceptibility mechanism for depression. *Nat. Neurosci.*, **8**, 828–834.

74 Cases, O., Seif, I., Grimsby, J., Gaspar, P., Chen, K., Pournin, S., Muller, U., Aguet, M., Babinet, C., Shih, J.C. *et al.* (1995) Aggressive behavior and altered amounts of brain serotonin and norepinephrine in mice lacking MAOA (see comments). *Science*, **268**, 1763–1766.

75 Brunner, H.G., Nelen, M., Breakefield, X.O., Ropers, H.H. and van Oost, B.A. (1993) Abnormal behavior associated with a point mutation in the structural gene for monoamine oxidase A. *Science*, **262**, 578–580.

76 McIntyre, M.H., Kantoff, P.W., Stampfer, M.J., Mucci, L.A., Parslow, D., Li, H., Gaziano, J.M., Abe, M. and Ma, J. (2007) Prostate cancer risk and ESR1 TA, ESR2 CA repeat polymorphisms. *Cancer Epidemiol. Biomarkers Prev.*, **16**, 2233–2236.

77 Kunnas, T.A., Laippala, P., Penttila, A., Lehtimaki, T. and Karhunen, P.J. (2000) Association of polymorphism of human alpha oestrogen receptor gene with coronary artery disease in men: a necropsy study. *BMJ*, **321**, 273–274.

78 Schuit, S.C., Oei, H.H., Witteman, J.C., Geurts van Kessel, C.H., van Meurs, J.B., Nijhuis, R.L., van Leeuwen, J.P., de Jong, F.H., Zillikens, M.C., Hofman, A., Pols, H.A. and Uitterlinden, A.G. (2004) Estrogen receptor alpha gene polymorphisms and risk of myocardial infarction. *JAMA*, **291**, 2969–2977.

79 Rokach, A., Pollak, A., Rosen, L., Friedlander, Y., Blumenfeld, A., Reznik, L. and Dresner-Pollak, R. (2005) Estrogen receptor alpha gene polymorphisms are associated with the angiographic extent of coronary artery disease. *J. Clin. Endocrinol. Metab.*, **90**, 6556–6560.

80 Lesch, K.P., Bengel, D., Heils, A., Sabol, S.Z., Greenberg, B.D., Petri, S., Benjamin, J., Muller, C.R., Hamer, D.H. and Murphy, D.L. (1996) Association of anxiety-related traits with a polymorphism in the serotonin transporter gene regulatory region. *Science*, **274**, 1527–1531.

81 Hariri, A.R., Mattay, V.S., Tessitore, A., Kolachana, B., Fera, F., Goldman, D., Egan, M.F. and Weinberger, D.R. (2002) Serotonin transporter genetic variation and the response of the human amygdala. *Science*, **297**, 400–403.

82 Rogers, G., Joyce, P., Mulder, R., Sellman, D., Miller, A., Allington, M., Olds, R., Wells, E. and Kennedy, M. (2004) Association of a duplicated repeat polymorphism in the 5'-untranslated region of the DRD4 gene with novelty seeking. *Am. J. Med. Genet.*, **126B**, 95–98.

83 Ebstein, R.P., Novick, O., Umansky, R., Priel, B., Osher, Y., Blaine, D., Bennett, E.R., Nemanov, L., Katz, M. and Belmaker, R.H. (1996) Dopamine D4 receptor (D4DR) exon III polymorphism associated with the human personality trait of novelty seeking. *Nat. Genet.*, **12**, 78–80.

84 Munafo, M.R., Yalcin, B., Willis-Owen, S.A. and Flint, J. (2008) Association of the dopamine D4 receptor (DRD4) gene and approach-related personality traits: meta-analysis and new data. *Biol. Psychiatry*, **63**, 197–206.

85 Faraone, S.V., Doyle, A.E., Mick, E. and Biederman, J. (2001) Meta-analysis of the association between the 7-repeat allele of the dopamine D(4) receptor gene and attention deficit hyperactivity disorder. *Am. J. Psychiatry*, **158**, 1052–1057.

86 Allen, N.C., Bagade, S., McQueen, M.B., Ioannidis, J.P., Kavvoura, F.K., Khoury, M.J., Tanzi, R.E. and Bertram, L. (2008) Systematic meta-analyses and field

synopsis of genetic association studies in
schizophrenia: the SzGene database.
Nat. Genet., **40**, 827–834.

87 Manor, I., Corbex, M., Eisenberg, J.,
Gritsenkso, I., Bachner-Melman, R.,
Tyano, S. and Ebstein, R.P. (2004)
Association of the dopamine D5 receptor
with attention deficit hyperactivity
disorder (ADHD) and scores on a
continuous performance test (TOVA).
Am. J. Med. Genet., **127B**, 73–77.

88 Maher, B.S., Marazita, M.L., Ferrell, R.E.
and Vanyukov, M.M. (2002) Dopamine
system genes and attention deficit
hyperactivity disorder: a meta-analysis.
Psychiatr. Genet., **12**, 207–215.

89 Lowe, N., Kirley, A., Mullins, C.,
Fitzgerald, M., Gill, M. and Hawi, Z.
(2004) Multiple marker analysis at the
promoter region of the DRD4 gene and
ADHD: evidence of linkage and
association with the SNP-616. *Am. J.
Med. Genet. B Neuropsychiatr. Genet.*,
131, 33–37.

90 Brookes, K.J., Mill, J., Guindalini, C.,
Curran, S., Xu, X., Knight, J., Chen,
C.K., Huang, Y.S., Sethna, V., Taylor, E.,
Chen, W., Breen, G. and Asherson, P.
(2006) A common haplotype of the
dopamine transporter gene associated
with attention-deficit/hyperactivity
disorder and interacting with maternal
use of alcohol during pregnancy. *Arch.
Gen. Psychiatry*, **63**, 74–81.

91 Brookes, K., Xu, X., Chen, W., Zhou, K.,
Neale, B., Lowe, N., Anney, R., Franke,
B., Gill, M., Ebstein, R., Buitelaar, J.,
Sham, P., Campbell, D., Knight, J.,
Andreou, P., Altink, M., Arnold, R.,
Boer, F., Buschgens, C., Butler, L.,
Christiansen, H., Feldman, F.,
Fleischman, K., Fliers, E., Howe-Forbes,
R., Goldfarb, A., Heise, A., Gabriels, I.,
Korn-Lubetzki, I., Johansson, L., Marco,
R., Medad, S., Minderaa, R., Mulas, F.,
Muller, U., Mulligan, A., Rabin, K.,
Rommelse, N., Sethna, V., Sorohan, J.,
Uebel, H., Psychogiou, L., Weeks, A.,
Barrett, R., Craig, I., Banaschewski, T.,
Sonuga-Barke, E., Eisenberg, J., Kuntsi,
J., Manor, I., McGuffin, P., Miranda, A.,
Oades, R.D., Plomin, R., Roeyers, H.,
Rothenberger, A., Sergeant, J.,

Steinhausen, H.C., Taylor, E.,
Thompson, M., Faraone, S.V. and
Asherson, P. (2006) The analysis of 51
genes in DSM-IV combined type
attention deficit hyperactivity disorder:
association signals in DRD4, DAT1 and
16 other genes. *Mol. Psychiatry*, **11**,
934–953.

92 Egan, M.F., Goldberg, T.E., Kolachana,
B.S., Callicott, J.H., Mazzanti, C.M.,
Straub, R.E., Goldman, D. and
Weinberger, D.R. (2001) Effect of COMT
Val108/158 Met genotype on frontal lobe
function and risk for schizophrenia.
Proc. Natl. Acad. Sci. U.S.A., **98**,
6917–6922.

93 Chamberlain, N.L., Driver, E.D. and
Miesfeld, R.L. (1994) The length and
location of CAG trinucleotide repeats in
the androgen receptor N-terminal
domain affect transactivation function.
Nucl. Acids Res., **22**, 3181–3186.

94 Giovannucci, E., Stampfer, M.J.,
Krithivas, K., Brown, M., Dahl, D.,
Brufsky, A., Talcott, J., Hennekens, C.H.
and Kantoff, P.W. (1997) The CAG
repeat within the androgen receptor
gene and its relationship to prostate
cancer [published erratum appears in
Proc. Natl. Acad. Sci. U.S.A. 1997, Jul
22; **94** (15), 8272]. *Proc. Natl. Acad.
Sci. U.S.A.*, **94**, 3320–3323.

95 Yaffe, K., Edwards, E.R., Lui, L.Y.,
Zmuda, J.M., Ferrell, R.E. and Cauley,
J.A. (2003) Androgen receptor CAG
repeat polymorphism is associated with
cognitive function in older men. *Biol.
Psychiatry*, **54**, 943–946.

96 Alevizaki, M., Cimponeriu, A.T.,
Garofallaki, M., Sarika, H.L., Alevizaki,
C.C., Papamichael, C., Philippou, G.,
Anastasiou, E.A., Lekakis, J.P. and
Mavrikakis, M. (2003) The androgen
receptor gene CAG polymorphism is
associated with the severity of coronary
artery disease in men. *Clin. Endocrinol.
(Oxford)*, **59**, 749–755.

97 Licinio, J. and Wong, M.L. (1999) The
role of inflammatory mediators in the
biology of major depression: central
nervous system cytokines modulate the
biological substrate of depressive
symptoms, regulate stress-responsive

systems, and contribute to neurotoxicity and neuroprotection. *Mol. Psychiatry*, **4**, 317–327.

98 Segman, R.H., Meltzer, A., Gross-Tsur, V., Kosov, A., Frisch, A., Inbar, E., Darvasi, A., Levy, S., Goltser, T., Weizman, A. and Galili-Weisstub, E. (2002) Preferential transmission of interleukin-1 receptor antagonist alleles in attention deficit hyperactivity disorder. *Mol. Psychiatry*, **7**, 72–74.

99 Misener, V.L., Schachar, R., Ickowicz, A., Malone, M., Roberts, W., Tannock, R., Kennedy, J.L., Pathare, T. and Barr, C.L. (2004) Replication test for association of the IL-1 receptor antagonist gene, IL1RN, with attention-deficit/hyperactivity disorder. *Neuropsychobiology*, **50**, 231–234.

100 Charmandari, E., Tsigos, C. and Chrousos, G. (2005) Endocrinology of the stress response. *Annu. Rev. Physiol.*, **67**, 259–284.

101 Wust, S., Federenko, I.S., van Rossum, E.F., Koper, J.W. and Hellhammer, D.H. (2005) Habituation of cortisol responses to repeated psychosocial stress-further characterization and impact of genetic factors. *Psychoneuroendocrinology*, **30**, 199–211.

102 Kumsta, R., Entringer, S., Hellhammer, D.H. and Wust, S. (2007) Cortisol and ACTH responses to psychosocial stress are modulated by corticosteroid binding globulin levels. *Psychoneuroendocrinology*, **32**, 1153–1157.

103 Porter, R.J., Gallagher, P., Watson, S. and Young, A.H. (2004) Corticosteroid-serotonin interactions in depression: a review of the human evidence. *Psychopharmacology (Berl.)*, **173**, 1–17.

104 Andrews, M.H. and Matthews, S.G. (2004) Programming of the hypothalamo-pituitary-adrenal axis: serotonergic involvement. *Stress*, **7**, 15–27.

105 Bingham, B. and Viau, V. (2008) Neonatal gonadectomy and adult testosterone replacement suggest an involvement of limbic arginine vasopressin and androgen receptors in the organization of the HPA axis. *Endocrinology*, **149**, 3581–3591.

106 Volpi, S., Rabadan-Diehl, C. and Aguilera, G. (2004) Vasopressinergic regulation of the hypothalamic pituitary adrenal axis and stress adaptation. *Stress*, **7**, 75–83.

107 Sullivan, R.M. and Dufresne, M.M. (2006) Mesocortical dopamine and HPA axis regulation: role of laterality and early environment. *Brain Res.*, **1076**, 49–59.

108 Serova, L.I., Maharjan, S. and Sabban, E.L. (2005) Estrogen modifies stress response of catecholamine biosynthetic enzyme genes and cardiovascular system in ovariectomized female rats. *Neuroscience*, **132**, 249–259.

109 Anisman, H. and Merali, Z. (2003) Cytokines, stress and depressive illness: brain-immune interactions. *Ann. Med.*, **35**, 2–11.

110 Johnson, W. (2007) Genetic and environmental influences on behavior: capturing all the interplay. *Psychol. Rev.*, **114**, 423–440.

111 Rutter, M., Moffitt, T.E. and Caspi, A. (2006) Gene-environment interplay and psychopathology: multiple varieties but real effects. *J. Child Psychol. Psychiatry*, **47**, 226–261.

112 Moffitt, T.E., Caspi, A. and Rutter, M. (2005) Strategy for investigating interactions between measured genes and measured environments. *Arch. Gen. Psychiatry*, **62**, 473–481.

113 Gicquel, C., El-Osta, A. and Le Bouc, Y. (2008) Epigenetic regulation and fetal programming. *Best Pract. Res. Clin. Endocrinol. Metab.*, **22**, 1–16.

114 Meaney, M.J., Szyf, M. and Seckl, J.R. (2007) Epigenetic mechanisms of perinatal programming of hypothalamic-pituitary-adrenal function and health. *Trends Mol. Med.*, **13**, 269–277.

115 Szyf, M., McGowan, P. and Meaney, M.J. (2008) The social environment and the epigenome. *Environ. Mol. Mutagen.*, **49**, 46–60.

116 Ptak, C. and Petronis, A. (2008) Epigenetics and complex disease: from etiology to new therapeutics. *Annu. Rev. Pharmacol. Toxicol.*, **48**, 257–276.

117 Nafee, T.M., Farrell, W.E., Carroll, W.D., Fryer, A.A. and Ismail, K.M. (2008)

Epigenetic control of fetal gene expression. *BJOG*, **115**, 158–168.

118 Oberlander, T.F., Weinberg, J., Papsdorf, M., Grunau, R., Misri, S. and Devlin, A.M. (2008) Prenatal exposure to maternal depression, neonatal methylation of human glucocorticoid receptor gene (NR3C1) and infant cortisol stress responses. *Epigenetics*, **3**, 97–106.

119 Traustadottir, T., Bosch, P.R. and Matt, K.S. (2005) The HPA axis response to

stress in women: effects of aging and fitness. *Psychoneuroendocrinology*, **30**, 392–402.

120 Carter, C.S. (2006) Sex differences in oxytocin and vasopressin: implications for autism spectrum disorders? *Behav. Brain Res.*, **176**, 170–186.

121 Holmes, T.H. and Rahe, R.H. (1967) The social readjustment rating scale. *J. Psychosom. Res.*, **11**, 213–218.

Part III
Cognition and Behavior

Stress – From Molecules to Behavior. Edited by Hermona Soreq, Alon Friedman, and Daniela Kaufer
Copyright © 2010 WILEY-VCH Verlag GmbH & Co. KGaA, Weinheim
ISBN: 978-3-527-32374-6

7
Corticosteroid Hormones in Stress and Anxiety –
Role of Receptor Variants and Environmental Inputs

Roel H. DeRijk, Efthimia Kitraki, and E. Ronald de Kloet

7.1
Introduction

Stressful events are very well remembered, but how does this work? Here we focus on the stress response mediators, in particular on the corticosteroids secreted from the adrenals as the endproduct of the hypothalamic–pituitary–adrenal (HPA) axis. Corticosteroids promote energy allocation if needed during stress and energy storage during rest. The best known action of corticosteroids is, however, that they are capable to restrain the initial stress reactions. For instance, tissue damage triggers inflammatory mediators, which are beneficial in the short run, but become damaging in the long run, and the overshoot is controlled by corticosteroids. Likewise psychological stressors trigger changes in the brain which are dampened by corticosteroids. Less well known is that the very same corticosteroids also can amplify the initial stress reactions. These complementary actions are mediated by receptors that act as ligand-driven transcription factors in gene regulation.

Brain corticosteroid receptors are central mediators of the stress response and thus indispensable for coping with every day challenges and preservation of the homeostatic balance. Two different types of nuclear receptors have been described: the widely expressed, low-affinity glucocorticoid receptor (GR) and the high-affinity mineralocorticoid receptor (MR), primarily found in higher limbic brain structures, that is, in the hippocampus, amygdala and parts of the cerebral cortex of rodents and primates [1]. The two receptor types mediate distinct actions of cortisol (primates) or corticosterone (rodents) in the HPA axis and the stress response. MRs are mainly involved in the ultradian [2] and circadian rhythm of hormone release and the maintenance of basal HPA axis activity. GRs modulate the processing of stressful information and facilitate behavioral adaptation, while promoting the termination of the stress response [3, 4]. Interestingly, in rodents recently a low-affinity membrane MR was discovered in hippocampus that can expand the actions mediated by the high-affinity nuclear MR. These membrane

Stress – From Molecules to Behavior. Edited by Hermona Soreq, Alon Friedman, and Daniela Kaufer
Copyright © 2010 WILEY-VCH Verlag GmbH & Co. KGaA, Weinheim
ISBN: 978-3-527-32374-6

MRs enhance excitability and are involved in the initial activation of the HPA axis during stress [5, 6].

In addition to the roles of MR and GR in HPA axis regulation, research in rodent models has delineated how these receptors mediate the actions of corticosteroids on learning and memory processes, as observed in spatial and fear conditioning paradigms [7]. The actions mediated by MR and GR are complementary in hippocampus, amygdala and the frontal cortex and affect distinct psychological domains of the perceived threat. Normally the corticosteroids secreted during such stressful experiences enhance cognitive, emotional and motivational processes. Local administration of selective agonists and antagonists [8] demonstrated that hippocampal GR and MR play a significant role in the transformation of short- into long-term memories and inputs from the amygdala can modify memory storage for fearful events [9]. The changes in structure and function in the brain during learning and memory have been termed allostasis [10].

According to the "MR:GR balance hypothesis" [11], the balanced activation of the two receptor types in the brain is critical for optimal function of HPA axis, as well as for optimal cognitive performance, neuroprotection and mood. Deviations from the proper functioning of the HPA axis under stress, in terms of hypo- or hyperactivation of the receptors, enhance vulnerability and promote the emergence of a wide spectrum of stress-related disorders such as depression and post-traumatic stress disorder (PTSD). Many depressed patients exhibit increased cortisol levels and adrenal hypertrophy, indicating a chronically activated HPA axis and reduced sensitivity to GR and MR signaling [12]. In contrast, increased GR sensitivity for negative feedback, accompanied by HPA axis hypoactivity and low basal cortisol levels characterize subjects with enhanced vulnerability to PTSD [13]. Under these conditions of chronically too high or too low concentrations of circulating corticosteroids, and thus imbalanced MR:GR activation, the cost to re-establish homeostasis may be too high. This then leads to wear and tear and ultimately to disease. The wear and tear has been termed allostatic load [10].

A large variability in the stress response has been observed in humans, as different subjects adopt different strategies to cope with a challenge, in order to achieve homeostasis at a minimal cost. These different strategies depend on the genetic background and life history of each individual. Variability in the stress response is therefore shaped during the life of an individual by complex and mutual interactions between genetic inputs and environmental factors, the latter translated by cognitive and emotional inputs. While the genetic inputs are determined by DNA sequence, the environmental inputs can cause lasting changes in gene expression by epigenetic processes. The outcome of this gene × environment interaction then somehow dictates the response to a certain challenge. A striking example of a lasting environmental input is early life experience: rats handled during the first week of life show reduced adrenocortical and emotional reactivity in later life. A dominant feature of a genetic input is the variation identified in the MR and GR genes, which has profound influences on stress responsiveness.

In this chapter, we focus on the role of corticosteroids in stress and anxiety. First, we address the question how the receptors for these hormones are involved in processing of fearful information and in modulation of stress responsiveness. Then, we highlight how genetic variants of the steroid receptors may modulate the action of the corticosteroids and how early life experiences cause lasting changes in stress responsiveness, cognitive functions and emotional reactivity. Finally, we discuss the implications of gene × environment interaction in the pathogenesis of stress-related anxiety disorders.

7.2
Corticosteroid Hormones, Fear and the Stress Response

7.2.1
Neuroanatomical Basis

Information of a perceived threat is conveyed through the sensory organs to the limbic brain areas, that is, hippocampus, amygdala and prefrontal cortex (see Figure 7.1). Neuronal networks in these areas are involved in the appraisal of

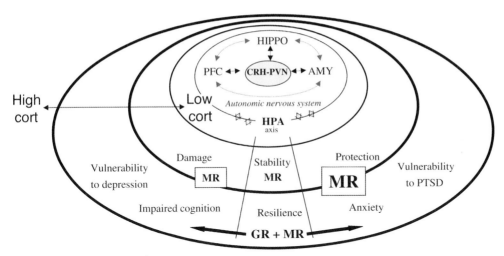

Figure 7.1 Corticosteroid levels and MR:GR balance determine health or disease. Emotional stressors mobilize CRH in hypothalamic and limbic areas, which signals the activation of autonomic and limbic-HPA axis systems. Corticosterone receptors (GR, MR), in cooperation with autonomic mediators and other neuropeptides, coordinate initial stress reactions with later adaptations to the stressor. Balanced MR:GR activation by corticosterone in time and space enables appraisal, encoding and proper extinction of the fearful memory, implying an effective coping response indicative for resilience. Imbalanced MR:GR and inadequate or excessive corticosterone action in control of the stress reaction have been related to cognitive impairment and enhanced vulnerability to depression, anxiety or PTSD. In this figure, corticosteroid levels are rising from the center toward the periphery and the larger case lettering for MR denotes relative excess or shortage of MR activation.

incoming information and in the organization of an appropriate response if the new information does not match with previous experience. In this limbic network the amygdala is important for the interpretation of emotions and its function has been associated with fear-related memory [14]. The medial prefrontal cortex exerts an inhibitory influence on amygdala-related fear-motivated behaviors. When the stressful event is emotionally arousing, the activation of the circuit comprising hippocampus, basolateral amygdala and prefrontal cortex is particularly important for HPA axis regulation [15]. Inputs from these limbic areas to the hypothalamus lead to stimulation of autonomic and HPA axis activity. Hippocampus and pre-frontal cortex exert opposing influences on HPA axis activation as compared to amygdala; the two former areas are inhibitory and the latter excitatory to the axis. Furthermore, brain stem catecholamine projections to the hypothalamus also have an excitatory influence on HPA axis function.

7.2.2
Role of Corticosteroids and Other Stress Mediators

The balance of opposing activities of corticosteroids on the HPA axis activation (negative feedback) or on the amygdala (further activation) during fearful experiences, though not fully elucidated, is at the core of stress hormone action underlying health versus psychopathology. Fearful experiences activate an amygdala projection to the hypothalamus which leads to the synthesis and release of CRH from the neurons of the paraventricular nucleus (PVN) [16]. CRH organizes the sympathetic and neuroendocrine response stimulating the release of peripheral catecholamines and the release of pro-opiomelanocortin peptides such as ACTH from anterior pituitary corticotrophs, which in turn activates the secretion of corticosteroids from the adrenal cortex. The amygdala itself also produces CRH. Cortisol feeds back on the amygdala in a feedforward mode, further enhancing amygdaloid synthesis and release of CRH, while activating the production of noradrenaline and other peptides which then further promote the processing of the fearful information [17].

How this positive feedback action is exerted on the level of the amygdala is not precisely known. The membrane MR may be involved, with a non-genomic enhancement of the local action of CRH and noradrenaline. Slower genomic GR-mediated actions are also involved, enhancing the synthesis and release of CRH from amygdala and the brain stem noradrenergic projections. Both the fast and slow corticosteroid actions enhance the positive feedforward loop in the amygdala, which contrasts with the well known negative feedback action at the level of hypothalamic CRH neurons and pituitary corticotrophs suppressing stress-induced HPA activity [15]. Moreover, cortisol blocks adrenergic action in the hippocampus at the post-synaptic receptor. The opposing activity of GR in the control of the CRH gene in amygdala and PVN is due to modulations exerted by different co-regulator cocktails [18]. Co-regulators may also be involved in activation via catecholamine synthesis and release versus suppression at the post-synaptic adrenergic level.

The function of different neuronal circuits all activated during a challenge requires coordination in time and space and corticosteroids provide a means to do so. Emotionally significant experiences usually form strong memories [19]. Stress hormones enhance the memory consolidation of these emotionally arousing experiences. This action exerted by the hormones requires the coordinated activation in time and space of the different components of the hippocampal–amygdala–prefrontal cortex circuit. Noradrenaline and CRH released from the amygdala strengthen synaptic contacts in the hippocampus, facilitating the storage of information in the hippocampus. In counterbalance, reciprocal connections from the prefrontal cortex to the amygdala inhibit fear responses by reducing amygdala excitation. Corticosteroids have receptors in all these regions which enable the hormone to coordinate these distinct regional functions in the limbic circuitry within time and space.

The concentration and time-dependent effects of corticosteroids on information processing are mediated by different types of corticosteroid receptors. The corticosteroids play an important role in the acquisition, consolidation and retrieval of several types of memory [7, 20, 21]. Using selective agonists and antagonists as well as transgenic approaches, the role of MR and GR in these distinct phases of information processing has been disentangled [3, 21]. First, stress-induced cortisol levels are thought to act via the low-affinity membrane MR. This fast action of cortisol converges in time with noradrenaline and CRH and amplifies their excitatory action to facilitate the acquisition of information. Next, the increased corticosteroid concentrations also activate slower genomic GR-mediated processes that facilitate memory consolidation [3, 17]. For this purpose the signaling cascades driven by corticosteroids and the other stress mediators need to converge in time and space for optimal information processing.

If corticosteroids act out of context, aberrant emotional processing and memory formation might take place. Elevated glucocorticoids (or acute stress) prior to, or out of context of the acquisition of new information can impair the learning process [22]. For instance if stress and corticosteroids are elevated one hour before learning, the genomic processes are underway and suppress the acquisition of new information. Likewise, memory retrieval is impaired if the individual previously has experienced an acute stressor and elevated corticosteroid concentrations [23–25]. One interpretation is that, in the face of the novel stressor, the previously learned response can be considered no longer relevant, and thus retrieval is disturbed.

In fear conditioning paradigms conducted in male animals, corticosteroid administration facilitates extinction of fear memory, while extinction is impaired in adrenalectomized animals lacking these hormones [20, 26]. This role of glucocorticoids in memory extinction was demonstrated long ago in forced extinction paradigms and in context-dependent fear conditioning [27].

There are also gender effects. In a first study on the impact of acute stress on classic conditioning in healthy humans, results indicated that exposure to a psychological stressor facilitated fear conditioning in men but not in women; the men most vulnerable to stress-induced facilitation were those with elevated cortisol levels [28]. Recent data underscore the significant gender differences in emotional memory [29] and hormonal stress responses [138] reported in humans.

7.2.3
Cognitive and Emotional Aspects of Stress

Cognitive processing is necessary to perceive and cope with a stressful event. At the same time, stress exposure and the concomitant level of anxiety can modify the cognitive process. Although, as mentioned above, emotionally arousing experiences may facilitate the consolidation of long term memory of this particular event [30], emotional arousal and high anxiety levels impair working memory and memory retrieval. Numerous studies in rodents have shown that high levels of exogenous or stress-induced corticosteroids before retention of inhibitory avoidance or spatial tasks impair memory recall 24h after training [23, 31–33]. The impairing effects on memory retrieval are transient and subside when corticosteroid levels are normalized. This impairment implicates GR activation in the hippocampus [34] with a fast onset of the effect [33] of the receptor, as well as interactions with a noradrenergic mechanism in the amygdala [9].

Likewise, administration of stress-level doses of cortisol to human subjects impairs delayed recall on episodic tasks [35, 36]. The glucocorticoid-induced memory impairment in humans can be blocked by the β-adrenoreceptor antagonist propranolol [37], further supporting a glucocorticoid–adrenergic interaction. This hippocampus–amygdala connection is also supported by imaging studies showing a greater activity in the hippocampus and amygdala of human subjects during the successful retrieval of emotionally arousing versus neutral information [38–40].

The impairing effect of stress hormones in cognitive function mentioned so far refers to situations of acute exposure. The reported effects of chronic stress, or chronically elevated glucocorticoids are variable. First of all, gender effects have to be taken into account. Impairing effects of chronic stress on spatial memory have been observed in most studies using male animals [41, 42]. However, the very same stress effects on females are rather memory-enhancing [43, 44]. This opposing effect of gender in memory performance implies that chronic anxiety may have a sexual dimorphic influence on cognitive function. This gender bias likely is due to the action of sex hormones.

Second, chronic stress may affect differently the various regions in the limbic circuitry with different outcome for cognitive performance. On the one hand, chronic stressors impair hippocampus-dependent spatial learning and memory as is commonly observed in rodent, but also in human studies [25]. Prolonged activation of the HPA axis and long-term increased cortisol levels in aged subjects are accompanied by reduced hippocampal volume and cognitive decline. Blocking the synthesis of endogenous cortisol with metyrapone can partially restore cognitive impairment [45]. On the other hand, if sensitization of amygdala function occurs in fear-conditioning tasks, the resulting high cortisol levels or chronic stress actually can enhance memory for adverse events [46].

In humans, high levels of stress can differently affect the various aspects of emotional memory. The retrieval of information on autobiographical events that implicate prefrontal cortex and hippocampus is impaired in healthy humans under stress, while the recall of memories supported by the amygdala is facilitated by stress [47]. The involvement of intact amygdala is a requirement for the enhanced memory recall for emotionally arousing experiences [48] and individuals exhibiting the largest increase in endogenous cortisol when viewing of emotionally arousing pictures show the greatest amygdala activation in PET studies [49]. These opposing effects of chronic stress on hippocampus- and amygdala-related memories can be explained by the differential impact it has on the neuronal structure and molecular functioning of the two systems. In rodents, the same emotional stress leads either to atrophy or to enhanced arborization of dendrites in the hippocampus and amygdala, respectively [50]. Furthermore, as stated in the previous section, the expression of GR and CRH mRNAs in the amygdala is increased following stress, as opposed to their decreasing levels in the hippocampus and hypothalamus.

In conclusion, corticosteroid actions in response to a stressful event are mediated by two receptor types (MR, GR) that are expressed abundantly in limbic brain areas subserving the cognitive and emotional processing of fearful experiences. The receptors coordinate in a complementary fashion both the initial reaction to a stressor and the management of later adaptive phases. First, still rather unexplored it seems that non-genomic MR-mediated actions participate in the appraisal of novel situations, risk assessment and in promoting appropriate coping behavior to deal with a challenge. Via this MR, corticosteroids amplify behavioral reactivity which may mount to violent attacks in social conflicts [51]. Peptidergic and autonomic mediators from amygdala cooperate with the corticosteroids in the formation of strong memories for emotionally arousing experiences. Second, genomic, mostly GR-mediated actions, are necessary to prepare the individual for the next encounter by promoting the storage of the fearful experience in the memory. Third, genomic MR-mediated actions are thought to be important for stability and integrity of neuronal circuitry, maintenance of homeostasis and in the behavioral realm in threshold or sensitivity of processes underlying the initial reactions to stressors.

7.3
Gene Variants and Early Life Experiences

Although the classic model for corticosteroid function is based on the translocation of activated receptors from the cytoplasm to the nucleus where they bind DNA-responsive elements, changing expression of the target gene, additional modes of action have been described. Monomers of GR interact with other transcription factors such a AP-1 and NF-κB, thereby changing their activity. MR has also been shown to affect these ubiquitous transcription factors [52–54]. In addition, fast non-genomic actions, occurring within minutes of corticosteroid mode of action mediated by MR and GR have been described [5, 6, 55]. Even ligand-independent effects of both MR and GR seem to occur *in vitro*, although their physiological significance is currently unclear [56, 57]. Furthermore, many levels of control of corticosteroid receptor function have been identified. These include chaperones (such as heat-shock proteins), post-translation modification of MR and GR and transcriptional coactivators and suppressors, among others [56, 58, 59]. DNA structure (methylation, histone acetylation) and the sequence and compositions of the glucocorticoid responsive element (GRE) may further modulate gene responsiveness to MR and or GR [60]. Together, these different modes of action of MR and GR may explain the enormous diversity in the effects of corticosteroids.

7.4
MR and GR Gene Structure

MR and GR gene-structure is comparable. The coding region of the genes is formed by exons 2–9, with 5′ the promoter region. The gene for MR is located on chromosome 4q31.1, while the GR gene is located on chromosome 5q31–32. From these genes several variants are derived. The "classic" GR protein consists of 777 amino acids (94 kDa), while the "classic" MR protein is composed of 984 amino acids (110 kDa).

For the human MR gene, two different promoter regions have been described, designated P1 and P2, located just upstream of exon 1α and 1β, respectively [61, 62]. Corticosteroids, aldosterone, estrogens and progesterone all regulate MR gene expression. Specifically, P1 responds to corticosterone while P2 is sensitive to aldosterone. In the rat, specific regulations of exon 1α, β and γ have been observed in hippocampal subregions [63]. Furthermore, conditions such as stress, ageing and antenatal corticosteroid treatment are known to affect hippocampal MR mRNA expression (see [64]).

In the coding region of the MR gene, a 12-bp insert just after exon 3, resulting from alternative splicing, does not affect DNA binding and its effect is currently unknown. However, tissue specific expression of this variant in human brain has been observed (see [64]). A splice variant skipping exons 5 and 6, resulting in a protein of 75 kDa, was found to be a ligand-independent transactivator capable of recruiting coactivators [65]. Finally, translational variants, resulting from alternative usage of the different ATG start codons in exon 2, give rise to MR-A and MR-B proteins with different transcriptional activities [66]. Similar to the MR gene, translational variants have also been found for the GR gene [67]. GR-A was found to be transcriptionally less active, resulting in an overall mild corticosteroid resistance [68].

GR expression is highly variable and is thought to underlie differences in effects of corticosteroids on metabolism, immune regulation and stress reactivity. The regulation of GR gene expression has extensively been studied in the brain of experimental animals and it has been shown that the receptor levels are sensitive to behavioral manipulations [69] and to almost all kinds of psychological, metabolic or immune stressors [70–74].

In line with the observed highly dynamic GR expression, the human GR gene 5′ promoter region consists of at least seven exon 1s, while exons 1A and 1C can both be further spliced in three alternative variants [75]. During different stages of development, tissue-specific regulation of GR has been shown. Interestingly, these differences during development appear permanent and seem to originate from the regulation of multiple alternative first exons or promoters of the GR gene, as shown in rats [76]. In human postmortem hippocampal tissue of Parkinson patients, Alzheimer patients and humans without any psychiatric diagnosis, no changes in methylation of this particular exon 1F was observed [77]. Also during ageing, GR regulation is altered with possible consequences for stress reactivity (see [64]). Furthermore, among the many factors that seem to influence GR expression are corticosteroids, neurotransmitters and cytokines (see [64]).

Several highly expressed splice variants of the GR gene include GR-P and GR-γ. GR-P (or GR-d) results from a splicing event in which exon 8 is replaced by intron G, giving rise to a truncated protein (676 amino acids). Ligand binding is absent in this variant due to a lack of exons 8–9; however, this variant could confer decreased corticosteroid sensitivity, possibly important in tumor cells, in which such expression is high (see [64]). The so-called GR-γ holds an insertion of an additional codon, GTA (Arg), between exons 3 and 4 at amino acid 452. Although hGR-γ seems rather ubiquitously expressed, its function is presently unknown.

GRβ is a variant of the 3′ UTR of exon 9, in which exon 8 joins exon 9β instead of exon 9α. Although originally defined as a non-ligand binding inhibitor of the GRα, more recent data indicate that GRβ binds RU486 (mifepristone, a GR antagonist), leading to translocation to the nucleus and regulation of gene expression. GRβ in the absence of ligand has effects on histone deacetylation, leading to the repression of transcription [78, 79]. Expression of the GRβ variant in human brain tissue was found to be very low (see [64]), although in immune cells several days of exposure to cytokines increases the GRβ/GRα ratio to levels in which the presumed dominant activity of GRβ over GRα could become important. A role for the GRβ isoform in the human brain still needs to be established.

The complexity of GR and MR gene expression allows for extensive tissue and context-specific regulation of corticosteroid sensitivity. Together with the different MR and GR modes of action and protein modification, the numerous and diverse physiological responses elicited by corticosteroid hormones can, at least partly, be explained.

7.5
Genetic Variants of the MR and GR Genes

Genetic variation in GR and MR genes can derive from classic mutations and single nucleotide polymorphisms (SNPs). Historically, mutations are associated with a severe clinical phenotype, and often consist of premature stop codons, deletions, inserts, abnormal splicing or amino acid changes. Mutations in the human MR gene resulting in mineralocorticoid resistance are associated with hypotension or pseudohypoaldosteronism (PHA1), further characterized by salt

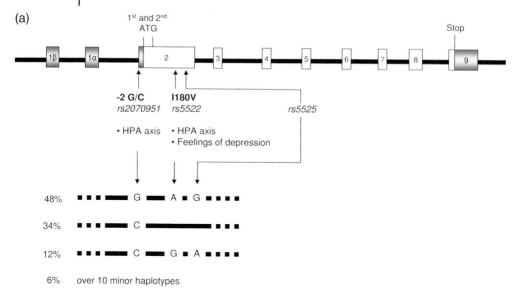

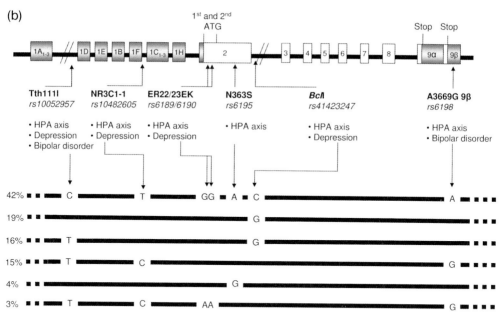

Figure 7.2 MR and GR gene genetic variants. The human MR and GR genes have a similar organization, with exon 2 to exon 9 being translated. The GR gene promoter region is highly complex, showing extensive tissue and developmental stage-specific alternative splicing; much less is known about the MR gene promoter region. Corticosteroid receptor proteins can roughly be divided in three functional parts: a transcriptional active domain (exon 2), the DNA binding domain (exons 3–4) and the ligand binding domain (exons 5–9), the latter also harboring transactivational capacity. Black arrows indicate important characterized single nucleotide polymorphisms (SNPs), as

loss, dehydration, vomiting and failure to thrive [80]. Mutations within the GR gene result in the corticosteroid resistance syndrome, which is associated with compensatory increases in circulating pituitary corticotropin (ACTH), causing not only excess of cortisol but also of androgens and steroids with salt-retaining activity. Consequently, patients suffer from clinical phenotypes which may vary (hypertension, chronic fatigue, hypokalemic alkalosis, hyperandrogenism) [81, 82].

Single nucleotide polymorphisms (SNPs) are one kind of variations in DNA sequences that are defined to have a frequency of at least 1%. SNPs mostly appear in combinations named haplotypes. In genes often blocks of combinations of several SNPs are found, giving rise to several haplotypes in a gene. Besides direct effects on the amino acid sequence, as often seen in overt "mutations", SNPs can influence gene expression by affecting promoter activity, transcription efficiency, gene splicing, mRNA stability, or translation efficacy. Information concerning functionality of a genetic variant *in vitro* gives an indication for a potential role *in vivo*. Below, we discuss some frequent and promising SNPs in the human MR and GR genes.

7.5.1
MR Gene

At position −2 (i.e., two nucleotides before the first ATG start codon), a G/C SNP (rs2070951) changes *in vitro* trans-activational activity [83] (see Figure 7.2a). However, the mechanism is currently unclear.

In exon 2 of the human MR gene, at codon 180, the GTT → ATT variation results in a isoleucine to valine change. Using a transfection assay with glucocorticoid responsive elements (GREs) upstream of luciferase, we and others showed that MR-180V shows *in vitro* loss of function using cortisol as a ligand, while aldosterone-induced transactivation was not affected [84]. This I180V is in almost 100% linkage with the synonymous GAT → GAC (rs5525) at codon 499.

7.5.2
GR Gene

Although many SNPs in the GR gene can be found on several SNPs public internet data bases and several have been tested in association studies, only a few SNPs in the GR gene have been tested for functionality *in vitro*. Also, more precise haplo-

---

discussed in the text. The name in bold is often historic/generic, while the rs-numbers are indicated in italics. Haplotypes are depicted, with the haplotype frequencies indicated at the right. The upper haplotype is the most frequent and therefore the so-called wild type.For MR (a), both −2G/C and I180V have been tested *in vitro* and found to affect

transactivation. For GR (b), all except *TthIIII* and *BclI* show changes in different *in vitro* assays. All described SNPs have been associated with several aspects of HPA axis reactivity, see text for details. Moreover, most GR and MR SNPs associate with some form of psychpathology.

type maps are provided by some research groups; for example, the group of Koper provided such a map obtained from a large Dutch population [85].

Between exon $1A_{1-3}$ and exon 1D, in a large intron of approximately 27 kb, the *TthIIII* restriction site (rs100529570) is located [82, 86] (see Figure 7.2b). No *in vitro* testing for this *TthIIII* restriction site has been reported. The region spanning approximately 4.5 kb before the first ATG in exon 2 harbors the alternative exon 1s: 1D, 1E, 1B, 1F, $1C_{1-3}$ and 1H. The NR3C1-1 (rs10482605) is 30 bp downstream of exon 1F [87]. Recently, Kumsta *et al.* found that the minor C-allele showed less transcriptional activity as tested in two different tumor cell lines [88].

In exon 2, between the first and second ATG, the codons 22 and 23 of exon 2 (GAG AGG [GluArg or ER] → GAA AAG [GlyLys or EK]; rs6189/rs6190) show 100% linked polymorphisms (haplotype). *In vitro*, the presence of this haplotype (i.e., the GAA AAG form) resulted in increased expression of GR-A, maybe as a result of changed secondary structure of the mRNA [89]. In COS cells, being devoid of endogenous MR and GR, the ER22/23EK displayed decreased transactivation, but no difference compared to wild type GR was seen on the NF-κB trans-repression [68]. This relative corticosteroid resistance of the ER22/23EK is in line with the observed associations of these SNPs with a more favorable metabolic profile in carriers [82, 90, 91].

At codon 363 (AAT → AGT, rs6195) of exon 2, a non-synonymous change from Asp to Ser was initially identified in a Dutch kindred with hypercortisolism. Functionality of this N363S was inferred using human peripheral blood monoclear lymphocytes of carriers, and dexamethasone induced GILZ/decreased IL-2 production. Homozygote carriers of the N363S displayed enhanced induction of GILZ, in line with the previously described increased corticosteroid sensitivity (using the 0.25 mg dexamethasone suppression test). In addition, a decreased inhibition of IL-2 production (not additionally stimulated) was found [92]. In contrast, using a whole blood assay, van Winsen *et al.* found a decreased sensitivity for dexamethasone using inhibition of lipopolysaccharide-induced TNFα production [93]. The extent of potential N363S effects was further tested, using a microarray, in both transiently and stably transfected cell lines [94]. This N363S has been associated with metabolic disfunction and a less favorable metabolic profile [82]. An association was also found with coronary artery disease (CAD) [95].

Inside intron B is the so-called *Bcl1* site (rs41423247) located 647 bp downstream of exon 2. For this *Bcl1* site, the wild type is a 5'-TGATCA-3' sequence and the variant is a 5'-TGATGA-3' sequence, the latter not being cut by *BclI*. In the wild type, the *BclI* enzyme cuts and generates a 2.5 kDa fragment, while the variant is not sensitive to *BclI*, generating a fragment of 4.5 kDa. This SNP has been associated with changes in cortisol sensitivity as tested, for example, with the beclomethasone skin-bleaching test. Originally increased corticosteroid sensitivity was found [96]. However, this was not observed on dexamethasone suppression of LPS-stimulated interleukin (IL)-6 production in whole blood [88]. Like the N363S, the *BclI* has been associated with metabolic dysfunction and a less favorable metabolic profile [64, 82, 97]. In a prospective study, female heterozygote carriers of the *BclI* showed increased fat deposition as compared to males [98].

Both the human GR exons 9α and 9β contain a large 3′ untranslated region (UTR) harboring several ATTTA motifs. These ATTTA motifs are well known to destabilize mRNA, providing a regulatory mechanism. We described an A to G change (A3669G; rs6198) at the first A in the ATTTA sequence resulting *in vitro* in stabilization of GRβ mRNA and an increased GRβ expression [99]. This variant, possible conferring corticosteroid resistance, was found to associate with rheumatoid arthritis and systemic lupus erythematosis (SLE) [99]. In line with this observation, homozygote carriers (approx. 2% of the population) seem to have an overactive immune system, with less colonization of *Staphylococcus aureus* in the nasal cavity [85]. Importantly, these homozygotes have also an almost three times increased incidence for CAD. However, also a more favorable metabolic profile has been associated with this GR A3669G polymorphism [100].

7.6
HPA Axis Regulation

7.6.1
Basal Cortisol Levels

MR-2 G/C associates with basal cortisol levels [101], being consistent with the finding that MR exerts tonic inhibition of the HPA axis under basal non-stress conditions [12]. In the GR gene, using a panel of six haplotype markers, Rautanen *et al.* showed that one particular haplotype was associated with higher cortisol during the day [102]. However, which GR gene SNPs contribute to this effect is unclear, although the *BclI* variant was included. GR *TthIIII* was found to associate with higher basal cortisol secretion in men [103]. These findings suggest that both MR and GR play a role in the regulation of non-stress levels of cortisol.

7.6.2
Feedback Sensitivity

Dexamethasone, a synthetic corticosteroid receptor agonist, has frequently been used to test different MR and GR genetic variants for the suppression of morning cortisol levels.

In a Dutch elderly population, large differences in morning plasma cortisol levels were measured after the ingestion of 0.25 mg dexamethasone on the previous morning [104]. In this study, GR N363S was found to associate with increased corticosteroid sensitivity, as inferred from the lower morning cortisol levels following 0.25 mg dexamethasone. GR *BclI* also associates with increased suppression by dexamethasone of cortisol in the morning (see [64]).

In contrast, GR ER22/23EK showed the opposite effect: a decreased sensitivity to 1 mg dexamethasone and thus higher post-dexamethasone cortisol levels [82]. Recently, both GR A3669G and NR3C1-1 were found to associate with significantly higher levels of post-dexamethasone (0.25 mg) plasma ACTH and cortisol in males, but with lowest levels in females [88, 105]. Importantly, the *TthIII*I-ER22/23EK haplotype and the *TthIII*I-*Bcl*I haplotype show linkage (see Figure 7.2b); therefore it is not clear which SNP exerts which effect. This linkage could also explain differences between studies, in addition to the effects of major confounders such as gender, age and other genes.

The combined dexamethasone-CRH challenge is a pharmacological test of the reactivity of the HPA axis, which further magnifies differences in dexamethasone sensitivity [106]. In a large German cohort of depressed patients ($n = 342$), ER22/23EK, N363S and *Bcl*I were tested for association with respect to cortisol and ACTH responses. Although trends were observed, no significant genotype effects were found for plasma cortisol and ACTH in these patients [107].

7.6.3
Psychosocial Challenges

The Trier social stress test (TSST) involves a public speaking and mental arithmetic task in front of an audience and camera. In an exclusive male twin cohort (both mono- and dizygotic twins), heritability for both cortisol and autonomic output during the TSST was observed [108]. Moreover, genetic effects changed following subsequent, identical sessions of the TSST, indicating the importance of context. GR N363S showed strong genetic effects, being associated with the highest levels of saliva cortisol in the aforementioned male twin cohort following the TSST [109]. For the tested GR *Bcl*I, heterozygotes had less-high responses, both for cortisol and ACTH, following the psychological challenge than the N363S carriers, but they were still higher than "wild-type" individuals. Unexpectedly, *Bcl*I homozygotes had lower cortisol responses than the controls, following both the psychological stressor and ACTH administration.

A recent follow-up study involved a new cohort with subjects selected based on GR genotype and included both males an females [105]. Woman were all oral contraceptive users, to eliminate effects of the menstural cycle, although other effects such as increases in plasma CBG may have occur. In males of this cohort, effects similar to GR *Bcl*I and N363S were found as in the (first) male twin cohort. In contrast, female GR 363S carriers displayed an almost blunted salivary cortisol response. The previously not-tested GR 9β A3669G was associated with higher levels, only in males, both for plasma cortisol and ACTH but not for saliva cortisol [105]. Similar results were obtained using GR NR3C1-1, which is in line with the fact that GR NR3C1-1 is in strong linkage with GR A3669G [88]. However, no effect of glucocorticoid receptor SNPs on autonomic function, measured as heart rate, was observed.

As tested in the (first) male twin cohort, MR I180V associated with increased saliva and plasma cortisol responses. Importantly, MR I180V also associated with increased autonomic output as measured by heartbeat [84], which the tested GR gene variants did not.

Clearly, MR and GR gene variants affect different aspects of HPA axis regulation, with strong gender effects appearing. Moreover, MR I180V modifies autonomic function during stress, with effects appearing within minutes.

7.7
Epigenetic Modifications Affecting GR and MR Actions

Apart from genetic variability, early life experiences can modify a organism's stress response in adulthood. In rodents, a programming effect of maternal corticosterone on fetal HPA axis has been shown, which has metabolic, cardiovascular and behavioral consequences for later life [76]. It appears that the mother–infant relationship and the nutritional status in the perinatal period can have programming effects on neuroendocrine reactivity, emotionality and cognitive performance exhibited in adulthood. Rodents like rats and mice are prematurely born animals, which have a period of hyporesponsiveness to common stressors (stress hyporesponsive period, SHRP) during the first two weeks of postnatal life. The SHRP is due to licking and grooming the pups (which keeps the central component of the stress system quiescent) and nutrition (which suppresses systemic stress responses from, e.g., the adrenal cortex and medulla).

During early postnatal life, brief maternal separation produces offspring with reduced adrenocortical and emotional reactivity in later life. These brief separations from the mother trigger enhanced maternal licking and grooming upon reunion. It is this sensory aspect of maternal care which has lasting effects on the development of the pup's stress system [76]. In contrast, prolonged maternal separation up to single periods of 24 h at postnatal day 3 or at postnatal day 9 has an opposite outcome. Generally, such early deprived animals show cognitive impairment and enhanced adrenocortical and emotional reactivity in later life [110]. Hence, the maternally deprived animal lacking feeding and care for some time is commonly used as a laboratory model for neglect, which is in turn a model for abuse.

Two experiments highlight either the importance of genes or the environment. In the first series of experiments, Rixt van der Veen [111] exposed two mice lines (DBA vs C57/Bl6) to foster mothers that displayed either high or low maternal care toward the pups. Subsequently, the mice were examined for cocaine intake and for a number of behavioral parameters. The results demonstrated that DBAs rather than C57 BL/6 were sensitive to the extent of maternal care, a finding which demonstrates that the susceptibility to maternal care is determined by genotype.

Danielle Champagne showed that the extent of maternal care also affects the adult HPA axis responsiveness and behavior. In agreement with previous data she demonstrated that the offspring of mothers providing high care have longer dendrites and more synaptic boutons in the hippocampal CA1 pyramidal cell field. The high-care offspring also shows enhanced synaptic plasticity and a better performance in tests on maze learning, under conditions in which the low-care offspring had short dendrites, poor LTP and poor performance [112]. Clear epigenetic modifications in the rat brain have been described as a consequence of the received maternal care: the degree of maternal care permanently altered the methylation pattern of GR gene promoters and subsequently the degree of receptor transactivational activity in the rat brain [113].

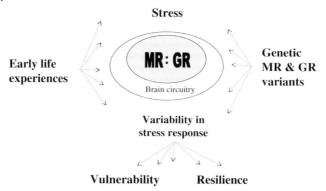

Figure 7.3 Balance MR:GR activity results from both early life experience and genetic make-up. Individual differences in brain corticosteroid receptor functionality are determined by genetic variants and epigenetic modifications triggered by early life experiences. The variability in receptor responsiveness largely dictates the variability in stress coping, leading to vulnerability or resilience.

At first glance the high-care offspring has better developed nerve cells and a stress-resistant phenotype. However, if the high-care group is exposed to a severe stressor they do not develop LTP and are unable to cope. Under those stressful conditions the low-care group performed much better. Apparently, the extent of maternal care prepares the mice for the life ahead.

The above-mentioned paradigms from animal research suggest the existence of analogous mechanisms in humans. Epigenetic modifications of GR functionality has not as yet been detected in humans. In preliminary studies the corresponding area in the human GR gene promoter was not found methylated in postmortem hippocampal tissue [77]. However, the epigenetics of GR could constitute a means for modifying the genetically determined stress responsivity by environmental challenges, leading to individual differences in stress susceptibility, as claimed for the rat.

In conclusion, the previously mentioned studies provide an excellent example of the importance of early life experiences in the programming of adult physiology and behavior. These stimuli translated into cognitive and emotional inputs can interact with the genetic inputs for each individual, enhancing or suppressing the expression of certain genetic traits, to sculpt the individual's stress coping styles (see Figure 7.3). The "three-hit model" predicts that the outcome of gene × environment interaction in the newborn prepares them for the life ahead. Offspring from an overprotected environment are predicted to fail in coping with severe stressors in later life, while the reverse seems the case for individuals that have suffered from adversity.

7.8
Stress-Related Psychopathology

Dysfunction of the limbic-HPA axis in stress has been associated with many psychiatric and psychosomatic disorders [12]. An impaired negative feedback mechanism, leading to hypercortisolaemia and increased HPA axis activity, is suggested to be central in the pathogenesis of depression and many antidepressants up-regulate limbic MR and GR levels toward normalization of the HPA axis function. Although there is an increasing body of evidence relating GR functionality with the emergence or relapse of the disease, the mechanism of their involvement is not clear. In contrast, the mechanism of the corticosteroid involvement in the emergence of anxiety-related disorders, like PTSD, is more clear. Below, we describe how increased GR sensitivity for negative feedback and subsequent HPA axis hypo-activation along with low cortisol levels can confer enhanced vulnerability to the development of PTSD following exposure to a traumatic life event.

Subjects suffering from PTSD have enhanced memories of certain features of a traumatic event. They vividly and selectively recall a past traumatic event in the form of intrusive memories or nightmares and show heightened arousal, accompanied in most of the cases with low basal cortisol levels [114]. However, in general they show impaired memory recall indicative of hippocampal dysfunction, often substantiated by a decreased hippocampal volume [115]. Apart from the hippocampus, dysfunction of additional brain areas subserving emotional memories, such as the amygdala and prefrontal cortex, has been suggested by brain imaging studies showing that, during exposure to traumatic reminders, PTSD patients are having abnormally low medial prefrontal cortex and abnormally high amygdala activation [116]. The loss of the inhibitory control of prefrontal cortex on amygdala, upon exposure to stress or stress reminders, has been proposed to participate in the increased anxiety observed in these subjects [117].

Based on the previously described mechanism of emotional memories formation, the earlier termination of HPA axis activation and the low cortisol levels in these individuals do not allow for an appropriate consolidation of the perceived information and the effective extinguishing of fear memories [118]. They thus cannot achieve a "normalization" of the system, allowing the storage of relevant-only information [22]. It has also been proposed that the low cortisol levels do not allow a proper inhibition of traumatic memory retrieval during the vicious cycles of re-experiencing and reconsolidating the traumatic memory [119]. Future studies will delineate this mechanism of this cortisol action.

7.8.1
Corticosteroid-Controlled Pathways as Targets for New Therapeutic Interventions

The inability of an individual to extinguish maladaptive fear memories can lead to persistent anxiety and related pathology. As anxiety-related disorders are associated with dysregulation of the HPA axis, modifying the effectiveness of stress hormone actions is a relevant target for new therapeutic interventions in the treatment of phobias and PTSD. Two main approaches are in use for the treatment of anxiety related disorders in humans: the disruption of the consolidation of a traumatic event and the facilitation of extinction of the aversive memories linked to the event.

Accumulating data from animal studies using pharmacological interventions or genetic modifications of GR function reveal the relevance of these molecules as targets for anxiety-related disorders. Blockade of the receptors during specific phases of memory processes (acquisition, consolidation, retrieval) showed a role of MR in the acquisition and retrieval phases, whereas GR is involved in the consolidation of memory [7]. GR overexpression increases anxiety-like behavior in mice, while decreased GR signaling in the brain is linked to depression-like phenotypes [120]. MR over-expression in the forebrain decreases anxiety-like behavior and modifies GR and serotonin 5HT-1a receptor expression in the hippocampus [121]. In contrast, forebrain MR[CaMKCre] knockout in mice that have the MR inactivated using the Cre/loxP-recombination system enhances emotional arousal after stress. Their fear acquisition and contextual fear memory are also enhanced. These mice show impaired behavioral and endocrine adaptation to stressors.

Animal models of PTSD, based on exposure to different stress paradigms, try to disentangle the role of corticosteroids in the neurobiology of the disease [122]. A strain- and corticosterone-dependent generalization of fear memories occurs in BALB/c and C57BL/6 mice. When corticosteroids were administered immediately after reactivation of a contextual fear memory, subsequent recall was significantly diminished [27, 123].

The first positive results of the use of cortisol treatment in PTSD were reported a few years ago by administering low doses of cortisol in patients for one month [124]. This treatment effectively blunted the recall of adverse memories without unwanted side effects. More recently, cortisone or cortisol administration was effective in reducing the self-reported fear of social and spider phobia-sufferers, respectively [125].

7.9
Early Experience and Psychopathology

In humans, maternal exposure to traumatic events like war or earthquakes during pregnancy has been associated with increased risk of their offspring to develop neuropsychiatric disorders in childhood or adulthood [126]. Although in humans the impact of mother's stress on the fetus is less robust, as the maternal cortisol is metabolized in the placenta by 11β HSD-2 to inactive cortisone, based on the animal literature one could hypothesize that maternal stress during gestation could similarly predispose the fetus to adult stress-related pathologies. In this context, the mother's genetic make-up concerning among other GR and MR polymorphisms and her subsequent stress vulnerability may be critical for the programming outcome and could provide an early mechanism for individual differences in stress responsivity. Individuals exposed during childhood in adverse conditions (abuse, malnutrition, violence) are at risk to develop stress-related diseases such as depression, obesity or PTSD [127]. However, caution should be exercised when generalizing the conclusions reached by these authors. Rinne *et al.* found that early adversity was associated with a dysregulated HPA axis activity rather than with depression and PTSD in a group of female borderline patients [128].

A recent study by Quellet-Morin *et al.* in twins addressed the question of the stress diathesis or the maternal mediation model in predicting the outcome of adverse experiences [129]. In the stress diathesis model following the studies of Caspi *et al.* [130] genetic factors program the newborn for a given cortisol secretion pattern with predetermined vulnerability to adverse experiences. In the maternal mediation model the genetic make-up for cortisol reactivity is abolished by an overriding impact of early familial adversity. The outcome of the study supports the maternal mediation hypothesis, which is in line with the animal studies on the low-care versus high-care offspring and the outcome of prolonged maternal deprivation.

7.10
Genetic Variants and Psychopathology

Epidemiological studies associating early adversity with MR:GR gene variants are lacking. Published studies rather associate healthy individuals or patient cohorts with some of these variants, but the available data base is insufficient for a definite conclusion. Below the outcome of recent studies is summarized.

In patients suffering from post-traumatic stress disorder (PTSD), a dysregulation of the HPA axis with low basal levels of cortisol and increased negative feedback has been observed [13]. In a relative small study, no association was found between GR N363S or *Bcl*I, although it has to be noted that also no differences were observed in corticosteroid sensitivity, as assessed using 0.25 mg DST [131].

In patients suffering from depression, West *et al.* also tested GR NR3C1-1, R23K, N363S and N766N in combination as haplotypes, for association with major depressive disorder (MDD, DSM-IV based) in two cohorts (in Belgium, $n = 180$; and in Sweden, $n = 134$) [87]. Although associations were not completely consistant between the two cohorts using single SNPs, when the combination of NR3C1-1 and ER22/23EK was tested significant associations were observed in both samples. In a German sample of 490 depressed patients, including mostly unipolar (86%) but also bipolar (13%) and dystymic patients (1%) and classified according to the DSM-IV, GR ER22/23EK and GR *Bcl*I were found to have significantly higher frequencies than control subjects [107], especially in patients with unipolar depression. Also a higher frequency for the GR N363S was observed, but this failed to reach significance ($P = 0.11$). Importantly, carriers of the GR ER22/23EK showed a much faster improvement on the HAM-D scale in response to antidepressant treatment. In addition, in a Dutch cohort, Brouwer *et al.* found GR *Bcl*I to be associated with worse treatment outcome [132]. In a recent study, the G/G genotype (minor allele) of the GR *Bcl*I polymorphism was found to be significantly more common in premenopausal women with major depression while these subjects also had greater abdominal obesity than non-GG carriers [133]. In a German cohort different from the one studied by van Rossum *et al.*, Zobel *et al.* found a significant association between GR *TthIII*I (decreased minor allele) and GR SNP in intron B (between exon 2 and exon 3) and the diagnosis of depression [86]. When testing different combinations of SNPs giving rise to putative haplotypes, she found the combination between *TthIII*I-*Bcl*I and a SNP in intron B to have a higher minor allele frequency in cases ($P = 0.0023$). These three markers are part of the same LD block and constitute a haplotype (see Figure 7.2). Moreover, the relation between these GR gene variants and intermediate phenotypes, including amygdala and hippocampal volume, was determined. The diagnosis-related alleles were indeed also associated with amygdala volume variation and hippocampal volume reduction. In a Dutch cohort of bipolar patients, a trend was found toward a protective effect of the GR A3669 (lower minor allele in patients) and a significant lower frequency of GR *TthIII*I, as compared to controls [134]. Moreover, GR A3669G was significant less frequent in patients showing fewer (hypo) manic episodes.

In two elderly cohorts an association has been performed between several GR and MR gene variants and measures of cognitive function. In the so-called Rotterdam Study, psychomoter speed was improved in carriers of the GR ER22/23EK, while no statistical significant effect was seen on global cognitive function and memory performance [135]. In another prospective study of subjects aged 85–90 years (Leiden 85-plus study) age-dependent decline of cognitive function was observed. However, no association with MR -2 G/C, MR I180V or GR ER22/23EK, N363S or *Bcl*I was found [101]. In this study, high morning cortisol levels ($\geq 47 \, \mu mol/l$) did predict lower attention and slower processing speed. As discussed above, C-allele carriers of MR-2 G/C had lower cortisol levels than non-carriers. Interestingly, in the Rotterdam study a protective effect of the ER22/23EK was suggested for dementia, although the effects were modest. White matter lesions in periventricular and subcortical areas (but not in the amygdala or hippocampus) were less prevalent in carriers, while progression of the subcortical lesions was 70% lower.

In conclusion, common GR gene variants have been found in several studies to associate not only with HPA axis reactivity but also with depression, further strengthening the link between functionality of GR and the disease (see Table 7.1). In addition, MR I180V modifies both the HPA axis and autonomic reactivity following a challenge and might be associated with depression. In addition, associations of MR and GR gene variants with HPA axis reactivity might be translated to individual differences in the levels of anxiety and arousal following stress.

Table 7.1 Phenotypes of MR and GR gene variants.

	Allele frequency	Corticosteroid sensitivity	HPA axis	Cardiovascular	Immune function	Metabolism	Psychopathology
GR *ThIII* 1 (*rs10052957*)	31%		Increased basal				Bipolar/unipolar
GR NR3C1-1 (*rs10482605*)	12%	Decreased	Increased				Unipolar
GR ER22/23EK (*rs6189/6190*)	3%	Decreased				Better	Unipolar
GR N363S (*rs6195*)	4%	Increased	Increased ♂ Decreased ♀			Worse	
GR *Bcl* site (*rs41423247*)	37%	Increased	Differential effects	Worse		Worse	Unipolar
GR A3669G (*rs6198*)	15%	Decreased	Increased	Worse	More active	Better	Bipolar
MR I180V (*rs5522*)	12%	Decreased	Increased	Increased			Elderly: feelings of depression

Specific phenotypes and individual frequencies of the different MR and GR gene variants. These phenotypes include associations with HPA axis regulation, cardiovascular and metabolic measures, immune system activity and depression. Blanks indicate not (properly) tested. Importantly, all SNPs both change certain aspects of HPA axis reactivity and associate with depression, suggesting a causal link.

7.11
Concluding Remarks

In this chapter, we address the role of corticosteroids and their receptors in the processing of fearful information and the stress response. Current evidence demonstrates that the two receptor types for the corticosteroids in the brain coordinate the initial stress reactions with the management of later adaptive phases (see Figure 7.1). MR are generally concerned with stability and integrity of the limbic circuitry, while its recently discovered membrane variant seems to amplify the initial stress and fear reactions. GR are generally concerned with recovery from stress and coping with the stressful experience, while learning how to deal with the event is stored for future use. The leading hypothesis is that a balance of MR:GR-mediated actions operating in concert with the other stress mediators serves homeostasis and health by enhancing plasticity in site- and context-dependent fashion.

Imbalance of these MR:GR-mediated processes is thought to be a significant factor in dysregulation of susceptibility pathways that underlie enhanced vulnerability to stress-related neuropathology. There is a higher risk for brain diseases that develop as a result of overexposure to corticosteroids, such as depression and psychoses, and diseases that prevail if insufficient corticosteroids are around to restrain the initial stress reactions, as is the case in PTSD. Such conditions are not only met by the actual corticosteroid concentrations, but also when the steroid concentration matches the demands dictated by the context and the time domain in which the stress reaction operates. Out of context and wrong timing of corticosteroid action are important factors in dysregulation of the stress system and the subsequent risk of cognitive impairments. If such conditions continue, the quality of life is seriously hampered, because these states of hyper- and hypocortisolism also have co-morbid metabolic, inflammatory and immune disorders.

The MR:GR imbalance can be caused by genetic variants in these receptors or early life experience or an interaction between genetic inputs × cognitive and emotional inputs. Associations of several receptor variants with basal (MR-2G/C) and stress-induced cortisol secretion have been described (MR I180V; GR *TthIII*I, NR3C1-1, ER22/23EK, N363S, *Bcl*1, A3669G 9β) and, in agreement with their role in HPA axis regulation, in some cases they have been associated with psychopathology (see Table 7.1). Since these genetic variants have relative high frequency, ranging from 5 to 55% carriers, they constitute a common vulnerability factor for stress-related disorders.

It should be noted that the findings of an association study are not always replicated in a next study published elsewhere. This is partially due to the different characteristics and homogeneity (age, sex, life history, nationality) of the cohorts used in these studies. Therefore, additional studies in well defined cohorts are needed. In accordance with the well known sex differences in corticosteroid action and receptor regulation, a sexually dimorphic profile of associations has also been referred for some of their genetic variants (e.g., GR N363S with HPA axis responses, GR *Bcl*I-site with metabolism). Also age could be a determinant, since the effects of the GR ER22/23EK variant on metabolic parameters seems to be age-dependent [136].

Early life experiences can modify the organism's stress response in adulthood. A programming effect of maternal corticosterone on fetal HPA axis has been shown in rodents, with a specific impact on metabolism, cardiovascular functions and behavior [137]. In humans, maternal exposure to traumatic events during pregnancy has been associated with increased risk of their offspring developing neuropsychiatric disorders. In this context, it would of interest to examine the offspring of mothers carrying GR and MR variants and displaying various degrees of maternal care.

We conclude with a selection of five pertinent research questions for stress aficionados:

1. How is the recently discovered membrane MR-mediated action in the same cell coordinated with the well known genomic actions exerted by a higher-affinity isoform of the very same MR?

2. To what extent is the processing of stressful information modulated by MR and GR variants and what is their relationship with psychopathology?

3. How does maternal care exert in the new born lasting effects on brain structure and function?

4. How does gender affect the outcome of MR:GR-mediated actions in the brain?

5. Which susceptibility pathways for pathogenesis of stress-related psycho-pathology can be identified in the brain?

Summary

Corticosteroids secreted after stress act on limbic brain circuits involved in the processing of fearful information. These actions are mediated by two receptor types abundantly expressed in limbic brain where they coordinate cognitive and emotional components of fear-motivated behavior. The two receptor types mediate the regulation of gene transcription, but recently membrane variants have been identified that can expand and integrate signaling of the steroid over different time domains. Hence MR is involved in appraisal of novel situations and selection of appropriate coping responses. To that effect cortisol enhances via MR in concert with other stress mediators the initial stress reaction. Via GR the hormone promotes recovery from stress, while facilitating the storage of the experience for adaptation and better coping in the future.

Variations in MR and GR gene function are predicted to affect the processing of stress and fearful information. The identification of gene variants in these receptors that modify the HPA axis and autonomic responses suggests that these genes indeed have the potential to determine susceptibility to stress. Moreover, epigenetic modification can affect the expression of MR and GR. These epigenetic modifications are prominent in response to early life experience. Recent research has demonstrated that the quality of maternal care has programming effects on cognitive performance, neuroendocrine and emotional reactivity.

These individual differences in stress responsiveness programmed by genetic variants and early life experience are thought to contribute to psychopathology. In this respect it is of relevance that the context of later life experiences matches the conditions for which the individual is programmed to respond. For designing novel pharmacotherapy of stress-related brain disorders it is therefore essential that the dynamics of stress reactions is considered from the perspective of genetic make-up and life history. Such novel treatments may be designed to promote resilience, still present in the diseased brain taking into account the timing and context of the drug application in relation to the experience of the stressful event.

Acknowledgment

The support of the Royal Netherlands Academy of Arts and Sciences is gratefully acknowledged.

References

1 Patel, P.D., Lopez, J.F., Lyons, D.M., Burke, S., Wallace, M. and Schatzberg, A.F. (2000) Glucocorticoid and mineralocorticoid receptor mRNA expression in squirrel monkey brain. *J. Psychiatr. Res.*, **34**, 383–392.

2 Atkinson, H.C., Wood, S.A., Castrique, E.S., Kershaw, Y.M., Wiles, C.C. and Lightman, S.L. (2008) Corticosteroids mediate fast feedback of the rat hypothalamic-pituitary-adrenal axis via the mineralocorticoid receptor. *Am. J. Physiol. Endocrinol. Metab.*, **294**, E1011–E1022.

3 Oitzl, M.S. and de Kloet, E.R. (1992) Selective corticosteroid antagonists modulate specific aspects of spatial orientation learning. *Behav. Neurosci.*, **106**, 62–71.

4 Oitzl, M.S., Reichard, H.M., Joels, M. and de Kloet, E.R. (2001) Point mutation in the mouse glucocorticoid receptor preventing DNA binding impairs spatial memory. *Proc. Natl. Acad. Sci. U.S.A.*, **98**, 12790–12795.

5 Karst, H., Berger, S., Turiault, M., Tronche, F., Schütz, G. and Joëls, M. (2005) Mineralocorticoid receptors are indispensible for non-genomic modulation of hippocampal glutamate transmission by corticosterone. *Proc. Natl. Acad. Sci. U.S.A.*, **102**, 19204–19207.

6 Joëls, M., Karst, H., DeRijk, R. and de Kloet, E.R. (2008) The coming out of the brain mineralocorticoid receptor. *Trends Neurosci.*, **31**, 1–7.

7 de Kloet, E.R., Oitzl, M.S. and Joels, M. (1999) Stress and cognition: are corticosteroids good or bad guys? *Trends Neurosci.*, **22**, 422–426.

8 Oitzl, M.S., Fluttert, M., Sutanto, W. and de Kloet, E.R. (1998) Continuous blockade of brain glucocorticoid receptors facilitates spatial learning and memory in rats. *Eur. J. Neurosci.*, **10**, 3759–3766.

9 Roozendaal, B., Barsegyan, A. and Lee, S. (2008) Adrenal stress hormones, amygdala activation, and memory for emotionally arousing experiences. *Prog. Brain Res.*, **167**, 79–97.

10 McEwen, B.S. (2004) Protection and damage from acute and chronic stress: allostasis and allostatic overload and relevance to the pathophysiology of

psychiatric disorders. *Ann. N. Y. Acad. Sci.*, **1032**, 1–7.

11 de Kloet, E.R., Vreugdenhil, E., Oitzl, M.S. and Joëls, M. (1998) Brain corticosteroid receptor balance in health and disease. *Endocr. Rev.*, **19**, 269–301.

12 de Kloet, E.R., Joëls, M. and Holsboer, F. (2005) Stress and the brain: from adaptation to disease. *Nat. Rev. Neurosci.*, **6**, 463–475.

13 Yehuda, R. (2002) Post-traumatic stress disorder. *N. Engl. J. Med.*, **346**, 108–114.

14 Pitkanen, A., Savander, V. and LeDoux, J.E. (1997) Organization of intra-amygdaloid circuitries in the rat: an emerging framework for understanding functions of the amygdala. *Trends Neurosci.*, **20**, 517–523.

15 Herman, J.P., Ostrander, M.M., Mueller, N.K. and Figueiredo, H. (2005) Limbic system mechanisms of stress regulation: hypothalamo-pituitary-adrenocortical axis. *Prog. Neuropsychopharmacol. Biol. Psychiatry*, **29**, 1201–1213.

16 Galvez, R., Mesches, M.H. and McGaugh, J.L. (1996) Norepinephrine release in the amygdala in response to footshock stimulation. *Neurobiol. Learn. Mem.*, **66**, 253–257.

17 Roozendaal, B., Schelling, G. and McGaugh, J.L. (2008) Corticotropin-Releasing Factor in the basolateral amygdala anhances memory consolidation via an interaction with the b-adrenoceptor – cAMP pathway: dependence on glucocorticoid receptor activation. *J. Neurosci.*, **28**, 6642–6651.

18 van der Laan, S., Lachize, S.B., Vreugdenhil, E., de Kloet, E.R. and Meijer, O.C. (2008) Nuclear receptor coregulators differentially modulate induction and glucocorticoid receptor-mediated repression of the corticotropin-releasing hormone gene. *Endocrinology*, **149**, 725–732.

19 Neisser, U., Winograd, E., Bergman, E.T., Schreiber, C.A., Palmer, S.E. and Weldon, M.S. (1996) Remembering the earthquake: direct experience vs. hearing the news. *Memory*, **4**, 337–357.

20 Bohus, B. and de Kloet, E.R. (1981) Adrenal steroids and extinction behavior: antagonism by progesterone, deoxycorticosterone and dexamethasone of a specific effect of corticosterone. *Life Sci.*, **28**, 433–440.

21 Sandi, C. and Rose, S.P. (1994) Corticosterone enhances long-term retention in one-day-old chicks trained in a weak passive avoidance learning paradigm. *Brain Res.*, **647**, 106–112.

22 Joëls, M., Pu, Z., Wiegert, O., Oitzl, M.S. and Krugers, H.J. (2006) Learning under stress: how does it work? *Trends Cogn. Sci.*, **10**, 152–158.

23 de Quervain, D.J.-F., Roozendaal, B. and McGaugh, J.L. (1998) Stress and glucocorticoids impair retrieval of long-term spatial memory. *Nature*, **394**, 787–790.

24 Diamond, D.M., Park, C.R., Campbell, A.M. and Woodson, J.C. (2005) Competitive interactions between endogenous LTD and LTP in the hippocampus underlie the storage of emotional memories and stress-induced amnesia. *Hippocampus*, **15**, 1006–1025.

25 Lupien, S.J., Maheu, F. and Weekes, N. (2005) Glucocorticoids: effects on human cognition, in *Handbook of Stress and the Brain* (eds T. Steckler, N.H. Kalin and J.M. Reul), Elsevier, Amsterdam, pp. 387–402.

26 Barrett, D. and Gonzalez-Lima, F. (2004) Behavioral effects of metyrapone on Pavlovian extinction. *Neurosci. Lett.*, **371**, 91–96.

27 Brinks, V., de Kloet, E.R. and Oitzl, M.S. (2008) Strain specific fear behaviour and glucocorticoid response to aversive events: modelling PTSD in mice. *Prog. Brain Res.*, **167**, 257–261.

28 Jackson, E.D., Payne, J.D., Nadel, L. and Jacobs, W.J. (2006) Stress differentially modulates fear conditioning in healthy men and women. *Biol. Psychiatry*, **59**, 516–522.

29 Cahill, L. (2003) Sex- and hemisphere-related influences on the neurobiology of emotionally influenced memory. *Prog. Neuropsychopharmacol. Biol. Psychiatry*, **27**, 1235–1241.

30 McGaugh, J.L. and Roozendaal, B. (2002) Role of adrenal stress hormones in forming lasting memories in the brain. *Curr. Opin. Neurobiol.*, **12**, 205–210.

31 Rashidy-Pour, A., Sadeghi, H., Taherain, A.A., Vafaei, A.A. and Fathollahi, Y. (2004) The effects of acute restraint stress and dexamethasone on retrieval of long-term memory in rats: an interaction with opiate system. *Behav. Brain Res.*, **154**, 193–198.

32 Roozendaal, B., McReynolds, J.R. and McGaugh, J.L. (2004) The basolateral amygdala interacts with the medial prefrontal cortex in regulating glucocorticoid effects on working memory impairment. *J. Neurosci.*, **24**, 1385–1392.

33 Sajadi, A.A., Samaei, S.A. and Rashidy-Pour, A. (2006) Intra-hippo-campal microinjections of anisomycin did not block glucocorticoid-induced impairment of memory retrieval in rats: an evidence for non-genomic effects of glucocorticoids. *Behav. Brain Res.*, **173**, 158–162.

34 Roozendaal, B., Griffith, Q.K., Buranday, J., Quervain, D.J.-F. and McGaugh, J.L. (2003) The hippocampus mediates glucocorticoid-induced impairment of spatial memory retrieval: dependence on the basolateral amygdala. *Proc. Natl. Acad. Sci. U. S. A.*, **100**, 1328–1333.

35 de Quervain, D.J., Roozendaal, B., Nitsch, R.M., McGaugh, J.L. and Hock, C. (2000) Acute cortisone administration impairs retrieval of long-term declarative memory in humans. *Nat. Neurosci.*, **3**, 313–314.

36 Wolf, O.T., Convit, A., McHugh, P.F., Kandil, E., Thorn, E.L., De, S.S., McEwen, B.S. and de Leon, M.J. (2001) Cortisol differentially affects memory in young and elderly men. *Behav. Neurosci.*, **115**, 1002–1011.

37 de Quervain, D.J., Aerni, A. and Roozendaal, B. (2007) Preventive effect of beta-adrenoceptor blockade on glucocorticoid-induced memory retrieval deficits. *Am. J. Psychiatry*, **164**, 967–969.

38 Dolan, R.J. (2000) Functional neuroimaging of the human amygdala during emotional processing and learning, in *The Amygdala* (ed. J.P.

Aggleton), Oxford University Press, New York, pp. 631–653.

39 Dolcos, F., Labar, K.S. and Cabeza, R. (2005) Remembering one year later: role of the amygdala and the medial temporal lobe memory system in retrieving emotional memories. *Proc. Natl. Acad. Sci. U. S. A.*, **102**, 2626–2631.

40 Smith, A.P., Stephan, K.E., Rugg, M.D. and Dolan, R.J. (2006) Task and content modulate amygdala-hippocampal connectivity in emotional retrieval. *Neuron*, **49**, 631–638.

41 McEwen, B.S. and Magarinos, A.M. (1997) Stress effects on morphology and function of the hippocampus. *Ann. N. Y. Acad. Sci.*, **821**, 271–284.

42 McLaughlin, K.J., Gomez, J.L., Baran, S.E. and Conrad, C.D. (2007) The effects of chronic stress on hippocampal morphology and function: an evaluation of chronic restraint paradigms. *Brain Res.*, **1161**, 56–64.

43 Luine, V.N., Beck, K.D., Bowman, R.E., Frankfurt, M. and Maclusky, N.J. (2007) Chronic stress and neural function: accounting for sex and age. *J. Neuroendocrinol.*, **19**, 743–751.

44 Kitraki, E., Kremmyda, O., Youlatos, D., Alexis, M.N. and Kittas, C. (2004) Gender-dependent alterations in corticosteroid receptor status and spatial performance following 21 days of restraint stress. *Neuroscience*, **125**, 47–55.

45 Lupien, S.J., Wilkinson, D.W., Brière, S., Ng Ying Kin, N.M.K., Meaney, M.J., and Nair, N.P.V. (2002) Acute modulation of aged human memory by pharma-cological manipulation of glucocorti-coids. *J. Clin. Endocrinol. Metab.*, **87**, 3798–3807.

46 Sandi, C., Merino, J.J., Cordero, M.I., Touyarot, K. and Venero, C. (2001) Effects of chronic stress on contextual fear conditioning and the hippocampal expression of the neural cell adhesion molecule, its polysialylation, and L1. *Neuroscience*, **102**, 329–339.

47 Mfe, J. and Jacobs, W.J. (1998) Emotional memory The effect of stress on "cool" and "hot" memory systems, in *The psychology of Learning and Motivaton Advances in Research and Theory* (ed. D.L.

Medin), Academic Press, San Diego, pp. 187–222.

48 Cahill, L., Babinsky, R., Markowitsch, H.J. and McGaugh, J.L. (1995) The amygdala and emotional memory. *Nature*, **377**, 295–296.

49 van Stegeren, A.H., Wolf, O.T., Everaerd, W., Scheltens, P., Barkhof, F. and Rombouts, S.A. (2007) Endogenous cortisol level interacts with noradrenergic activation in the human amygdala. *Neurobiol. Learn. Mem.*, **87**, 57–66.

50 Vyas, A., Mitra, R., Shankaranarayana Rao, B.S., and Chattarji, S. (2002) Chronic stress induces contrasting patterns of dendritic remodeling in hippocampal and amygdaloid neurons. *J. Neurosci.*, **22**, 6810–6818.

51 Kruk, M.R., Halász, J., Meelis, W. and Haller, J. (2005) Fast positive feedback between the adrenocortical stress response and a brain mechanism involved in aggressive behavior. *Behav. Neurosci.*, **118**, 1062–1070.

52 Fiebeler, A., Schmidt, F., Muller, D.N., Park, J.K., Dechend, R., Bieringer, M., Shagdarsuren, E., Breu, V., Haller, H. and Luft, F.C. (2001) Mineralocorticoid receptor affects AP-1 and nuclear factor-kappab activation in angiotensin II-induced cardiac injury. *Hypertension*, **37**, 787–793.

53 Pascual-Le Tallec, L. and Lombes, M. (2005) The mineralocorticoid receptor: a journey exploring its diversity and specificity of action. *Mol. Endocrinol.*, **19**, 2211–2221.

54 Meijer, O.C., van der Laan, S., Lachize, S., Steenbergen, P.J. and de Kloet, E.R. (2006) Steroid receptor coregulator diversity: what can it mean for the stressed brain? *Neuroscience*, **138**, 891–899.

55 Tasker, J.G., Di, S. and Malcher-Lopes, R. (2006) Minireview: rapid glucocorticoid signaling via membrane-associated receptors. *Endocrinology*, **147**, 5549–5556.

56 Heitzer, M.D., Wolf, I.M., Sanchez, E.R., Witchel, S.F. and Defranco, D.B. (2007) Glucocorticoid receptor physiology. *Rev. Endocr. Metab. Disord.*, **8**, 321–330.

57 Christ, M., Wehling, M., Kirsch, E., Viengchareun, S., Zennaro, M.C. and Lombes, M. (2005) Enhancement of beta-adrenergic cAMP-signaling by the mineralocorticoid receptor. *Mol. Cell. Endocrinol.*, **231**, 23–31.

58 Viengchareun, S., Le, M.D., Martinerie, L., Munier, M., Pascual-Le, T.L. and Lombes, M. (2007) The mineralocorticoid receptor: insights into its molecular and (patho) physiological biology. *Nucl. Recept. Signal.*, **5**, e012.

59 van der Laan, S. and Meijer, O.C. (2008) Pharmacology of glucocorticoids: beyond receptors. *Eur. J. Pharmacol.*, **585**, 483–491.

60 Meijer, O.C. (2006) Understanding stress through the genome. *Stress*, **9**, 61–67.

61 Zennaro, M.-C., Le Menuet, D., and Lombes, M. (1996) Characterization of the human mineralocorticoid receptor gene 5′-regulatory region: evidence for differential hormonal regulation of two alternative promoters via nonclassical mechanisms. *Mol. Endocrinol.*, **10**, 1549–1560.

62 Castren, M. and Damm, K. (1993) A functional promoter directing expression of a novel type of rat mineralocorticoid receptor mRNA in brain. *J. Neuroendocrinol.*, **5**, 461–466.

63 Vazquez, D.M., Lopez, J.F., Morano, M.I., Kwak, S.P., Watson, S.J. and Akil, H. (1998) Alpha, beta, and gamma mineralocorticoid receptor messenger ribonucleic acid splice variants: differential expression and rapid regulation in the developing hippocampus. *Endocrinology*, **139**, 3165–3177.

64 DeRijk, R.H. and de Kloet, E.R. (2008) Corticosteroid Receptor polymorphisms: determinants of vulnerability and resilience. *Eur. J. Pharmacol.*, **853**, 303–311.

65 Zennaro, M.-C., Souque, A., Viengchareun, S., Poisson, E. and Lombes, M. (2001) A new human MR splice variant is a ligand-independent transactivator modulating corticosteroid action. *Mol. Endocrinol.*, **15**, 1586–1598.

66 Pascual-Le Tallec, L., Demange, C., and Lombes, M. (2004) Human mineralocorticoid receptor A and B protein forms produced by alternative translation sites display different transcriptional activities. *Eur. J. Endocrinol.*, **150**, 585–590.

67 Yudt, M.R. and Cidlowski, J.A. (2002) The glucocorticoid receptor: coding a diversity of proteins and responses through a single gene. *Mol. Endocrinol.*, **16**, 1719–1726.

68 Russcher, H., Smit, P., van den Akker, E.L.T., van Rossum, E.F.C., Brinkman, A.O., de Jong, F.H., Lamberts, S.W. and Koper, J.W. (2005) Two polymorphisms in the glucocorticoid receptor gene directly affect glucocorticoid-regulated gene expression. *J. Clin. Endocrinol. Metab.*, **90**, 5804–5810.

69 Szyf, M., Weaver, I.C.G., Champagne, F.A., Diorio, J. and Meaney, M.J. (2005) Maternal programming of steroid receptor expression and phenotype through DNA methylation in the rat. *Front. Neuroendocrinol.*, **26**, 139–162.

70 Soulis, G., Kitraki, E. and Gerozissis, K. (2008) Early neuroendocrine alterations in female rats following a diet moderately enriched in fat. *Cell. Mol. Neurobiol.*, **25**, 869–880.

71 Herman, J.P. and Spencer, R. (1998) Regulation of hippocampal glucocorticoid receptor gene transcription and protein expression *in vivo*. *J. Neurosci.*, **18**, 7462–7473.

72 Holmes, M.C., French, K.L. and Seckl, J.R. (1997) Dysregulation of diurnal rhythms of serotonin 5-HT2C and corticosteroid receptor gene expression in the hippocampus with food restriction and glucocorticoids. *J. Neurosci.*, **17**, 4056–4065.

73 Hill, M.R., Stith, R.D. and McCallum, R.E. (1988) Human recombinant IL-1 alters glucocorticoid receptor function in Reuber hepatoma cells. *J. Immunol.*, **141**, 1522–1528.

74 Sterlemann, V., Ganea, K., Liebl, C., Harbich, D., Alam, S., Holsboer, F., Muller, M.B. and Schmidt, M.V. (2008) Long-term behavioral and neuro-endocrine alterations following chronic social stress in mice: implications for stress-related disorders. *Horm. Behav.*, **53**, 386–394.

75 Turner, J.D. and Muller, C.P. (2005) Structure of the glucocorticoid receptor (NR3C1) gene 5′ untranslated region: identification, and tissue distribution of multiple human exon 1. *J. Mol. Endocrinol.*, 283–289.

76 Meaney, M.J., Szyf, M. and Seckl, J.R. (2007) Epigenetic mechanisms of perinatal programming of hypothalamic-pituitary-adrenal function and health. *Trends Mol. Med.*, **13**, 269–277.

77 Moser, D., Moliter, A., Kumsta, R., Tatschner, T., Riederer, P. and Meyer, J. (2007) The glucocorticoid receptor gene exon 1-F promoter is not methylated at the NGFI-A binding site in human hippocampus. *World J. Biol. Psychiatry*, **8**, 262–268.

78 Kelly, A., Bowen, H., Jee, Y.K., Mahfiche, N., Soh, C., Lee, T., Hawrylowicz, C. and Lavender, P. (2008) The glucocorticoid receptor beta isoform can mediate transcriptional repression by recruiting histone deacetylases. *J. Allergy Clin. Immunol.*, **121**, 203–208.

79 Lewis-Tuffin, L.J., Jewell, C.M., Bienstock, R.J., Collins, J.B. and Cidlowski, J.A. (2007) Human glucocorticoid receptor beta binds RU-486 and is transcriptionally active. *Mol. Cell. Biol.*, **27**, 2266–2282.

80 Zennaro, M.-C. and Lombes, M. (2004) Mineralocorticoid resistance. *TEM*, **15**, 264–270.

81 Chrousos, G.P., Detera-Wadleigh, S. and Karl, M. (1993) Syndromes of glucocorticoid resistance. *Ann. Intern. Med.*, **119**, 1113–1124.

82 van Rossum, E.F.C. and Lamberts, S.W. (2004) Polymorphisms in the gluco-corticoid receptor gene and their associations with metabolic parameters and body composition. *Recent. Prog. Horm. Res.*, **59**, 333–357.

83 van Leeuwen, N., Wust, S., Meijer, O.C., de Kloet, E.R., Zitman, F.G. and DeRijk, R.H. (2006) The role of single nucleotide polymorphisms (SNPs) in the mineralocorticoid receptor (MR) in stress response and psychopathology.

Am. J. Med. Genet. B Neuropsychiatr. Genet., **141B**, 752.

84 DeRijk, R.H., Wüst, S., Meijer, O.C., Zennaro, M.-C., Federenko, I.S., Hellhammer, D.H., Giacchetti, G., Vreugdenhil, E., Zitman, F.G. and de Kloet, E.R. (2006) A common polymorphism in the mineralocorticoid receptor modulates stress responsiveness. *J. Clin. Endocrinol. Metab.*, **91**, 5083–5089.

85 van den Akker, E.L., Nouwen, J.L., Melles, D.C., van Rossum, E.F., Koper, J.W., Uitterlinden, A.G., Hofman, A., Verbrugh, H.A., Pols, H.A., Lamberts, S.W. and van Belkum, A. (2006) Staphylococcus aureus nasal carriage is associated with glucocorticoid receptor gene polymorphisms. *J. Infect. Dis.*, **194**, 814–818.

86 Zobel, A., Jessen, F., von Widdern, O., Schuhmacher, A., Hofels, S., Metten, M., Rietschel, M., Scheef, L., Block, W., Becker, T., Schild, H.H., Maier, W., and Schwab, S.G. (2008) Unipolar depression and hippocampal volume: impact of DNA sequence variants of the glucocorticoid receptor gene. *Am. J. Med. Genet. B Neuropsychiatr. Genet.*, **147**, 836–843.

87 van West, D., Van Den, E.F., Del Favero, J., Souery, D., Norrback, K.F., van Duijn, C., Sluijs, S., Adolfsson, R., Mendlewicz, J., Deboutte, D., Van Broeckhoven, C., and Claes, S. (2006) Glucocorticoid receptor gene-based SNP analysis in patients with recurrent major depression. *Neuropsychopharmacology*, **31**, 620–627.

88 Kumsta, R., Entringer, S., Koper, J.W., van Rossum, E.F., Hellhammer, D.H. and Wust, S. (2008) Glucocorticoid receptor gene polymorphisms and glucocorticoid sensitivity of subdermal blood vessels and leukocytes. *Biol. Psychol.*, **79**, 179–184.

89 Russcher, H., van Rossum, E.F.C., de Jong, F.H., Brinkman, A.O., Lamberts, S.W. and Koper, J.W. (2005) Increased expression of the glucocorticoid receptor – a translational isoform as a result of the ER22/23EK polymorphism. *Mol. Endocrinol.*, **19**, 1687–1696.

90 Finken, M.J.J., Meulenbelt, I., Dekker, F.W., Frölich, M., Romijn, J.A., Slagboom, E.P. and Wit, J.M. (2007) The 23K variant of the R23K polymorphism in the glucocorticoid receptor gene protects against postnatal growth failure and insulin resistance after preterm birth. *J. Clin. Endocrinol. Metab.*, **92**, 4777–4782.

91 Kuningas, M., Mooijaart, S.P., Slagboom, P.E., Westendorp, R.G. and van Heemst, H.D. (2006) Genetic variants in the glucocorticoid receptor gene (NR3C1) and cardiovascular disease risk. The Leiden 85-plus study. *Biogerontology*, **7**, 231–238.

92 van den Akker, E.L., Russcher, H., van Rossum, E.F., Brinkmann, A.O., de Jong, F.H., Hokken, A., Pols, H.A., Koper, J.W. and Lamberts, S.W. (2006) Glucocorticoid receptor polymorphism affects transrepression but not transactivation. *J. Clin. Endocrinol. Metab.*, **91**, 2800–2803.

93 van Winsen, L.L., Hooper-van Veen, T., van Rossum, E.F., Polman, C.H., van den Berg, T.K., Koper, J.W., and Uitdehaag, B.M. (2005) The impact of glucocorticoid receptor gene polymorphisms on glucocorticoid sensitivity is outweighted in patients with multiple sclerosis. *J. Neuroimmunol.*, **167**, 150–156.

94 Jewell, C.M. and Cidlowski, J.A. (2007) Molecular Evidence for a Link between the N363S Glucocorticoid Receptor Polymorphism and Altered Gene Expression. *J. Clin. Endocrinol. Metab.*, **92**, 3268–3277.

95 Lin, R.C.Y., Wang, X.L. and Morris, B.J. (2003) Association of coronary artery disease with the glucocorticoid receptor N363S variant. *Hypertension*, **41**, 404–407.

96 Panarelli, M., Holloway, C.D., Fraser, R., Connell, J.M.C., Ingram, M., Anderson, N.N. and Kenyon, C.J. (1998) Glucocorticoid receptor polymorphism, skin vasoconstriction and other metabolic intermediate phenotypes in normal subjects. *J. Clin. Endocrinol. Metab.*, **83**, 1846–1852.

97 Syed, A.A., Halpin, C.G., Irving, J.A.E., Unwin, N.C., White, M., Bhopal, R.S., Redfern, C.P.F. and Weaver, J.U. (2008) A common intron 2 polymorphism of the glucocorticoid receptor gene is associated with insulin resistance in men. *Clin. Endocrinol. (Oxf.)*, **68**, 879–884.

98 Tremblay, A., Bouchard, L., Bouchard, C., Despres, J.P., Drapeau, V. and Perusse, L. (2003) Long-term adiposity changes are related to a glucocorticoid receptor polymorphism in young females. *J. Clin. Endocrinol. Metab.*, **88**, 3141–3145.

99 DeRijk, R.H., Schaaf, M., Turner, G., Datson, N.A., Vreugdenhil, E., Cidlowski, J.A., de Kloet, E.R., Emery, P., Sternberg, E.M. and Detera-Wadleigh, S. (2001) A glucocorticoid receptor variant that increases the stability of the glucocorticoid receptor β-isoform is associated with rheumatoid arthritis. *J. Rheum.*, **28**, 2383–2388.

100 Syed, A.A., Irving, J.A., Redfern, C.P., Hall, A.G., Unwin, N.C., White, M., Bhopal, R.S. and Weaver, J.U. (2006) Association of glucocorticoid receptor polymorphism A3669G in exon 9beta with reduced central adiposity in women. *Obesity (Silver Spring)*, **14**, 759–764.

101 Kuningas, M., DeRijk, R.H., Westendorp, R.G., Jolles, J., Slagboom, E.P. and van Heemst, D. (2007) Mental performance in old age is dependent on cortisol and genetic variance in the mineralocorticoid and glucocorticoid receptors. *Neuropsychopharmacology*, **32**, 1–7.

102 Rautanen, A., Eriksson, J.G., Kere, J., Andersson, S., Osmond, C., Tienari, P., Sairanen, H., Barker, D.J., Phillips, D.I., Forsen, T. and Kajantie, E. (2006) Associations of body size at birth with late-life cortisol concentrations and glucose tolerance are modified by haplotypes of the glucocorticoid receptor gene. *J. Clin. Endocrinol. Metab.*, **91**, 4544–4551.

103 Rosmond, R., Chagnon, Y.C., Chagnon, M., Pérusse, L., Bouchard, C. and Björntorp, P. (2000) A polymorphism of the 5′-flanking region of the glucocorticoid receptor gene locus is associated with basal cortisol secretion in men. *Metabolism*, **49**, 1197–1199.

104 Huizenga, N.A.T.M., Koper, J.W., de Lange, P., Pols, H.A., Stolk, R.P., Burger, H., Grobbee, D.E., Brinkman, A.O., de Jong, F.H. and Lamberts, S.W. (1998) A polymorphism in the glucocorticoid receptor gene may be associated with an increased sensitivity to glucocorticoids *in vivo. J. Clin. Endocrinol. Metab.*, **83**, 144–151.

105 Kumsta, R., Entringer, S., Koper, J.W., van Rossum, E.F.C., Hellhammer, D.H. and Wüst, S. (2007) Sex specific associations between common glucocorticoid receptor gene variants and hypothalamus-pituitary-adrenal axis responses to psychological stress. *Biol. Psychiatry*, **62**, 863–869.

106 Holsboer, F., von Bardeleben, U., Wiedemann, K., Muller, O.A. and Stalla, G.K. (1987) Serial assessment of corticotropin-releasing hormone response after dexamethasone in depression. Implications for pathophysiology of DST nonsuppression. *Biol. Psychiatry*, **22**, 228–234.

107 van Rossum, E.F.C., Binder, E.B., Majer, M., Koper, J.W., Ising, M., Modell, S., Salyakina, D., Lamberts, S.W. and Holsboer, F. (2006) Polymorphisms of the glucocorticoid receptor gene and major depression. *Biol. Psychiatry*, **59**, 681–688.

108 Federenko, I.S., Nagamine, M., Hellhammer, D.H., Wadhwa, P.D. and Wust, S. (2004) The heritability of hypothalamus pituitary adrenal axis responses to psychosocial stress is context dependent. *J. Clin. Endocrinol. Metab.*, **89**, 6244–6250.

109 Wüst, S., van Rossum, E.F.C., Federenko, I., Koper, J.W., Kumsta, R. and Hellhammer, D. (2004) Common polymorphisms in the glucocorticoid receptor gene are associated with adrenocortical responses to psychosocial stress. *J. Clin. Endocrinol. Metab.*, **89**, 563–564.

110 Lehmann, J., Pryce, C.R., Jongen-Relo, A.L., Stohr, T., Pothuizen, H.H.J. and

Feldon, J. (2002) Comparison of maternal separation and early handling in terms of their neurobehavioral effects in aged rats. *Neurobiol. Aging*, **23**, 457–466.

111 van der Veen, R., Koehl, M., Abrous, D.N., de Kloet, E.R., Piazza, P.V. and Deroche-Gamonet, V. (2008) Maternal environment influences cocaine intake in adulthood in a genotype-dependent manner. *PLoS One*, **3**, 1–9.

112 Champagne, D.L., Bagot, R.C., van Hasselt, F., Ramakers, G., Meaney, M.J., de Kloet, E.R., Joels, M. and Krugers, H. (2008) Maternal care and hippocampal plasticity: evidence for experience-dependent structural plasticity, altered synaptic functioning, and differential responsiveness to glucocorticoids and stress. *J. Neurosci.*, **28**, 6037–6045.

113 Weaver, I.C.G., Cervoni, N., Champagne, F.A., D'Allessio, A.C., Sharma, S., Seckl, J.R., Dymov, S., Szyf, M. and Meaney, M.J. (2004) Epigenetic programming by maternal behavior. *Nat. Neurosci.*, **7**, 847–854.

114 Yehuda, R. (2006) Advances in understanding neuroendocrine alterations in PTSD and their therapeutic implications. *Ann. N. Y. Acad. Sci.*, **1071**, 137–166.

115 Bremner, J.D., Krystal, J.H., Charney, D.S. and Southwick, S.M. (1996) Neural mechanisms in dissociative amnesia for childhood abuse: relevance to the current controversy surrounding the "false memory syndrome". *Am. J. Psychiatry*, **153**, 71–82.

116 Shin, L.M., Whalen, P.J., Pitman, R.K., Bush, G., Macklin, M.L., Lasko, N.B., Orr, S.P., McInerney, S.C. and Rauch, S.L. (2001) An fMRI study of anterior cingulate function in posttraumatic stress disorder. *Biol. Psychiatry*, **50**, 932–942.

117 Akirav, I. and Maroun, M. (2007) The role of the medial prefrontal cortex-amygdala circuit in stress effects on the extinction of fear. *Neuronal Plast.*, article ID 30873, 1–11.

118 Cai, W.H., Blundell, J., Han, J., Greene, R.W. and Powell, C.M. (2006) Postreactivation glucocorticoids impair recall of established fear memory. *J. Neurosci.*, **26**, 9560–9566.

119 de Quervain, D.J. (2008) Glucocorticoid-induced reduction of traumatic memories: implications for the treatment of PTSD. *Prog. Brain Res.*, **167**, 239–247.

120 Tronche, F., Kellendonk, C., Kretz, O., Gass, O., Anlag, K., Orban, P.C., Bock, R., Klein, R. and Schutz, G. (1999) Disruption of the glucocorticoid receptor gene in the nervous system results in reduced anxiety. *Nat. Genet.*, **23**, 99–103.

121 Rozeboom, A.M., Akil, H. and Seasholtz, A.F. (2007) Mineralocorticoid receptor overexpression in forebrain decreases anxiety-like behavior and alters the stress response in mice. *Proc. Natl. Acad. Sci. U. S. A.*, **104**, 4688–4693.

122 Miller, M.M. and McEwen, B.S. (2006) Establishing an agenda for translational research on PTSD. *Ann. N.Y. Acad. Sci.*, **1071**, 294–312.

123 Brinks, V., van der Mark, M.H., de Kloet, E.R. and Oitzl, M.S. (2007) Differential MR/GR activation in mice results in emotional states beneficial or impairing for cognition. *Neural Plast.*, **2007**, 90163.

124 Aerni, A., Traber, R., Hock, C., Roozendaal, B., Schelling, G., Papassotiropoulos, A., Nitsch, R.M., Schnyder, U. and de Quervain, D.J. (2004) Low-dose cortisol for symptoms of posttraumatic stress disorder. *Am. J. Psychiatry*, **161**, 1488–1490.

125 Soravia, L.M., Heinrichs, M., Aerni, A., Maroni, C., Scheling, G., Ehlert, U., Roozendaal, B. and de Quervain, D.J.-F. (2006) Glucocorticoids reduce phobic fear in humans. *Proc. Natl. Acad. Sci. U.S.A.*, **103**, 5585–5590.

126 Van Os, J. and Selten, J.P. (1998) Prenatal exposure to maternal stress and subsequent schizophrenia. The May 1940 invasion of the Netherlands. *Br. J. Psychiatry*, **172**, 324–326.

127 Heim, C., Newport, D.J., Heit, S., Graham, Y.P., Wilcox, M., Bonsall, R., Milller, A.H. and Nemeroff, C.B. (2001) Pituitary-adrenal and autonomic responses to stress in woman after

sexual and physical abuse in childhood. *JAMA*, **284**, 592–597.

128 Rinne, T., de Kloet, E.R., Wouters, L., Goekoop, J.G., DeRijk, R.H. and Brink, W. (2002) Hyperresponsiveness of hypothalamic-pituitary-adrenal axis to combined dexamethasone/corticotropin-releasing hormone challenge in female borderline personality disorder subjects with a history of sustained childhood abuse. *Biol. Psychiatry*, **52**, 1102–1112.

129 Ouellet-Morin, I., Boivin, M., Dionne, G., Lupien, S.J., Arseneault, L., Barr, R.G., Perusse, D. and Tremblay, R.E. (2008) Variations in heritability of cortisol reactivity to stress as a function of early familial adversity among 19-month-old twins. *Arch. Gen. Psychiatry*, **65**, 211–218.

130 Caspi, A., Sugden, K., Moffit, T.E., Taylor, A., Craig, I.W., Harrington, H., McClay, J., Mill, J., Martin, J., Braithwaite, A. and Poulton, R. (2003) Influence of life stress on depression: moderation by a polymorphism in the 5-HTT gene. *Science*, **301**, 386–389.

131 Bachmann, A.W., Sedgley, T.L., Jackson, R.V., Gibson, J.N., Young, R.M. and Torpy, D.J. (2005) Glucocorticoid receptor polymorphisms and post-traumatic stress disorder. *Psychoneuroendocrinology*, **30**, 297–306.

132 Brouwer, J.P., Appelhof, B.C., van Rossum, E.F., Koper, J.W., Fliers, E., Huyser, J., Schene, A.H., Tijssen, J.G., Van Dyck, R., Lamberts, S.W., Wiersinga, W.M. and Hoogendijk, W.J. (2006) Prediction of treatment response by HPA-axis and glucocorticoid receptor polymorphisms in major depression. *Psychoneuroendocrinology*, **31**, 1154–1163.

133 Krishnamurthy, P., Romagni, P., Torvik, S., Gold, P.W., Charney, D.S., Detera-Wadleigh, S. and Cizza, G. (2008) Glucocorticoid receptor gene polymorphisms in premenopausal women with major depression. *Hom. Metab. Res.*, **40**, 194–198.

134 Spijker, A.T., Hoencamp, E., van Rossum, E.F.C., DeRijk, R.H., Haffmans, J., Blom, M., Manenschijn, L., Koper, J.W., Lamberts, S.W.J. and Zitman, F.G. (2009) Functional polymorphism of the glucocorticoid receptor gene associates with (hypo)mania in bipolar disorder, *Bipolar Disord.*, **11**, 95–101.

135 van Rossum, E.F., de Jong, F.J., Koper, J.W., Uitterlinden, A.G., Prins, N.D., van Dijk, E.J., Koudstaal, P.J., Hofman, A., de Jong, F.H., Lamberts, S.W. and Breteler, M.M. (2008) Glucocorticoid receptor variant and risk of dementia and white matter lesions. *Neurobiol. Aging*, **29**, 716–723.

136 van Rossum, E.F.C., Voorhoeve, P.G., te Velde, S.J., Koper, J.W., Delamarre-van de Waal, H.A., Kemper, H.C.G. and Lamberts, S.W. (2004) The ER22/23EK polymorphism in the glucocorticoid receptor gene is associated with a beneficial body composition and muscle strength in young adults. *J. Clin. Endocrinol. Metab.*, **89**, 4004–4009.

137 Holmes, M.C. and Seckl, J.R. (2006) The role of 11beta-hydroxysteroid dehydrogenase in the brain. *Mol. Cell. Endocrinol.*, **248**, 9–14.

138 Kudielka, B.M., Hellhammer, D.H. and Wüst, S. (2009) Why do we respond so differently? Reviewing determinants of human salivary cortisol responses to challenge. *Psychoneuroendocrinology*, **34**, 2–18.

8

Corticotropin-Releasing Factor (CRF) and CRF-Related Peptides – a Linkage Between Stress and Anxiety

Thomas Blank and Joachim Spiess

8.1
The CRF Family, Its Receptors and Ligands

The CRF family of neuropeptides has undergone considerable expansion during recent years. CRF, a 41-amino-acid (aa) peptide, also called corticotropin-releasing hormone (CRH) or corticoliberin, is the major regulator of the adaptive response to internal or external stressors [1, 2]. CRF was the first peptide isolated of a family of mammalian CRF-related peptides that now includes urocortin 1 (UCN_1), urocortin 2 (UCN_2; also known as stresscopin-related peptide) and urocortin 3 (UCN_3; also known as stresscopin) [3] (Table 8.1).

Consistent with a role for this family of peptides in survival and adaptation, the structures of CRF and UCN_1 are highly conserved across mammalian species and during evolution, as shown by its 45–48% similarity with sauvagine and urotensin I, a fish neuropeptide present in the urophysis, a caudal neuro-secretory organ of the common suckerfish *Catostomus commersoni* [4–6]. The first component of its name UCN_1 is derived from its homology to the teleost hormone urotensin I, while the second component stems from its homology to CRF. Like CRF, UCN_1 can stimulate ACTH production from anterior pituitary corticotrophs *in vitro* and *in vivo*. UCN_1 is more potent than CRF with regard to other biological effects, including suppression of appetite, but is less potent in generating anxiety.

The biological actions of CRF and UCN are mediated via two types of G protein-coupled receptors, CRF_1 and CRF_2, which have different expression patterns and physiological functions. Recently, a third CRF receptor, termed CRF_3, was cloned from catfish. Interestingly, CRF_3 receptor expression is restricted to the catfish pituitary, a tissue that normally expresses CRF_1 in other species [7]. CRF_1 exhibits high affinity toward CRF and UCN_1 but low affinity toward UCN_2 and no affinity toward UCN_3. The CRF_1 is mainly expressed in CNS and the anterior pituitary. Indeed, CRF_1 is expressed in the entire paraventricular-infundibular system, the septum, bed nuclei of stria terminalis nucleus accumbens, the cerebral cortex and the limbic system (amygdala, hippocampus) [8, 9].

Stress – From Molecules to Behavior. Edited by Hermona Soreq, Alon Friedman, and Daniela Kaufer
Copyright © 2010 WILEY-VCH Verlag GmbH & Co. KGaA, Weinheim
ISBN: 978-3-527-32374-6

Table 8.1 Amino acid sequences of CRF-related peptides.

Peptide	Sequences	Identity
oCRF	SQE**PPI**SL**DLTFHLLR**EVLEMTKADQLAQQA**HSNRK**LLDIA-NH2	83%
r/hCRF	SEE**PPI**SL**DLTFHLLR**EVLEMARAEQLAQQA**HSNRK**LMEII-NH2	100%
Sauvagin	ZG**PPI**SI**DLSLELLR**KMIEIEKQEKEKQQA**ANNR**LLLDTI-NH2	45%
Urotensin I	NDD**PPI**SI**DLTFHLLR**NMIEMARNENQREQAG**LNRK**YLDEV-NH2	54%
rUrocortin I	DD**PPLS**I**DLTFHLLR**TLLELARTQSQRERAEQ**NRI**IFDSV-NH2	45%
hUrocortin I	DN**PPLS**I**DLTFHLLR**TLLELARTQSQRERAEQ**NRI**IFDSV-NH2	45%
mUrocortin I	DD**PPLS**I**DLTFHLLR**TLLELARTQSQRERAEQ**NRI**IFDSV-NH2	45%
oUrocortin I	DD**PPLS**I**DLTFHLLR**TLLELARTQSQRERAEQ**NRI**IFDSV-NH2	45%
hUrocortin II	IVL**SLD**VPIG**LLQ**ILLEQARARAAREQ**ATTN**ARILARV-NH2	34%
mUrocortin II	VIL**SLD**VPIG**LLR**ILLEQARYKAARNQ**AATN**AQILAHV-NH2	34%
hUrocortin III	FTL**SLD**VPTNIMNLLFNIAKAKNLRAQ**AAAN**AHLMAQI-NH2	32%
mUrocortin III	FTL**SLD**VPTNIMNILFNIDKAKNLRAK**AAAN**AQLMAQI-NH2	26%

In addition, alternative splicing of the primary transcript encoding CRF_1 can lead to a number of other variants, named CRF_{1b-n}, all of which display impaired signaling. The functional significance of these other transcripts is still poorly characterized, but some splice variants (such as CRF_{1d}) have been reported to modulate CRF and UCN_1 signaling in transfected cells by acting as a "decoy receptor" able to compete with CRF_{1a} for agonist binding [10]. Three different splice variants of CRF_2 have been identified in mammals: CRF_{2a}, CRF_{2b} and CRF_{2c} [11]. The CRF_2 variants have a common carboxy-terminal region and structurally distinct amino-terminal extracellular domains (the region involved in ligand binding) that contain 34 aa for CRF_{2a}, 61 aa for CRF_{2b} and 20 aa for CRF_{2c} [3, 12]. Recently, a novel soluble splice variant was identified in the mouse brain and shown to encode only the first extracellular domain of CRF_{2a} and to function as a soluble binding protein for CRF and UCN_1 [13]. Despite sharing 70% aa sequence similarity, CRF_1 and CRF_2 display distinct characteristic affinities for the CRF family of peptides. All species homologs of CRF, UCN_1, urotensin-I and sauvagine bind to and activate CRF_1 with high affinity. UCN_1, urotensin-I and sauvagine also bind to CRF_2 with high affinity, whereas CRF binds with significantly lower affinity to this receptor subtype [3]. UCN_2 and UCN_3 are selective activators of CRF_2 [14–16].

It is well documented that stimulation of CRF_{1a}, CRF_{2a} and CRF_{2b} activates adenylyl cyclase/cAMP signaling pathways through the coupling and activation of $G\alpha_s$ proteins. However, several recent reports indicate that the nature of a trimeric complex (CRF agonist, CRF receptor, G protein) influences the pattern of intracellular signaling in a tissue-specific manner [10]. Finally, CRF-BP, a 37-kDa glycoprotein, is a soluble protein that binds CRF. It was first identified in a soluble form in human plasma. The human and rat CRF-BP cDNAs encode highly homologous

322-aa glycoproteins [17]. CRF-BP has been detected in anthropoid primates and rodents but not in evolutionarily earlier species, suggesting that a complex regulatory mechanism involving CRF-BP developed later in evolution [17–19]. In the circulation and in the interstitial spaces, CRF-BP binds to either CRF or UCNs with high affinity. In primates, the CRF-BP gene is expressed in several tissues, including liver, placenta and brain [20, 21]. By binding to CRF and UCNs, CRF-BP reduces their bioavailability and prevents their binding to the biologically active CRF receptors, thus controlling in a fast and transient manner the concentration of the available ligands.

8.2
The Role of CRF and CRF-Related Peptides in Neuroendocrine Aspects of Stress

CRF and its receptors (CRF_1 and CRF_2) are important regulators of the hypothalamic–pituitary–adrenal (HPA) axis and central stress responses. Studies on CRF_1- and CRF_2-deficient mice indicate that these receptors play different roles in the HPA axis. CRF_1 and CRF_2 seem to act in an antagonistic manner, such that CRF_1 activates and CRF_2 attenuates the stress response. During a stress response, CRF is released from the paraventricular nucleus of the hypothalamus (PVN) and activates CRF_1 on anterior pituitary corticotropes to stimulate the release of ACTH. It was shown that CRF_1-deficient mice are unable to mount a stress-induced HPA response in terms of circulating adrenocorticotrophic hormone (ACTH) and corticosterone, but their baseline ACTH levels are normal and baseline corticosterone levels virtually undetectable [22, 23]. From these findings it was concluded that CRF_1 is crucial for stress-induced HPA responsiveness but not for the baseline hypothalamic-pituitary drive. ACTH then enters the blood stream and acts at MC2 receptors in the cortex of the adrenal gland to stimulate the synthesis and release of glucocorticoids (GCs). GCs enter into the cytoplasm of cells throughout the body and the brain, where they interact with their receptors. A prominent negative-feedback system acts to inhibit further CRF production and release from the hypothalamus. Hypothalamic vasopressin (AVP) has been shown to act synergistically with CRF to augment the release of ACTH in rodents and humans, suggesting that AVP may also play a physiologic role in the neuroendocrine stress response [24–27]. Stress input also stimulates central CRF release, capable of activating the autonomic nervous system, producing a classic "fight or flight" response and a stimulated release of norepinephrine and epinephrine into the circulation. Interestingly, this centrally responsive CRF is also influenced by the glucocorticoid feedback of the HPA axis, but in the opposing direction such that limbic CRF, especially within the central nucleus of the amygdala (CeA), is increased with glucocorticoid exposure [28, 29]. Alterations in CRF expression in the CeA have been linked to increased emotionality and stress responsivity, a likely factor contributing to an increased vulnerability for disease onset [30, 31]. Conceivably, chronic stress or even acute stress in an individual genetically predisposed to stress sensitivity could cause such a dysregulation in homeostatic balance by disrupting normal feedback mechanisms. Studies examining effects of chronic stress or models of elevated glucocorticoid levels have demonstrated a potential mechanism of reduced sensitivity to the negative feedback resulting in an HPA axis that is hyper-responsive to new stimuli [32, 33]. This decrease in negative-feedback sensitivity may be due to effects on hippocampal tone and its normally inhibitory role on PVN activity [33–35]. Unlike CRF, peripheral UCN_2 or UCN_3 administration does not increase corticosterone secretion, consistent with the relative absence of CRF_2 on ACTH-secreting pituitary corticotroph cells [11]. In contrast, intracerebroventricular or intravenous administration of UCN_1 activates the pituitary–adrenal axis (as or more potently than CRF), via a CRF_1-dependent mechanism, stimulating ACTH release and proopiomelanocortin synthesis in pituitary corticotrophs [36]. However, UCN_1 is not a likely physiological regulator of the HPA axis. Unlike CRF-deficient mice, UCN_1-deficient mice exhibit normal basal and stress-induced HPA hormone levels [37, 38]. Similarly, unlike CRF antisera, peripheral administration of specific UCN_1 antisera does not modify basal, stress-induced or adrenalectomy-induced ACTH levels. It remains possible, however, that the type 2 UCNs modulate HPA-axis activity at the hypothalamic level in paracrine or autocrine fashion. Indeed, UCN_2 and UCN_3 mRNA are increased in the parvocellular PVN following immobilization/restraint stress [39]. A recent study demonstrated a stress-induced response that is temporally differentiated in hypophysiotropic neurons of the PVN [40]. FOS protein expression in CRF cells in the PVN after a single 30-min episode of acute restraint was slightly decreased in animals killed 2h after the stress. In contrast, non-preganglionic Edinger–Westphal (npEW) nucleus-UCN_1 neurons never responded to stressful stimuli within 1h. In CRF-overexpressing mice, npEW-UCN_1 mRNA is strongly down-regulated [41], whereas this mRNA is up-regulated in CRF-null mice [42]. It was furthermore demonstrated that the stress response of the EW-UCN_1 system does not habituate, which is in contrast with the habituating response of the hypothalamo–pituitary adrenal axis [40]. These results point to an intriguing possibility that CRF/UCN_1 neuronal circuits comprise two separate, but functionally interrelated entities, which are coordinately regulated by acute stressors, but are inversely coupled during chronic stress. Thus, it appears that npEW-UCN_1 neurons specifically respond to stress, and npEW-UCN_1 and PVN-CRF neuronal systems possess complementary dynamics that would serve to adequately terminate the central stress responses (Figure 8.1). Such collaboration between the two systems would implicate a very important role of UCN_1 in adaptation to stress and, as a consequence, in stress-related disorders like anxiety, major depression and the abuse of drugs.

8.3
The Role of CRF and CRF-Related Peptides in Behavioral Aspects of Stress

Reacting to psychological stressors requires appraisal by higher brain structures such as the cingulate cortex and the orbital/medial prefrontal cortex [43, 44]. Threat appraisal also involves subcortical structures such as the bed nucleus of the stria terminalis and the hippocampus, as well as further integration by hypothalamic

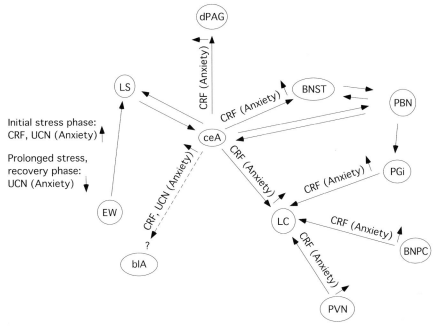

Figure 8.1 Brain region-specific contribution of the CRF system to generate stress-induced anxiety (the diagram is not intended to reflect the true anatomic localization of any structure). CRF, corticotropin-releasing factor; UCN, urocortin; LC, locus coeruleus; PGi, nucleus paragigantocellularis; PBN, parabrachial nuclei; PVN, hypothalamic paraventricular nucleus; BN PC, Barrington's nucleus and pericoerulear regions; BNST, bed nucleus of stria terminals; dPAG, dorsal periaqueductal gray; ceA, central nucleus of the amygdala; blA, basolateral amygdala; LS, lateral septum; EW, Edinger–Westphal nucleus.

and brain stem structures [45]. CRF receptors in all of these regions affect components of the stress response [46]. For example, CRF infused into the locus coeruleus in rodents intensifies anxiety-related behaviors, and neurons in the locus coeruleus are sensitized to CRF after being exposed to psychological stressors (Figure 8.1) [47]. Further evidence that brain CRF systems play an important role in the regulation of the behavioral response to stress comes from studies using competitive CRF receptor antagonists, as these antagonists attenuate both CRF- and stress-induced behavioral changes. α-Helical CRF$_{(9-41)}$, a non-selective peptide CRF receptor antagonist, decreases the CRF-enhanced acoustic startle response in rats when centrally injected [48]. Transgenic mice overexpressing CRF reveal lifelong elevations of CRF and corticosterone levels, resulting in sensorimotor gating deficits. Surprisingly, pharmacological experiments using CRF$_1$ antagonists demonstrated that sustained overactivation of CRF$_1$ rather than excessive glucocorticoid receptor stimulation underlies impaired sensorimotor gating in CRF-overexpressing mice [49]. The non-peptidergic CRF$_1$ antagonist CP-154,526 prevents stress-induced impairment in context-dependent fear conditioning and hippocampal synaptic plasticity [50]. However, CRF_1 and CRF_2 are differentially involved in the modulation of conditioned fear. Injection of human/rat (h/r)CRF into the dorsal hippocampus enhances conditioned fear by activation of CRF_1. In contrast, conditioned fear is reduced by h/rCRF acting through CRF_2 in the lateral septum to enhance anxiety-like behavior [51, 52]. Thus, depending on the brain region and the receptor subtype involved, CRF enhances or reduces conditioned fear. CRF-overproducing mice also exhibit increased exploratory behavior in the elevated plus maze following injection with α-helical $CRF_{(9-41)}$ compared with the response after receiving vehicle [53]. A second non-selective peptide CRF receptor antagonist, D-Phe-$CRF_{(12-41)}$, reduces CRF-induced increases in locomotor activation [54]. D-Phe-$CRF_{(12-41)}$ also reverses increases in CRF-induced decreases in intracranial self-stimulation reward thresholds induced by central CRF administration [55]. Astressin, a third non-selective peptide CRF receptor antagonist, decreases CRF-induced locomotor activation and closed arm exploration in the elevated plus maze in rats [56]. In primates, antalarmin treatment blocked CRF_1 and thereby dramatically decreased stress-induced behaviors such as body tremors and grimacing, while increasing exploratory behaviors normally suppressed during stress [57]. Studies that focused on the neurobiological substrates underlying stress-induced reinstatement of cocaine, heroin and alcohol-seeking behavior suggest that an increased release of CRF is a likely mediator of this behavior. It was reported that intraventricular administration of the non-specific CRF receptor antagonist d-Phe $CRF_{(12-41)}$ or systemic administration of the specific CRF_1 antagonist CP-154,526 attenuates stress-induced reinstatement of alcohol-seeking behavior [58]. In a very recent study it was also suggested that CRF receptor antagonism could be an efficacious treatment for stress-induced relapse to tobacco smoking behavior [59, 60]. Again, different roles for both CRF receptors in stress-induced relapse to drug seeking behavior can be assumed. Evidence comes from the observation that footshock-induced reinstatement of cocaine seeking can be blocked by selective blockade of ventral tegmental area CRF_2 but not CRF_1 [61].

8.4
The Role of CRF and CRF-Related Peptides in Anxiety

CRF plays an important role in the regulation of anxiety-related behavior and is implicated in anxiety and depressive disorders. Several lines of evidence point to the participation of CRF_1 in mediating the effects of CRF. First, CRF_1, but not CRF_2, binds CRF with high affinity. Second, CRF_2-deficient mice show reduced anxiety-related behavior [22]. Third, central administration of CRF_1 antisense oligodeoxynucleotides restrain CRF-evoked and social-defeat-evoked anxiety-related behaviors and elicit anxiolytic-like effects in certain anxiety tests, whereas CRF_2 antisense did not exert any significant effects in these tests [62–64]. Studies in rodents have also found supportive evidence for CRF_1 antagonists to be useful in alleviation of stress-induced maladaptive coping responses in the forced swim and learned helplessness models [65]. Treatment with the small molecule CRF_1 antagonist, antalarmin, in male rats diminished plasma elevations of HPA axis stress hormones in models of acute and chronic stress as well as decreased CRF-induced anxiety-like behaviors. *In vivo* monitoring of the CRF_1 in the living brain could soon become possible as a result of the recent accomplishments in the development of non-peptidergic CRF_1 ligands for single photon emission computed tomography (SPECT) and positron emission tomography (PET). This should reveal any changes in CRF_1 receptor expression that occur in emotional states. Although there is robust evidence that CRF_1 is highly involved in anxiety-related behavior, a role for CRF_2 cannot be excluded. However, the functional involvement of CRF_2 in the behavioral stress response is not as clear as initially anticipated and there is currently no general consensus as the results of various studies have yielded conflicting results. CRF_2 knockout mice show an anxiogenic-like phenotype when examined in the elevated plus maze, open field and light/dark emergence tests [66, 67]. In contrast, other studies report anxiogenic effects of CRF_2 agonists in the elevated plus maze and the acoustic startle test in mice [68, 69]. Central infusion of CRF_2 antisense does not produce anxiolytic-like effects [62] but rather decreases stress coping behaviors [63] and induces an anxiogenic-like response. Also, intracerebroventricular injections of antisauvagine 30, a CRF_2 receptor antagonist, produce increased anxiety-like behavior in the mouse [67]. It is critical to note that studies reporting anxiolytic effects of CRF_2 agonists tested animals under relatively low-stress conditions, in which the animals were handled by the experimenter and habituated to the testing environment. In contrast, reports of anxiogenic effects of CRF_2 activation are from subjects tested in high-stress environments, for example, during mild restraint in the acoustic startle chamber while exposed to loud noises, shock-induced freezing, or immobilization. However, the current hypothesis is that during the acute (early) phase of the stress response the increase in emotionality (anxiety) is evoked by CRF-mediated CRF_1 activation and UCN- or UCN_3-mediated CRF_2 activation, presumably in the amygdala, bed nucleus of the stria terminalis (BNST) and/or intermediate lateral septal nucleus (iLS; Figure 8.1). As part of the recovery phase CRF_2, following activation by UCN_1, UCN_2 and/ or UCN_3, participates in reducing emotionality some hours after the stressful experience. Thus, CRF_2 has a dual mode of action on anxiety-related behavior. Interestingly, in some systems – for instance, the suppression of feeding behavior – activation of CRF_2 has an additive effect with the CRF_1-mediated orexigenic effect. Central UCN_1 administration shares many neurochemical and behavioral properties of CRF treatment, reflecting their pharmacological similarity. These include behavioral arousing properties in familiar environments and proconvulsant and anxiogenic-like effects [70]. The anxiogenic-like properties of central UCN_1 infusion, mediated at least partly by CRF_1, have been shown in several paradigms, including the open field [71], elevated plus maze [72], light/dark box [71], defensive withdrawal [73] and social interaction tests [74]. Unlike CRF_1 agonists, i.c.v. administration of type 2 UCNs does not consistently have anxiogenic-like effects. Rather, UCN_2 had delayed, anxiolytic-like effects under high baseline anxiety conditions in the plus maze [75, 76] and opposed the anxiogenic-like effects of CRF in the open field [77]. Finally, whereas CRF_1 agonists increased activity in familiar environments, type 2 UCNs had mild motor suppressing effects and opposed the activating effects of CRF [72]. Currently, the endogenous anxiety-related roles of UCNs remain unclear as one UCN_1-deficient mouse model exhibited normal anxiety-like behavior and autonomic responses to stress [38], whereas another UCN_1-deficient mouse model showed increased anxiety-like behavior in the plus maze and open field tests [37]. Recently generated UCN_2-deficient mice also did not exhibit altered anxiety-like behavior in the plus maze, light/dark box or conditioned fear tests [78].

8.5
Crosstalk Between the CRF and Serotonin Systems

The behavioral effects induced by CRF seem to be mediated, in part, by CRF actions on 5-HT systems within the brain. The intersection of CRF pathways with serotonin (5-HT) neurotransmission is a likely site where stress influences mood. Electrophysiological, biochemical and anatomical localization studies have shown direct CRF input to the 5-HT producing dorsal raphe nucleus (DRN). Recent studies demonstrated that relatively low doses of CRF (administered i.c.v.) inhibited 5-HT release in two different terminal regions of the DRN (i.e., striatum, lateral septum), suggesting that the site of action of CRF effects on 5-HT release is at the level of the cell bodies in the DRN [79, 80]. Both CRF_1 and CRF_2 have been detected in the raphe and proposed to have opposing roles for 5-HT release. As CRF has a 10-fold higher affinity for CRF_1 than CRF_2, low doses of CRF in the raphe could preferentially activate CRF_1, where higher doses could potentially stimulate neurons expressing both receptors. One can hypothesize that activation of CRF_1 inhibits 5-HT release while activation of CRF_2 may augment its release [81, 82]. A putative influence of CRF/urocortin on serotonergic neurotransmission is underscored by the results of several neuroanatomical studies. Moreover, receptor binding and *in situ* hybridization studies have forwarded evidence for the presence of CRF receptors in serotonergic cell body regions. More recently, evidence has shown a direct effect of antalarmin blockade of CRF_1 in the DR, preventing swim stress-elicited c-fos expression in the dorsolateral dorsal raphe [83]. As both swim stress and CRF decrease 5-HT release, these results suggest that certain stresses engage CRF inputs to the raphe that function to inhibit 5-HT neurons and release. The immunohistochemical co-localization of CRF with either $vGlut_1$ or $vGlut_2$ suggests that the co-release of CRF and glutamate may function to regulate post-synaptic targets in the DRN [84]. Infusion of CRF into the dorsal raphé nucleus has dose-dependent effects on extracellular 5-HT levels in the nucleus accumbens. Infusion of 100 ng CRF into the dorsal raphé decreases extracellular 5-HT concentrations in the nucleus accumbens. In contrast, a higher dose of 500 ng CRF infused into the dorsal raphé resulted in an increase in accumbal 5-HT [85]. Furthermore, the results suggest that the decrease in accumbal 5-HT release induced by 100 ng CRF is dependent on CRF_1 in the dorsal raphé nucleus, while the increase in 5-HT release in the nucleus accumbens following 500 ng CRF infusions into the dorsal raphé appears to be mediated by CRF_2. A decrease in serotonergic activity in the nucleus accumbens has been associated with increased aggression and impulsivity [86, 87]. In contrast, an increase in serotonergic activity in the nucleus accumbens is associated with stressful conditions and showed a significant increase with helpless behavior [88]. Therefore, CRF effects on the raphé serotonergic output to nucleus accumbens has the potential for modifying a suite of behaviors that include the development of learned helplessness after stress exposure.

8.6
Future Directions

The two prominent CRF receptors, CRF_1 and CRF_2, tend to mediate opposing actions. CRF_1 appears to mediate many of the anxiety-related actions of CRF, while CRF_2 mediates more of the stress effects on vegetative functions. Consistent with this distinction, CRF_1 is more abundant in cortical–limbic pathways that mediate fear and anxiety-related behaviors, whereas CRF_2 is predominantly found in sub-cortical brain regions [89, 90]. It is unfortunate for students of human development that CRF cannot be non-invasively measured. Furthermore, although CRF can be assayed in samples of cerebral spinal fluid (CSF), CSF concentrations do not allow differentiation of the brain locus of production.

In order to better understand the mediating role of CRF between stress and anxiety and to potentially target the CRF system for therapeutic interventions future studies have to determine the specific expression pattern of newly identified CRF receptor splice variants in native and stressed tissue and reveal their signal transduction efficacy under *in vivo* conditions. Further emphasis should be put on the characterization of differential brain region-specific CRF receptor expression levels, different turn-on conditions of parallel pathways that crosstalk, or alternative coupling between CRF receptors and different G proteins, as was recently found [91]. An unresolved issue is whether CRF receptors exert their behavioral effects by acting through signaling pathways associated with scaffold proteins. Such a setup would facilitate spatial and temporal control of specific interactions between CRF receptors and other receptor systems by means of protein–protein interactions and, at the same time, could influence the conformation of specific arrays of signaling components downstream of CRF and other partner receptors. The complex interactions among all of these mechanisms are highly likely to result in a cell-specific final outcome of the CRF-induced activation of intracellular pathways and, if adaptation fails, to ultimately contribute to stress-induced anxiety disorders.

Summary

With the characterization of the 41-amino-acid CRF and other peptide members of the mammalian CRF family, urocortins 1–3 and the cloning of CRF(1) and

CRF(2) receptors, which display distinct affinity for CRF ligands, combined with the development of selective CRF receptor antagonists, it has become possible to unravel the importance of the CRF(1) receptor in stress-related endocrine (activation of pituitary–adrenal axis), behavioral (anxiety/depression, altered feeding), autonomic (activation of sympathetic nervous system) and immune responses. Particular attention is paid to CRF and serotonin which, closely linked to stress, neuroendocrine regulation, social behavior and learning/memory, play critical roles in the regulation of anxiety-related behavior.

References

1 Vale, W., Spiess, J. and Rivier, C. (1981) Characterization of a 41-residue ovine hypothalamic peptide that stimulates secretion of corticotropin and beta-endorphin. *Science*, **213**, 1394–1397.

2 Spiess, J., Rivier, J., Rivier, C. and Vale, W. (1981) Primary structure of corticotropin-releasing factor from ovine hypothalamus. *Proc. Natl. Acad. Sci. U.S.A.*, **78**, 6517–6521.

3 Hauger, R.L. *et al.* (2003) International union of pharmacology. XXXVI. Current status of the nomenclature for receptors for corticotropin-releasing factor and their ligands. *Pharmacol. Rev.*, **55**, 21–26.

4 Montecucchi, P.C., Anastasi, A., de Castiglione, R. and Erspamer, V. (1980) Isolation and amino acid composition of sauvagine. An active polypeptide from methanol extracts of the skin of the South American frog phyllomedusa. *Int. J. Peptide Protein Res.*, **16**, 191–199.

5 Lovejoy, D.A. and Balment, R.J. (1999) Evolution and physiology of the corticotropin-releasing factor (CRF) family of neuropeptides in vertebrates. *Gen. Comp. Endocrinol.*, **115**, 1–22.

6 Lederis, K., Vale, W.W., Rivier, J.E., MacCannell, K.L., McMaster, D., Kobayashi, Y., Suess, U. and Lawrence, J. (1982) Urotensin I – a novel CRF-like peptide in catostomus commersoni urophysis. *Proc. West. Pharmacol. Soc.*, **25**, 223–227.

7 Arai, M., Assil, I.Q. and Abou-Samra, A.B. (2001) Characterization of three corticotropin- releasing factor receptors in catfish: a novel third receptor is predominantly expressed in pituitary and urophysis. *Endocrinology*, **142**, 446–454.

8 Merchenthaler, I. (1984) Corticotropin releasing factor (CRF)-like immunoreactivity in the rat central nervous system. Extrahypothalamic distribution. *Peptides*, **1**, 53–69.

9 Dautzenberg, F.M. and Hauger, R.L. (2002) The CRF peptide family and their receptors, yet more partners discovered. *Trends Pharmacol. Sci.*, **23**, 71–77.

10 Hillhouse, E.W. and Grammatopoulos, D.K. (2006) The molecular mechanisms underlying the regulation of the biological activity of corticotropin-releasing hormone receptors: implications for physiology and pathophysiology. *Endocr. Rev.*, **27**, 260–286.

11 Lovenberg, T.W., Chalmers, D.T., Liu, C.G. and Desouza, E.B. (1995) $CRF_{2\alpha}$ and $CRF_{2\beta}$ receptor mRNAs are differentially distributed between the rat central nervous system and peripheral tissues. *Endocrinology*, **136**, 4139–4142.

12 Catalano, R.D., Kyriakou, T., Chen, J., Easton, A. and Hillhouse, E.W. (2003) Regulation of corticotropin-releasing hormone type 2 receptors by multiple promoters and alternative splicing: identification of multiple splice variants. *Mol. Endocrinol.*, **17**, 395–410.

13 Chen, A.M. *et al.* (2005) A soluble mouse brain splice variant of type 2alpha corticotropin- releasing factor (CRF) receptor binds ligands and modulates their activity. *Proc. Natl. Acad. Sci. U. S. A.*, **102**, 2620–2625.

14 Lewis, K., Li, C., Perrin, M.H., Blount, A., Kunitake, K., Donaldson, C., Vaughan, J.,

Reyes, T.M. *et al.* (2001) Identification of urocortin III, an additional member of the corticotropin-releasing factor (CRF) family with high affinity for the CRF2 receptor. *Proc. Natl. Acad. Sci. U. S. A.*, **98**, 7570–7575.

15 Reyes, T.M., Lewis, K., Perrin, M.H., Kunitake, K.S., Vaughan, J., Arias, C.A., Hogenesch, J.B. *et al.* (2001) Urocortin II: a member of the corticotropin-releasing factor (CRF) neuropeptide family that is selectively bound by type 2 CRF receptors. *Proc. Natl. Acad. Sci. U. S. A.*, **98**, 2843–2848.

16 Hsu, S.Y. and Hsueh, A.J. (2001) Human stresscopin and stresscopin-related peptide are selective ligands for the type 2 corticotropin-releasing hormone receptor. *Nat. Med.*, **7**, 605–611.

17 Cortright, D.N., Goosens, K.A., Lesh, J.S. and Seasholtz, A.F. (1997) Isolation and characterization of the rat corticotropin-releasing hormone (CRH)-binding protein gene: transcriptional regulation by cyclic adenosine monophosphate and CRH. *Endocrinology*, **138**, 2098–2108.

18 Behan, D.P., Heinrichs, S.C., Troncoso, J.C., Liu, X.J., Kawas, C.H., Ling, N. and De Souza, E.B. (1995) Displacement of corticotropin releasing factor from its binding protein as a possible treatment for Alzheimer's disease. *Nature*, **378**, 284–287.

19 Seasholtz, A.F., Burrows, H.L., Karolyi, I.J. and Camper, S.A. (2001) Mouse models of altered CRH-binding protein expression. *Peptides*, **22**, 743–751.

20 Chatzaki, E., Margioris, A.N. and Gravanis, A. (2002) Expression and regulation of corticotropin- releasing hormone binding protein (CRH-BP) in rat adrenals. *J. Neurochem.*, **80**, 81–90.

21 Baigent, S.M. and Lowry, P.J. (2000) MRNA expression profiles for corticotrophin-releasing factor (CRF), urocortin, CRF receptors and CRF-binding protein in peripheral rat tissues. *J. Mol. Endocrinol.*, **25**, 43–52.

22 Timpl, P., Spanagel, R., Sillaber, I., Kresse, A., Reul, J.M.H.M., Stalla, G.K., Blanquet, V., Steckler, T., Holsboer, F. and Wurst, W. (1998) Impaired stress response and reduced anxiety in mice lacking a functional corticotropin-

releasing hormone receptor 1. *Nat. Genet.*, **19**, 162–166.

23 Smith, G.W., Aubry, J.-M., Dellu, F., Contarino, A., Bilezikijan, L.M., Gold, L.H., Chen, R., Marchuk, Y. *et al.* (1998) Corticotropin releasing factor receptor 1-deficient mice display decreased anxiety, impaired stress response, and aberrant neuroendocrine development. *Neuron*, **20**, 1093–1102.

24 Bilezikjian, L.M. and Vale, W.W. (1987) Regulation of ACTH secretion from corticotrophs: the interaction of vasopressin and CRF. *Ann. N. Y. Acad. Sci.*, **512**, 85–96.

25 DeBold, C.R., Sheldon, W.R., DeCherney, G.S., Jackson, R.V., Alexander, A.N., Vale, W., Rivier, J. and Orth, D.N. (1984) Arginine vasopressin potentiates adrenocorticotropin release induced by ovine corticotropin-releasing factor. *J. Clin. Invest.*, **73**, 533–538.

26 Liu, J.H., Muse, K., Contreras, P., Gibbs, D., Vale, W., Rivier, J. and Yen, S.S. (1983) Augmentation of ACTH-releasing activity of synthetic corticotropin releasing factor (CRF) by vasopressin in women. *J. Clin. Endocrinol. Metab.*, **57**, 1087–1089.

27 Rivier, C., Rivier, J., Mormede, P. and Vale, W. (1984) Studies of the nature of the interaction between vasopressin and corticotropin-releasing factor on adrenocorticotropin release in the rat. *Endocrinology*, **115**, 882–886.

28 Altemus, M., Smith, M.A., Diep, V., Aulakh, C.S. and Murphy, D.L. (1994) Increased mRNA for corticotrophin releasing hormone in the amygdala of fawn-hooded rats: a potential animal model of anxiety. *Anxiety*, **6**, 251–257.

29 Schulkin, J., Gold, P.W. and McEwen, B.S. (1998) Induction of corticotropin-releasing hormone gene expression by glucocorticoids: implication for understanding the states of fear and anxiety and allostatic load. *Psychoneuroendocrinology*, **23**, 219–243.

30 Adamec, R.E. (2003) Stress effects on limbic function and behavior. *Prog. Neuropsychopharmacol. Biol. Psychiatry.*, **27**, 1173–1175.

31 Adamec, R.E. and McKay, D. (1993) Amygdala kindling, anxiety, and

corticotrophin releasing factor (CRF). *Physiol. Behav.*, **54**, 423–431.

32 Murakami, K., Akana, S.F. and Dallman, M.F. (1997) Dopamine-beta-hydroxylase activity is necessary for hypothalamo-pituitary-adrenal (HPA) responses to ether, and stress-induced facilitation of subsequent HPA responses to acute ether emerges as HPA responses are inhibited by increasing corticosterone (B). *J. Neuroendocrinol.*, **9**, 601–608

33 Young, E.A., Akana, S. and Dallman, M.F. (1990) Decreased sensitivity to gluco-corticoid fast feedback in chronically stressed rats. *Neuroendocrinology*, **51**, 536–542.

34 Herman, J.P., Schafer, K.H., Sladek, C.D., Day, R., Young, E.A., Akil, H. and Watson, S.J. (1989) Chronic electro-convulsive shock treatment elicits up-regulation of CRF and AVP mRNA in select populations of neuroendocrine neurons. *Brain Res.*, **501**, 235–246.

35 Herman, J.P., Schafer, M.K., Young, E.A., Thompson, R., Douglass, J., Akil, H. and Watson, S.J. (1989) Evidence for hippo-campal regulation of neuroendocrine neurons of the hypothalamo-pituitary-adrenocortical axis. *J. Neurosci.*, **9**, 3072–3082.

36 Vaughan, J., Donaldson, C., Bittencourt, J., Perrin, M.H., Lewis, K., Sutton, S., Chan, R., Turnbull, A.V. *et al.* (1995) Urocortin, a mammalian neuropeptide related to fish urotensin I and to corticotropin-releasing factor. *Nature*, **378**, 287–292.

37 Vetter, D.E., Li, C., Zhao, L., Contarino, A., Liberman, M.C., Smith, G.W., Marchuk, Y., Koob, G.F., Heinemann, S.F., Vale, W. and Lee, K.F. (2002) Urocortin-deficient mice show hearing impairment and increased anxiety-like behavior. *Nat. Genet.*, **31**, 363–369.

38 Wang, X., Su, H., Copenhagen, L.D., Vaishnav, S., Pieri, F., Shope, C.D., Brownell, W.E., De Biasi, M., Paylor, R. and Bradley, A. (2002) Urocortin-deficient mice display normal stress- induced anxiety behavior and autonomic control but an impaired acoustic startle response. *Mol. Cell. Biol.*, **22**, 6605–6610.

39 Tanaka, Y., Makino, S., Noguchi, T., Tamura, K., Kaneda, T. and

Hashimoto, K. (2003) Effect of stress and adrenalectomy on urocortin II mRNA expression in the hypothalamic paraventricular nucleus of the rat. *Neuroendocrinology*, **78**, 1–11.

40 Viau, V. and Sawchenko, P.E. (2002) Hypophysiotropic neurons of the paraventricular nucleus respond in spatially, temporally, and phenotypically differentiated manners to acute vs. repeated restraint stress: rapid publication. *J. Comp. Neurol.*, **445**, 293–307.

41 Kozicz, T., Korosi, A., Korsman, C., Tilburg-Ouwens, D., Groenink, L., Veening, J., van der Gugten, G.J., Roubos, E. and Olivier, B. (2004) Urocortin expression in the Edinger-Westphal nucleus is down-regulated in transgenic mice over-expressing neuronal corticotropin-releasing factor. *Neuroscience*, **123**, 589–594.

42 Weninger, S.C., Peters, L.L. and Majzoub, J.A. (2000) Urocortin expression in the Edinger-Westphal nucleus is up-regulated by stress and corticotropin-releasing hormone deficiency. *Endocrinology*, **141**, 256–263.

43 Barbas, H. (1995) Anatomic basis of cognitive-emotional interactions in the primate prefrontal cortex. *Neurosci. Biobehav. Rev.*, **19**, 499–510.

44 Diorio, D., Viau, V. and Meaney, M.J. (1993) The role of the medial prefrontal cortex (cingulate gyrus) in the regulation of hypothalamic-pituitary-adrenal responses to stress. *J. Neurosci.*, **13**, 3839–3847.

45 Davis, M., Walker, D.L. and Lee, Y. (1997) Roles of the amygdala and bed nucleus of the stria terminalis in fear and anxiety measured with the acoustic startle reflex. *Ann. N. Y. Acad. Sci.*, **821**, 305–331.

46 Bale, T.L. and Vale, W.W. (2004) CRF and CRF receptors: role in stress responsivity and other behaviors. *Annu. Rev. Pharmacol. Toxicol.*, **44**, 525–557.

47 Butler, P.D., Weiss, J.M., Stout, J.C. and Nemeroff, C.B. (1990) Corticotropin-releasing factor produces fear-enhancing and behavioral activating effects following infusion into the locus coeruleus. *J. Neurosci.*, **10**, 176–183.

48 Swerdlow, N.R., Britton, K.T. and Koob, G.F. (1989) Potentiation of acoustic startle by corticotropin-releasing factor (CRF) and by fear are both reversed by alpha-helical CRF (9-41). *Neuropsychopharmacology*, **2**, 285–292.

49 Groenink, L., Dirks, A., Verdouw, P.M., de Graaff, M., Peeters, B.W., Millan, M.J. and Olivier, B. (2008) CRF(1) not glucocorticoid receptors mediate prepulse inhibition deficits in mice overexpressing CRF. *Biol. Psychiatry.*, **63**, 360–368.

50 Blank, T., Nijholt, I., Vollstaedt, S. and Spiess, J. (2003) The corticotropin-releasing factor receptor 1 antagonist CP-154,526 markedly enhances associative learning and paired-pulse facilitation immediately after a stressful experience. *Behav. Brain Res.*, **138**, 207–213.

51 Radulovic, J., Ruhmann, A., Liepold, T. and Spiess, J. (1999) Modulation of learning and anxiety by corticotropin-releasing factor (CRF) and stress: differential roles of CRF receptors 1 and 2. *J. Neurosci.*, **19**, 5016–5025.

52 Todorovic, C., Radulovic, J., Jahn, O., Radulovic, M., Sherrin, T., Hippel, C. and Spiess, J. (2007) Differential activation of CRF receptor subtypes removes stress-induced memory deficit and anxiety. *Eur. J. Neurosci.*, **25**, 3385–3397.

53 Stenzel-Poore, M.P., Heinrichs, S.C., Rivest, S. *et al.* (1994) Overproduction of corticotropin-releasing factor in transgenic mice: a genetic model of anxiogenic behavior. *J. Neurosci.*, **14**, 2579–2584.

54 Menzaghi, F., Howard, R.L., Heinrichs, S.C. *et al.* (1994) Characterization of a novel and potent corticotropin-releasing factor antagonist in rats. *J. Pharmacol. Exp. Ther.*, **269**, 564–572.

55 Macey, D.J., Koob, G.F. and Markou, A. (2000) CRF and urocortin decreased brain stimulation reward in the rat: reversal by a CRF receptor antagonist. *Brain Res.*, **866**, 82–91.

56 Spina, M.G., Basso, A.M., Zorrilla, E.P. *et al.* (2000) Behavioral effects of central administration of the novel CRF antagonist astressin in rats. *Neuropsychopharmacology*, **22**, 230–239.

57 Habib, K.E., Weld, K.P., Rice, K.C., Pushkas, J., Champoux, M., Listwak, S., Webster, E.L., Atkinson, A.J. *et al.* (2000) Oral administration of a corticotropin-releasing hormone receptor antagonist significantly attenuates behavioral, neuroendocrine, and autonomic responses to stress in primates. *Proc. Natl. Acad. Sci. U. S. A.*, **97**, 6079–6084.

58 Le, A.D., Harding, S., Juzytsch, W., Watchus, J., Shalev, U. and Shaham, Y. (2000) The role of corticotrophin-releasing factor in stress-induced relapse to alcohol-seeking behavior in rats. *Psychopharmacology (Berl.)*, **150**, 317–324.

59 Zislis, G., Desai, T.V., Prado, M., Shah, H.P. and Bruijnzeel, A.W. (2007) Effects of the CRF receptor antagonist d-Phe CRF((12-41)) and the alpha2-adrenergic receptor agonist clonidine on stress-induced reinstatement of nicotine-seeking behavior in rats. *Neuropharmacology*, **53**, 958–966.

60 George, O., Ghozland, S., Azar, M.R., Cottone, P., Zorrilla, E.P., Parsons, L.H., O'Dell, L.E., Richardson, H.N. and Koob, G.F. (2007) CRF-CRF$_1$ system activation mediates withdrawal-induced increases in nicotine self-administration in nicotine-dependent rats. *Proc. Natl. Acad. Sci. U. S. A.*, **104**, 17198–17203.

61 Wang, B., You, Z.B., Rice, K.C. and Wise, R.A. (2007) Stress-induced relapse to cocaine seeking: roles for the CRF(2) receptor and CRF-binding protein in the ventral tegmental area of the rat. *Psychopharmacology (Berl.)*, **193**, 283–294.

62 Heinrichs, S.C., Lapansky, J., Lovenberg, T.W., De Souza, E.B. and Chalmers, D.T. (1997) Corticotropin-releasing factor CRF1, but not CRF2, receptors mediate anxiogenic-like behavior. *Regul. Pept.*, **71**, 15–21.

63 Liebsch, G., Landgraf, R., Engelmann, M., Lörscher, P. and Holsboer, F. (1999) Differential behavioural effects of chronic infusion of CRH 1 and CRH 2 receptor antisense oligonucleotides into the rat brain. *J. Psychiatr. Res.*, **33**, 153–163.

64 Liebsch, G., Landgraf, R., Gerstberger, R., Probst, J.C., Wotjak, C.T., Engelmann, M., Holsboer, F. and Montkowski, A. (1995) Chronic infusion of a CRH1 receptor antisense oligodeoxynucleotide into the central nucleus of the amygdala reduced anxiety-related behavior in socially defeated rats. *Regul. Pept.*, **59**, 229–239.

65 Arborelius, L., Skelton, K.H., Thrivikraman, K.V., Plotsky, P.M., Schulz, D.W. and Owens, M.J. (2000) Chronic administration of the selective corticotropin-releasing factor 1 receptor antagonist CP-154,526: behavioral, endocrine and neurochemical effects in the rat. *J. Pharmacol. Exp. Ther.*, **294**, 588–597.

66 Bale, T.L., Contarino, A., Smith, G.W. *et al.* (2000) Mice deficient for evidence corticotropin- releasing hormone receptor-2 display anxiety-like behaviour and are hypersensitive to stress. *Nat. Genet.*, **24**, 410–414.

67 Kishimoto, T., Radulovic, J., Radulovic, M. *et al.* (2000) Deletion of crhr2 reveals an anxiolytic role for corticotropin- releasing hormone receptor-2. *Nat. Genet.*, **24**, 415–419.

68 Pelleymounter, M.A., Joppa, M., Ling, N. and Foster, A.C. (2002) Pharmacological evidence supporting a role for central corticotropin-releasing factor(2) receptors in behavioral, but not endocrine, response to environmental stress. *J. Pharmacol. Exp. Ther.*, **302**, 145–152.

69 Pelleymounter, M.A., Joppa, M., Ling, N. and Foster, A.C. (2004) Behavioral and neuroendocrine effects of the selective CRF2 receptor agonists urocortin II and urocortin III. *Peptides*, **25**, 659–666.

70 Zorrilla, E.P. and Koob, G.F. (2005) The roles of urocortins 1, 2 and 3 in the brain, in *Handbook of Stress and the Brain, Techniques in the Behavioral and Neural Sciences*, Vol. **15** (eds T. Steckler, N.H. Kalin and J.M.H.M. Reul), Elsevier Science, New York, pp. 179–203.

71 Moreau, J.L., Kilpatrick, G. and Jenck, F. (1997) Urocortin, a novel neuropeptide with anxiogenic-like properties. *Neuroreport*, **8**, 1697–1701.

72 Jones, D.N., Kortekaas, R., Slade, P.D., Middlemiss, D.N. and Hagan, J.J. (1998) The behavioural effects of corticotropin-releasing factor-related peptides in rats. *Psychopharmacology*, **138**, 124–132.

73 Spina, M.G., Merlo-Pich, E., Akwa, Y., Balducci, C., Basso, A.M., Zorrilla, E.P., Britton, K.T., Rivier, J. *et al.* (2002) Time-dependent induction of anxiogenic-like effects after central infusion of urocortin or corticotropin-releasing factor in the rat. *Psychopharmacology*, **160**, 113–121.

74 Sajdyk, T.J. and Gehlert, D.R. (2000) Astressin, a corticotropin releasing factor antagonist, reverses the anxiogenic effects of urocortin when administered into the basolateral amygdala. *Brain Res.*, **877**, 226–234.

75 Valdez, G.R., Inoue, K., Koob, G.F., Rivier, J., Vale, W.W. and Zorrilla, E.P. (2002) Human urocortin II: mild locomotor suppressive and delayed anxiolytic-like effects of a novel corticotropin-releasing factor related peptide. *Brain Res.*, **943**, 142–150.

76 Valdez, G.R., Zorrilla, E.P., Rivier, J., Vale, W.W. and Koob, G.F. (2003) Locomotor suppressive and anxiolytic-like effects of urocortin 3, a highly selective type 2 corticotropin-releasing factor agonist. *Brain Res.*, **980**, 206–212.

77 Venihaki, M., Sakihara, S., Subramanian, S., Dikkes, P., Weninger, S.C., Liapakis, G., Graf, T. and Majzoub, J.A. (2004) Urocortin, III a brain neuropeptide of the corticotropin- releasing hormone family: modulation by stress and attenuation of some anxiety-like behaviours. *J. Neuroendocrinol.*, **16**, 411–422.

78 Chen, A., Zorrilla, E., Smith, S., Rousso, D., Levy, C., Vaughan, J., Donaldson, C., Roberts, A., Lee, K.F. and Vale, W. (2006) Urocortin 2-deficient mice exhibit gender-specific alterations in circadian hypothalamus-pituitary-adrenal axis and depressive-like behaviour. *J. Neurosci.*, **26**, 5500–5510.

79 Price, M.L., Curtis, A.L., Kirby, L.G., Valentino, R.J. and Lucki, I. (1998) Effects of corticotropin-releasing factor on brain serotonergic activity. *Neuropsychopharmacology*, **18**, 492–502.

80 Price, M.L. and Lucki, I. (2001) Regulation of serotonin release in the lateral septum and striatum by corticotropin-releasing factor. *J. Neurosci.*, **21**, 2833–2841.

81 Hammack, S.E., Richey, K.J., Schmid, M.J., LoPresti, M.L., Watkins, L.R. and Maier, S.F. (2002) The role of corticotropin-releasing hormone in the dorsal raphe nucleus in mediating the behavioral consequences of uncontrollable stress. *J. Neurosci.*, **22**, 1020–1026.

82 Valentino, R.J., Liouterman, L. and Van Bockstaele, E.J. (2001) Evidence for regional heterogeneity in corticotropin-releasing factor interactions in the dorsal raphe nucleus. *J. Comp. Neurol.*, **435**, 450–463.

83 Roche, M., Commons, K.G., Peoples, A. and Valentino, R.J. (2003) Circuitry underlying regulation of the serotonergic system by swim stress. *J. Neurosci.*, **23**, 970–977.

84 Waselus, M. and Van Bockstaele, E.J. (2007) Co-localization of corticotropin-releasing factor and vesicular glutamate transporters within axon terminals of the rat dorsal raphe nucleus. *Brain Res.*, **1174**, 53–65.

85 Lukkes, J.L., Forster, G.L., Renner, K.J. and Summers, C.H. (2008) Corticotropin-releasing factor 1 and 2 receptors in the dorsal raphé differentially affect serotonin release in the nucleus accumbens. *Eur. J. Pharmacol.*, **578**, 185–193.

86 Haney, M., Noda, Y., Kream, R. and Miczek, K. (1990) Regional serotonin and dopamine activity; sensitivity to amphetamine and aggressive behavior in mice. *Aggress. Behav.*, **16**, 259–270.

87 Cardinal, R.N., Pennicott, D.R., Sugathapala, C.L., Robbins, T.W. and Everitt, B.J. (2001) Impulsive choice induced in rats by lesions of the nucleus accumbens core. *Science*, **292**, 2499–2501.

88 Petty, F., Kramer, G., Wilson, L. and Jordan, S. (1994) *In vivo* serotonin release and learned helplessness. *Psychiatry Res.*, **52**, 285–293.

89 Sanchez, M.M., Young, L.J., Plotsky, P.M. and Insel, T.R. (2000) Distribution of corticosteroid receptors in the rhesus brain: relative absence of glucocorticoid receptors in the hippocampal formation. *J. Neurosci.*, **20**, 4657–4668.

90 Vythilingam, M., Heim, C., Newport, J., Miller, A.H., Anderson, E. *et al.* (2002) Childhood trauma associated with smaller hippocampal volume in women with major depression. *Am. J. Psychiatry*, **159**, 2072–2080.

91 Blank, T., Nijholt, I., Grammatopoulos, D.K., Randeva, H.S., Hillhouse, E.W. and Spiess, J. (2003) Corticotropin-releasing factor receptors couple to multiple G-proteins to activate diverse intracellular signaling pathways in mouse hippocampus: role in neuronal excitability and associative learning. *J. Neurosci.*, **23**, 700–707.

9
Stress, Emotion and Memory: the Good, the Bad and the Intriguing

Marie-France Marin, Tania Elaine Schramek, Françoise S. Maheu, and Sonia J. Lupien

9.1
Introduction

Prior to becoming part of our day-to-day conversations, the term "stress" was used by engineers to explain forces that can put strain on a structure in such a way that a piece of metal can shatter like glass if its stress level is reached. In 1936, Hans Selye (reproduced in [1]) borrowed the term and viewed stress as being a non-specific phenomenon representing the intersection of symptoms produced by a wide variety of noxious agents. For many years, Selye tested various conditions (e.g., fasting, extreme cold, operative injuries, drug adminis-tration) that gave rise to the stress response and the release of stress hormones therein. These in turn could produce morphological changes in the body that were representative of a stress response (e.g., enlargement of the adrenal glands, atrophy of the thymus, gastric ulceration). Selye's view of the concept of stress was that the determinants of the stress response were non-specific, that is, many unspecific conditions can put strain on the organism and ultimately lead to disease.

Not all researchers agreed with Selye's model, particularly with the notion that the determinants of the stress response were non-specific. The reason for this was simple. While Selye spent his entire career working on physical stres-sors (e.g., heat, cold, pain), we all know that some of the worst stressors we encounter in life are psychological in nature and are potent inducers of the stress response. In addition, researchers sought to be able to reconcile the highly individual nature of stress (i.e., what is stressful for some is not for others) with the universal nature of the stress response. In other words, there must be specific characteristics of situations that render them stressful for whoever would be exposed to them. Accordingly, a psychologist named John Mason spent many years measuring stress hormone levels in humans in various jobs or conditions deemed to be stressful (e.g., air traffic controllers, parachute jumping) and summarized his results and those of others in a seminal paper published in the late 1960s.

Stress – From Molecules to Behavior. Edited by Hermona Soreq, Alon Friedman, and Daniela Kaufer
Copyright © 2010 WILEY-VCH Verlag GmbH & Co. KGaA, Weinheim
ISBN: 978-3-527-32374-6

<u>N</u>ovelty

<u>U</u>npredictability

<u>T</u>hreat to the ego

<u>S</u>ense of control decreased

Figure 9.1 In order for a situation to be perceived as stressful, it must be interpreted as either *n*ovel and/or *u*npredictable and/or *t*hreatening to the ego of the person experiencing it and/or decreasing the perceived control that the person has over the situation. Interestingly enough, the more characteristics present in a given situation, the more stressful the experience. A user-friendly acronym that one can use in order to remember those characteristics is N.U.T.S.

Mason was able to describe three main psychological determinants that when present within a situation would give rise to a stress response in a majority of individuals. He showed that stress hormone release occurred when a situation is interpreted as being either novel and/or unpredictable and/or decreases the individual's sense of control over the situation [2]. Although this work led to a series of debates between Selye and Mason, further studies confirmed that the determinants of the stress response were indeed highly specific and, therefore, were potentially predictable and measurable. Since then, we have come to know that socio-evaluative threat (i.e., ego involvement) is also a potent inducer of the stress response in humans (Figure 9.1) [3].

9.2
The Relativity of Stress

From this short historical background, we can now define stress as being a threat, real or implied, to the physical (i.e., homeostasis) or psychological integrity of an individual. In this sense, stress can be absolute (a real threat induced by an earthquake for instance, leading to a significant stress response in every person facing this threat), or it can be relative (an implied threat induced by the interpretation of a situation as being novel and/or unpredictable and/or uncontrollable and/or threatening for the ego; e.g., a public speaking task).

Absolute stressors serve adaptive purposes and are those events or situations that necessarily lead to a stress response in the majority (if not the totality) of individuals when they are first confronted with it. Examples of absolute stressors include being in or witnessing an accident, confronting a dangerous animal, being submitted to extreme cold or heat. These extreme and particular situations constitute absolute stressors in that, due to their aversive nature, a stress response has to be elicited for one's survival and/or well-being. In westernized societies, absolute stressors are rare, but are nonetheless those that elicit the greatest physiological response and quite often vivid memories [4]. Conversely, relative stressors are those events or situations that elicit a stress response only in a certain proportion of individuals depending on their interpretation of a situation as being novel and/or unpredictable and/or threatening to their ego and/or decreasing their sense of control. As such, there are considerable individual differences with respect to the amount of stress hormones released in the face of a psychological challenge. For example, having to unexpectedly deliver a speech may be very stressful for a given individual and not at all for another – it all depends of the interpretation of the situation.

9.3
Important Characteristics of Stress Hormones

When a situation is interpreted as being stressful, it triggers the activation of the sympathetic–adrenal–medullary and the hypothalamic–pituitary–adrenal (HPA) axes. The former, under sympathetic control, results in the release of catecholamine (epinephrine and norepinephrine, also known as adrenalin and noradrenalin) by the adrenal glands that sit atop the kidneys; the later results in the secretion of a class of steroid hormones known as the glucocorticoids (GCs). More specifically, neurons in the hypothalamus release a hormone called corticotropin-releasing hormone (CRH), which triggers the subsequent secretion and release of a polypeptide hormone called adrenocorticotropin (ACTH) from the pituitary gland. The ACTH then leaves the central nervous system via the bloodstream and binds to its receptors on the adrenal glands, which in turn releases the GCs. This top-down control of adrenocortical function is complemented by negative feedback mechanisms exerted by GCs through receptors at the pituitary and hypothalamic levels and at the level of the hippocampus [5]. Thus there are two main classes of stress hormones released by the adrenals, the catecholamines and the GCs (corticosterone in animals, cortisol in humans).

Stress hormones have a variety of different effects in target systems throughout the organism, which can be summarized as aiming to increase the availability of energy substrates in different parts of the body and allowing for optimal adaptation to the changing demands of the environment. In humans, under basal conditions, GC secretion exhibits a 24-h circadian profile in which GC concentrations are at the highest in the morning (the circadian peak) and slowly decline over the course of the day reaching their lowest levels in late afternoon, evening and nocturnal period (the circadian trough) and show an abrupt elevation after the first few hours of sleep [6].

9.3.1
Stress Hormone Receptors

Circulating GCs bind with high affinity to two receptor subtypes; the mineralocorticoid (MR or type I) and GC (GR or type II) receptors. Although both receptor types have been implicated in mediating GC feedback effects [7], there are two major differences between MR and GR receptors. First, MRs bind GCs with an affinity that is about six to ten times higher than that of GRs [8]. This differential affinity results in a striking difference in occupation of the two receptor types under different conditions and time of day. Thus, during the circadian trough (the PM phase in humans, the AM phase in rats), the endogenous hormone occupies more than 90% of MRs, but only 10% of GRs. However, during stress and/or the circadian peak of GC secretion (the AM phase in humans and the PM phase in rats), MRs become saturated and there is approximately 67–74% occupation of GRs [9].

Given that they are liposoluble, the GCs can easily cross the blood–brain barrier and access the brain. This brings us to the second major difference between the two GC receptor types, namely, their distribution in the brain. The MR is exclusively present in the limbic system, with a preferential distribution in the hippocampus, parahippocampal gyrus, entorhinal and insular cortices. The GR on the other hand, is present in both subcortical (paraventricular nucleus, other hypothalamic nuclei, hippocampus, parahippocampal gyrus) and cortical structures, with a preferential distribution in the prefrontal cortex [8, 10]. As we can see in the following sections, the impact of GCs on cognitive function can be best understood in terms of the differential effects of MR and GR activation within these brain regions.

With respect to the catecholamines, while peripheral epinephrine is not lipophilic and therefore cannot cross the blood–brain barrier, it nonetheless impacts the brain by activating β-adrenergic receptors on vagal nerve afferents that terminate in the nucleus of the solitary tract, which either project directly to adrenergic neurons in the amygdala or indirectly by increasing basal noradrenergic firing in the locus coeruleus. This in turn results in increased norepinephrine output in the amygdala [11, 12] by far, the most important brain area for the processing of fear- and emotion-laden information [13].

9.3.2
The Impact of Stress Hormones on Cognition

The presence of stress hormone receptors in the hippocampus, amygdala and prefrontal cortex is interesting in light of their critical roles in multiple cognitive processes, particularly in learning and memory [14, 15]. The prefrontal cortex is involved in higher cognitive functions such as reasoning, planning, selective attention, working memory and some aspects of language [16, 17]. Importantly, the prefrontal cortex is the crucial instance that determines which information will be considered and which information will be inhibited during the performance of goal-directed behaviors (e.g., selective attention [18]). The amygdala is critically involved in assessing the emotional significance of events [19]. It is especially important for the processing of the emotions of anger, aggression and fear. The hippocampal formation is involved in learning and memory. It specifically subserves the formation and retrieval of episodic memories, that is, the human ability to travel back in time and remember where (spatial component) and when (temporal component) a specific event occurred [20]. During the acquisition or encoding of new material the hippocampal formation is engaged in relational learning mechanisms that bind the different components of an event into an integrated memory trace [21]. Roughly speaking, the described brain structures are critically involved in evaluating a situation and selecting an appropriate response (prefrontal cortex) with regard to the emotional content of the situation (amygdala) and in relation to past experiences (hippocampal formation). The presence of stress hormone receptors in these brain areas lead researchers to ask whether stress hormones also played a role in learning and memory.

9.4
Effects of GCs on Learning and Memory

Indeed, direct cause–effect relationships between learning, memory and GCs have been demonstrated in both animals and humans. Most studies in rodents examining these relationships have done so using tasks that involve either associative learning (e.g., contextual fear conditioning where the animal must remember the context in which it received a shock) or spatial memory (e.g., water maze in which the animals must learn the location of a hidden platform that will allow them to escape the water by using spatial cues). Common approaches of manipulating GC levels in rodents include the use of receptor agonists and antagonists, GC synthesis inhibitors and/or exposure to an acute stressor that represents a relevant threat to the animal.

In humans, measuring the acute effects of GCs on memory processing typically consists of administering pharmacological agents that either increase or decrease circulating levels of GCs before learning or before retention testing. Manipulations effected before or shortly after learning tend to affect acquisition and consolidation processes while those effected before recalling influence the retrieval of the newly acquired material.

It is important to note that the consolidation of material in memory occurs in two phases, the first known as the early phase takes place in the first 3h after encoding. The second or late phase begins about three hours post-learning and can continue for up to 24h, owing to the fact that it involves gene transcription and protein synthesis [22]. Thus, memory tested within the first 24h of an event (or stimuli) is assessing events surrounding encoding and consolidation processes while retention testing occurring after 24h is assessing long-term memory. This brings to light the importance of taking into account the time-course for GC actions when manipulating GC levels in memory tasks. If exogenous GCs are administered (or if stress is induced) immediately before encoding then it is likely that encoding will not be impacted upon, given that GCs will not have the time to access the brain. Instead, GCs will likely be acting upon consolidation processes. However, if there is a delay (fitting the pharmacokinetics of the compound used for instance) between GC manipulation and encoding, then GCs are more likely to exert their effect at the level of encoding. Finally, GC manipulation effected prior to retention testing is likely to affect retrieval processes.

It is also possible to induce endogenous increases in GC secretion using laboratory-based psychological stressors. By far the most commonly used and most reliable at inducing an increase in GC levels is the Trier Social Stress Test (TSST), in which subjects have to deliver a free speech and perform mental arithmetic in front of an audience [23]. Subjects can be exposed to the "to be remembered" information before TSST administration or undergo the TSST and then go onto encoding. In both types of studies, experimenters must verify the efficacy of their drug/psychological manipulations by measuring GC levels in either the blood or saliva. The latter is most commonly used, as it is non-invasive, reliable and provides a measure of the free unbound fraction of GCs, that is, the bioactive form that can cross the blood–brain barrier and affect the brain regions that subserve learning and memory.

9.4.1
Direct Effects of GCs

In the water maze, when circulating levels of endogenous GCs are acutely elevated with the addition of exogenous GCs prior to learning, increased latency to find the platform and altered swim patterns, and thus impaired performance is observed. Removal of the adrenal glands and therefore the source of GCs also results in memory impairments in the water maze [24, 25]. Similarly, deficits in contextual fear conditioning are observed after the central administration of selective GR receptor antagonists such that when a rat is placed in the context where it received a shock, it displays less of the characteristic freezing response, which shows that the animal has poor retention of the shock experience [26]. The administration of metyrapone, which suppresses endogenous GCs by inhibiting the conversion of 11-deoxycortisol to cortisol by 11βhydroxylase and thus inhibiting further synthesis of GCs, dose-dependently impairs acquisition in the water maze [27]. Taken together, these findings indicate that in rodents, either an excess or absence of GCs affects learning and memory most often impairing these processes. Moreover, the findings of impaired performance due to the absence of GCs indicate that the GCs, to a certain extent, are necessary for optimal memory processing.

Much like in rodents, in humans, both an excess and the absence of GCs can interfere with learning and memory functioning. Maheu and collaborators demonstrated that GC synthesis inhibition with metyrapone results in decreased declarative (i.e., episodic) memory performance [28]. Kirschbaum and colleagues found that the administration of high doses of synthetic GCs resulted in impaired declarative memory in a task involving the cued-recall of a list of 24 concrete nouns. Deficits were also found when participants performed a spatial memory task involving the reproduction of a series of spatial navigations within a complex visual scene. They further showed that TSST-induced elevations of endogenous GCs resulted in impaired declarative memory performance on word list learning and spatial memory [29].

A recent meta-analysis [30] showed that overall, when there is an increase in GC levels before retrieval, GCs have an impairing effect. With the aim of examining the impact of exogenous GC administration on encoding, consolidation and retrieval, de Quervain and colleagues (2000) [31] gave a medium dose of synthetic GCs either before the acquisition of a word list, immediately after, or just before the retrieval of the list. The results revealed significant impairments in memory when the drug was administered just before retrieval, thus suggesting specific effects of glucocorticoids on the retrieval of previously learned information [31].

As noted above, memory processes mediated by prefrontal cortex regions are also affected by GC administration. In 1999, we reported data showing impairments in working memory function with a high dose of synthetic GCs in young male subjects [32]. In this study, participants were infused for 100 min. with one of three doses of synthetic GCs or placebo and working memory function was tested during the infusion period. The results revealed that performance on the working memory task decreased significantly at the highest dose of hydrocortisone thus suggesting that in young individuals, GCs can have negative effects on frontal lobe-dependent functions. The results of Wolf and colleagues confirm this notion as they have shown that performance on a prefrontal cortex-dependent task was negatively affected by the administration of synthetic GCs [33]. Interestingly, a recent study showed that endogenous increase in GCs, induced by exposure to a psychosocial stressor, can also impair working memory [34].

9.4.2
Modulatory Effects of GCs

It is important to note however, that significant increases or decreases of circulating GC levels do not always lead to impaired performance. In fact, there is a great paradox in the field of stress research that relates to the fact that stress hormones are associated with both positive and negative effects on cognition. In the animal studies described above, one important feature common to these tasks is that their aversive/stressful nature induces an increase in *endogenous* GC levels. Thus, any GCs administered to the animals will have an additive effect and likely increase GC concentrations to supraphysiological levels. Interestingly, in the absence of exogenous GC manipulations, memory performance is often enhanced by the endogenous release induced by the stressful/aversive nature of these tasks suggesting that GCs can also exert indirect or modulatory effects on learning and memory.

Sandi and collaborators demonstrated that decreasing the water temperature in the water maze optimizes the acquisition and retention of the platform's location [35]. Similar results have been observed in adrenalectomized rats in contextual fear conditioning when low to moderate doses of GCs were given soon after learning thus mimicking the endogenous release that would normally take place as a function of the task [36]. Low to moderate doses of GCs given to rats in the water maze also results in enhanced performance [35].

In humans, Beckwith and collaborators (1986) showed that exogenous GC administration enhanced memory performance [37]. When considering the effects of exogenous GC administration however, one must keep in mind the circadian rhythm of GC secretion. Recall that in humans in the AM phase GC levels are at their peak, whereas they are at their lowest in the PM phase. Lupien and collaborators showed that the administration of synthetic GCs in the AM phase prior to learning impaired performance in a hippocampus-dependent declarative memory task, whereas the same dose given in the afternoon, enhanced memory performance [38]. These results demonstrate that there exists an inverted U-shaped function between circulating levels of GCs and learning and memory processes such that up to a given optimal level, GCs enhance performance, while anything above this optimal range results in impaired performance.

These findings also highlight the importance of the ratio of MR/GR GC receptor occupancy in determining the *direction* of GC-induced cognitive changes, a notion proposed in the literature under the GC receptor balance hypothesis [39, 40]. Recall that the differential binding affinities of the two receptor types result in MR receptors having to be saturated before GR activation occurs. Thus, low to moderate doses of GCs result in a high MR/GR ratio (hypothetically: MR receptors 100% saturated and GR receptors partially occupied 50% or a ratio of 2.0) and enhanced memory performance, whereas high doses of GCs results in a low MR/GR ratio (hypothetically: 100% saturation of both MR receptors and GR receptors or a ratio of 1.0) and impaired memory. In hippocampal cells, the induction of long-term potentiation (LTP), the electrophysiological correlate of synaptic reinforcement thought to underlie learning and memory, is optimal when GC levels are mildly elevated (high MR/GR ratio). The induction of LTP is blocked, however, after adrenalectomy or when high doses of GCs are administered, both of which result in a low MR/GR ratio [9, 41].

In humans, memory deficits, achieved through the administration of the GC synthesis inhibitor metyrapone, that is, low MR/GR ratio, were reversed by returning GCs to baseline levels with hydrocortisone, that is, increasing the MR/GR ratio [38]. Taken together, these findings indicate that <u>acute</u> manipulations of GCs can have both direct and biphasic modulatory effects (i.e., inverted U-shaped function) on learning and memory processes in both animals and humans.

Thus, at this point we have seen that GC-induced effects on learning and memory largely depend upon the circulating levels of the endogenous hormone (i.e., circadian rhythm) at the time of testing, the nature of the task, as well as the dose of exogenous GCs given. Another important factor to consider is the duration of GC exposure.

9.4.3
Chronic Effects of GCs on Cognition and Brain Integrity

While short-term responses of the brain to novel and potentially threatening situations may be adaptive and result in new learning and acquired behavioral strategies for coping, as may be the case for absolute stress, repeated stress can cause both cognitive impairments and structural changes in the brain and in the hippocampus in particular.

Animal models of chronic stress in which rats are physically restrained over varying numbers of days highlight that chronic elevations of GCs result in cognitive deficits in functions ranging from working memory to spatial memory [42, 43]. The cognitive effects of long-term elevations of GCs in human populations were studied in disorders affecting GC levels and using the exogenous administration of synthetic GCs to healthy subjects. Depression was associated with hypercortisolism in some [44–46] but not all studies [47]. Studies revealed that depressed individuals show decrements in simple and complex attentional tasks as well as verbal and visual memory tasks [48–51]. Similar cognitive deficits were found in individuals with steroid-induced psychosis (typically after long-term steroid use) [52]. The same was also reported in patients suffering from Cushing's disease [53, 54], a medical condition in which endogenous levels of GCs are chronically elevated. Finally, we showed that a significant proportion of elderly individuals show an endogenous increase of GC levels over time that is also linked to impaired memory performance [55]. Thus, overall, extended elevations in GCs appear to be detrimental to several cognitive processes.

Chronic exposure to stress also was shown to affect the integrity of brain structures and more specifically to be associated with significant changes in the cell populations within the hippocampus. For example, the treatment of adult rats with synthetic GCs decreases the proliferation of granule cell precursors [56] while the removal of adrenal steroids stimulates the generation of new granule cells [57]. There is also evidence that the exogenous administration of GCs suppresses hippocampal neurogenesis. Acute elevations of endogenous GCs decrease the number of adult-generated neurons in the dentate gyrus in various species, including the rat [58], tree shrew [59] and marmoset [60], and repeated stress results in prolonged suppression of cell proliferation in the dentate gyrus in the adult tree shrew [61]. Stress-induced decreases in dentate gyrus cell proliferation could also contribute, along with atrophy of the pyramidal cells of the hippocampus, to changes in hippocampal volume observed after chronic exposure to stress. In addition, GC-induced structural changes in the hippocampus have also been shown in humans. Over the past two decades, many studies revealed that psychiatric conditions known to lead to dysregulations in endogenous GC levels such as depression, PTSD and schizophrenia are also associated with a smaller hippocampal volume [62–64]. Moreover, hippocampal atrophy associated with chronic exposure to high levels of GCs was also reported in Cushing patients [53] and elderly individuals [65].

Thus in summary, in both animals and humans, the impact of acute changes (increases or decreases) in GC levels on learning and memory processing is relative in that it depends on the magnitude of the resulting change in endogenous levels, which is related to either the dose of exogenous GCs given or to the nature of the stressor, all of which will in turn vary as a function of the time of day at which these manipulations took place owing to the circadian rhythm of GC secretion. Chronic elevations of GC levels on the other hand, have a clear negative impact on learning and memory functioning and the integrity of brain regions that subserve these processes.

While we began the section on stress hormones and cognition stating that GCs affect the functioning of brain areas dense with stress hormone receptors, the reader has perhaps taken note that no attention has been paid to cognitive processes subserved by the amygdala. The rationale for this deliberate omission relates to the fact that GC effects in the amygdala cannot be considered in isolation. As we show below, the catecholamines play a central role in GC-mediated effects on memory for emotionally laden stimuli, situations, or events.

9.5
Stress, Emotion and Memory

It is well known that what we encode and remember from an event primarily depends on the attention that is devoted to this event and its components. This is because the more attention given to an event, the higher the probability that this event will be elaborated upon (relating the information from this event to other situations and related concepts stored in memory) at the time of encoding. In turn, the level of attention devoted to an event at the time of encoding greatly depends on the emotional salience of this event.

Most of us remember what we were doing and who we were with when we learned of the World Trade Center attacks, but the majority of us may have difficulty remembering what we were doing and who we were with two weeks before or after the attacks. The emotions elicited by the event (e.g., surprise, fear, anger, etc.) resulted in a "flashbulb memory" most likely because the totality of our attention was directed to the event, leading to a deeper elaboration and thus, to an optimization of our memory for this event. This highly adaptive tendency is inherent to the cognitive focus that occurs at the time of a stress response; a notion that is confirmed when we look at findings obtained in studies of trauma victims.

9.5.1
Central Focus Phenomenon

Studies show that trauma victims report vivid memories for the traumatic event but cannot recall information surrounding the traumatic event. An analogous phenomenon is well known in the field of law enforcement. Witnesses to violent crimes demonstrate a weapon focus such that the weapon captures most of the victim's attention, resulting in a reduced ability to recall other details of the scene and to recognize the assailant at a later time [66]. In an earlier study Christianson and Loftus (1991) showed subjects a series of slides in which the emotional valence of one critical slide was varied. In the neutral version, the critical slide depicted a woman riding a bicycle. In the emotional version, the same woman was seen lying injured on the street near the bicycle. In both versions, a peripheral car was seen in the distant background. The results showed that the central information (woman and bicycle) was well retained in the emotional condition, while the peripheral information (car in background) was better retained in the neutral condition [67]. These findings are consistent with a myriad of others showing that emotional material is better retained relative to neutral information [68, 69]. Further still, memory for central information is also less susceptible to the effects of misleading information than is memory for peripheral details [70]. In this study, retention of a short movie was tested at two intervals; three weeks after encoding and four weeks after encoding. In the first recall session, half the participants were asked misleading questions and the other half were asked unbiased questions. At the next recall session all questions were the same for all participants. The results suggested that individuals asked misleading questions where more likely to provide incorrect answers for questions concerning the peripheral details than for the central details, thus, providing good support for the strength of the central versus peripheral effect.

9.5.2
Retention Intervals and Emotional Memory

A second important characteristic of memory for emotional events concerns time-dependent memory processes (i.e., the retention intervals at which memory is assessed). Studies in humans show that emotional and/or traumatic information is better recalled after a delay compared to immediately after encoding. In contrast, memory for neutral events decreases over longer intervals [71–73]. A good example of this is provided by Kleinsmith and Kaplan (1963), who presented subjects with pairs of words to learn. Some words were neutral, while others had a strong emotionally negative component (e.g., "rape", "mutilation"). Memory of the word pairs was tested 2 min or 1 week after encoding. They showed that memory for the emotional stimuli was better at longer intervals. These results have been replicated many times and the majority of studies have shown enhanced memory for emotionally arousing material, provided that memory is tested at longer retention intervals [74].

Walker [101] explains this retention interval effect in emotional memory within the context of Hebb's (1949) [102] concept of reverberating circuits. He suggests that emotional information, because it creates a high level of arousal at the time of learning, produces a more active consolidation process and leads to a stronger memory trace and better recall at longer intervals. In contrast, because it creates a high level of arousal, emotional information also acts by inhibiting recall at shorter interval, due to the fact that the process of consolidation is active (and thus inaccessible) at this time. In support of this view a meta-analysis [75] looking at 48 studies revealed that at short delays such as 2 min low arousal leads to better immediate memory than high arousal. At delays longer than 20 min the reverse is observed, with high arousal leading to better delayed memory than low arousal. These findings provide further evidence that emotional memory is best at longer delays and that arousal is an important mediator in emotional memory processing.

9.5.3
Stress Hormones and Emotional Memory

While we know that elevated GCs at encoding can result in impaired retention performance, however when emotional stimuli are used the opposite is true. Elevated GCs enhance the encoding and consolidation of emotional material. What might explain these differential effects?

Enter the catecholamines. The effects of GCs on memory for emotional material are best understood by examining the biological mechanisms that underlie emotional memory processing. Extensive data indicate that emotionally arousing experiences produce a retrograde enhancement of memory for that experience, owing to the actions of the stress hormones released during these experiences. Emotionally arousing events often result in the release of the peripheral catecholamines epinephrine and norepinephrine by the adrenal medulla, the secretion of central norepinephrine by the locus coeruleus and the release of GCs by the adrenal glands. Recall that the peripheral catecholamines activate β-adrenergic receptors on vagal nerve afferents that terminate in the nucleus of the solitary tract, which either project directly to adrenergic neurons in the amygdala or indirectly by increasing basal noradrenergic firing in the locus coeruleus. This in turn results in increased norepinephrine output in the amygdala and the activation of adrenergic receptors throughout the brain [11, 12] and is also thought to modulate hippocampal activity in such a way that it enhances the consolidation of emotionally arousing material [76].

9.5.4
Stress Hormone Effects of the Encoding and Consolidation of Emotional Material

Animal and human studies confirm the role of the catecholamines and the GCs in the memory-modulating effects of emotionally arousing events. In rodents, post-learning stimulation of the noradrenergic system enhances, whereas post-learning blockade inhibits, long-term memory (LTM) in an inhibitory avoidance task. Likewise, post-training exposure to psychosocial stress enhances retention of inhibitory avoidance in animals, whereas pre-training corticosteroid synthesis inhibition blocks this enhancement [77]. In humans, pre-learning blockade of central β-adrenergic receptors or pre-learning corticosteroid synthesis inhibition impairs long-term declarative memory for emotionally arousing material [28, 78, 79], whereas pre- or post-learning stimulation of the noradrenergic or corticosteroid systems enhances it [80–82]. These findings suggest that both classes of hormones are critically involved in the encoding and consolidation of emotional material.

As mentioned earlier, GC secretion follows a circadian rhythm; and, thus, studies varying the levels of GCs either endogenously or exogenously must take this into consideration when exerting their manipulations. We recently took advantage of the circadian rhythm of the GC secretion to modulate the effects of GCs on consolidation [83]. Participants were exposed either in the morning or the afternoon to a psychosocial stressor before viewing a movie containing both neutral and emotional material. In contrast to most studies showing a GC-mediated enhancement of consolidation we found that, when tested one week later, the group exposed to the stressor in the morning showed impaired memory for emotional material. No difference was found for the neutral material. Given that basal levels of GCs are elevated in the morning, stressing participants in the AM phase likely brought GC levels to a point where they were no longer beneficial for the consolidation of the emotional material, that is, the right side of the inverted U-shape function relating GCs and memory performance. Thus overall, the effects of GCs and the catecholamines on emotional memory also appear to be both direct and modulatory as was the case for neutral material.

9.5.5
Stress Hormone Effects of the Retrieval of Emotional Material

Also extensively studied are the effects of GCs on the retrieval of emotional material. What these data show us is that elevated GCs at the time of retrieval result in impaired retention of emotional material. For instance, rats exposed to an acute stressor (foot shock) or administered corticosterone before being tested on the retention of a water maze spatial task show impaired performance compared to controls. However, the administration of metyrapone (an inhibitor of GC synthesis) prevents this stress-induced retention impairment. Therefore in this aversive testing paradigm, GC levels elevated prior to retention testing seem to impair memory retrieval [84].

In humans, Kuhlmann and colleagues (2005) found significant impairments in memory performance when they gave GCs prior to the retrieval of a word list encoded 5 h prior to the retention test. The retrieval impairment was observed for the negative words, whereas the recall for the neutral words was intact [85]. These results go along with the notion that GCs have a greater impact on the processing of emotionally arousing material.

Endogenous elevations induced by a physiological stress task (the cold pressor task in which participants keep their arm in very cold water for as long as possible) also affect the retrieval of emotional information. In one study, the authors exposed half of their participants to the cold pressor task 1 h after having learned both neutral and negative words. The other half of the participants was exposed to warm water. Although all participants exposed to the physiological stressor showed an increase in sympathetic activity, they did not all show an increase in GCs. The authors then subdivided the stressed participants into GC responders and non-responders. Interestingly, the responders (those who showed an elevation in GC levels) recalled significantly fewer words than the non-responders and the controls, and this effect was mostly pronounced for the mildly arousing words compared to the highly arousing and neutral words [86]. While certainly not conclusive, these findings do point toward a necessity for the concerted actions of the catecholamines and the GCs for the impairing effects of GCs on the retrieval of emotional memories to be observed.

de Quervain and colleagues (2007) wished to address precisely this question. They conducted a double-blind, placebo-controlled study in which they presented emotional and neutral words to their participants and asked them to learn them for later recall. At 24 h later, 1 h prior to retention testing, one group of participants received cortisone (25 mg), the second received 40 mg of propranolol (a β-adrenergic receptor antagonist) and the third group received both drugs. The authors found that propanolol alone did not affect the recall of either emotional or neutral words. Consistent with the literature, the individuals that received cortisone showed impaired retrieval of the emotionally arousing words (keep in mind that the catecholamines are not affected in this group, i.e., they act as they typically would by being elevated). Most interestingly, no impairment was observed in those individuals that received the concurrent administration of cortisone and propranolol, that is, resulting in elevated GCs in the absence of β-adrenergic stimulation [87]. These findings confirmed previously obtained animal data showing that the impairing effects of GCs on the retrieval of emotionally arousing material depend upon the coactivation of the noradrenergic system and the GCs.

9.6
Contextual Effects

As we can see in the above sections, a certain level of arousal at acquisition or at memory testing needs to be present in order for GCs to modulate memory performance. In support of this view, a study by Okuda and colleagues [88] contrasted the effects of GCs on memory consolidation by administering synthetic GCs or vehicle after acquisition to rats that were previously habituated to the learning environment (low arousal at the time of acquisition) and rats that were not habituated (high arousal at the time of acquisition). The results showed that only the non-habituated group showed enhanced memory performance when tested 24 h later, while the habituated group did not. In humans, a recent study by Abercrombie *et al.* [89] reported that stress-induced increases in GCs were associated with enhanced memory consolidation, only in participants who reported high arousal and negative affect at the time of testing.

Similarly, Kuhlmann and Wolf [90] asked participants to learn a list of words and 4 h later they gave exogenous GCs. Retention of the word list was assessed 1 h after GC administration. The authors modified the experimental setting during the 1-h wait between the drug administration and memory recall such that in half of the participants, the experimental setting was formal (e.g., participants waiting in a noisy area, alone, etc.), while in the other half of the participants, the experimental setting was more relaxing (e.g., the participant spent time chatting with the research assistant in his/her office). The results showed that the exogenous dose of GCs impaired retrieval only in the participants tested in the formal experimental setting, while it had no effect in the participants tested in the relaxed condition. Altogether, these results show that one can eliminate the GC-induced enhancement of consolidation or impairment of retrieval by inducing slight changes in the test setting and thereby likely decreasing arousal.

How can we explain the influence of the environmental setting on the effects of GCs on memory consolidation and retrieval? Two existing models can be used to understand these effects. The first, put forward by Roozendaal [91], suggested that under stressful conditions there is an enhancement of the consolidation of novel information *related* to the stressful situation, while there is a decrease in the retrieval of information *not related* to the stressful situation. Consequently, the differential effects of GCs on memory consolidation and retrieval do not relate to the same type of information pertaining to the stressful situation. The impairing effect on retrieval occurs because, in order to facilitate the consolidation of the threatening information, the system prevents the retrieval of unrelated types of information. This model would help to explain the increased memory of central events and the impaired memory of peripheral events discussed earlier.

DeKloet and colleagues [9] suggest that stress does not impair the retrieval of previously consolidated information, but rather it *interferes* with the encoding of the new situation that is created by the application of stress (or elevated GCs), right before retrieval. The threat posed by the new event supersedes the retrieval of previously learned, less important information in order to encode the new situation. Consequently, in the model of DeKloet and collaborators, the effects of stress on retrieval are best explained by the fact that stress applied before retrieval creates a *new* event to be encoded by the brain, thus preventing the retrieval of less important information encoded previously. The model of De Kloet and collaborators may help to explain why one needs to test participants in an aroused state in order to observe the modulatory effects of GC on memory retrieval. Indeed, an arousing situation at the time of retrieval would represent a new and different situation that would necessitate full attention from the subjects, thus explaining the impairing effects of stress on retrieval. Although we have no clear answers as to which hypothesis holds true, what we can state with certainty is that the dynamic nature of the interactions between the two classes of stress hormones, the brain, cognitive function and the environment are rich and complex and promise more interesting things to come, some of which we now address.

9.7
New Directions

For many years, the effects of stress on different memory processes have been studied with the idea that consolidation was the last step in these processes. Recall that memory formation is a time-dependent process in which new memories are initially in a dynamic or "labile" state (i.e., in short-term memory) but that over time become "fixed" or consolidated (long-term memory) and, thus, insensitive to disruption. The consolidation model thus implies that any information that has been fully consolidated should be insensitive to treatment, be it stress or any other amnestic treatment. However, animal data and more recently human data suggest that, once in the long-term memory system, memories are not necessarily immune to modifications. In fact, by reactivating a fixed memory (e.g., by recalling it), it brings it back into an unstable state where it is subject to modifications the same way that it was during the first round of consolidation. In 1968, an important study demonstrated that electroconvulsive shocks could impair previously consolidated memories but only if they were reactivated prior to shock administration [92]. Since then, important findings have demonstrated evidence for *reconsolidation* for different types of memories and in different species [93–95]. In humans, to date there is evidence of reconsolidation for procedural [96] and neutral declarative memories [97].

At a time where a certain level of convergence was starting to arise in the field of stress and the different memory processes, the question is now open again as to whether GCs could also have an impact on memory *reconsolidation* in the same way that they have the capacity to affect memory consolidation. Further still, we suggest that reactivating a stressful/emotional memory may in fact also lead to a reactivation of the HPA axis [98]. These elevated GCs may therefore then have the potential to affect the memory trace being reconsolidated as they have been shown to do so during consolidation.

It would also be interesting to broaden our understanding of some of the characteristics of emotional memory (i.e., the effect of central vs peripheral details). From what we see here, this effect is generally accepted as one of the features of emotional memory processing. We propose instead that the central versus peripheral focus effect is likely a characteristic of negative and arousing emotional memories that have *also elicited GC release*. To our knowledge there is little or no data addressing whether this effect also applies to positive emotions. Intuitively, the answer to this question seems to be no because we can all think of a positive emotional event/experience in our lives for which we are able to recall many if not all of the elements that comprised the entire event, both the central and peripheral details (most likely due to the arousal and the catecholamines). Indeed, little is known about the impact of GCs on positive emotional memory. While positive emotional stimuli have been used in experiments looking at stress hormone effects [99, 100], few if any have looked at positive emotion alone given that most studies present positive stimuli within the same context as negative stimuli.

9.8
Conclusion

After more than a century of research on the effects of GCs on human cognitive function, we come to view the actions of these steroids as being modulatory rather than direct. History taught us two important things: first, that these steroids can access the brain and lead to changes in cognitive function; second, we learned that GC actions on cognitive processes can also be positive and that the positive or negative effects of GCs depend on the balance of the two GC receptor types known to exist today.

The story was supposed to become simpler as new discoveries were being made about the actions of GCs and the catecholamines on the animal and human brain. Yet, it became more complex, dynamic and difficult to grasp. Indeed, we now suspect that important variables related to the testing environment itself can have an important impact on the stress response and, consequently, can predict its impact on cognitive performance. Here, the time of day, valence and nature of the information to be learned, as well as the context of the experiment are summarized as examples of the types of variables that have to be controlled when assessing the effects of GCs on cognition in humans, since these factors have been shown to impact on either the stress response itself and/or on cognitive performance in humans. However, as complex as the story gets, it tells us that the stress hormones are very important players in the interface between the brain and the environment. This renders the study of their actions more exciting than ever.

Summary

This chapter summarizes the effects of stress hormones on cognitive function in humans. The stress response generated by the brain and body in the face of a real or implied threat (i.e., absolute or relative stressors) leads to the secretion of stress hormones that cross the blood–brain barrier and readily access the brain, where-upon they can influence learning and memory by binding to stress hormone receptors in the very brain regions that subserve these processes. The effects of stress hormones on cognition in humans are relative, in that they can be positive or negative, depending on a number of factors such as time of day, dose and the nature and valence of the to-be-remembered material, as well as the context of the experiment.

References

1 Selye, H. (1998) A syndrome produced by diverse nocuous agents. 1936. *J. Neuropsychiatry Clin. Neurosci.*, **10**, 230–231.

2 Mason, J.W. (1968) A review of psychoendocrine research on the sympathetic-adrenal medullary system. *Psychosom. Med.*, **30** (Suppl.), 631–653.

3 Dickerson, S.S. and Kemeny, M.E. (2004) Acute stressors and cortisol responses: a theoretical integration and synthesis of laboratory research. *Psychol. Bull.*, **130**, 355–391.

4 Lupien, S.J., Ouellet-Morin, I., Hupbach, A., Walker, D., Tu, M.T. and Buss, C. (2006) Beyond the stress concept: allostatic load – a developmental biological and cognitive perspective, in *Developmental Psychopathology Volume 2: Developmental Neuroscience* (ed. D. Cicchetti and D.J. Cohen), Hoboken, New Jersey, pp. 578–628.

5 Brown, R.E. (1999) *An Introduction to Neuroendocrinology*, 5th edn, Cambridge University Press, Cambridge.

6 Nelson, R.J. (2000) *An Introduction to Behavioral Endocrinology*, 2nd Edn., Sinauer Associates, Sunderland, MA.

7 Bradbury, M.J., Akana, S.F. and Dallman, M.F. (1994) Roles of type I and II corticosteroid receptors in regulation

of basal activity in the hypothalamo-pituitary-adrenal axis during the diurnal trough and the peak: evidence for a nonadditive effect of combined receptor occupation. *Endocrinology*, **134**, 1286–1296.

8 Reul, J.M. and de Kloet, E.R. (1985) Two receptor systems for corticosterone in rat brain: microdistrisbution and differential occupation. *Endocrinology*, **117**, 2505–2511.

9 de Kloet, E.R., Oitzl, M.S. and Joels, M. (1999) Stress and cognition: are corticosteroids good or bad guys? *Trends Neurosci.*, **22**, 422–426.

10 Birmingham, M.K., Sar, M. and Stumpf, W.E. (1984) Localization of aldosterone and corticosterone in the central nervous system assessed by quantitative autoradiography. *Neurochem. Res.*, **9**, 333–350.

11 Clayton, E.C. and Williams, C.L. (2000) Adrenergic activation of the nucleus tractus solitarius potentiates amygdala norepinephrine release and enhances retention performance in emotionally arousing and spatial memory tasks. *Behav. Brain Res.*, **112**, 151–158.

12 Fallon, J.L. and Ciofi, P. (1992) Distribution of monoamines within the amygdala, in *The Amygdala: Neurobiological Aspects of Emotion, Memory, and Mental Dysfunction* (ed. J. Aggleton), John Wiley & Sons, Inc., New York, pp. 94–114.

13 McGaugh, J.L., Cahill, L. and Roozendaal, B. (1996) Involvement of the amygdala in memory storage: interaction with other brain systems. *Proc. Natl. Acad. Sci. U. S. A.*, **93**, 13508–13514.

14 Buckner, R.L., Kelley, W.M. and Petersen, S.E. (1999) Frontal cortex contributes to human memory formation. *Nat. Neurosci.*, **2**, 311–314.

15 Squire, L.R. (1982) The neuropsychology of human memory. *Annu. Rev. Neurosci.*, **5**, 241–273.

16 Connolly, J.D., Goodale, M.A., Menon, R.S. and Munoz, D.P. (2002) Human fMRI evidence for the neural correlates of preparatory set. *Nat. Neurosci.*, **5**, 1345–1352.

17 Owen, A.M., Doyon, J., Petrides, M. and Evans, A.C. (1996) Planning and spatial working memory: a positron emission tomography study in humans. *Eur. J. Neurosci.*, **8**, 353–364.

18 Bunge, S.A., Ochsner, K.N., Desmond, J.E., Glover, G.H. and Gabrieli, J.D. (2001) Prefrontal regions involved in keeping information in and out of mind. *Brain*, **124**, 2074–2086.

19 Anderson, A.K. and Phelps, E.A. (2001) Lesions of the human amygdala impair enhanced perception of emotionally salient events. *Nature*, **411**, 305–309.

20 Tulving, E. (2002) Episodic memory: from mind to brain. *Annu. Rev. Psychol.*, **53**, 1–25.

21 Davachi, L. and Wagner, A.D. (2002) Hippocampal contributions to episodic encoding: insights from relational and item-based learning. *J. Neurophysiol.*, **88**, 982–990.

22 Kandel, E.R. (2001) The molecular biology of memory storage: a dialogue between genes and synapses. *Science*, **294**, 1030–1038.

23 Kirschbaum, C., Pirke, K.M. and Hellhammer, D.H. (1993) The 'Trier Social Stress Test'–a tool for investigating psychobiological stress responses in a laboratory setting. *Neuropsychobiology*, **28**, 76–81.

24 Oitzl, M.S. and de Kloet, E.R. (1992) Selective corticosteroid antagonists modulate specific aspects of spatial orientation learning. *Behav. Neurosci.*, **106**, 62–71.

25 Veldhuis, H.D. and De Kloet, E.R. (1983) Antagonistic effects of aldosterone on corticosterone-mediated changes in exploratory behavior of adrenalectomized rats. *Horm. Behav.*, **17**, 225–232.

26 Donley, M.P., Schulkin, J. and Rosen, J.B. (2005) Glucocorticoid receptor antagonism in the basolateral amygdala and ventral hippocampus interferes with long-term memory of contextual fear. *Behav. Brain Res.*, **164**, 197–205.

27 Roozendaal, B., Bohus, B. and McGaugh, J.L. (1996) Dose-dependent suppression of adrenocortical activity with metyrapone: effects on emotion and

memory. *Psychoneuroendocrinology*, **21**, 681–693.

28 Maheu, F.S., Joober, R., Beaulieu, S. and Lupien, S.J. (2004) Differential effects of adrenergic and corticosteroid hormonal systems on human short- and long-term declarative memory for emotionally arousing material. *Behav. Neurosci.*, **118**, 420–428.

29 Kirschbaum, C., Wolf, O.T., May, M., Wippich, W. and Hellhammer, D.H. (1996) Stress- and treatment-induced elevations of cortisol levels associated with impaired declarative memory in healthy adults. *Life Sci.*, **58**, 1475–1483.

30 Het, S., Ramlow, G. and Wolf, O.T. (2005) A meta-analytic review of the effects of acute cortisol administration on human memory. *Psychoneuro-endocrinology*, **30**, 771–784.

31 de Quervain, D.J., Roozendaal, B., Nitsch, R.M., McGaugh, J.L. and Hock, C. (2000) Acute cortisone administration impairs retrieval of long-term declarative memory in humans. *Nat. Neurosci.*, **3**, 313–314.

32 Lupien, S.J., Gillin, C.J. and Hauger, R.L. (1999) Working memory is more sensitive than declarative memory to the acute effects of corticosteroids: a dose-response study in humans. *Behav. Neurosci.*, **113**, 420–430.

33 Wolf, O.T., Convit, A., McHugh, P.F., Kandil, E., Thorn, E.L., De Santi, S., McEwen, B.S. and de Leon, M.J. (2001) Cortisol differentially affects memory in young and elderly men. *Behav. Neurosci.*, **115**, 1002–1011.

34 Schoofs, D., Preuss, D. and Wolf, O.T. (2008) Psychosocial stress induces working memory impairments in an n-back paradigm. *Psychoneuro-endocrinology*, **33**, 643–653.

35 Sandi, C., Loscertales, M. and Guaza, C. (1997) Experience-dependent facilitating effect of corticosterone on spatial memory formation in the water maze. *Eur. J. Neurosci.*, **9**, 637–642.

36 Pugh, C.R., Tremblay, D., Fleshner, M. and Rudy, J.W. (1997) A selective role for corticosterone in contextual-fear conditioning. *Behav. Neurosci.*, **111**, 503–511.

37 Beckwith, B.E., Petros, T.V., Scaglione, C. and Nelson, J. (1986) Dose-dependent effects of hydrocortisone on memory in human males. *Physiol. Behav.*, **36**, 283–286.

38 Lupien, S.J., Wilkinson, C.W., Briere, S., Menard, C., Ng Ying Kin, N.M. and Nair, N.P. (2002) The modulatory effects of corticosteroids on cognition: studies in young human populations. *Psychoneuroendocrinology*, **27**, 401–416.

39 De Kloet, E.R., Vreugdenhil, E., Oitzl, M.S. and Joels, M. (1998) Brain corticosteroid receptor balance in health and disease. *Endocr. Rev.*, **19**, 269–301.

40 Oitzl, M.S., van Haarst, A.D., Sutanto, W. and de Kloet, E.R. (1995) Corticosterone brain mineralocorticoid receptors (MRs) and the activity of the hypothalamic-pituitary-adrenal (HPA) axis: the Lewis rat as an example of increased central MR capacity and a hyporesponsive HPA axis. *Psychoneuroendocrinology*, **20**, 655–675.

41 Pavlides, C., Ogawa, S., Kimura, A. and McEwen, B.S. (1996) Role of adrenal steroid mineralocorticoid and glucocorticoid receptors in long-term potentiation in the CA1 field of hippocampal slices. *Brain Res.*, **738**, 229–235.

42 Cerqueira, J.J., Mailliet, F., Almeida, O.F., Jay, T.M. and Sousa, N. (2007) The prefrontal cortex as a key target of the maladaptive response to stress. *J. Neurosci.*, **27**, 2781–2787.

43 Luine, V., Villegas, M., Martinez, C. and McEwen, B.S. (1994) Repeated stress causes reversible impairments of spatial memory performance. *Brain Res.*, **639**, 167–170.

44 Holsboer, F. (2000) The corticosteroid receptor hypothesis of depression. *Neuropsychopharmacology*, **23**, 477–501.

45 Nemeroff, C.B. (1996) The corticotropin-releasing factor (CRF) hypothesis of depression: new findings and new directions. *Mol. Psychiatry*, **1**, 336–342.

46 Pariante, C.M. and Miller, A.H. (2001) Glucocorticoid receptors in major depression: relevance to pathophysiology and treatment. *Biol. Psychiatry*, **49**, 391–404.

47 Brouwer, J.P., Appelhof, B.C., Hoogendijk, W.J., Huyser, J., Endert, E., Zuketto, C., Schene, A.H., Tijssen, J.G., Van Dyck, R., Wiersinga, W.M. and Fliers, E. (2005) Thyroid and adrenal axis in major depression: a controlled study in outpatients. *Eur. J. Endocrinol.*, **152**, 185–191.

48 Beats, B.C., Sahakian, B.J. and Levy, R. (1996) Cognitive performance in tests sensitive to frontal lobe dysfunction in the elderly depressed. *Psychol. Med.*, **26**, 591–603.

49 Roy-Byrne, P.P., Weingartner, H., Bierer, L.M., Thompson, K. and Post, R.M. (1986) Effortful and automatic cognitive processes in depression. *Arch. Gen. Psychiatry*, **43**, 265–267.

50 Schatzberg, A.F., Posener, J.A., DeBattista, C., Kalehzan, B.M., Rothschild, A.J. and Shear, P.K. (2000) Neuropsychological deficits in psychotic versus nonpsychotic major depression and no mental illness. *Am. J. Psychiatry*, **157**, 1095–1100.

51 Tancer, M.E., Brown, T.M., Evans, D.L., Ekstrom, D., Haggerty, J.J. Jr., Pedersen, C. and Golden, R.N. (1990) Impaired effortful cognition in depression. *Psychiatry Res.*, **31**, 161–168.

52 Varney, N.R., Alexander, B. and MacIndoe, J.H. (1984) Reversible steroid dementia in patients without steroid psychosis. *Am. J. Psychiatry*, **141**, 369–372.

53 Starkman, M.N., Gebarski, S.S., Berent, S. and Schteingart, D.E. (1992) Hippocampal formation volume, memory dysfunction and cortisol levels in patients with Cushing's syndrome. *Biol. Psychiatry*, **32**, 756–765.

54 Forget, H., Lacroix, A., Somma, M. and Cohen, H. (2000) Cognitive decline in patients with Cushing's syndrome. *J. Int. Neuropsychol. Soc.*, **6**, 20–29.

55 Lupien, S., Lecours, A.R., Lussier, I., Schwartz, G., Nair, N.P. and Meaney, M.J. (1994) Basal cortisol levels and cognitive deficits in human aging. *J. Neurosci.*, **14**, 2893–2903.

56 Cameron, H.A. and Gould, E. (1994) Adult neurogenesis is regulated by adrenal steroids in the dentate gyrus. *Neuroscience*, **61**, 203–209.

57 Cameron, H.A. and McKay, R.D. (1999) Restoring production of hippocampal neurons in old age. *Nat. Neurosci.*, **2**, 894–897.

58 Tanapat, P., Galea, L.A. and Gould, E. (1998) Stress inhibits the proliferation of granule cell precursors in the developing dentate gyrus. *Int. J. Dev. Neurosci.*, **16**, 235–239.

59 Gould, E., McEwen, B.S., Tanapat, P., Galea, L.A. and Fuchs, E. (1997) Neurogenesis in the dentate gyrus of the adult tree shrew is regulated by psychosocial stress and NMDA receptor activation. *J. Neurosci.*, **17**, 2492–2498.

60 Gould, E., Tanapat, P., McEwen, B.S., Flugge, G. and Fuchs, E. (1998) Proliferation of granule cell precursors in the dentate gyrus of adult monkeys is diminished by stress. *Proc. Natl. Acad. Sci. U. S. A.*, **95**, 3168–3171.

61 Fuchs, E., Flugge, G., McEwen, B.S., Tanapat, P. and Gould, E. (1997) Chronic subordination stress inhibits neurogenesis and decreases the volume of the granule cell layer. *Soc. Neurosci.*, **23**, 317.

62 Bremner, J.D., Narayan, M., Anderson, E.R., Staib, L.H., Miller, H.L. and Charney, D.S. (2000) Hippocampal volume reduction in major depression. *Am. J. Psychiatry*, **157**, 115–118.

63 Gurvits, T.V., Shenton, M.E., Hokama, H., Ohta, H., Lasko, N.B., Gilbertson, M.W., Orr, S.P., Kikinis, R., Jolesz, F.A., McCarley, R.W. and Pitman, R.K. (1996) Magnetic resonance imaging study of hippocampal volume in chronic combat-related posttraumatic stress disorder. *Biol. Psychiatry*, **40**, 1091–1099.

64 Nelson, M.D., Saykin, A.J., Flashman, L.A. and Riordan, H.J. (1998) Hippocampal volume reduction in schizophrenia as assessed by magnetic resonance imaging: a meta-analytic study. *Arch. Gen. Psychiatry*, **55**, 433–440.

65 Lupien, S.J., de Leon, M., de Santi, S., Convit, A., Tarshish, C., Nair, N.P., Thakur, M., McEwen, B.S., Hauger, R.L. and Meaney, M.J. (1998) Cortisol levels during human aging predict hippocampal atrophy and memory deficits. *Nat. Neurosci.*, **1**, 69–73.

66 Christianson, S.A. (1992) Emotional
stress and eyewitness memory: a critical
review. *Psychol. Bull.*, **112**, 284–309.

67 Christianson, S.A. and Loftus, E.F.
(1991) Remembering emotional events:
the fate of detailed information. *Cogn.
Emot.*, **5**, 81–108.

68 Cahill, L. and McGaugh, J.L. (1995) A
novel demonstration of enhanced
memory associated with emotional
arousal. *Conscious. Cogn.*, **4**,
410–421.

69 Heuer, F. and Reisberg, D. (1990) Vivid
memories of emotional events: the
accuracy of remembered minutiae. *Mem.
Cognit.*, **18**, 496–506.

70 Roebers, C.M. and Schneider, W. (2000)
The impact of misleading questions on
eyewitness memory in children and
adults. *Appl. Cognit. Psychol.*, **14**,
509–526.

71 Kleinsmith, L.J. and Kaplan, S. (1963)
Paired-associate learning as a function of
arousal and interpolated interval. *J. Exp.
Psychol.*, **65**, 190–193.

72 Kleinsmith, L.J., Kaplan, S. and Tarte,
R.D. (1963) The relationship of arousal
to short- and longterm verbal recall. *Can.
J. Psychol.*, **17**, 393–397.

73 Quevedo, J., Sant'Anna, M.K., Madruga,
M., Lovato, I., de-Paris, F., Kapczinski,
F., Izquierdo, I. and Cahill, L. (2003)
Differential effects of emotional arousal
in short- and long-term memory in
healthy adults. *Neurobiol. Learn. Mem.*,
79, 132–135.

74 Lupien, S.J., Fiocco, A., Wan, N., Maheu,
F., Lord, C., Schramek, T. and Tu, M.T.
(2005) Stress hormones and human
memory function across the lifespan.
Psychoneuroendocrinology, **30**, 225–242.

75 Park, J. (2005) Effect of arousal and
retention delay on memory: a meta-
analysis. *Psychol. Rep.*, **97**, 339–355.

76 Roozendaal, B., Okuda, S., Van der Zee,
E.A. and McGaugh, J.L. (2006)
Glucocorticoid enhancement of memory
requires arousal-induced noradrenergic
activation in the basolateral amygdala.
Proc. Natl. Acad. Sci. U. S. A., **103**,
6741–6746.

77 Liu, L., Tsuji, M., Takeda, H., Takada, K.
and Matsumiya, T. (1999) Adrenocortical
suppression blocks the enhancement of
memory storage produced by exposure
to psychological stress in rats. *Brain Res.*,
821, 134–140.

78 Cahill, L., Prins, B., Weber, M. and
McGaugh, J.L. (1994) Beta-adrenergic
activation and memory for emotional
events. *Nature*, **371**, 702–704.

79 van Stegeren, A.H., Everaerd, W., Cahill,
L., McGaugh, J.L. and Gooren, L.J.
(1998) Memory for emotional events:
differential effects of centrally versus
peripherally acting beta-blocking agents.
Psychopharmacology (Berl.), **138**,
305–310.

80 Abercrombie, H.C., Kalin, N.H.,
Thurow, M.E., Rosenkranz, M.A. and
Davidson, R.J. (2003) Cortisol variation
in humans affects memory for
emotionally laden and neutral
information. *Behav. Neurosci.*, **117**,
505–516.

81 Buchanan, T.W. and Lovallo, W.R.
(2001) Enhanced memory for emotional
material following stress-level cortisol
treatment in humans. *Psychoneuroendocrinology*, **26**, 307–317.

82 Cahill, L. and Alkire, M.T. (2003)
Epinephrine enhancement of human
memory consolidation: interaction with
arousal at encoding. *Neurobiol. Learn.
Mem.*, **79**, 194–198.

83 Maheu, F.S., Collicutt, P., Kornik, R.,
Moszkowski, R. and Lupien, S.J. (2005)
The perfect time to be stressed: a
differential modulation of human
memory by stress applied in the
morning or in the afternoon. *Prog.
Neuropsychopharmacol. Biol. Psychiatry*,
29, 1281–1288.

84 de Quervain, D.J., Roozendaal, B. and
McGaugh, J.L. (1998) Stress and
glucocorticoids impair retrieval of
long-term spatial memory. *Nature*, **394**,
787–790.

85 Kuhlmann, S., Kirschbaum, C. and
Wolf, O.T. (2005) Effects of oral cortisol
treatment in healthy young women on
memory retrieval of negative and neutral
words. *Neurobiol. Learn. Mem.*, **83**,
158–162.

86 Buchanan, T.W., Tranel, D. and
Adolphs, R. (2006) Impaired memory
retrieval correlates with individual
differences in cortisol response but not

autonomic response. *Learn. Mem.*, **13**, 382–387.

87 de Quervain, D.J., Aerni, A. and Roozendaal, B. (2007) Preventive effect of beta-adrenoceptor blockade on glucocorticoid-induced memory retrieval deficits. *Am. J. Psychiatry*, **164**, 967–969.

88 Okuda, S., Roozendaal, B. and McGaugh, J.L. (2004) Glucocorticoid effects on object recognition memory require training-associated emotional arousal. *Proc. Natl. Acad. Sci. U. S. A.*, **101**, 853–858.

89 Abercrombie, H.C., Speck, N.S. and Monticelli, R.M. (2006) Endogenous cortisol elevations are related to memory facilitation only in individuals who are emotionally aroused. *Psychoneuro-endocrinology*, **31**, 187–196.

90 Kuhlmann, S. and Wolf, O.T. (2006) A non-arousing test situation abolishes the impairing effects of cortisol on delayed memory retrieval in healthy women. *Neurosci. Lett.*, **399**, 268–272.

91 Roozendaal, B. (2002) Stress and memory: opposing effects of glucocorticoids on memory consolidation and memory retrieval. *Neurobiol. Learn. Mem.*, **78**, 578–595.

92 Misanin, J.R., Miller, R.R. and Lewis, D.J. (1968) Retrograde amnesia produced by electroconvulsive shock after reactivation of a consolidated memory trace. *Science*, **160**, 554–555.

93 Child, F.M., Epstein, H.T., Kuzirian, A.M. and Alkon, D.L. (2003) Memory reconsolidation in Hermissenda. *Biol. Bull.*, **205**, 218–219.

94 Nader, K., Schafe, G.E. and Le Doux, J.E. (2000) Fear memories require protein synthesis in the amygdala for reconsolidation after retrieval. *Nature*, **406**, 722–726.

95 Pedreira, M.E., Perez-Cuesta, L.M. and Maldonado, H. (2002) Reactivation and reconsolidation of long-term memory in the crab Chasmagnathus: protein synthesis requirement and mediation by NMDA-type glutamatergic receptors. *J. Neurosci.*, **22**, 8305–8311.

96 Walker, M.P., Brakefield, T., Hobson, J.A. and Stickgold, R. (2003) Dissociable stages of human memory consolidation and reconsolidation. *Nature*, **425**, 616–620.

97 Hupbach, A., Gomez, R., Hardt, O. and Nadel, L. (2007) Reconsolidation of episodic memories: a subtle reminder triggers integration of new information. *Learn. Mem.*, **14**, 47–53.

98 Lupien, S.J. and Schramek, T.E. (2006) The differential effects of stress on memory consolidation and retrieval: a potential involvement of reconsolidation? Theoretical comment on beckner *et al.* (2006). *Behav. Neurosci.*, **120**, 735–738.

99 Kuhlmann, S., Piel, M. and Wolf, O.T. (2005) Impaired memory retrieval after psychosocial stress in healthy young men. *J. Neurosci.*, **25**, 2977–2982.

100 Smeets, T., Jelicic, M. and Merckelbach, H. (2006) The effect of acute stress on memory depends on word valence. *Int. J. Psychophysiol.*, **62**, 30–37.

101 Walker, E.L. (1958) Action decrement and its relation to learning. *Psychol. Rev.*, **65**, 129–142.

102 Hebb, D.O. (1949) *Organization of Behaviour: a Neuropsychological Theory*, Wiley, New York.

10
Contribution of Early Life Stress to Anxiety Disorder

Marta Weinstock

10.1
Introduction

Anxiety disorders are serious medical conditions estimated to affect approximately 20% of the population [1]. Fear is a biologically adaptive physiological and behavioural response to an actual or anticipated threat, while anxiety involves uncertainty as to the expectancy of the threat and is triggered by more generalized cues. Anxiety is characterized by a more diffuse state of distress with symptoms of hyper-arousal and worry in which psychosocial and/or physical function is often compromised, as manifested in somatic symptoms of tension, dysphoria and depression [2]. In subjects with anxiety disorders stimuli that may be only mildly aversive for non-anxious individuals can elicit hyper-arousal, emotional distress and attempts to escape from or avoid the anxiety provoking object.

Observations from several sources show that early life stress is a major risk factor for anxiety and/or depressive disorder in later life [3–8]. Further analyses indicate that stressful life events involving danger and threat of loss are more often related to anxiety disorder, whereas experience of humiliation like sexual abuse or neglect are more potent predictors of depression [9].

The healthy individual is able to maintain a dynamic equilibrium or "homeostasis" when challenged by extrinsic or intrinsic adverse forces or stressors. In response to such forces, the mind and body activate a complex series of reactions in the central and peripheral nervous systems [10]. These reactions are short-lived and enable the individual to respond in the appropriate manner to environmental challenge. If the mechanisms underlying these reactions are disrupted by exposure to excess levels of stress hormones during fetal development, excessive or prolonged activation of these systems may result. Such hormonal exposure can also compromise neuronal function resulting in reduced ability to cope with subsequent stress. Such impaired coping occurs in pathological anxiety and depressive disorder.

The prenatal and early postnatal life are periods of rapid neurological development, characterized by a high turnover of neuronal connections termed "programming", thus making the organism particularly vulnerable to both organizing and disorganizing influences [11]. Brain programming slows down during childhood and puberty and may be disrupted by hypoxia, glucocorticoids, adrenaline or cytokines released during maternal stress, viral or bacterial infection. This chapter discusses the experimental findings in support of a role of an adverse prenatal or early postnatal environment on neuronal and hormonal processes that could make the organism susceptible to anxiety disorder.

10.2
Organization of the Stress Response

The response of the organism to stress involves the activation of several interacting components in the central and peripheral nervous systems. For a more detailed account of the normal control of these systems and the changes that occur during gestation, the reader is referred to the reviews in [12] and [13]. The peripheral components of the stress system include the efferent limbs of the sympathetic nervous system, adrenal medulla and adrenal cortex. The main central components of the hypothalamic–pituitary–adrenal (HPA) axis are corticotrophin-releasing hormone (CRH) and arginine vasopressin (AVP) found in paraventricular nuclei (PVN) of the hypothalamus and noradrenergic cell bodies found in the brain stem. CRH acts as the principal regulator of the pituitary adrenal axis and is released into the portal circulation together with AVP in response to perceived stress. CRH promotes the release adrenocorticotropic hormone (ACTH) and β-endorphin from the anterior pituitary, and together with ACTH, activates the adrenal gland to release cortisol (corticosterone, CORT, in rodents) from the cortex and catecholamines from the medulla [14]. Both rapid activation and de-activation of the HPA axis is critical to permit effective handling of acute stress and to prevent adverse effects due to prolonged activity of glucocorticoids. Such adverse effects include alterations in brain neurotransmitter function and actions on the reproductive, catabolic and immune systems. The neuroendocrine response to stress is under stringent regulation through the action of multiple neurotransmitters and feedback loops. The release of CRH from the hypothalamus is activated among others by 5-HT, acetylcholine (ACh), noradrenaline (NA) and cytokines, and is inhibited by γ-amino butyric acid (GABA) and opioid peptides. CRH release is also under negative feedback control by cortisol acting on specific receptors in the hippocampus and hypothalamus [10].

10.2.1
Glucocorticoid Receptors

Glucocorticoids act via two intracellular receptors, mineralocorticoid (MR) and glucocorticoid (GR), with high and low affinity respectively. MR and GR are co-localized in neurons in the hippocampus, amygdala and prefrontal cortex (PFC) [15], while GRs are also found in the cingulate cortex, nucleus accumbens, thalamus, hypothalamic CRH neurons and pituitary, in which they exert their feedback control on the HPA axis [16]. GRs are activated as levels of glucocorticoids increase either by stress or during the appropriate period of the circadian rhythm (morning in humans, afternoon and evening in rodents). MRs are activated by resting levels of glucocorticoids and during the trough of the circadian rhythm. This has been demonstrated by the use of specific MR and GR antagonists that has enabled the identification of the role of these receptors in the actions of glucocorticoids on the brain and in the regulation of HPA axis activity. Administration of an MR antagonist, RU 28318, prior to exposure of a rat to a novel environment caused a rise in basal and peak levels of HPA activity and enhanced the response to novelty. The GR antagonist mifepristone had no effect on basal HPA axis activity but attenuated and prolonged the response to novelty [15]. Although glucocorticoids inhibit CRH release from the PVN, they stimulate its synthesis and release from neurons in other brain areas, like the central nucleus of the amygdala (ceA) [17].

In the human, GR and MRs are present in the hippocampus from at least 24 weeks of gestation [18]. It is not known when these receptors first appear in this and other brain regions. In the rat, GR mRNA can first be detected in the hippocampus, hypothalamus and pituitary from day 13 of gestation [19] and in the amygdala on day 17 [20]. MR mRNA appears in the hippocampus only on day 16 of gestation [21]. The paraventricular nucleus (PVN) develops in the rat between days 13–15 and is able to react to maternal COR on day 15 of gestation [22].

10.2.2
CRH and Its Receptors

CRH is a 41-amino-acid peptide that produces its biological effects through activation of at least two receptors CRH1 and CRH2. In addition to the PVN, CRH is found in specific brain regions, ceA, bed nucleus of the stria terminalis (BNST) and in the locus coeruleus (LC) [23]. CRH-containing neurons project from the ceA to the LC [24], from which noradrenergic neurons innervate multiple cortical and subcortical regions, including the amygdala, hippocampus and hypothalamus [25]. Stress increases CRH expression in the LC, and injection of the peptide into the LC increases the release of NA in its terminal areas [26]. The CRH system, including the peptide, its binding protein and CRH1 receptors, continue to develop after birth and become established at different rates in various brain regions throughout the first three weeks of age in the rat [27]. The normal pattern of development of CRH, CRH-bp and CRH1 in different brain regions can be altered by stress during the early postnatal period [27]. It is not known whether gestational stress can also alter the development of CRH and its receptors in the fetus.

CRH receptors are seven transmembrane, G protein-coupled that are predominantly linked to the activation of adenylate cyclase through Gs. In contrast to classic neurotransmitters that are released in a frequency-dependent fashion, CRH and other neuropeptides have a longer half-life of activity after release, enabling them to produce long-lasting effects on synaptic transmission and gene transcription. Thus, an increase in the release and activity of neuropeptides could have significant effects on neuronal signaling and consequentially on behavior. At the molecular level, expression of the CRH gene is determined by a complex of transcriptional regulatory factors with specific DNA sequences. Modulation of the interactions of these factors with their corresponding DNA-regulatory elements enables the rapid expression of CRH genes in response to environmental stimuli. In order to activate transcription of the CRH gene, the cyclic AMP response element must be bound to its binding protein and then be phosphorylated. In PVN neurons this process is inhibited by glucocorticoids [28].

10.2.3
Role of the Amygdala and CRH in Mediating Anxiogenic Behavior

Functional MRI studies have shown increased activation of the amygdala in association with anxiety states [29]. Within the amygdala the basolateral (blA) and ceA nuclei are believed to regulate affective responses. The ceA is considered to be the primary efferent site for eliciting conditioned fear responses, and lesions in the ceA prevent many of the behavioural effects associated with stress and the development of conditioned emotional responses [30]. CRH cell bodies in the ceA and BNST project to the blA and other brain areas. The blA receives multimodal information from several cortical regions and processes them to allocate emotional salience [31].

Central administration of CRH causes anxiety-like behavior through activation of CRH1 receptors in the septo-hippocampal system, BNST, LC and raphé and in amygdala nuclei [31, 32]. It is proposed that higher CRH activity in the amygdala induced by stress could alter the balance toward greater excitability and increase resistance to GABA-induced inhibition, thereby reducing the ability to cope with the stressful stimulus [33]. Activation of CRH1 receptors in the BNST causes suppression of social interaction but not of increased fear as measured in the EPM [34]. By contrast, microinjection of a CRH antagonist into the amygdala blocks the anxiogenic effects of social defeat and ethanol withdrawal [35, 36].

10.3
Alterations in Circulating Hormonal Levels Induced by Chronic Gestational Stress

10.3.1
Experimental Animals

In rats, chronic maternal stress was shown to increase the level and duration of circulating CORT in both the mother and fetus [37, 38]. After acute stress, CORT released into the maternal blood is sequestered by CORT binding protein (CORT-bp), thereby limiting its effects. Chronic stress results in a reduction of CORT-bp and also in impairment of the feedback regulation via GR in the maternal HPA axis, thereby prolonging the elevation of maternal free CORT [37]. Thus, the fetuses of chronically stressed mothers are frequently exposed to the steroid for relatively long periods compared to those after a single stress exposure. The rat fetus can respond to maternal stress on day 15 of gestation by an increase in the expression of CRH mRNA in the PVN [22]. The more intense the maternal stress, as indicated by the magnitude of the increase in circulating CORT, the greater the number of Fos-immunoreactive cells in the maternal and fetal PVN. It is not known whether the fetal response is intensified after chronic maternal stress.

10.3.2
Human Subjects

During gestation in primates the placenta becomes an additional source of CRH from which it can be released by cortisol, NA, oxytocin, PGE_2 and PGF_2 and IL-1 [13]. In a normal human pregnancy, plasma levels of CRH rise dramatically from the 25th week, reaching levels at 30 weeks which are 25-fold greater than those prior to 19 weeks [39]. However, the biological activity of CRH normally remains very low since it is largely sequestered by CRH-bp. Close to term the levels of CRH-bp fall and those of free CRH rise thereby initiating the process of labor [40]. Maternal levels of cortisol ACTH and β-endorphin also increase over this period, but only two- to threefold. In contrast to its inhibitory influence on the promoter region of the CRH gene in the PVN, cortisol activates the same gene in the placenta thereby increasing CRH synthesis [41].

While in rats or monkeys a well defined stress can be applied during a given period in gestation and its effect on the offspring evaluated under controlled conditions, human studies have mainly relied on self reports of perceived stress or high levels of anxiety during pregnancy. Such perceived stress was associated with a significant increase in circulating CRH [42]. Although much of the released CRH is bound to CRH-bp, some may be secreted into the fetal circulation and activate CRH1 receptors in the pituitary and adrenal to release cortisol. High circulating levels of free CRH may serve as a signal to the fetus that the maternal environment is threatened, potentially compromising its survival and ultimately shortening gestation. This was supported by the finding that CRH in excess of normal levels in the 28th to 31st week of gestation significantly predicted pre-term labor [43]. Moreover, a significant correlation was found between the magnitude of plasma CRH at 31 weeks of gestation and the presence of excess levels of cortisol at 15 weeks, thereby testifying to a relation between maternal stress and circulating CRH [39]. Studies performed in women exposed during pregnancy to a major earthquake in California showed that the length of gestation was shortest when the mother was in the first trimester at the time of the earthquake and increased as gestation progressed [44]. This observation provides an indirect indication of the influence of stress during early gestation on CRH levels and birth weight.

Under normal conditions, the fetus is protected from high levels of maternal cortisol through its conversion to inactive cortisone by 11β-hydroxysteroid dehydrogenase 2 (11β-HSD2) in the placenta [45]. Reduced intrauterine growth is associated with a reduction in mRNA and activity of 11β-HSD2, thereby exposing the fetus to excess levels of the active glucocorticoid [46]. In addition to elevation of cortisol, maternal stress could impair fetal brain programming through excess sympatho-adrenal activation resulting in a decrease in uteroplacental perfusion [47, 48].

10.4
Alterations in the Regulation of the HPA Axis as a Result of Prenatal Stress

10.4.1
Experimental Animals

10.4.1.1 Effects on Basal HPA Axis Activity
Maternal stress induced in rats by restraining them in small containers thrice daily for 45-min periods during the last week of gestation altered the circadian rhythm of CORT in the offspring. Prenatally stressed (PS) male and female rats were shown to secrete higher amounts of total and free COR at the end of the light period compared to controls, but only PS females secreted more COR over the whole diurnal cycle [49]. The majority of other studies did not find differences in morning resting levels of CORT in adult PS and control offspring [50].

10.4.1.2 Effect of Prenatal Stress on the Response to Stress of the HPA Axis
Many studies have shown that subjection of pregnant rats or mice to various forms of chronic stress like restraint, unpredictable noise or electric shocks once or thrice daily during the last semester (days 14–21) of gestation, causes a permanent altera-tion in the regulation of the HPA axis response to stress in the adult offspring (reviewed in [50]). This is seen in a greater increase and/or longer duration of CORT release than in control offspring [51–56]. Other studies [52, 54, 55, 57] found that the prolongation of CORT release was associated with down-regulation of hippocampal MR and GR and a higher number of Fos immunoreactive neurons in the LC [58]. However, in rats specifically bred for high (HAB) or low (LAB) levels of anxiety, maternal psychosocial stress during days 4–18 of gestation, had a dif-ferential effect on the response to stress in adulthood. HAB but not LAB rats showed a smaller increase in plasma CORT than controls [59]. Prenatal stress also reduced CRH mRNA in the PVN of HAB but not LAB rats [59]. Also no altered response to stress was found in the offspring of Fisher rats stressed chronically during the last week of gestation [60]. This shows that there is a genetic basis for differences in the effect of prenatal stress on the programming of the HPA axis. There is also a gender difference, as it has been shown that adult PS female rats show increased CORT elevation after subsequent stress but the response in their male siblings did not differ from that of controls [55, 57, 61]. In support of a gender difference in the effect of prenatal stress we have recently observed two- to threefold higher levels of CRH mRNA in the PVN and in the amygdala of PS female rats than in controls ($P < 0.01$) but no increase in their male siblings (Zohar and Weinstock, unpublished observations).

When pregnant rats were stressed only in the second week (days 7–13) of gestation before GR or MR appear in the fetus, no alteration was found in the response to stress or in the number of MR and GR in adulthood [53]. On the other hand, a single 2-h period of restraint stress on day 16 of gestation resulted in a decrease in resting levels of CORT and a suppression of the HPA axis response to electric shock in the adolescent rat offspring [62]. These findings indicate that maternal stress of sufficient intensity and chronicity applied after the appearance of GR in the fetus can induce permanent alterations in the regulation of the HPA axis and in CRH mRNA in the PVN and amygdala, particularly in the female offspring.

In mice, maternal restraint stress from day 8.5 to day 20.0 of gestation caused a delayed recovery of CORT levels in adult males after immobilization stress and a decrease in the expression of GR mRNA (but not MR mRNA) in the hippocampus, hypothalamus and amygdala [51]. However, no significant changes were detected in the content of CRH in the hypothalamus or amygdala. The difference from our finding of an increase in CRH mRNA in PS rats may be due to the species, gender and type of measurement, gene or protein.

In rhesus monkeys, daily stress in midgestation from day 45 to day 90 [63], or from day 90 to day 145 [64] caused a larger increase in both basal plasma levels of ACTH and cortisol and in the response to stress in the juvenile offspring than in controls. It is not known when GR and MR appear in this species in relation to the timing of the stress or whether there is a gender difference in the response of the HPA axis of PS monkeys to subsequent stress.

Hypoxia before and during parturition and in the postnatal period, is a common causes of perinatal distress, morbidity, and mortality. The 6-month-old male offspring of rats subjected to hypoxia from prenatal day 19 to postnatal day 14 showed an increase in CORT release in response to stress and higher basal CRH mRNA levels in the PVN than control offspring. Maternal hypoxia had no effect on the levels of vasopressin mRNA in the PVN, CRH receptor-1 mRNA in the anterior pituitary or GR mRNA in the hippocampus [65].

10.4.2
Human Subjects

10.4.2.1 Effects of Prenatal Stress on Basal HPA Axis Activity

In prospective studies in humans the presence of stress during pregnancy was inferred either from maternal self reports of high levels of anxiety or distress, or because their infants had a low birth weight, which had been shown to be well correlated with increased circulating levels of CRH [43]. Others have used measures of maternal salivary cortisol as an indicator of stress [66, 67].

Based on the above indications of the presence of maternal stress, measurements were made of the basal activity of the HPA axis and/or its response to stress in children and adolescent offspring. In pre-pubertal children (aged 7–9 years), morning peak cortisol was found to be negatively correlated to birth weight in girls, while evening cortisol was positively correlated to birth weight in boys [68]. The 15-year-old male and female offspring of mothers with self reported anxiety at 12–22 weeks of pregnancy had lower than normal cortisol secretion on awakening, but this was higher than expected in the evening. In the girls only, the altered profile of diurnal cortisol was correlated with depressive symptoms. Maternal anxiety between 23–31 or 32–40 weeks of pregnancy had no effect on diurnal cortisol in the offspring [69]. These data suggest that human females like rats may show a greater sensitivity to the disrupting effects of gestational stress on the circadian release of cortisol.

10.4.2.2 Effect of Prenatal Stress on the Response to Stress of the HPA Axis

Measurements made in the 7–9-year-old children of the increase in salivary cortisol in response to a psychological stress test showed that they were negatively correlated to birth weight in boys but not in girls. Salivary cortisol was also increased in response to acute psychological stress in teen-aged male twins of low birth weight in comparison with that in boys of normal birth weight [70]. The increase in salivary cortisol in response to the first day of school after a long break in 5–6-year-old boys and girls was correlated both with maternal cortisol during pregnancy and measures of perceived maternal stress [67]. No other observations were reported on the behavioural reactions to the stress or behavior in general in the children in these studies.

In summary, the data from studies in experimental animals and humans demonstrate that chronic maternal stress that is associated with a significant elevation of circulating glucocorticoids and CRH (in humans). If gestational stress occurs during a particular period of gestation it can permanently alter the response to stress of the HPA axis in the offspring. In order to produce these changes in rats the mothers must be stressed on several days during the last week of gestation when GR receptors are present. In humans, the influence of the timing of stress on the HPA axis regulation in the offspring is less clear and can be as early as 12–22 weeks when the fetal HPA axis is responsive to maternal stress.

10.5
Alterations in the Regulation of the HPA Axis as a Result of Postnatal Stress

10.5.1
Experimental Animals

In the rat, the HPA axis matures toward adult activity during the pre-weaning period and can be still be modulated by neonatal manipulations. These include short periods of daily handling of 15 min or less, which speeds the recovery of plasma CORT after exposure to stress [71], or prolonged maternal deprivation (MD) that increases and prolongs stress-induced elevation of circulating CORT [72]. Maternal deprivation for several hours daily is a complex psychological and physiological stress involving both a lack of maternal care, hunger and thirst. When pups are maternally deprived for at least 3 hr daily from the age of 3–12 days an increase in circulating ACTH and CORT is seen in response to stress both in juvenile and adult rats compared to non-deprived rats [72–74], resembling that seen in PS rats described in a previous section. By contrast, MD of only 30 min daily over days 7–13 caused no alteration in the basal levels of CORT or in the magnitude or duration of its increase in response to stress in adulthood [75]. However, when the rat pups were immobilized individually for 30 min each day, CORT levels increased more than in controls and remained elevated for a longer time in adulthood. This alteration in the regulation of the HPA axis may result from repeated elevation of circulating glucocorticoid after each daily immobilization and reduction of GR and MR, as described for chronic prenatal stress.

10.5.2
Human Subjects

Enduring alterations in the regulation of the HPA axis were also shown in those who endured early life adversity, parental neglect and sexual abuse, all of which have been identified as risk factors for depression and anxiety disorder in adulthood [76, 77]. Women that suffered sexual abuse during childhood or adolescence showed a greater increase in circulating cortisol in response to a psychological stress test than unstressed women [78, 79]. However, a blunted increase in ACTH and cortisol was found in response to a standardized psychological stressor in asymptomatic adult male and female subjects reporting significant childhood maltreatment [80]. These differences may result from the nature, severity and duration of the early trauma and from genetic factors. Taken together, these and other studies [81, 82] demonstrate that childhood trauma can induce long-lasting effects on the regulation of the HPA axis, manifested by either hypo- or hyperactivity in response to emotional stress.

10.6
Association Between Prenatal Stress and Anxiety Disorder

10.6.1
Experimental Animals

In experimental animals it much easier than in human subjects to control the time of stress exposure during gestation when investigating its potential effect on development of anxiogenic behavior in the offspring. Thus, it was found that the offspring of Rhesus monkeys whose mothers were stressed either in early (days 45–90) [63] or in late gestation, days 90–145 (term 165 days) [83] showed more signs of anxiety than controls when exposed to novelty challenge. Anxiety symptoms were indicated by irritability, clinging to companions, less exploration and fewer social interactions. The stress paradigms used in these studies and their timing during gestation also increased the response of the HPA axis to stress in the offspring.

An inability to perform life-sustaining activities in aversive conditions is characteristic of a chronic anxiety disorder. We found that adult PS rats showed such impaired coping in a test of maternal (pup-retrieval) and food-seeking behavior. Under normal conditions, PS female rats showed the same latencies as controls to retrieve their neonatal pups from which they had been separated. However, more than 50% of PS females failed to retrieve their pups under an aversive condition of cold air puffs directed onto the alley through which they had to pass, compared to less than 5% of the controls [84]. Food seeking behavior after food deprivation was also significantly disrupted by mild foot shocks in PS females but not in controls [85]. While control rats no longer released CORT after the third exposure to the same novel environment PS rats continued to release significant amount of the glucocorticoid and void faecal pellets even after eight exposures indicating that they were much slower to learn that the environment was no longer threatening [85].

An anxiety state was modeled in rats and mice in tests involving fear-like reactions, which can be elicited in unfamiliar open spaces like the "open field" or in the elevated plus maze (EPM) and accompanied by CORT release. The EPM presents the rodent with a conflict between the desire to explore a novel situation and its fear of height and open spaces [86]. When pregnant rats were stressed once daily throughout gestation [87, 88], or by variable forms of stress [89] or restraint three times daily [90], over days 17–21 the time spent by male and female offspring in the open arms of the EPM was significantly reduced compared that in controls, indicating that they were more fearful. PS rats also displayed more fear than controls in a novel environment like a brightly lit open field [91, 92] or in the cage emergence test [60].

By using a shorter duration of daily restraint stress over days 14–21 of gestation it was possible to induce anxiogenic behavior in the EPM in the female but not in the male offspring [93]. Thus it appears, at least in rats, that females have a lower threshold than males for the induction of anxiogenic behavior and of dysregulation of the HPA axis. Others found that milder forms of maternal stress did not induce a difference in the cage emergence test between PS and control males [94, 95]. However, subjection of the PS and control rats to restraint stress prior to the cage emergence test resulted in the detection of greater fear-related behavior in PS rats, which was associated with higher concentrations of CRH in the amygdala [94]. Moreover, the difference in the latency to enter an aversive environment between PS and control subjects was abolished by intracerebral injection of a CRH antagonist [94, 95]. These findings accord with our recent observation of an association between anxiogenic behavior of PS rats to elevation of CRH mRNA in the amygdala (Zohar and Weinstock, unpublished observations). In PS rats, others have reported expansion of the lateral nucleus of the amygdala, an area in which learned fear is encoded [96], and a decrease in the number of benzodiazepine binding sites in ceA, supporting the finding of increased anxiogenic behavior in the EPM [97].

Several of the maternal stress procedures described above that increased anxiety in the offspring also altered the reactivity to stress of the HPA axis [54, 56, 63, 83].

Moreover a single period of maternal restraint for 2 h only on day 16 of gestation that resulted in a decrease in CORT release in response to stress in the offspring also induced lower levels of anxiety than in control offspring [62]. These data suggest that maternal stress can cause an alteration in the programming of defined brain regions responsible for fear-related behavior and for regulating the reactivity of the HPA axis to stress.

A critical role of excess maternal CORT release in the changes in programming of the fetal HPA axis [98] and anxiogenic behavior [93] was demonstrated by their prevention by adrenalectomy and administration of maintenance levels of CORT prior to subjecting the pregnant rats to daily stress. Injection of CORT to the adrenalectomized mothers to mimic stress levels re-instated anxiogenic behavior [99] and HPA axis dysregulation [98].

The anxiogenic behavior induced by prenatal stress in rats could be prevented by separation of the pups from the nest for brief periods of 3 min daily over days 2–10 of age (neonatal handling). This procedure had no effect on the behavior of controls subjected to normal animal husbandry of cage cleaning and handling [100]. The short handling procedure also prevented a significant number of changes induced by prenatal stress in hippocampal genes that play a role in presynaptic organization and function thereby highlighting the importance of this processes in the mediation of anxiogenic behavior [101].

In spite of the clear association described above between altered HPA axis regulation and anxiogenic behavior, other studies were able to obtain alterations in the stress response without increased anxiety and *vice versa* [60, 61]. Gestational stress in mice on days 8.5–19.5 [51], 10–18 [102] or 15–21 [103] did not induce any significant anxiogenic behavior in the male or female offspring, but nevertheless altered the regulation of the HPA axis in association with down regulation of GR receptors in several brain regions [51, 104]. The discrepancy from the findings cited earlier may depend on genetic factors, timing and intensity of the maternal stress in relation to the development of the HPA axis and critical brain areas like the amygdala for expression of heightened anxiety.

10.6.2
Human Subjects

Early emotional problems in infants and anxiogenic behavior in children are predictors of anxiety and depressive disorders in adults [105]. This observation has prompted the assessment of behavior in young children in relation to antenatal maternal anxiety in the hope of averting major anxiety or depressive disorders in adulthood. Most of such investigations based their indication of maternal stress on reports of self-assessed perceived stress, anxiety and recall of adverse events. Although the reports may reflect the women's emotional response to adverse circumstances during pregnancy, they may also signify emotional attributes unrelated to such events. These in turn, could influence the mother's perception of the infant's or children's emotional state and may confound the interpretation of the results [106].

In order to control for the presence of maternal trait anxiety, a study was performed on a large number of mother and infant pairs that assessed the incidence of emotional problems in four-year-old children in relation to maternal antenatal anxiety at 18 and 32 weeks and postnatal anxiety or depression at 8, 21 and 33 months. A significant link was found between antenatal anxiety during late gestation (32 weeks) and emotional problems in boys and girls and attention deficits in boys. By an appropriate factor analysis the authors were able to exclude postnatal anxiety and depression as causative factors and considered the behavioural outcome in the offspring to result from a mechanism operating in the prenatal period [107]. A later study conducted on 8 to 9-year-old children found that antenatal anxiety at 12–22 weeks, which was significantly higher than at 32–40 weeks, was a more significant predictor of anxiety [108]. Others also found that the associations with childhood emotional behavior were strongest when maternal anxiety occurred in early and mid-gestation [109, 110]. The difference in the effect of the timing of maternal stress in the study of O'Connor *et al.* [107] from that of Van den Bergh and Marcoen [108], was present even though maternal anxiety was assessed in a similar manner by questionnaires that measured state and trait anxiety [107], or also by increases in salivary cortisol [111]. These discrepancies reflect the difficulty encountered in human studies when attempting to relate a particular behavioural outcome to the timing of gestational stress. More studies are needed using larger numbers of pregnant women and their children to clarify how the time of occurrence of maternal stress, its chronicity and intensity affects the behavioural outcome in addition to or independent of postnatal care.

In summary, the data suggest that maternal stress during a critical phase of fetal brain development may increase the likelihood of anxiety and depressive disorders in rats and monkeys. However, the appearance of such disorders depends on the timing and intensity of the maternal stress in relation to development of critical neuronal systems, gender, conditions of the test and genetic factors. There do not appear to be any data from prospective studies in humans or non-human primates indicating such a gender difference in association with prenatal stress, although retrospective studies indicate that the incidence of a chronic anxiety state and depression is higher in women than in men [112], particularly if associated with prior stressful events [113].

10.7
Association Between Early Life Stress and Anxiety Disorder

10.7.1
Experimental Animals

A number of early studies by Harlow and his colleagues showed that maternal separation in monkeys at one week of age lead in later life to disrupted social interaction and increased anxious and depressive behaviours [114–116]. More recently it was also shown that maternal separation at this age resulted in a significant decrease in the guanylate cyclase 1α (1GUCY1A3) mRNA expression in the amygdala, the levels of which are strongly related to social behavior. 1GUCY1A3 mRNA is found in several limbic structures of the rodent brain, with highest levels in the amygdala [117].

In rats, most studies determined the effect of early maternal separation for shorter or longer periods on the regulation of the HPA axis, but a few of them also investigated their effects on offspring behavior. Periodic separation of the pups from the rat dam during the stress hypo-responsive period of the first two weeks of life is considered to be a potent naturally occurring stress. Maternal deprivation (MD) of pups for 3h daily during days 2–21 of age caused them to spend more time than control rats in the open arms of the EPM and show less neophobia toward a novel snack (particularly the females). This behavior resembled that of rats subjected as neonates to brief separation of only 15 min or less [118]. However, the controls in this study were not handled at all throughout the pre-weaning period, a procedure that is also known to increases their anxiety in comparison to those subjected to routine animal husbandry and cage cleaning. Maternal separation for 3h also resulted in lower levels of anxiety in several different mouse strains that used the same non-handled control [119]. However, when rat pups were separated 3h daily at 3–12 days and their behavior was compared to that of pups subjected to normal levels of cage cleaning and handling, anxiety-related behavior was significantly increased in both males and females in adulthood [74, 120–122]. The anxiogenic behavior was associated with a higher release of ACTH in MD males on exposure to the EPM but not in females [74], while the release of ACTH was lower than in controls in both sexes in the study by Daniels *et al.* [120]. MD for 24h only on day 9 had no significant effect on anxiogenic behavior of the adult offspring in the EPM but after prior exposure to electric shocks for 5 min, MD males but not females were more anxious than controls [123]. These MD males but not females showed more intense staining of CRH mRNA in the amygdala, but not in the PVN. No information was given about the response of the HPA axis to stress in these male rats.

The data in the foregoing section show that early life stress at a critical time during development can reduce the threshold for the precipitation of anxious behavior in later life and supports the findings in PS rats cited above. The lower threshold for fear-related behavior in males may be due to greater activity of CRH in the ceA induced in the pups by 24h of maternal deprivation on a given day in the pre-weaning period. The selective effect of this form of early life stress on the behavior and CRH signaling in males appears to be at variance with the greater susceptibility of PS females than males to show anxiogenic behavior. However, it is more likely that the a selective effect in a given gender depends on the relative amounts of stress hormones that reach critical brain areas at different stages of brain development, and the age at which the behavior of the animals is assessed.

10.7.2
Human Subjects

A large number of studies provide epidemiological evidence that stress during childhood can result in significant adult psychopathology. These include substance abuse [124], depression [125] chronic anxiety or panic disorder [126]. The problem with most of these studies is that they depend on the recollection of events that occurred many years previously by people with significant psychopathology. Nevertheless the number of such reports suggests that early life stress (ELS) may be able to cause long-lasting influences on emotional regulation.

In order to provide more critical information about the nature of such ELS on subsequent emotionality, a multinational study was performed on subjects without significant psychopathology. This examined the presence of ELS like severe family conflict, emotional abuse, death of parent or sibling, poverty and neglect and natural disaster on their emotional experiences as adults [5]. It was found that levels of anxiety were most strongly correlated with sustained family conflict, social rejection, physical or emotional abuse and separation from family. Another study showed that re-exposure of prenatally stressed individuals to an adverse event later in childhood resulted in a significantly greater incidence of anxiety and depressive symptoms around puberty in girls than in those not experiencing early life stress [127]. The lower threshold for the development of anxiety and depressive disorder after an additional trauma in those that had experienced prenatal or ELS could result from the presence of higher levels of CRH in key areas of the amygdala mRNA, as shown in experimental animals.

Summary

The data from the foregoing studies allow the inference that psychological stress or trauma occurring during prenatal period or during early life can induce enduring changes in key brain areas involved in the regulation of the HPA axis and of fear-related behavior. These alterations may be present already in childhood or become manifest after a subsequent stressful experience in later life, that could precipitate an anxiety state, depression or other mental health illness according to the nature and timing of the original psychological stress in the pregnant mother or during early childhood. Studies in experimental animals have provided a greater understanding of the nature of the changes in hormonal and neurotransmitter activity induced in the developing brain by prenatal and early postnantal stress. These include the role of glucocorticoids and CRH and their receptors in determining the regulation of the HPA axis as well as in discrete brain regions subserving fear-related behavior. Together with more focussed studies in human subjects this could lead to more effective strategies for the prevention and treatment of affective disorders induced by early life stress.

References

1 Greenberg, P.E., Sisitsky, T., Kessler, R.C. *et al.* (1999) *J. Clin. Psychiatry*, **60**, 427–435.

2 Bishop, S.J. (2007) *Trends Cogn. Sci.*, **11**, 307–316.

3 Bergman, K., Sarkar, P., O'Connor, T.G. *et al.* (2007) *J. Am. Acad. Child Adolesc. Psychiatry*, **46**, 1454–1463.

4 Brown, G.W. and Harris, T.O. (1993) *Psychol. Med.*, **23**, 143–154.

5 Cohen, R.A., Paul, R.H., Stroud, L. *et al.* (2006) *Int. J. Psychiatry Med.*, **36**, 35–52.

6 Gutman, D.A. and Nemeroff, C.B. (2003) *Physiol. Behav.*, **79**, 471–478.

7 Kessler, R.C., Davis, C.G. and Kendler, K.S. (1997) *Psychol. Med.*, **27**, 1101–1119.

8 Phillips, N.K., Hammen, C.L., Brennan, P.A. *et al.* (2005) *J. Abnorm. Child Psychol.*, **33**, 13–24.

9 Hettema, J.M. (2008) *Depress. Anxiety*, **25**, 300–316.

10 Charmandari, E., Kino, T., Souvatzoglou, E. and Chrousos, G.P. (2003) *Horm. Res.*, **59**, 161–179.

11 Barker, D.J. (1998) *Clin. Sci.*, **95**, 115–128.

12 Habib, K.E., Gold, P.W. and Chrousos, G.P. (2001) *Endocrino. Metab. Clin. North Am.*, **30**, 695–728, vii–viii.

13 Weinstock, M. (2005) *Brain Behav. Immun.*, **19**, 296–308.

14 Chrousos, G.P. and Gold, P.W. (1992) *JAMA*, **267**, 1244–1252.

15 De Kloet, E.R. (2004) *Ann. N. Y. Acad. Sci.*, **1018**, 1–15.

16 De Kloet, E.R., Vreugdenhil, E., Oitzl, M.S. and Joels, M. (1998) *Endocr. Rev.*, **19**, 269–301.

17 Makino, S., Gold, P.W. and Schulkin, J. (1994) *Brain Res.*, **640**, 105–112.

18 Noorlander, C.W., De Graan, P.N., Middeldorp, J. *et al.* (2006) *J. Comp. Neurol.*, **499**, 924–932.

19 Cintra, A., Solfrini, V., Bunnemann, B. *et al.* (1993) *Neuroendocrinology*, **57**, 1133–1147.

20 Kitraki, E., Alexis, M.N., Papalopoulou, M. and Stylianopoulou, F. (1996) *Neuroendocrinology*, **63**, 305–317.

21 Diaz, R., Brown, R.W. and Seckl, J.R. (1998) *J. Neurosci.*, **18**, 2570–2580.

22 Fujioka, T., Fujioka, A., Endoh, H. *et al.* (2003) *Neurosci.*, **118**, 409–415.

23 Carrasco, G.A. and Van de Kar, L.D. (2003) *Eur. J. Pharmacol.*, **463**, 235–272.

24 Van Bockstaele, E.J., Bajic, D., Proudfit, H. and Valentino, R.J. (2001) *Physiol. Behav.*, **73**, 273–283.

25 Foote, S.L., Bloom, F.E. and Aston-Jones, G. (1983) *Physiol. Rev.*, **63**, 844–914.

26 Lavicky, J. and Dunn, A.J. (1993) *J. Neurochem.*, **60**, 602–612.

27 Vazquez, D.M., Bailey, C., Dent, G.W. *et al.* (2006) *Brain Res.*, **1121**, 83–94.

28 Legradi, G., Holzer, D., Kapcala, L.P. and Lechan, R.M. (1997) *Neuroendocrinology*, **66**, 86–97.

29 Anand, A. and Shekhar, A. (2003) *Ann. N. Y. Acad. Sci.*, **985**, 370–388.

30 LeDoux, J.E. (1994) *Semin. Neurosci.*, **6**, 231–237.

31 Shekhar, A., Truitt, W., Rainnie, D. and Sajdyk, T. (2005) *Stress*, **8**, 209–219.

32 Sahuque, L.L., Kullberg, E.F., McGeehan, A.J. *et al.* (2006) *Psychopharmacology*, **186**, 122–132.

33 Maroun, M. and Richter-Levin, G. (2003) *J. Neurosci.*, **23**, 4406–4409.

34 Lee, Y., Fitz, S., Johnson, P.L. and Shekhar, A. (2008) *Neuropsychopharmacology*, **33**, 2586–2594.

35 Heinrichs, S.C., Pich, E.M., Miczek, K.A. *et al.* (1992) *Brain Res.*, **581**, 190–197.

36 Rassnick, S., Heinrichs, S.C., Britton, K.T. and Koob, G.F. (1993) *Brain Res.*, **605**, 25–32.

37 Takahashi, L.K., Turner, J.G. and Kalin, N.H. (1998) *Psychoneuroendocrinology*, **23**, 571–581.

38 Weinstock, M., Fride, E. and Hertzberg, R. (1988) *Prog. Brain Res.*, **73**, 319–331.

39 Sandman, C.A., Glynn, L., Schetter, C.D. *et al.* (2006) *Peptides*, **27**, 1457–1463.

40 Perkins, A.V., Linton, E.A., Eben, F. *et al.* (1995) *Br. J. Obstet. Gynaecol.*, **102**, 118–122.

41 King, B.R., Smith, R. and Nicholson, R.C. (2001) *Peptides*, **22**, 1941–1947.

42 Wadhwa, P.D., Dunkel-Schetter, C., Chicz-DeMet, A. *et al.* (1996) *Psychosom. Med.*, **58**, 432–446.

43 Inder, W.J., Prickett, T.C., Ellis, M.J. et al. (2001) *J. Clin. Endocrinol. Metab.*, **86**, 5706–5710.

44 Glynn, L.M., Wadhwa, P.D., Dunkel-Schetter, C. et al. (2001) *Am. J. Obstet. Gynecol.*, **184**, 637–642.

45 Yang, K. (1997) *Rev. Reprod.*, **2**, 129–132.

46 Dy, J., Guan, H., Sampath-Kumar, R. et al. (2008) *Placenta*, **29**, 193–200.

47 Cosmi, E.V., Luzi, G., Gori, F. and Chiodi, A. (1990) *Eur. J. Obstet. Gynecol. Reprod. Biol.*, **36**, 239.

48 Myers, R.E. (1975) *Am. J. Obstet. Gynecol.*, **122**, 47–59.

49 Koehl, M., Darnaudery, M., Dulluc, J. et al. (1999) *J. Neurobiol.*, **40**, 302–315.

50 Weinstock, M. (2008) *Neurosci. Biobehav. Rev.*, **32**, 1073–1086.

51 Chung, S., Son, G.H., Park, S.H. et al. (2005) *Endocrinology*, **146**, 3202–3210.

52 Henry, C., Kabbaj, M., Simon, H. et al. (1994) *J. Neuroendocrinol.*, **6**, 341–345.

53 Koenig, J.I., Elmer, G.I., Shepard, P.D. et al. (2005) *Behav. Brain Res.*, **156**, 251–261.

54 Maccari, S., Darnaudery, M., Morley-Fletcher, S. et al. (2003) *Neurosci. Biobehav. Rev.*, **27**, 119–127.

55 Weinstock, M., Matlina, E., Maor, G.I. et al. (1992) *Brain Res.*, **595**, 195–200.

56 Weinstock, M., Poltyrev, T., Schorer-Apelbaum, D. et al. (1998) *Physiol. Behav.*, **64**, 439–444.

57 McCormick, C.M., Smythe, J.W., Sharma, S. and Meaney, M.J. (1995) *Brain Res. Dev. Brain Res.*, **84**, 55–61.

58 Viltart, O., Mairesse, J., Darnaudery, M. et al. (2006) *Psychoneuroendocrinology*, **31**, 769–780.

59 Bosch, O.J., Kromer, S.A. and Neumann, I.D. (2006) *Eur. J. Neurosci.*, **23**, 541–551.

60 Van den Hove, D.L., Blanco, C.E., Aendekerk, B. et al. (2005) *Dev. Neurosci.*, **27**, 313–320.

61 Richardson, H.N., Zorrilla, E.P., Mandyam, C.D. and Rivier, C.L. (2006) *Endocrinology*, **147**, 2506–2517.

62 Cannizzaro, C., Plescia, F., Martire, M. et al. (2006) *Behav. Brain Res.*, **169**, 128–136.

63 Schneider, M.L., Moore, C.F., Kraemer, G.W. et al. (2002) *Psychoneuro-endocrinology*, **27**, 285–298.

64 Clarke, A.S., Wittwer, D.J., Abbott, D.H. and Schneider, M.L. (1994) *Dev. Psychobiol.*, **27**, 257–269.

65 Raff, H., Jacobson, L. and Cullinan, W.E. (2007) *J. Neuroendocrinol.*, **19**, 907–912.

66 de Weerth, C., van Hees, Y. and Buitelaar, J.K. (2003) *Early Hum. Dev.*, **74**, 139–151.

67 Gutteling, B.M., de Weerth, C. and Buitelaar, J.K. (2005) *Psychoneuroendocrinology*, **30**, 541–549.

68 Jones, A., Godfrey, K.M., Wood, P. et al. (2006) *J. Clin. Endocrinol. Metab.*, **91**, 1868–1871.

69 Van den Bergh, B.R., Van Calster, B., Smits, T. et al. (2008) *Neuropsychopharmacology*, **33**, 536–545.

70 Wust, S., Entringer, S., Federenko, I.S. et al. (2005) *Psychoneuroendocrinology*, **30**, 591–598.

71 Ader, R. (1970) *Physiol. Behav.*, **5**, 837–839.

72 Levine, S., Huchton, D.M., Wiener, S.G. and Rosenfeld, P. (1991) *Dev. Psychobiol.*, **24**, 547–558.

73 van Oers, H.J., de Kloet, E.R. and Levine, S. (1998) *Brain Res. Dev. Brain Res.*, **111**, 245–252.

74 Wigger, A. and Neumann, I.D. (1999) *Physiol. Behav.*, **66**, 293–302.

75 Yoshihara, T. and Yawaka, Y. (2008) *Physiol. Behav.*, **93**, 322–326.

76 Kendler, K.S., Kuhn, J.W. and Prescott, C.A. (2004) *Psychol. Med.*, **34**, 1475–1482.

77 Weiss, E.L., Longhurst, J.G. and Mazure, C.M. (1999) *Am. J. Psychiatry*, **156**, 816–828.

78 Heim, C., Newport, D.J., Bonsall, R. et al. (2001) *Am. J. Psychiatry*, **158**, 575–581.

79 Heim, C., Newport, D.J., Heit, S. et al. (2000) *JAMA*, **284**, 592–597.

80 Carpenter, L.L., Carvalho, J.P., Tyrka, A.R. et al. (2007) *Biol. Psychiatry*, **62**, 1080–1087.

81 Cicchetti, D. and Rogosch, F.A. (2001) *Dev. Psychopathol.*, **13**, 677–693.

82 Rohleder, N., Joksimovic, L., Wolf, J.M. and Kirschbaum, C. (2004) *Biol. Psychiatry*, **55**, 745–751.

83 Clarke, A.S. and Schneider, M.L. (1993) *Dev. Psychobiol.*, **26**, 293–304.

84 Fride, E., Dan, Y., Gavish, M. and Weinstock, M. (1985) *Life Sci.*, **36**, 2103–2109.

85 Fride, E., Dan, Y., Feldon, J. *et al.* (1986) *Physiol. Behav.*, **37**, 681–687.

86 Handley, S.L. and Mithani, S. (1984) *Naunyn Schmiedebergs Arch. Pharmacol.*, **327**, 1–5.

87 Estanislau, C. and Morato, S. (2005) *Behav. Brain Res.*, **163**, 70–77.

88 Fride, E. and Weinstock, M. (1988) *Life Sci.*, **42**, 1059–1065.

89 Murmu, M.S., Salomon, S., Biala, Y. *et al.* (2006) *Eur. J. Neurosci.*, **24**, 1477–1487.

90 Vallee, M., Mayo, W., Dellu, F. *et al.* (1997) *J. Neurosci.*, **17**, 2626–2636.

91 Dickerson, P.A., Lally, B.E., Gunnel, E. *et al.* (2005) *Physiol. Behav.*, **86**, 586–593.

92 Poltyrev, T., Keshet, G.I., Kay, G. and Weinstock, M. (1996) *Dev. Psychobiol.*, **29**, 453–462.

93 Zagron, G. and Weinstock, M. (2006) *Behav. Brain Res.*, **175**, 323–328.

94 Cratty, M.S., Ward, H.E., Johnson, E.A. *et al.* (1995) *Brain Res.*, **675**, 297–302.

95 Ward, H.E., Johnson, E.A., Salm, A.K. and Birkle, D.L. (2000) *Physiol. Behav.*, **70**, 359–366.

96 Salm, A.K., Pavelko, M., Krouse, E.M. *et al.* (2004) *Brain Res. Dev. Brain Res.*, **148**, 159–167.

97 Barros, V.G., Rodriguez, P., Martijena, I.D. *et al.* (2006) *Synapse*, **60**, 609–618.

98 Barbazanges, A., Piazza, P.V., Le Moal, M. and Maccari, S. (1996) *J. Neurosci.*, **16**, 3943–3949.

99 Salomon, S. and Weinstock, M. (2007) *Neural Plast.*, 95.

100 Wakshlak, A. and Weinstock, M. (1990) *Physiol. Behav.*, **48**, 289–292.

101 Bogoch, Y., Biala, Y.N., Linial, M. and Weinstock, M. (2007) *J. Neurochem.*, **101**, 1018–1030.

102 Nishio, H., Tokumo, K. and Hirai, T. (2006) *Int. J. Dev. Neurosci.*, **24**, 263–268.

103 Pallares, M.E., Scacchi Bernasconi, P.A., Feleder, C. and Cutrera, R.A. (2007) *Physiol. Behav.*, **92**, 951–956.

104 Ishiwata, H., Shiga, T. and Okado, N. (2005) *Neurosci.*, **133**, 893–901.

105 Pine, D.S. (2007) *J. Child Psychol. Psychiatry*, **48**, 631–648.

106 DiPietro, J.A., Novak, M.F., Costigan, K.A. *et al.* (2006) *Child Dev.*, **77**, 573–587.

107 O'Connor, T.G., Heron, J., Golding, J. *et al.* (2002) *Br. J. Psychiatry*, **180**, 502–508.

108 Van den Bergh, B.R. and Marcoen, A. (2004) *Child Dev.*, **75**, 1085–1097.

109 Huizink, A.C., de Medina, P.G., Mulder, E.J. *et al.* (2002) *J. Am. Acad. Child Adolesc. Psychiatry*, **41**, 1078–1085.

110 Huizink, A.C., Robles de Medina, P.G., Mulder, E.J. *et al.* (2003) *J. Child Psychol. Psychiatry*, **44**, 810–818.

111 Davis, E.P., Glynn, L.M., Schetter, C.D. *et al.* (2007) *J. Am. Acad. Child Adolesc. Psychiatry*, **46**, 737–746.

112 Wauterickx, N. and Bracke, P. (2005) *Soc. Psychiatry Psychiatr. Epidemiol.*, **40**, 691–699.

113 Sherrill, J.T., Anderson, B., Frank, E. *et al.* (1997) *Depress. Anxiety*, **6**, 95–105.

114 Harlow, H.F. and Suomi, S.J. (1971) *J. Autism Child. Schizophr.*, **1**, 246–255.

115 Harlow, H.F. and Zimmermann, R.R. (1959) *Science.*, **130**, 421–432.

116 Suomi, S.J., Eisele, C.D., Grady, S.A. and Harlow, H.F. (1975) *J. Abnorm. Psychol.*, **84**, 576–578.

117 Sabatini, M.J., Ebert, P., Lewis, D.A. *et al.* (2007) *J. Neurosci.*, **27**, 3295–3304.

118 McIntosh, J., Anisman, H. and Merali, Z. (1999) *Brain Res. Dev. Brain Res.*, **113**, 97–106.

119 Millstein, R.A. and Holmes, A. (2007) *Neurosci. Biobehav. Rev.*, **31**, 3–17.

120 Daniels, W.M., Pietersen, C.Y., Carstens, M.E. and Stein, D.J. (2004) *Metab. Brain Dis.*, **19**, 3–14.

121 Diaz Lujan, V.E., Castellanos, M.M., Levin, G. and Suarez, M.M. (2008) *Int. J. Dev. Neurosci.*, **26**, 415–422.

122 Veenema, A.H., Reber, S.O., Selch, S. *et al.* (2008) *Endocrinology*, **149**, 2727–2736.

123 Barna, I., Balint, E., Baranyi, J. *et al.* (2003) *Brain Res. Bull.*, **62**, 85–91.

124 Harrison, P.A., Fulkerson, J.A. and Beebe, T.J. (1997) *Child Abuse Negl.*, **21**, 529–539.

125 Infrasca, R. (2003) *J. Affect. Disord.*, **76**, 103–111.

126 Friedman, S., Smith, L., Fogel, D. *et al.* (2002) *J. Anxiety Disord.*, **16**, 259–272.

127 Costello, E.J., Worthman, C., Erkanli, A. and Angold, A. (2007) *Arch. Gen. Psychiatry*, **64**, 338–344.

Part IV
Immune Responses

Stress – From Molecules to Behavior. Edited by Hermona Soreq, Alon Friedman, and Daniela Kaufer
Copyright © 2010 WILEY-VCH Verlag GmbH & Co. KGaA, Weinheim
ISBN: 978-3-527-32374-6

11
Stress Effects on Immunity in Vertebrates and Invertebrates

Michael Shapira

11.1
Introduction

The general notion is that stress is immunosuppressive. This goes back to Hans Selye's seminal 1936 paper [1], which described shrinking of the thymus and other lymphoid organs in diversely stressed rats. Whereas the importance of the thymus as the organ in which T lymphocytes differentiate and mature was not known at the time, it became apparent not much later that stress hormones have anti-inflammatory effects [2]. This forms the basis for many current-day anti-inflammatory drugs. In accordance with this, studies both in men and in rodents show that stress increases the susceptibility to various infectious diseases [3–5]. At the same time, however, it was observed that stress also causes enhanced inflammatory responses representing uncontrolled activation of the immune system [6]. This, in turn, causes a whole different array of additional pathologies, that is, autoimmune, inflammatory and allergic disorders. Two-faceted in its effects, stress emerged as uniquely detrimental to our immunity. This view is now beginning to change as more recent studies raise the distinction between acute, or short-term, and chronic stress responses. Data accumulated in detailed experiments shows that the acute stress response has an adaptive advantage in preparing the body for wound healing and pathogen invasion, in line with the fight or flight response [7]. Similar to other aspects of the fight or flight response, immune responses should be terminated. Termination is embedded within the stress response itself. Thus, depending on the phase of the stress response it could either activate specific branches of the immune system or repress them, leading to variable outcomes under different stress regimes. In this chapter, I attempt to address this complexity and for the sake of some simplification I focus more on systemic stress effects, leaving stress-immune interaction in the central nervous system not addressed. This is the focus of Chapters 12 and 13 in this book.

To better understand the different effects of stress on the immune system it is therefore necessary to consider the timeline of the stress response, the stress

Stress – From Molecules to Behavior. Edited by Hermona Soreq, Alon Friedman, and Daniela Kaufer
Copyright © 2010 WILEY-VCH Verlag GmbH & Co. KGaA, Weinheim
ISBN: 978-3-527-32374-6

mediators acting in different phases and the different components of the immune system affected.

Additional insights about the effects of stress on immunity may come from simpler organisms. Molecules highly similar to vertebrate stress hormones were identified in invertebrates such as annelids, insects and mollusks. Moreover, exposure of these creatures to such hormones induces what was termed a proto-stress response, similar in many ways to the vertebrate response (reviewed in [8]). Although the extent of this similarity is yet not clear, the prospect of using model invertebrates such as *Drosophila* or *Caenorhabditis elegans* to understand some of the molecular mechanisms underlying stress effects on immunity merits discussion.

In addition to classic model invertebrates, a new source of relevant data is recent studies of marine invertebrates, driven by a growth in aquaculture of various crustaceans and mollusks, which is limited to a great extent by infectious diseases [9, 10]. Such studies provide a constantly increasing knowledge about immune functions and their regulation by stress in simple organisms, and through comparative analysis could shed more light on the effects of stress of the vertebrate immune system.

11.2
The Neuroendocrine Stress Response in Vertebrates and Invertebrates

Time line for stress responses is variable. Its onset and duration depend on the stressor. However, a basic framework could be described which holds true for a classic fight or flight response (occurring for example in a prey being hunted by a predator; see [11]). Two principal systems contribute to this response: (i) the hypothalamic–pituitary–adrenal (HPA) axis, which utilizes a neuroendocrine cascade; and (ii) the autonomous sympathetic nervous system, which sends processes from the brain to the spinal cord and from spinal cord ganglia to internal organs, and uses mostly the neurotransmitter norepinephrine, excluding neurons innervating the adrenal gland, which use acetylcholine.

Acute Stress Within seconds of exposure to a stressor, the HPA axis is activated [11]. It starts with secretion of corticotropine-releasing hormone (CRH) from the hypothalamus, the major site of neuroendocrine control. CRH activates secretion of the pro-opiomelanocortin (POMC)-derived adrenocorticotropic hormone (ACTH, aka corticotropine) from the pituitary gland into the blood stream. Minutes later, ACTH reaches the adrenal cortex where it activates secretion of the gluco-corticoid hormone (GCH), the main stress hormone in vertebrates. This activation lasts more than 1 h following the initial stimulus and is turned off by a negative feedback loop in which GCH reaching the brain shuts down ACTH secretion. In parallel, autonomous centers, also in the hypothalamus, are activated and further activate chromaffin cells in the adrenal medulla to secrete catecholamines (CA), mostly epinephrine (EPI). EPI secretion is also activated by GCH.

Chronic Stress GCH secretion from the adrenal is subjected to several layers of negative feedback regulation. Secretion in response to acute stress turns itself off within 1–2 h, by directly inhibiting both steady-state and stress-response ACTH secretion from the pituitary gland. This repression may last up to 24 h [12]. Repeated surges in GCH secretion result in a delayed inhibition, which may last for days and weeks. This is caused by inhibition of the synthesis of both CRH in the hypothalamus and ACTH in the pituitary gland. Repeated sessions of acute mild stress such as immobilization were shown to cause an increase in GCH steady state, elevating significantly GCH levels in the nadir of the diurnal cycle, but as a consequence of the latter, dampening the cycle. In addition, a repeated stress regime attenuates the acute stress response, leading to GCH surges that are both delayed and weaker [13].

A typical situation of chronic stress is usually not associated with repeated cycles of acute stress, but rather a continuation of a moderate stress (e.g., social stress). A typical example is the stress associated with caring for a sick spouse; another is being a part of hostile marriage [3, 14, 15]. Experimental models of social stress were also developed to enable detailed analysis and manipulation of stress parameters. The chronic subordinate colony housing protocol (CSC) is one such model, established both in rats and in mice, which focuses on males that were defeated by a dominant peer and remain housed with it, enduring harassments [16, 17]. One such study carried out for 19 days showed that chronic stress caused hypertrophy of the adrenal, appearing within 2 days of stress, and was associated with increased EPI steady state levels in the blood. GCH blood levels, measured at their highest point in the diurnal cycle, were only transiently increased and stabilized at a lower level. At the end of the 19 days adrenal cells showed reduced responsiveness to ACTH [17]. A similar study showed that, while peak GCH levels were lower after chronic stress, the overall steady-state levels were higher [18]. Similar to the latter, a study in men subjected to chronic confinement stress of several months in a spaceflight simulator showed increased urine norepinephrine (NE) and a dampened GCH diurnal cycle [19].

It thus appears that while acute stress is manifested in high levels of NE/EPI and GCH, chronic stress is associated with slightly higher levels of EPI and slightly higher, or unchanged levels of GCH.

11.2.1
Stress Responses and Hormones in Invertebrates

Both CRH- and ACTH-like molecules, as well as glucocorticoids can be identified by immunhistochemistry in invertebrate hemocytes, cells that are circulating in the invertebrate hemolymph or coelom. Protein sequence determination for some of these molecules showed high similarity to their human counterparts [20]. The catecholamines (CA) dopamine (DA), NE and EPI are also common in invertebrates, and EPI is found in the hemolymph of arthropods and mollusks. Moreover, incubation of snail hemocytes with CRH or ACTH resulted in a rapid release of NE (reviewed in [8]), suggesting conservation of the basic hormonal cascade.

In a series of experiments Ottaviani *et al.*, examined the function of these hormones in the freshwater snails *Planorbarius corneus* and *Viviparus ater*, using the human hormone homologs. Within minutes of exposing snail hemocytes to

ACTH, levels of DA, EPI and NE within the hemocytes increased, followed by their release to the hemolymph. The peak of release was 15 min after the initial exposure and it lasted for 45 min [21]. EPI release, which could be induced also with CRH exposure, was attenuated when cells were treated with anti-ACTH antibodies. Furthermore, in the heart of the oyster *Crassostrea gigas* cells were identified that are similar to vertebrate chromaffin cells and responded to ACTH injection with release of NE [22]. These studies suggest a sequence of events similar to that described in vertebrate stress response where CRH regulates ACTH secretion, which in turn causes a transient increase in the release of CA [8].

As for the natural conditions that induce stress responses in invertebrate, it was shown that at least some stressors relate to changes in the organism's environment. Among those, prolonged exposure to high salinity stress was found to cause an increase in hemocyte CRH immunoreactivity in mussels [23]. Similarly, in a study performed in oysters, changes in salinity and temperature caused a long-lasting (up to 72 h) increase in hemolymph CA [24]. In contrast to these long lasting changes, mechanical disturbance, representing an acute stressor, was found to cause a rapid and transient increase in hemolymph EPI levels in abalone [25], and mechanical agitation of oysters caused a transient increase in both NE and DA hemolymph levels [22]. Such studies describe a relatively simple system that could be used to assess the extent to which stress causes an adaptive response and when it becomes a burden. At this point, however, the distinction between acute and chronic stress in most studies of mollusk stress responses is yet not emphasized.

Although stress responses in insects are less well described than those of mollusks, they do show some similarities. In general, the insect equivalent of EPI is octopamine. Together with DA, these biogenic amines are the main stress hormones in insects. Their levels increase in response to both short-term stress, such as handling, and to long term stress such as social stress, immobilization, temperature changes, high population density, vibration, starvation and insecticides [26]. Such responses were described in locusts, cockroaches and *Drosophila*. In adult *Drosophila*, it was shown that the initial increase in biogenic amine levels appears within 15 min of exposure to a stressor accompanied by an initial decrease in the levels of their synthesizing enzymes. As stress continues, the levels of the synthesizing enzymes increase to contribute to maintenance of elevated biogenic amine levels [27]. Later in the response, levels of the juvenile hormone increase. This is probably initiated by the increase in biogenic amines [28]. Changes in this important developmental hormone, together with the function of ecdysteroid hormones, delays larval development and metamorphosis and promotes diapause [29].

11.3
The Immune System in Vertebrates and Invertebrates

The immune system in vertebrates consists of two branches, innate and adaptive immunity. The adaptive immune system was until not too long ago the better-understood branch, with qualities that are most associated with immunity–

specificity and memory. These qualities rely on the recognition of unique molecular determinants, antigens, by membrane receptors on lymphocytes, the cells responsible for most functions of the adaptive immune system. However, adaptive immunity is a relatively recent invention in evolution, appearing first in its current form in jawed fish [30]. Furthermore, as was first pointed out by Charles Janeway, it is not a stand-alone defense mechanism and needs the innate immune system for its activation [31–33]. Innate immunity is the more ancient of the two branches and exists in all animals, from amoebas to humans. Invertebrates rely solely on this system for their antimicrobial defense [34, 35]. Numerous studies are now focused on identifying and characterizing the conserved mechanisms that make innate immunity in vertebrates and invertebrates and the way such mechanisms affect the function of vertebrate adaptive immunity. In this section, I describe the basics of the two systems focusing on elements that are most affected by stress.

Vertebrate Adaptive Immunity This branch provides a very specific defense mechanism, which targets molecular determinants (antigens) that are recognized as "non-self", or foreign. The adaptive immune system takes long to be activated upon first exposure to a foreign determinant (in the range of days), but responds much faster and stronger in subsequent exposures, hence the term adaptive (acquired is another term used). Furthermore, once activated the adaptive immune system maintains a long-lasting memory of a pathogen. This property forms the basis for vaccination.

The providers of adaptive immune functions are the B and T lymphocytes. B lymphocytes are responsible for humoral responses (disseminated by the circulatory system). Upon activation they proliferate and differentiate to plasma cells, which secrete antibodies directed against a specific antigen. Such responses are most efficient against extracellular pathogens or secreted toxins. Adorning such foreign agents with antibodies impairs their activity and marks them for phagocytosis and subsequent hydrolysis (see below).

T lymphocytes are important components of cell-mediated immunity. Once activated, they proliferate and differentiate to cytotoxic cells. These cells bind to cells presenting the antigen they recognize and secrete pore-forming cytotoxins and apoptosis-inducing enzymes, leading to its death. They are most useful against intracellular pathogens (e.g., viruses) which expose their associated antigens on the surface of infected cells, or against cells bearing antigens representing "altered self" (e.g., tumor cells).

A second type of T lymphocytes are Helper T cells (T_H), which are important for activating both cytotoxic T cells and antibody-secreting B cells. The choice between one kind of immunity or another (cell-mediated vs humoral) is determined to a large extent by the differentiation path taken by naïve T_H cells. This path could lead to two major fates, Th1 (pro-inflammatory, supporting cell-mediated immunity) and Th2 (anti-inflammatory, supporting humoral immunity); a third important cell fate, Th17, is not discussed in this chapter. The differentiation path is initiated by signaling proteins called cytokines (or interleukins). IL-12 and interferon (IFN)γ initiate differentiation to Th1 cells; IL-4 drives differentiation to Th2 cells. Cytokines are also the critical messengers secreted by mature T_H cells to mediate their influence on other cells. IFNγ is the signature cytokine expressed by Th1 cells, whereas IL-4 secretion is a signature of Th2 cells [36, 37]. With these cytokines T_H cells orchestrate much of the adaptive immune responses and are of crucial importance. This is the reason for the devastating consequences caused by infection and destruction of these cells by the HIV virus.

Vertebrate Innate Immunity   Innate immunity is a more ancient defense mechanism, existing in one form or another in all animals and providing a rapid response (in the range of minutes to hours) to invading pathogens, not affected by previous experience. The responses produced by this system are stereotypical, in the sense that they are not extremely specific (as the adaptive immune system is). Innate immune responses are activated by recognition of conserved molecular patterns typical of certain classes of pathogens, for example, by lipopolysaccharide (LPS), which is a major component of the envelope of Gram-negative bacteria, or by dsRNA typically produced in virally infected cells [38]. However, these stereotypical responses are quite efficient in dealing with most pathogens, as evident by the evolutionary success of many invertebrate phyla including more than a few species with a long lifespan (e.g., the lobster, which can live according to some reports up to 100 years). Moreover, these responses are of utmost importance for telling "self" from "non-self" and for determining subsequent responses of the adaptive immune system [33].

The innate immune system includes physical barriers such as the skin epithelium and mucosal surfaces in the lungs and intestine. In addition, similar to the adaptive immune system, it presents both humoral and cell-mediated responses (reviewed in [39]). Humoral responses of the innate immune system include secretion of various antimicrobial pore-forming peptides and enzymes that hydrolyze the microbial envelope (e.g., lysozymes). They also include the complement system, which is activated by a proteolytic signal-amplifying cascade, causing damage to microbial envelopes as well as marking pathogens for further treatment [40, 41]. This variety of humoral responses is produced by different cells and in many cases is initiated by cells of the tissue first in contact with the pathogen, such as epithelial cells. The same cells also secrete cytokines, and together with cell attractant molecules called chemokines, recruit white blood cells to the site of infection and activate them to generate the necessary response.

The cell-mediated component of innate immune responses includes several specialized cell types that similar to lymphocytes are grouped under the umbrella of white blood cells. Cytotoxic cells include granulocytes (neutrophils, eosinophils, basophils), cells that obtain their name from the granules in which they store antimicrobial peptides and reactive oxygen species which they release upon contact with a pathogen. Natural killer (NK) cells are also cytotoxic cells and a special subgroup of T lymphocytes. Unlike most lymphocytes, however, they do not need to be activated before they become cytotoxic, hence the name "natural killers". They also do not alter their responses in subsequent exposure to a pathogen making them a part of the innate immune system rather than the adaptive.

Another type of innate immune cells are phagocytic cells (macrophages, neutrophils, dendritic cells) that are either resident to epithelial tissues or that, by following chemotactic signals, leave blood capillaries to migrate to sites of wound or infection. They engulf cellular debris and invading pathogens, internalize and degrade them in special organelles. Once degraded, molecular components of the pathogen can serve as antigens, presented on membrane receptors where they fulfill an important role, together with secreted cytokines, in activating cells of the adaptive immune system.

Invertebrate Immunity As mentioned above, all animals have an innate immune system. However, these differ between organisms, with the different body plans constraining the development of various parts of this system. Invertebrates that have a circulatory system (mainly coelomates, e.g., echinoderms (sea urchin, star fish), insects, crustaceans and mollusks) have both cell-mediated and humoral immunity; pseudocoelomates (e.g., *C. elegans*), which have a separated body cavity but no circulatory system, rely mostly on local humoral responses; acoelomates, such as cnidaria (e.g., corals), mostly do not have a circulatory system and rely on cell-mediated immunity.

Cell-mediated immunity in invertebrates depends on hemocytes as they are called in insects or in mullosks. Echinoderms have cells with similar properties that are called coelomocytes and cnidarians use similar cells called amoebocytes (*C. elegans* also have coelomocytes in their pseudocoelum, but their relation to immunity is still unknown). As a general rule such cells are capable of pathogen recognition, attachment followed by agglutination, phagocytosis and elimination of invaders by respiratory burst or exocytosis of antimicrobial factors. Those that are circulating, such as hemocytes, are also capable of chemotaxis and infiltration to sites of infection [42, 44].

The humoral component of invertebrate immunity includes reactive oxygen species, a diverse array of antimicrobial peptides effective against different classes of pathogens [45, 46] and various hydrolyzing enzymes (lysozymes, proteases pholapholipases, etc.). Components of the complement system perform vital roles in echinoderms immunity and genes homologous to vertebrate complement genes were identified even in cnidaria [43, 47]. In addition, invertebrates use another proteolytic cascade, the prophenoloxidase (proPO) system, which generates toxic intermediates and drives melanisation and encapsulation of pathogens, restricting their movement inside the host and aiding in eliminating them [48].

The invertebrate humoral response originates in three locations: cells local to the site of infection (mostly of epithelial origin), specialized organs, such as the insect fat body [46], and in hemocytes, the same cells described above as the main source of invertebrate stress responses. The contribution of these cells to both stress and immunity raises the possibility that hemocytes integrate stress signals and immune responses and thus form a functional parallel to the vertebrate HPA axis. There are great implications in such a possibility for simplifying research of the principles governing stress effects on immunity and the cellular mechanisms underlying it.

11.3.1
Effects of Acute Stress on Immunity

11.3.1.1 Vertebrates

The effects of stress on immunity in vertebrates were believed for many years to be generally suppressive. Hallmarks in this suppression included decreased numbers of circulating leukocytes, suggested to be associated with reduced proliferation; killing of developing T lymphocytes with the subsequent shrinkage of the thymus; and a shift from Th1 (inflammatory, cell-mediated) responses to Th2 (anti-inflammatory, humoral) responses (reviewed in [36, 49]), accompanied by elevated production of antibodies from B cells [50, 51]. Excluding the increase in antibody production, that is thought to be mediated by CA secretion, much of stress effects have been attributed to the action of adrenal GCH acting through the low affinity glucocorticoid receptor (GR), a nuclear hormone receptor that is expressed in most cells including those of the immune system [52, 53]. The immune suppressive qualities of GCH are well known and are used in clinical settings to treat autoimmune diseases, inflammation and for preventing organ rejection after transplantation [54].

Studies of the effects of GCH and its activated receptor found multi-level effects on the immune system. GCH blocks the initial steps necessary for signaling the identification of a "non-self". It rapidly lowers circulating levels of macrophages, eosinophils, basophils and lymphocytes by changing expression of cell adhesion molecules ([55]; reviewed in [11, 56]); conversely, circulating neutrophil numbers increase as are various antimicrobial proteins that they express [57]). GCH also inhibits synthesis of pro-inflammatory cytokines in macrophages (IL-1β, IL-6, IL-12) [58–61] and interferes with the migration and maturation of antigen-presenting dendritic cells (DCs), thus impairing the presentation of antigens acquired in the periphery to T cells in lymph nodes and the initiation of Th1 differentiation [62]. GCH is also effective in directing the shift from Th1 pro-inflammatory responses, such as IL-2 and IFNγ secretion, to Th2 anti-inflammatory responses, such as IL-10 secretion [63–65]. The anti-inflammatory effects of GCH occur in part through GR-regulated transcription of anti-inflammatory genes (e.g., annexin-1, IL-1 receptor antagonist), but mostly through transrepression, which is either dependent on direct binding of the hormone-receptor complex to promoters, or more frequently, on interference with the function of AP-1 and NF-κB, important transcription factor regulators of pro-inflammatory responses (reviewed in [66]).

The immunosuppressive effects of stress and stress hormones, discussed above, described a paradox in which one facet of an evolutionary adaptive feature, the fight-or-flight response appeared to be non-adaptive, suppressing a system that one would think might be useful in case of wounding. However, examples deviating from this trend were also reported. Studies in humans showed that specific lymphocyte subpopulations, mainly NK and cytotoxic T cells, increased rapidly and transiently following a mild acute mental stress [67, 68]. This increase was activated by β-adrenergic receptors (βAR)[69], which are expressed on most immune cells, although in significantly different levels [53]. Reports of additional βAR-dependent immune changes occurring after acute stress include attenuation of pro-inflammatory cytokine secretion (IL-1β, IL-12, TNFα) from phagocytic cells [70, 71], suppression of Th1 responses, such as IFNγ and IL-2 secretion [72, 73] and enhancement of anti-inflammatory (Th2) responses, such as IL-10 secretion [70]. It is yet unknown whether these effects are attributed to adrenally secreted EPI or to direct sympathetic innervation of lymphoid organs (for the latter, see [74]).

Further support for the adaptive nature of acute stress immune changes comes from studies performed in rats by Dhabhar *et al.* [13, 75, 76]. They found that acute stress, unlike chronic stress, has adaptive effects on the immune system, both of the innate and adaptive branches. Acute stress caused a rapid increase in blood GCH levels accompanied by a decrease in the number of circulating T cells, B cells, NK cells and monocytes (precursors of phagocytic cells). However, the decrease in white blood cells did not simply represent decreased immunity, as these leukocytes were not gone but redistributed from the blood to other tissues, to organs such as lymph nodes and to the skin. There, they were suggested to be more available for wound healing and antimicrobial defense [76]. Leukocyte redistribution is controlled mainly by the HPA axis and mediated by adrenal hormones, particularly by GCH – adrenalectomy was sufficient to abolish both redistribution and the resulting "peripheral immunity"; administration of GCH resulted in redistribution similar to that occurring during stress [77]. It should be noted, as the authors did, that such cell redistribution, while advantageous in anticipation of wound healing, might expose individuals to blood-borne pathogens.

In summary, new themes emerge in studies of the effects of acute stress on the immune system. The most prominent of those is enhancement of cell-mediated immunity. It includes a transient increase in circulating cells as well as redeployment of cells from the blood to the periphery, both of the innate and adaptive branches. On the mechanistic level, both CAs and GCH contribute to acute stress responses. It appears though that adrenal hormone secretion is more prominent than direct sympathetic innervation.

Redistribution of leukocytes to the skin occurs not only in rodents, but also in people following acute social stress [78]. In addition, an acute psychological stress test was shown to increase the activation of the complement system [79]. Altogether, this data demonstrates how ancient adaptive responses are mistakingly activated by perceived threats associated with modern life.

Additional similarities were shown as part of a meta-analysis of almost 300 papers published by 2001 [80]. This analysis found that acute time-limited stress (e.g., public speaking) was largely reported to increase NK cells and cytotoxic T cells (or cell-mediated immunity) in the blood. This is similar to the very initial and transient CA-dependent phase observed in rodents (see above). B and helper T cells did not show any changes, but their proliferative capacity decreased. IL-6 and IFNγ secretion increased, representing increased Th1 responses. Other studies examining cases of a pre-exam stress showed no marked effect on cell numbers, but did show a Th1 to Th2 shift (decreased IFNγ secretion; increased IL-10). Although the results in the two types, or protocols, of acute stress should have been identical some discrepancies exist. Whether those are artifacts of sampling or that they represent real differences in the stress conditions is not clear. However, such discrepancies demonstrate the complexity, which is the inherent problem in studying stress and immunity in vertebrates.

11.3.1.2 Invertebrates

In theory, it should be simpler to control stress protocols in invertebrates, which respond more to environmental conditions than to psychological factors (if at all), enabling a better distinction between acute and chronic stress. In practice, however, little attention has been given so far to doing so. The result is sometimes contradictory reports as to the effects of stress on the invertebrate immune system.

The general trend is, similar to the old dogma in vertebrates, immunosuppression and increased susceptibility to infectious diseases (reviewed for the case of abalone in [81]). Exposure to biogenic amines causes a decrease in most immune parameters in various invertebrates. In shrimp, both NE and DA injections caused a transient and rapid (within 2 h) decrease in phenoloxidase activity, respiratory burst and phagocytic activity [82, 83]. In oyster hemocytes, NE, acting through the βAR was found to inhibit hemocyte phagocytic activity and respiratory burst [84, 85].

However, several reports that employed acute stress protocols showed that, in mollusks as in the more recent reports coming from vertebrates, stress initially enhances immunity. A 15-min shaking of abalone resulted in a transient increase of CAs returning to baseline levels as soon as the stress ended. The number of circulating hemocytes and their migratory and phagocytic activities decreased while shaking was ongoing, but climbed rapidly to four times higher-than-normal levels once it ended, returning to baseline levels 6 h later [25]. A similar protocol applied to oysters showed the same outcome for all immune parameters including this time also enhanced respiratory burst [84]. Mussels also showed an initial increase in the number of circulating hemocytes following several types of stress including salinity, mechanical stress and air exposure [86].

Studies in insects of the connection between stress and immunity are scarce, but so far support the adaptive nature of the acute stress response. Flight or fight responses in insects include the secretion of the neurohormone octopamine, an NE equivalent. Octopamine was shown to increase hemocyte phagocytic activity in the cockroach, resulting in increased resistance to bacterial infection. Further analysis of the molecular mechanisms underlying these effects showed that octopamine function was mediated by its receptor and that downstream to receptor activation production of the second messanger inositol triphosphate was essential [87]. Interestingly, octopamine receptors are also expressed on cells of the main source of antimicrobial peptides, the fat body [88].

Reports as those described above demonstrate that the protective nature of stress initial effects on immune functions is an evolutionarily conserved feature. This suggests that these effects contribute to adaptation. At the molecular level, biogenic amines appear to mediate acute stress effects on the immune system both in vertebrates and invertebrates. This opens the door to using invertebrates as a tool to study the molecular mechanisms underlying the interface between stress and the immune system. In addition to biogenic amines, it is possible that glucocorticoid hormones participate in invertebrate stress responses. However, their roles are yet to be determined.

11.3.2
Effects of Chronic Stress

11.3.2.1 Vertebrates
Most of the deleterious effects of stress on the immune system, some of which are described above, are probably the consequences of chronic stress, a constant preparedness for catastrophic events that takes its toll on homeostatic processes. Dhabhar *et al.* showed how some potentially beneficial effects of acute stress are lost when the same acute stress regimen is repeated many times [13]. Chronic stress dampens the diurnal cycle of GCH and its stress-induced responses. It also attenuates the decrease in circulating leukocytes and the cell-mediated skin immune response.

The CA-dependent increase in circulating cytotoxic T lymphocytes and NK cells, occurring through activation of the βAR (see above) are also lost following chronic stress, leading to lymphopenia. This is caused both by direct effect of chronic CA treatment on T cell proliferation but also by attenuating levels of IL-2, which is necessary for co-stimulation of T cell proliferation [36].

The major themes in chronic stress are adrenal hypertrophy, accompanied by elevated steady-state EPI and GCH levels and a dampened GCH diurnal cycle, and shrinkage of the thymus [17, 19]. Thymus shrinkage is thought to be caused by GCH-dependent decrease in proliferation of naïve T cells, further leading to reduced numbers of circulating lymphocytes [89]. In contrast, chronic stress, both in rodents and in humans, enhances pro-inflammatory responses, including elevated levels of plasma cytokines such as IL-6, IFNγ and TNFα [90–92] and increased number of circulating granulocytes [19, 93]. Whereas elevated cytokines are caused by reduced sensitivity to GCH (see more below), the increase in granulocyte numbers is thought to be caused by increased steady state levels of CAs [93]. The enhancement of the pro-inflammatory axis was shown in humans to be associated with slower wound healing [14, 90]. Additionally, together with Th1 cytokines, this enhancement is probably what forms the basis for the detrimental effects of stress on inflammatory, autoimmune and allergic pathologies. Unlike acute stress that has the potential to increase the number of B lymphocytes and humoral immunity, chronic stress attenuates B lymphocyte responses and antibody production together with cell-mediated immunity [94, 95]. This general suppression is reflected, in chronically-stressed individuals (e.g., those caring for demented spouses), in reduced humoral and cellular responses to vaccination, either against viral or bacterial pathogens [53].

In rodents, social stress causes reduced GCH sensitivity in spleen macrophages and increased release of the pro-inflammatory cytokine IL-6. This may be a general mechanism through which chronic stress reverses some of the effects of acute stress. The mechanism(s) responsible for decreased GCH sensitivity are not fully comprehended. Several reports showed that in macrophages it occurs through reduced nuclear translocation of GR [91, 92, 96]. Upregulation of the dominant negative GR isoform GRβ is another possible mechanism [56].

In summary, chronic stress is the "bad stress" (or distress, as coined by Dhabhar *et al.*). The combination of increased CA levels and GCH resistance seems to be sufficient for most, if not all, of the pathological effects of stress. If the purpose of stress hormones is to direct the immune system to provide protection from possible wounding and to keep in check possibly detrimental inflammatory responses, their chronic activation seems to activate negative feedback mechanisms that prevent them from doing so, leading to rampant and improper immune system activation.

11.3.2.2 Invertebrates

As discussed above, the distinction between acute and chronic stress studies in invertebrates is, to date, not strongly emphasized. However, several studies focusing on the effects of environmental conditions on the immune system and the well-being of such organisms can provide some information on the effects of chronic stress on immunity. At the same time, such studies may lead to understanding how stress may select different modes of immunity.

In abalone, increased ammonia levels, which typically follow increased animal density, caused within 24 h a decrease in most measured immune parameters, including total hemocyte counts, phenoloxidase activity, phagocytic activity and clearance efficiency of the bacterial pathogen *Vibrio parahaemolyticus*, but not the release of reactive oxygen species in a respiratory burst [97]. This is similar to the effects of chronic stress on the vertebrate immune system and suggests that crowding causes stress, which ultimately leads to increased susceptibility to infectious diseases. However, under some conditions such infectious diseases may also select for better-adapted individuals that are more resistant to infectious diseases. This was supported by a study in locusts showing that animals reared under crowding conditions had elevated antimicrobial hemolymph activity and were more resistant to a natural fungal pest [98]. It is not clear what is the reason for the differences shown in these two studies between mollusks and insects. One possibility is that they arise due to different selective forces acting on the different organisms. Using invertebrate system, it should be possible to conduct studies of the core questions regarding how immune modulation by stress could lead to adaptation.

Pollution, such as mercury or crude oil contamination, may also represent a situation of chronic stress. A microarray analysis performed in mussels showed that such stress caused an increase in the expression of metalothionein genes, which encode metal sequestering genes, and a decrease in genes encoding lysozymes and antimicrobial peptides [99]. Water temperature is another factor that may effect immunity in marine invertebrates and is naturally of growing interest as global warming becomes a greater issue. A recent report shows that corals have increased number and density of amoebocytes, their immune cells, in higher temperatures [100]. What are the limits of this beneficial effects of water temperature on coral immunity is yet unknown.

Summary and Future Prospects

It is apparent from this chapter that the effects of stress on immunity in diverse organisms share some common features. At the basis are the adaptive consequences of the stress response. In its essence stress appears to lead to better adaptation, preparing the organism for possible wounding and infection. It is the toll of prolonged activation that makes stress detrimental, but the point when good becomes bad is still unknown. In fact, determining this point will be very important for future attempts to diminish the detrimental effects of stress responses while maintaining the protective ones. This is complicated in vertebrates. Instead, one could start with the much simpler invertebrates.

Similarities are not limited to the basics of the phenomenon. Both mollusks, and to a lesser degree insects, show similarities in the hormonal milieu that mediates stress responses and affects the immune system. This would certainly contribute to the relevance of invertebrate models in elucidating the cellular and molecular underpinnings of stress hormones effects on immune functions.

With invertebrate model organisms in mind, it might be useful to think about the classic model organisms *Drosophila melanogaster* and *Caenorhabditis elegans* as systems to approach questions regarding the interface between stress and immunity. I have discussed these two organisms only briefly in this chapter, the reason being that their stress responses are not as understood as those of mollusks. Nevertheless, as discussed above, similarities in stress hormone function as well as the principles of the stress response are demonstrated in insects and thus suggest its usefulness for studying stress and immunity. The genetic tractability of *Drosophila* and the extensive biological tool kit it brings with it make it a potentially powerful system for such studies.

The worm *C. elegans* offers its own technical advantages for research, with the efficiency of *in vivo* RNA silencing being one of several. However, much less is known about its hormonal responses in general and during stress in particular. *C. elegans* does not have a circulatory system; the impressive nervous system it does have (out of a total of 959 somatic cells in the adult hermaphrodite, 302 are neurons) might be an alternative for maintaining coordination and homeostasis. Yet, more and more hormones in *C. elegans* are being characterized, some of which are steroids mediating various environmental signals [101, 102]. Moreover, the nuclear hormone receptor gene family in *C. elegans* has more members than in any other animal (encoding ca. 150 proteins [103]), suggesting that hormones play an important function in its physiology.

C. elegans provides an accessible system to follow genome scale responses both to stress and to immune chellanges and, as we found, such analyses expose overlaps suggesting ancient wiring of the two responses [104]. Expanding such studies could provide valuable clues about the interplay between stress and immunity.

References

1 Selye, H. (1936) A syndrome produced by diverse nocuous agents. *Nature*, **138**, 32.

2 Selye, H. (1949) Effect of ACTH and cortisone upon an anaphylactoid reaction. *Can. Med. Assoc. J.*, **61**, 553–556.

3 Kiecolt-Glaser, J.K., Glaser, R., Gravenstein, S., Malarkey, W.B. and Sheridan, J. (1996) Chronic stress alters the immune response to influenza virus vaccine in older adults. *Proc. Natl. Acad. Sci. U. S. A.*, **93**, 3043–3047.

4 Konstantinos, A.P. and Sheridan, J.F. (2001) Stress and influenza viral infection: modulation of proinflammatory cytokine responses in the lung. *Respir. Physiol.*, **128**, 71–77.

5 Godbout, J.P. and Glaser, R. (2006) Stress-induced immune dysregulation: implications for wound healing, infectious disease and cancer. *J. Neuroimmune Pharmacol.*, **1**, 421–427.

6 Padgett, D.A. and Glaser, R. (2003) How stress influences the immune response. *Trends Immunol.*, **24**, 444–448.

7 Dhabhar, F.S. and McEwen, B.S. (1999) Enhancing versus suppressive effects of stress hormones on skin immune function. *Proc. Natl. Acad. Sci. U.S.A.*, **96**, 1059–1064.

8 Ottaviani, E. and Franceschi, C. (1996) The neuroimmunology of stress from invertebrates to man. *Prog. Neurobiol.*, **48**, 421–440.

9 Bachere, E., Gueguen, Y., Gonzalez, M., de Lorgeril, J., Garnier, J. and Romestand, B. (2004) Insights into the anti-microbial defense of marine invertebrates: the penaeid shrimps and the oyster Crassostrea gigas. *Immunol. Rev.*, **198**, 149–168.

10 Lacoste, A., Jalabert, F., Malham, S.K., Cueff, A. and Poulet, S.A. (2001) Stress and stress-induced neuroendocrine changes increase the susceptibility of juvenile oysters (Crassostrea gigas) to Vibrio splendidus. *Appl. Environ. Microbiol.*, **67**, 2304–2309.

11 Sapolsky, R.M., Romero, L.M. and Munck, A.U. (2000) How do glucocorticoids influence stress responses? Integrating permissive, suppressive, stimulatory, and preparative actions. *Endocr. Rev.*, **21**, 55–89.

12 Keller-Wood, M.E. and Dallman, M.F. (1984) Corticosteroid inhibition of ACTH secretion. *Endocr. Rev.*, **5**, 1–24.

13 Dhabhar, F.S. and McEwen, B.S. (1997) Acute stress enhances while chronic stress suppresses cell-mediated immunity *in vivo*: a potential role for leukocyte trafficking. *Brain Behav. Immun.*, **11**, 286–306.

14 Kiecolt-Glaser, J.K., Loving, T.J., Stowell, J.R., Malarkey, W.B., Lemeshow, S., Dickinson, S.L. and Glaser, R. (2005) Hostile marital interactions, proinflammatory cytokine production, and wound healing. *Arch. Gen. Psychiatry*, **62**, 1377–1384.

15 Bauer, M.E., Vedhara, K., Perks, P., Wilcock, G.K., Lightman, S.L. and Shanks, N. (2000) Chronic stress in caregivers of dementia patients is associated with reduced lymphocyte sensitivity to glucocorticoids. *J. Neuroimmunol.*, **103**, 84–92.

16 Stefanski, V., Knopf, G. and Schulz, S. (2001) Long-term colony housing in long evans rats: immunological, hormonal, and behavioral consequences. *J. Neuroimmunol.*, **114**, 122–130.

17 Reber, S.O., Birkeneder, L., Veenema, A.H., Obermeier, F., Falk, W., Straub, R.H. and Neumann, I.D. (2007) Adrenal insufficiency and colonic inflammation after a novel chronic psycho-social stress paradigm in mice: implications and mechanisms. *Endocrinology*, **148**, 670–682.

18 Sterlemann, V., Ganea, K., Liebl, C., Harbich, D., Alam, S., Holsboer, F., Muller, M.B. and Schmidt, M.V. (2008) Long-term behavioral and neuroendocrine alterations following chronic social stress in mice: implications for stress-related disorders. *Horm. Behav.*, **53**, 386–394.

19 Chouker, A., Smith, L., Christ, F., Larina, I., Nichiporuk, I., Baranov, V., Bobrovnik, E., Pastushkova, L.,

Messmer, K., Peter, K. and Thiel, M. (2002) Effects of confinement (110 and 240 days) on neuroendocrine stress response and changes of immune cells in men. *J. Appl. Physiol.*, **92**, 1619–1627.

20 Salzet, M., Salzet-Raveillon, B., Cocquerelle, C., Verger-Bocquet, M., Pryor, S.C., Rialas, C.M., Laurent, V. and Stefano, G.B. (1997) Leech immunocytes contain proopiomelanocortin: nitric oxide mediates hemolymph proopiomelanocortin processing. *J. Immunol.*, **159**, 5400–5411.

21 Ottaviani, E., Caselgrandi, E., Petraglia, F. and Franceschi, C. (1992) Stress response in the freshwater snail Planorbarius corneus (L.) (Gastropoda, Pulmonata): interaction between CRF, ACTH, and biogenic amines. *Gen. Comp. Endocrinol.*, **87**, 354–360.

22 Lacoste, A., Malham, S.K., Cueff, A., Jalabert, F., Gelebart, F. and Poulet, S.A. (2001) Evidence for a form of adrenergic response to stress in the mollusc Crassostrea gigas. *J. Exp. Biol.*, **204**, 1247–1255.

23 Malagoli, D., Di Paolo, I. and Ottaviani, E. (2007) Presence of and stress-related changes in urocortin-like molecules in neurons and immune cells from the mussel Mytilus galloprovincialis. *Peptides*, **28**, 1545–1552.

24 Lacoste, A., Malham, S.K., Cueff, A. and Poulet, S.A. (2001) Stress-induced catecholamine changes in the hemolymph of the oyster Crassostrea gigas. *Gen. Comp. Endocrinol.*, **122**, 181–188.

25 Malham, S.K., Lacoste, A., Gelebart, F., Cueff, A. and Poulet, S.A. (2003) Evidence for a direct link between stress and immunity in the mollusc Haliotis tuberculata. *J. Exp. Zoolog. A Comp. Exp. Biol.*, **295**, 136–144.

26 Hirashima, A., Sukhanova, M. and Rauschenbach, I. (2000) Biogenic amines in Drosophila virilis under stress conditions. *Biosci. Biotechnol. Biochem.*, **64**, 2625–2630.

27 Gruntenko, N., Chentsova, N.A., Bogomolova, E.V., Karpova, E.K., Glazko, G.V., Faddeeva, N.V., Monastirioti, M. and Rauschenbach, I.Y. (2004) The effect of mutations altering biogenic amine metabolism in Drosophila on viability and the response to environmental stresses. *Arch. Insect Biochem. Physiol.*, **55**, 55–67.

28 Woodring, J. and Hoffmann, K.H. (1994) The effects of octopamine, dopamine and serotonin on juvenile hormone synthesis, *in vitro*, in the cricket, Gryllus bimaculatus. *J. Insect Physiol.*, **40**, 797–802.

29 Jones, G. (1995) Molecular mechanisms of action of juvenile hormone. *Annu. Rev. Entomol.*, **40**, 147–169.

30 Kasahara, M., Suzuki, T. and Pasquier, L.D. (2004) On the origins of the adaptive immune system: novel insights from invertebrates and cold-blooded vertebrates. *Trends Immunol.*, **25**, 105–111.

31 Janeway, C.A., Jr. (1989) Approaching the asymptote? Evolution and revolution in immunology. *Cold Spring Harb Symp Quant Biol*, **54** (1), 1–13.

32 Medzhitov, R., Preston-Hurlburt, P. and Janeway, C.A. Jr. (1997) A human homologue of the Drosophila Toll protein signals activation of adaptive immunity. *Nature*, **388**, 394–397.

33 Janeway, C.A. Jr. and Medzhitov, R. (2002) Innate immune recognition. *Annu. Rev. Immunol.*, **20**, 197–216.

34 Hoffmann, J.A., Kafatos, F.C., Janeway, C.A. and Ezekowitz, R.A. (1999) Phylogenetic perspectives in innate immunity. *Science*, **284**, 1313–1318.

35 Mylonakis, E., Casadevall, A. and Ausubel, F.M. (2007) Exploiting amoeboid and non-vertebrate animal model systems to study the virulence of human pathogenic fungi. *PLoS Pathog.*, **3**, e101.

36 Elenkov, I.J. and Chrousos, G.P. (2006) Stress system – organization, physiology and immunoregulation. *Neuroimmunomodulation*, **13**, 257–267.

37 Reiner, S.L. (2007) Development in motion: helper T cells at work. *Cell*, **129**, 33–36.

38 Kabelitz, D. and Medzhitov, R. (2007) Innate immunity – cross-talk with adaptive immunity through pattern

recognition receptors and cytokines. *Curr. Opin. Immunol.*, **19**, 1–3.

39 Eckmann, L. (2004) Innate immunity and mucosal bacterial interactions in the intestine. *Curr. Opin. Gastroenterol.*, **20**, 82–88.

40 Fujita, T., Matsushita, M. and Endo, Y. (2004) The lectin-complement pathway–its role in innate immunity and evolution. *Immunol. Rev.*, **198**, 185–202.

41 Zipfel, P.F., Mihlan, M. and Skerka, C. (2007) The alternative pathway of complement: a pattern recognition system. *Adv. Exp. Med. Biol.*, **598**, 80–92.

42 Ottaviani, E. and Franceschi, C. (1997) The invertebrate phagocytic immunocyte: clues to a common evolution of immune and neuroendocrine systems. *Immunol. Today*, **18**, 169–174.

43 Gross, P.S., Al-Sharif, W.Z., Clow, L.A. and Smith, L.C. (1999) Echinoderm immunity and the evolution of the complement system. *Dev. Comp. Immunol.*, **23**, 429–442.

44 Meister, M. and Lagueux, M. (2003) Drosophila blood cells. *Cell Microbiol.*, **5**, 573–580.

45 Bulet, P., Stocklin, R. and Menin, L. (2004) Anti-microbial peptides: from invertebrates to vertebrates. *Immunol. Rev.*, **198**, 169–184.

46 Lemaitre, B. and Hoffmann, J. (2007) The host defense of Drosophila melanogaster. *Annu. Rev. Immunol.*, **25**, 697–743.

47 Hemmrich, G., Miller, D.J. and Bosch, T.C. (2007) The evolution of immunity: a low-life perspective. *Trends Immunol.*, **28**, 449–454.

48 Cerenius, L., Lee, B.L. and Soderhall, K. (2008) The proPO-system: pros and cons for its role in invertebrate immunity. *Trends Immunol.*, **29**, 263–271.

49 Cohen, S. and Herbert, T.B. (1996) Health psychology: psychological factors and physical disease from the perspective of human psychoneuro-immunology. *Annu. Rev. Psychol.*, **47**, 113–142.

50 Sanders, V.M. and Munson, A.E. (1985) Norepinephrine and the antibody response. *Pharmacol. Rev.*, **37**, 229–248.

51 Sanders, V.M., Baker, R.A., Ramer-Quinn, D.S., Kasprowicz, D.J., Fuchs, B.A. and Street, N.E. (1997) Differential expression of the beta2-adrenergic receptor by Th1 and Th2 clones: implications for cytokine production and B cell help. *J. Immunol.*, **158**, 4200–4210.

52 Karin, M. (1998) New twists in gene regulation by glucocorticoid receptor: is DNA binding dispensable? *Cell*, **93**, 487–490.

53 Glaser, R. and Kiecolt-Glaser, J.K. (2005) Stress-induced immune dysfunction: implications for health. *Nat. Rev. Immunol.*, **5**, 243–251.

54 McEwen, B.S., Biron, C.A., Brunson, K.W., Bulloch, K., Chambers, W.H., Dhabhar, F.S., Goldfarb, R.H., Kitson, R.P., Miller, A.H., Spencer, R.L. and Weiss, J.M. (1997) The role of adrenocorticoids as modulators of immune function in health and disease: neural, endocrine and immune interactions. *Brain Res. Brain Res. Rev.*, **23**, 79–133.

55 Cronstein, B.N., Kimmel, S.C., Levin, R.I., Martiniuk, F. and Weissmann, G. (1992) A mechanism for the antiinflammatory effects of corticosteroids: the glucocorticoid receptor regulates leukocyte adhesion to endothelial cells and expression of endothelial-leukocyte adhesion molecule 1 and intercellular adhesion molecule 1. *Proc. Natl. Acad. Sci. U. S. A.*, **89**, 9991–9995.

56 Webster, J.I., Tonelli, L. and Sternberg, E.M. (2002) Neuroendocrine regulation of immunity. *Annu. Rev. Immunol.*, **20**, 125–163.

57 Weber, P.S., Madsen-Bouterse, S.A., Rosa, G.J., Sipkovsky, S., Ren, X., Almeida, P.E., Kruska, R., Halgren, R.G., Barrick, J.L. and Burton, J.L. (2006) Analysis of the bovine neutrophil transcriptome during glucocorticoid treatment. *Physiol. Genomics*, **28**, 97–112.

58 Waage, A., Slupphaug, G. and Shalaby, R. (1990) Glucocorticoids inhibit the production of IL6 from monocytes, endothelial cells and fibroblasts. *J. Immunol.*, **20**, 2439–2443.

59 Lee, S.W., Tsou, A.P., Chan, H., Thomas, J., Petrie, K., Eugui, E.M. and

Allison, A.C. (1988) Glucocorticoids selectively inhibit the transcription of the interleukin 1 beta gene and decrease the stability of interleukin 1 beta mRNA. *Proc. Natl. Acad. Sci. U. S. A.*, **85**, 1204–1208.

60 Beutler, B., Krochin, N., Milsark, I.W., Luedke, C. and Cerami, A. (1986) Control of cachectin (tumor necrosis factor) synthesis: mechanisms of endotoxin resistance. *Science*, **232**, 977–980.

61 Blotta, M.H., DeKruyff, R.H. and Umetsu, D.T. (1997) Corticosteroids inhibit IL-12 production in human monocytes and enhance their capacity to induce IL-4 synthesis in CD4+ lymphocytes. *J. Immunol.*, **158**, 5589–5595.

62 Banchereau, J. and Steinman, R.M. (1998) Dendritic cells and the control of immunity. *Nature*, **392**, 245–252.

63 Ramirez, F., Fowell, D.J., Puklavec, M., Simmonds, S. and Mason, D. (1996) Glucocorticoids promote a TH2 cytokine response by CD4+ T cells *in vitro*. *J. Immunol.*, **156**, 2406–2412.

64 Brinkmann, V. and Kristofic, C. (1995) Regulation by corticosteroids of Th1 and Th2 cytokine production in human CD4+ effector T cells generated from CD45RO- and CD45RO+ subsets. *J. Immunol.*, **155**, 3322–3328.

65 Matyszak, M.K., Citterio, S., Rescigno, M. and Ricciardi-Castagnoli, P. (2000) Differential effects of corticosteroids during different stages of dendritic cell maturation. *Eur. J. Immunol.*, **30**, 1233–1242.

66 Newton, R. and Holden, N.S. (2007) Separating transrepression and transactivation: a distressing divorce for the glucocorticoid receptor? *Mol. Pharmacol.*, **72**, 799–809.

67 Naliboff, B.D., Benton, D., Solomon, G.F., Morley, J.E., Fahey, J.L., Bloom, E.T., Makinodan, T. and Gilmore, S.L. (1991) Immunological changes in young and old adults during brief laboratory stress. *Psychosom. Med.*, **53**, 121–132.

68 Bachen, E.A., Manuck, S.B., Marsland, A.L., Cohen, S., Malkoff, S.B., Muldoon, M.F. and Rabin, B.S. (1992) Lymphocyte subset and cellular immune responses to a brief experimental stressor. *Psychosom. Med.*, **54**, 673–679.

69 Benschop, R.J., Nieuwenhuis, E.E., Tromp, E.A., Godaert, G.L., Ballieux, R.E. and van Doornen, L.J. (1994) Effects of beta-adrenergic blockade on immunologic and cardiovascular changes induced by mental stress. *Circulation*, **89**, 762–769.

70 Elenkov, I.J., Papanicolaou, D.A., Wilder, R.L. and Chrousos, G.P. (1996) Modulatory effects of glucocorticoids and catecholamines on human interleukin-12 and interleukin-10 production: clinical implications. *Proc. Assoc. Am. Physicians*, **108**, 374–381.

71 Meltzer, J.C., MacNeil, B.J., Sanders, V., Pylypas, S., Jansen, A.H., Greenberg, A.H. and Nance, D.M. (2004) Stress-induced suppression of *in vivo* splenic cytokine production in the rat by neural and hormonal mechanisms. *Brain Behav. Immun.*, **18**, 262–273.

72 Fleshner, M., Nguyen, K.T., Cotter, C.S., Watkins, L.R. and Maier, S.F. (1998) Acute stressor exposure both suppresses acquired immunity and potentiates innate immunity. *Am. J. Physiol.*, **275**, R870–R878.

73 Sanders, V.M. and Straub, R.H. (2002) Norepinephrine, the beta-adrenergic receptor, and immunity. *Brain Behav. Immun.*, **16**, 290–332.

74 Nance, D.M. and Sanders, V.M. (2007) Autonomic innervation and regulation of the immune system (1987–2007). *Brain Behav. Immun.*, **21**, 736–745.

75 Dhabhar, F.S., Miller, A.H., Stein, M., McEwen, B.S. and Spencer, R.L. (1994) Diurnal and acute stress-induced changes in distribution of peripheral blood leukocyte subpopulations. *Brain Behav. Immun.*, **8**, 66–79.

76 Dhabhar, F.S. and McEwen, B.S. (1996) Stress-induced enhancement of antigen-specific cell-mediated immunity. *J. Immunol.*, **156**, 2608–2615.

77 Dhabhar, F.S. (2002) Stress-induced augmentation of immune function – the role of stress hormones, leukocyte trafficking, and cytokines. *Brain Behav. Immun.*, **16**, 785–798.

78 Cole, S.W., Kemeny, M.E., Weitzman, O.B., Schoen, M. and Anton, P.A. (1999)

Socially inhibited individuals show heightened DTH response during intense social engagement. *Brain Behav. Immun.*, **13**, 187–200.

79 Burns, V.E., Edwards, K.M., Ring, C., Drayson, M. and Carroll, D. (2008) Complement cascade activation after an acute psychological stress task. *Psychosom. Med.*, **70**, 387–396. doi: 10.1097/PSY.0b013e31816ded22.

80 Segerstrom, S.C. and Miller, G.E. (2004) Psychological stress and the human immune system: a meta-analytic study of 30 years of inquiry. *Psychol. Bull.*, **130**, 601–630.

81 Hooper, C., Day, R., Slocombe, R., Handlinger, J. and Benkendorff, K. (2007) Stress and immune responses in abalone: limitations in current knowledge and investigative methods based on other models. *Fish Shellfish Immunol.*, **22**, 363–379.

82 Cheng, W., Chieu, H.-T., Ho, M.-C. and Chen, J.-C. (2006) Noradrenaline modulates the immunity of white shrimp Litopenaeus vannamei. *Fish Shellfish Immunol.*, **21**, 11–19.

83 Chang, C.-C., Wu, Z.-R., Kuo, C.-M. and Cheng, W. (2007) Dopamine depresses immunity in the tiger shrimp Penaeus monodon. *Fish Shellfish Immunol.*, **23**, 24–33.

84 Lacoste, A., Malham, S.K., Cueff, A. and Poulet, S.A. (2001) Noradrenaline modulates oyster hemocyte phagocytosis via a beta-adrenergic receptor-cAMP signaling pathway. *Gen. Comp. Endocrinol.*, **122**, 252–259.

85 Lacoste, A., Malham, S.K., Cueff, A. and Poulet, S.A. (2001) Noradrenaline modulates hemocyte reactive oxygen species production via beta-adrenergic receptors in the oyster Crassostrea gigas. *Dev. Comp. Immunol.*, **25**, 285–289.

86 Malagoli, D., Casarini, L., Sacchi, S. and Ottaviani, E. (2007) Stress and immune response in the mussel Mytilus galloprovincialis. *Fish Shellfish Immunol.*, **23**, 171–177.

87 Baines, D. and Downer, R.G. (1994) Octopamine enhances phagocytosis in cockroach hemocytes: involvement of inositol trisphosphate. *Arch. Insect Biochem. Physiol.*, **26**, 249–261.

88 Roeder, T. (2005) Tyramine and octopamine: ruling behavior and metabolism. *Annu. Rev. Entomol.*, **50**, 447–477.

89 Engler, H. and Stefanski, V. (2003) Social stress and T cell maturation in male rats: transient and persistent alterations in thymic function. *Psychoneuroendocrinology*, **28**, 951–969.

90 Kiecolt-Glaser, J.K., Preacher, K.J., MacCallum, R.C., Atkinson, C., Malarkey, W.B. and Glaser, R. (2003) Chronic stress and age-related increases in the proinflammatory cytokine IL-6. *Proc. Natl. Acad. Sci. U. S. A.*, **100**, 9090–9095.

91 Quan, N., Avitsur, R., Stark, J.L., He, L., Lai, W., Dhabhar, F. and Sheridan, J.F. (2003) Molecular mechanisms of glucocorticoid resistance in splenocytes of socially stressed male mice. *J. Neuroimmunol.*, **137**, 51–58.

92 Scheinman, R.I., Gualberto, A., Jewell, C.M., Cidlowski, J.A. and Baldwin, A.S. Jr. (1995) Characterization of mechanisms involved in transrepression of NF-kappa B by activated gluco-corticoid receptors. *Mol. Cell. Biol.*, **15**, 943–953.

93 Benschop, R.J., Rodriguez-Feuerhahn, M. and Schedlowski, M. (1996) Catecholamine-induced leukocytosis: early observations, current research, and future directions. *Brain Behav. Immun.*, **10**, 77–91.

94 Murray, S.E., Lallman, H.R., Heard, A.D., Rittenberg, M.B. and Stenzel-Poore, M.P. (2001) A genetic model of stress displays decreased lymphocytes and impaired antibody responses without altered susceptibility to streptococcus pneumoniae. *J. Immunol.*, **167**, 691–698.

95 Jarillo-Luna, A., Rivera-Aguilar, V., Garfias, H.R., Lara-Padilla, E., Kormanovsky, A. and Campos-Rodriguez, R. (2007) Effect of repeated restraint stress on the levels of intestinal IgA in mice. *Psychoneuroendocrinology*, **32**, 681–692.

96 Stark, J.L., Avitsur, R., Padgett, D.A., Campbell, K.A., Beck, F.M. and Sheridan, J.F. (2001) Social stress

induces glucocorticoid resistance in macrophages. *Am. J. Physiol. Regul. Integr. Comp. Physiol.*, **280**, R1799–R1805.

97 Cheng, W., Hsiao, I.S. and Chen, J.C. (2004) Effect of ammonia on the immune response of Taiwan abalone Haliotis diversicolor supertexta and its susceptibility to Vibrio parahaemolyticus. *Fish Shellfish Immunol.*, **17**, 193–202.

98 Wilson, K., Thomas, M.B., Blanford, S., Doggett, M., Simpson, S.J. and Moore, S.L. (2002) Coping with crowds: density-dependent disease resistance in desert locusts. *Proc. Natl. Acad. Sci. U.S.A.*, **99**, 5471–5475.

99 Dondero, F., Piacentini, L., Marsano, F., Rebelo, M., Vergani, L., Venier, P. and Viarengo, A. (2006) Gene transcription profiling in pollutant exposed mussels (Mytilus spp.) using a new low-density oligonucleotide microarray. *Gene*, **376**, 24–36.

100 Mydlarz, L.D., Holthouse, S.F., Peters, E.C. and Harvell, C.D. (2008) Cellular responses in sea fan corals: granular amoebocytes react to pathogen and climate stressors. *PLoS ONE*, **3**, e1811.

101 Motola, D.L., Cummins, C.L., Rottiers, V., Sharma, K.K., Li, T., Li, Y., Suino-Powell, K., Xu, H.E., Auchus, R.J., Antebi, A. and Mangelsdorf, D.J. (2006) Identification of ligands for DAF-12 that govern dauer formation and reproduction in C. elegans. *Cell*, **124**, 1209–1223.

102 Butcher, R.A., Fujita, M., Schroeder, F.C. and Clardy, J. (2007) Small-molecule pheromones that control dauer development in Caenorhabditis elegans. *Nat. Chem. Biol.*, **3**, 420–422.

103 Antebi, A. (2006) Nuclear hormone receptors in C. elegans. *WormBook*, **2006**, 1–13.

104 Shapira, M., Hamlin, B.J., Rong, J., Chen, K., Ronen, M. and Tan, M.W. (2006) A conserved role for a GATA transcription factor in regulating epithelial innate immune responses. *Proc. Natl. Acad. Sci. U. S. A.*, **103**, 14086–14091.

12
Immunity to Self Maintains Resistance to Mental Stress: Boosting Immunity as a Complement to Psychological Therapy

Gil M. Lewitus, Osnat Schwartz-Stav, and Michal Schwartz

12.1
Introduction

Acute stress can be defined as a series of events, comprising a stimulus (stressor), that generates a reaction in the brain (stress perception), leading to a emotional and a physiologic fight-or-flight response in the body (stress response) [1]. This evolutionarily adaptive response endows the organism with the ability to deal with the stressor by temporarily adapting homeostasis to the new situation. These stress-response mechanisms are well regulated and, under physiological conditions, the body returns to its normal pre-stress homeostasis shortly after removal of the stressor. However, when the dysregulatoin is not resolved and homeostasis is not restored, long-lasting changes often result in post-traumatic stress disorder (PTSD) [2].

An important system that is affected by stress is immune function; from the evolutionary point of view, when the organism is confronted with a stressful situation (such as a predator) there is a high risk of injury or infection. Therefore it stands to reason that the immune system should be on alert so that it may be rapidly activated. It has been generally accepted that stress suppresses the immune system [3–6]; however, in recent years, an increasing number of studies have shown that stress can boost as well as suppress the immune response. In some cases, stress suppresses immunity, thereby increasing susceptibility to infection and cancer [7], while in other cases, stress is thought to heighten immunity to the extent that it can even cause inflammatory and autoimmune diseases [8]. Possible reasons for this apparent dichotomy are the different types of stress models (mild vs severe) that were tested, the duration of the stress (acute vs chronic) and the ability of the individual to control immunity, particularly immunity to self, as discussed below.

Data from our laboratory and that of Cohen show an association between T cells and adaptation to stress [9, 10]. In these experiments, a short-term exposure of mice to predator odor resulted in a long-term effect on both behavioral (reflecting psychological) and physiological functioning, reminiscent of PTSD in humans

Stress – From Molecules to Behavior. Edited by Hermona Soreq, Alon Friedman, and Daniela Kaufer
Copyright © 2010 WILEY-VCH Verlag GmbH & Co. KGaA, Weinheim
ISBN: 978-3-527-32374-6

[11]. In these studies, it was shown that animals that suffer from an immune deficit are impaired in their ability to cope with stress and thus develop a high incidence of PTSD. In contrast, transgenic mice that overexpress T cells reactive to the CNS-related self-antigen, myelin basic protein (MBP), have reduced incidence of PTSD-like symptoms than their matched controls [9]. These results suggest not only that stress influences the ability of the immune system to fight off invading pathogens, but that activation of the immune system following stress is necessary to optimize coping abilities and maintenance of brain homeostasis.

It has been known for decades that stress affects the immune system, yet very little if any effort has been devoted to the issue as to whether the immune system affects the ability to cope with stressful condition and, if so, whether immune-based manipulations are effective in ameliorating psychotic conditions. These issues are the focus of the present chapter.

12.2
Stress and the Blood–Brain Barrier

Previously, the CNS was considered to be an immune privileged site, mainly due to the blood–brain barrier (BBB), an anatomical and physiological barrier that maintains the integrity of the CNS [12]. This barrier is composed of the blood vessel endothelia, joined by tight junctions, and is supplemented by the perivascular space and a basement membrane that separates it from the CNS parenchyma [13, 14]. An additional mechanism is an immunological barrier which creates an immune suppressive environment [15], with reduced antigen presentation in the CNS [16, 17]. This unique CNS state is thought to protect it against disruptions in the complex neuronal network. However, the notion of the CNS as an immune privileged site is being undermined by the observation that in non-pathogenic conditions, activated T cells can cross the BBB and constantly survey the CNS [18]. Furthermore, it was shown that activated or resting memory T cells, regardless of antigenic specificity, can penetrate the CNS [13, 19]. Interestingly, acute stress can regulate BBB permeability; this regulation is achieved by activation of brain mast cells by corticotrophin-releasing hormone (CRH) [20, 21], thereby increasing brain surveillance by T cells. Stress-induced BBB permeability was mainly investigated in the context of pathologies such as multiple sclerosis and brain metastases [22, 23]. However, the physiological relevance of these phenomena was never investigated. In order to study the relevance of trafficking to the CNS under stress conditions, it is important to understand the process underlying lymphocyte trafficking to the resting healthy CNS.

12.3
Lymphocyte Surveillance of the CNS

Lymphocyte trafficking/surveillance of the CNS was mostly studied in disease or injury models where substantial neuroinflammation occurs. Therefore, the mechanisms of brain surveillance in non-inflamed brain are poorly understood. There are several ways by which lymphocytes from the blood can penetrate the CNS, and this migration is governed by a variety of chemokine and adhesion molecules (for a review, see [24]).

One of the possible lymphocyte entry points is from the blood to the cerebrospinal fluid (CSF) across the choroid plexus. This route is likely to be of physiological relevance, as CSF of healthy individuals contains ~2500 leukocytes/ml. The majority of the lymphocytes in the CSF are CD3+ T cells, with low numbers of B and NK cells. Most T cells have a phenotype of memory/activated cells, including both CD4+ and CD8+ subpopulations [25]. Kivisakk *et al.* characterized the surface phenotype of CSF cells and defined the expression of selected adhesion molecules on vasculature in the choroid plexus from patients without inflammatory disorders of the CNS [26–28]. They found that the CD3+ T cells in the CSF express CXCR3, CCR5 and CCR6. Most of the CD3+ positive cells were CD4$^+$/CD45RA$^-$/CD27$^+$/CD69$^+$-activated central memory T cells expressing high levels of CCR7, L-selectin and P-selectin glycoprotein ligand 1. They also identified CD3+ T cells in the choroid plexus stroma in CNS tissue sections from individuals that died without known neurological disorders. Moreover, they detected P-, E-selectin and ICAM-1 in large venules in the choroid plexus and subarachnoid space, but not in parenchymal microvessels. In light of their findings, they suggested that interaction between P-selectin and its ligand and an ICAM-1/ LFA1 interaction may be required for the immuno surveillance of T cells[27, 28]. It was further shown that, for the migration of activated CD8+ and Th1 CD4+ cells to the CSF, expression of P-selectin on choroid plexus cells is needed, but this is not the case for Th2 CD4+ cells [29]. It is important to note that both these experiments were done by a passive transfer of activated cells; therefore it may represent an inflammatory condition more than a physiological one.

12.4
Stress-Induced Lymphocyte Mobilization

Dhabhar *et al.* [30] were the first to suggest that stress induces redistribution of leukocytes and not their loss. They further showed that during acute stress in rats there is a significant redistribution in peripheral blood lymphocytes and that the cessation of stress results in a rapid return of leukocytes numbers to baseline levels [31]. Further evidence for the stress-induced mobilization of immune cells comes from experiments done on human subjects, examining the expression of adhesion molecules and chemokine receptors on PBMCs following acute stress. In these experiments, subjects were asked to perform a public speaking task that was shown to activate the HPA axis and the sympathetic nervous system and can be considered as a form of mild acute stress. The speaking task significantly affected the circulating lymphocytes phenotype. There was a decrease in the lymph node homing receptor, L-selectin, and increased density of LFA-1 [32]. Furthermore, subjects exhibited a selective increase in the number of circulating T cells expressing CXCR2, CXCR3 and CCR5 [33]. The lymphocytes after the stress express features that endow them with favorable migration toward tissue rather than to lymphoid organs. Thus, the redistribution of lymphocytes and their phenotype suggests an enhancement in immune surveillance toward primary immune defense areas, such as the skin, the lungs and the brain. Indeed acute stress in rats and mice appears to enhance immune function of the skin. Acute restraint stress significantly enhances delayed-type hypersensitivity (DTH) responses in the skin by inducing rapid redeployment of immune cells out of the blood and into the skin [34, 35]. Moreover, epinephrine-mediated stress increases lymphocyte traffic to the mucosal surface of the lungs [36]. As the brain is one of the organs affected by stress, we proposed that acute stress may enhance brain immune surveillance.

Recently, we tested this hypothesis and found an association between enhanced trafficking of T cells to the brain and an enhanced ability to adapt to stress. For example, the Balb/c strain of mouse that demonstrates enhanced T-cell recruitment to the brain following stress has lower levels of anxiety and reduced acoustic startle response, relative to the C57BL/6J strain with limited brain lymphocyte recruitment [10]. The increase in T cells in the CNS was seen in the choroid plexus surrounding the hippocampus. The infiltration of T cells was associated with a selective increase in ICAM-1 expression (the receptor for LFA-1) by choroid plexus cells [10]. Reyes *et al.* [37] looked at the difference in the gene expression in the paraventricular nucleus (PVN) after acute restraint stress and after immunological stress (injection of LPS). To their surprise, they found that acute restraint stress induced up-regulation of immunological molecules different from those elicited by the LPS. The immune-associated molecules observed were related to lymphocyte migration, attraction and adhesion, such as PECAM and CCL27, suggesting that stress may enhance lymphocyte recruitment to the PVN as part of the mechanism used to cope with mental stress, to facilitate speedy protection of the brain from deviated levels of neurotransmitters and other compounds whose abnormal levels represent threats.

12.5
Stress Hormones and Lymphocyte Trafficking

The mobilization of lymphocytes after stress is induced by the stress hormones. There are two main systems by which the CNS can regulate the immune system following stress. One is through the activation of the hypothalamic–pituitary–adrenal (HPA) axis. In this pathway, CRH from the PVN activates the anterior pituitary gland to release adrenocorticotrophic hormone (ACTH), which induces the secretion of glucocorticoids from the adrenal cortex. The second is through the activation of the sympathetic autonomic nervous system (SNS). The main mediators of the SNS are the catecholamines released from the nerve terminals and the adrenal medulla. Both glucocorticoids and epinephrine are involved in lymphocyte mobilization, and exogenous administration of these hormones can mimic the effect of stress [38]. Conversely, adrenalectomy reduces the redistribution of leukocytes induced by stress [31]. Epinephrine induces the release of lymphocytes from the spleen [39] and increases circulating CD8 and natural killer cells [40]. Acute administration of epinephrine increases memory CD4 T cell surveillance of the lungs [36]. Cortisol augments the expression of CXCR4 on T lymphocytes and enhances their migration toward CXCL12 [41], as well as up-regulating LFA-1 (the ligand for ICAM-1) expression on lymphocytes [42]. Moreover, we observed that corticosterone regulates ICAM-1 expression on the choroid plexus [10].

12.6
The Physiological Relevance of Lymphocytes in the Brain

Although it is commonly accepted that psychological stress can lead to neurodegeneration, the underlying mechanisms for these processes are poorly understood. Traumatic stress can be associated with long-lasting behavioral changes as well as morphological changes in different parts of the brain. There are several brain areas implicated in the traumatic stress response, including the amygdala, prefrontal cortex and hippocampal formation, of which the latter is the most extensively studied. The hippocampus is considered to be important in learning and memory [43], as well as in regulating the HPA axis [44, 45] and in other stress-related functions. In PTSD patients, a wide range of memory problems including defects in declarative as well as non-declarative memory are observed [46]. Moreover, an asymmetry of the limbic stricture [47] and a decrease in the volume of the hippocampus [48] were observed in patients with PTSD. The reduction in the hippocampus volume in response to severe stress can be attributed to several factors, such as regression of dendritic processes [49], inhibition of neurogenesis [50] and apoptosis [51] (reviewed in [52]). The main mediators of the above phenomena are the glucocorticoids, which may act directly or indirectly [53]. However, there are several other mediators such as cytokines [54], free radicals [55] and stress-induced changes in the secretion of brain-derived neurotransmitters and neurotrophic factors [56]. Therefore, stress can be considered to be an acute cause of neurodegeneration, especially if leading to long-lasting abnormal behavior such as PTSD or depression. One of the responses that were found to be effective both physiologically and therapeutically in containing degenerative conditions is a well controlled immune response that recognizes self-antigens in the CNS; such responses are collectively termed "protective autoimmunity".

12.7
Protective Autoimmunity

The concept of protective autoimmunity refers to the protection and maintenance of injured CNS tissue by T cells, recognizing self-antigens residing in the CNS [57]. The occurrence of protective autoimmunity is contrary to the previously accepted dogma in which all autoimmune T cells were viewed as potentially disease-causing. The protective autoreactive T cell response is antigen-specific: thus, to induce healing of axonal injury, the protective T cells should react to myelin-associated antigen; and for injury in the gray matter, the T cells should react to protein expressed at that site [58–60]. The neuroprotective capacity of autoreactive T cells was shown in several models of CNS trauma, such as mechanical injury to the optic nerve and spinal cord [58, 59], chemical toxicity [60] and preventing maladaptation to mental stress [9, 10] and mental dysfunction resulting from exposure to unbalanced levels of neurotransmitters [61]. This neuroprotective response is a physiological response that is spontaneously evoked following CNS injury; however, the degree of protection observed is genetically controlled [62]. Nevertheless boosting the immune response by active or passive vaccination against CNS antigens in all strains enhances the neuroprotective immune response [58, 60, 63]. In addition, we and others have found that the adaptive immune system has an important role in supporting hippocampal-dependent cognitive ability [61, 64, 65]. Thus, T cell-deficient mice exhibit impaired spatial cognitive abilities, as assayed with the Morris water maze learning and memory test. However, the cognitive impairment can be restored by the passive transfer of T cells from immunocompetent mice. Moreover, we have found that the adaptive immune system has an important role in regulating neurogenesis as well as BDNF levels in the hippocampus [65].

Although the mechanisms by which autoreactive T cells protect the damaged CNS are not fully elucidated, in recent years several possibilities were proposed. One of the first mechanisms suggested is the ability of T cells to directly and indirectly produce neurotrophic factors that are required for the maintenance and survival of neurons; these factors include NGF, BDNF, GDNF and others [66, 67]. Furthermore, microglia that are activated by CNS-specific T cells have enhanced buffering capacity for the extracellular neurotoxin, glutamate and reduce the secretion of NO and TNF-alpha, in contrast to the microglia activated with lipopolysaccharide (LPS), which acquire a cytotoxic phenotype [68].

Since stress enhances lymphocyte mobilization to the brain and the presence of lymphocytes can reduce the induction of behavioral abnormalities, we investigated the possibility of using immune-based therapies for the prevention of PTSD-like symptoms and depression. Using the same immunization protocol used to rescue neurons from mechanical injury, we were able to reduce the anxiety levels and the startle response in mice pre-exposed to a predator odor, possibly through supporting the recovery of BDNF in the dente gyrus of the hippocampus [10]. In addition, immunization with a weak agonist of myelin-derived peptide reduced depressive-like behaviors (such as anhedonia) in rats subjected to chronic mild stress by regulating hippocampal neurogenesis and BDNF levels [79]. These findings suggest new possibilities for the development of a T cell-based therapy for stress-induced pathologies such as PTSD and depression, by increasing the body's physiological coping ability in response to stress.

12.8
Behavioral Immunization

The view that prior stressful experiences may strengthen an organism's resistance to subsequent stressors has long been recognized [69–72]. For example, in children, past mildly stressful life events are associated with reduced emotional distress during hospital admission [73], attenuated fearfulness in a day care setting [70] and decreased cardiovascular responses to psychologically stressful laboratory tests [74]. Animal studies in rodents and in non-human primates support this notion; experiments in squirrel monkeys exposed to brief periods of maternal separation stress exhibit diminished anxiety and attenuated HPA axis responses to subsequent stressors, compared with unstressed monkeys [72]. This phenomenon was termed "behavioral immunization" [75, 76]. Importantly, the process of post-natal development that builds the mental self involves this "behavioral immunization" to traumatic experiences and affect brain architecture and mental resilience. Thus building the mental self is a reflection of being exposed to modest stressful episodes that leave behind memory. Psychoanalytic theories suggest that the mental "self" is built and reinforced by "optimal frustrations" that produce tolerable mental stresses. In cases of repetitive exposure to intolerable stresses (trauma) the "self" might become fragmented and unable to cope with future stresses [77, 78]. The psychobiological mechanisms that promote the development of stress resistance by a previous mild controllable stress are not fully elucidated.

12.9
Immune Memory to Self Antigens Underlies Resilience to Mental Stress

We suggest that exposure to modest stress activates an immune response to self antigens that constitutively exist in the brain. The successful resolution of a stressful experience elicits cognitive processing as well as immunological memory, which makes subsequent coping efforts more efficient. To distinguish between the cognitive memories of an experience and the stress-induced immunological memory, we therapeutically induced immune memory by vaccination with a self antigen., Mice were inoculated with a CNS-derived peptide emulsified in adjuvant one week prior to their exposure to predator odor; control animals were injected with the adjuvant alone. Mice that were vaccinated with the CNS-related peptide showed reduced anxiety levels and a reduced acoustic startle response.

These data further suggest that, similar to immunological challenge (natural or by vaccination), behavioral challenge by a stressful experience can activate specific immune cells leading to the development of memory cells, that can be quickly recruited upon subsequent exposure to the stressor. According to our view, living in a psychologically sterile environment results in reduced immune activation and thus in reduced immunological memory; such an individual copes poorly upon exposure to even moderate stress levels. In contrast, an overwhelmingly stress-

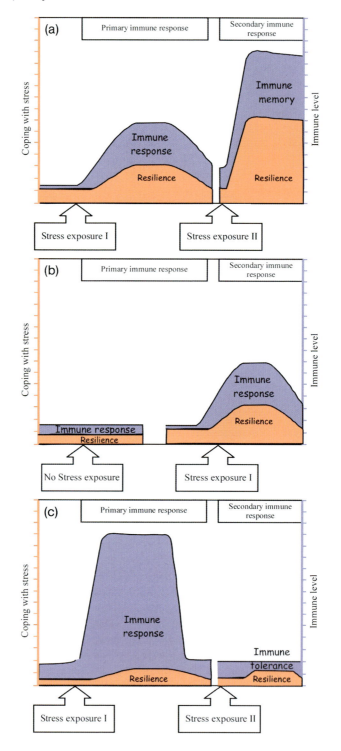

ful experience, as well as chronic stress, can overwhelm the immune system's response to CNS antigens, leading to immune tolerance. These extreme effects will result in reduced resilience to future stressful episodes (Figure 12.1).

In conclusion, we propose that a mildly stressful experience stimulates CNS-specific lymphocytes in a manner similar to inoculation with an antigen and induces immunological memory. As a result, the secondary response to the same challenge (psychological or immunological) is more efficient and effective. Thus, we suggest that it may be possible to therapeutically induce the immunological repertoire needed for mental resilience, thereby providing an adjuvant to psychotherapy, to protect individuals from stress-induced pathologies.

Summary

A controlled adaptive immune response recognizing brain antigen is one of the physiological mechanisms protecting the brain from consequences of injury and is a key player in the maintenance of brain plasticity under non-pathological condition. Immune activation and lymphocytes trafficking are part of the body's defense mechanism following acute psychological stress, enhancing the natural ability of the organism to fight off the damage caused by the stressful condition and helping to restore brain homeostasis. This immune response is amenable to boosting, thereby offering new directions for the development of a therapy for stress-induced pathologies such as post-traumatic stress disorders (PTSD) and depression, in the form of a T cell-based vaccination, which increases the body's physiological ability to cope with stress. In this chapter we discuss brain immune-surveillance, how a healthy immune system maintains psychological resilience as manifested by the ability to cope with stress and depression.

Figure 12.1 The relationship between a previous stressful experience, immunological memory and behavioral memory manifested by resilience to a subsequent stress exposure. (a) First exposure to a mildly stressful situation induces trafficking and activation of CNS-specific T cells. The autoreactive T cells participate in restoration of homeostasis and eventually contribute to the overall accumulation of memory cells, which leads to enhanced resilience to a second exposure to stress. (b) A psychologically sterile environment results in a lack of autoimmune activation and thus in reduced immunological memory; such an individual will cope poorly upon exposure to even moderate stress levels. (c) An overwhelmingly stressful experience or chronic stress may lead to an overwhelming immune response to CNS antigens, possibly leading to immune tolerance and reduced resilience to future stressful episodes.

References

1 Dhabhar, F.S. and McEwen, B.S. (1997) Acute stress enhances while chronic stress suppresses cell-mediated immunity *in vivo*: a potential role for leukocyte trafficking. *Brain Behav. Immun.*, **11**, 286–306.

2 Yehuda, R. and McFarlane, A.C. (1995) Conflict between current knowledge about posttraumatic stress disorder and its original conceptual basis. *Am. J. Psychiatry*, **152**, 1705–1713.

3 Borysenko, M. and Borysenko, J. (1982) Stress, behavior, and immunity: animal models and mediating mechanisms. *Gen. Hosp. Psychiatry*, **4**, 59–67.

4 Cohen, S., Tyrrell, D.A. and Smith, A.P. (1991) Psychological stress and susceptibility to the common cold. *N. Engl. J. Med.*, **325**, 606–612.

5 Kort, W.J. (1994) The effect of chronic stress on the immune response. *Adv. Neuroimmunol.*, **4**, 1–11.

6 Maier, S.F., Watkins, L.R. and Fleshner, M. (1994) Psychoneuroimmunology. The interface between behavior, brain, and immunity. *Am. Psychol.*, **49**, 1004–1017.

7 Fife, A., Beasley, P.J. and Fertig, D.L. (1996) Psychoneuroimmunology and cancer: historical perspectives and current research. *Adv. Neuroimmunol.*, **6**, 179–190.

8 Gold, S.M. and Heesen, C. (2006) Stress and disease progression in multiple sclerosis and its animal models. *Neuroimmunomodulation*, **13**, 318–326.

9 Cohen, H., Ziv, Y., Cardon, M., Kaplan, Z., Matar, M.A., Gidron, Y., Schwartz, M. and Kipnis, J. (2006) Maladaptation to mental stress mitigated by the adaptive immune system via depletion of naturally occurring regulatory CD4+CD25+ cells. *J. Neurobiol.*, **66**, 552–563.

10 Lewitus, G.M., Cohen, A. and Schwartz, M. (2008) Reducing post-traumatic anxiety by, immunization. *Brain Behav. Immun.*, **22**, 1108–1114.

11 Adamec, R., Walling, S. and Burton, P. (2004) Long-lasting, selective, anxiogenic effects of feline predator stress in mice. *Physiol. Behav.*, **83**, 401–410.

12 Reese, T.S. and Karnovsky, M.J. (1967) Fine structural localization of a blood-brain barrier to exogenous peroxidase. *J. Cell Biol.*, **34**, 207–217.

13 Owens, T., Tran, E., Hassan-Zahraee, M. and Krakowski, M. (1998) Immune cell entry to the CNS – a focus for immuno-regulation of EAE. *Res. Immunol.*, **149**, 781–789, 844–846, 855–860.

14 Prat, A., Biernacki, K., Wosik, K. and Antel, J.P. (2001) Glial cell influence on the human blood-brain barrier. *Glia*, **36**, 145–155.

15 Neumann, H. (2000) The immunological microenvironment in the CNS: implications on neuronal cell death and survival. *J. Neural. Transm. Suppl.*, **59**, 59–68.

16 Hailer, N.P., Heppner, F.L., Haas, D. and Nitsch, R. (1998) Astrocytic factors deactivate antigen presenting cells that invade the central nervous system. *Brain Pathol.*, **8**, 459–474.

17 Sun, D., Coleclough, C. and Whitaker, J.N. (1997) Nonactivated astrocytes downregulate T cell receptor expression and reduce antigen-specific proliferation and cytokine production of myelin basic protein (MBP)-reactive T cells. *J. Neuroimmunol.*, **78**, 69–78.

18 Flugel, A., Matsumuro, K., Neumann, H., Klinkert, W.E., Birnbacher, R., Lassmann, H., Otten, U. and Wekerle, H. (2001) Anti-inflammatory activity of nerve growth factor in experimental autoimmune encephalomyelitis: inhibition of monocyte transendothelial migration. *Eur. J. Immunol.*, **31**, 11–22.

19 Zamvil, S.S. and Steinman, L. (1990) The T lymphocyte in experimental allergic encephalomyelitis. *Annu. Rev. Immunol.*, **8**, 579–621.

20 Esposito, P., Chandler, N., Kandere, K., Basu, S., Jacobson, S., Connolly, R., Tutor, D. and Theoharides, T.C. (2002) Corticotropin-releasing hormone and brain mast cells regulate blood-brain-barrier permeability induced by acute stress. *J. Pharmacol. Exp. Ther.*, **303**, 1061–1066.

21 Esposito, P., Gheorghe, D., Kandere, K., Pang, X., Connolly, R., Jacobson, S. and Theoharides, T.C. (2001) Acute stress increases permeability of the blood-brain-barrier through activation of brain mast cells. *Brain Res.*, **888**, 117–127.

22 Hasegawa, H., Ushio, Y., Hayakawa, T., Yamada, K. and Mogami, H. (1983) Changes of the blood-brain barrier in experimental metastatic brain tumors. *J. Neurosurg.*, **59**, 304–310.

23 Kermode, A.G., Thompson, A.J., Tofts, P., MacManus, D.G., Kendall, B.E., Kingsley, D.P., Moseley, I.F., Rudge, P. and McDonald, W.I. (1990) Breakdown of the blood-brain barrier precedes symptoms and other MRI signs of new lesions in multiple sclerosis. Pathogenetic and clinical implications. *Brain*, **113** (Pt 5), 1477–1489.

24 Ransohoff, R.M., Kivisakk, P. and Kidd, G. (2003) Three or more routes for leukocyte migration into the central nervous system. *Nat. Rev. Immunol.*, **3**, 569–581.

25 Svenningsson, A., Andersen, O., Edsbagge, M. and Stemme, S. (1995) Lymphocyte phenotype and subset distribution in normal cerebrospinal fluid. *J. Neuroimmunol.*, **63**, 39–46.

26 Kivisakk, P., Liu, Z., Trebst, C., Tucky, B., Wu, L., Stine, J., Mack, M., Rudick, R.A., Campbell, J.J. and Ransohoff, R.M. (2003) Flow cytometric analysis of chemokine receptor expression on cerebrospinal fluid leukocytes. *Methods*, **29**, 319–325.

27 Kivisakk, P., Mahad, D.J., Callahan, M.K., Trebst, C., Tucky, B., Wei, T., Wu, L., Baekkevold, E.S., Lassmann, H., Staugaitis, S.M., Campbell, J.J. and Ransohoff, R.M. (2003) Human cerebrospinal fluid central memory CD4+ T cells: evidence for trafficking through choroid plexus and meninges via P-selectin. *Proc. Natl. Acad. Sci. U.S.A.*, **100**, 8389–8394.

28 Kivisakk, P., Trebst, C., Liu, Z., Tucky, B.H., Sorensen, T.L., Rudick, R.A., Mack, M. and Ransohoff, R.M. (2002) T-cells in the cerebrospinal fluid express a similar repertoire of inflammatory chemokine receptors in the absence or presence of CNS inflammation: implications for CNS

trafficking. *Clin. Exp. Immunol.*, **129**, 510–518.

29 Carrithers, M.D., Visintin, I., Kang, S.J. and Janeway, C.A. Jr. (2000) Differential adhesion molecule requirements for immune surveillance and inflammatory recruitment. *Brain*, **123** (Pt 6), 1092–1101.

30 Dhabhar, F.S., Miller, A.H., Stein, M., McEwen, B.S. and Spencer, R.L. (1994) Diurnal and acute stress-induced changes in distribution of peripheral blood leukocyte subpopulations. *Brain Behav. Immun.*, **8**, 66–79.

31 Dhabhar, F.S., Miller, A.H., McEwen, B.S. and Spencer, R.L. (1995) Effects of stress on immune cell distribution. Dynamics and hormonal mechanisms. *J. Immunol.*, **154**, 5511–5527.

32 Goebel, M.U. and Mills, P.J. (2000) Acute psychological stress and exercise and changes in peripheral leukocyte adhesion molecule expression and density. *Psychosom. Med.*, **62**, 664–670.

33 Bosch, J.A., Berntson, G.G., Cacioppo, J.T., Dhabhar, F.S. and Marucha, P.T. (2003) Acute stress evokes selective mobilization of T cells that differ in chemokine receptor expression: a potential pathway linking immunologic reactivity to cardiovascular disease. *Brain Behav. Immun.*, **17**, 251–259.

34 Dhabhar, F.S. and McEwen, B.S. (1999) Enhancing versus suppressive effects of stress hormones on skin immune function. *Proc. Natl. Acad. Sci. U.S.A.*, **96**, 1059–1064.

35 Dhabhar, F.S., Satoskar, A.R., Bluethmann, H., David, J.R. and McEwen, B.S. (2000) Stress-induced enhancement of skin immune function: a role for gamma interferon. *Proc. Natl. Acad. Sci. U. S. A.*, **97**, 2846–2851.

36 Kradin, R., Rodberg, G., Zhao, L.H. and Leary, C. (2001) Epinephrine yields translocation of lymphocytes to the lung. *Exp. Mol. Pathol.*, **70**, 1–6.

37 Reyes, T.M., Walker, J.R., DeCino, C., Hogenesch, J.B. and Sawchenko, P.E. (2003) Categorically distinct acute stressors elicit dissimilar transcriptional profiles in the paraventricular nucleus of the hypothalamus. *J. Neurosci.*, **23**, 5607–5616.

38 Shephard, R.J. (2003) Adhesion molecules, catecholamines and leucocyte redistribution during and following exercise. *Sports Med.*, **33**, 261–284.

39 Crary, B., Hauser, S.L., Borysenko, M., Kutz, I., Hoban, C., Ault, K.A., Weiner, H.L. and Benson, H. (1983) Epinephrine-induced changes in the distribution of lymphocyte subsets in peripheral blood of humans. *J. Immunol.*, **131**, 1178–1181.

40 Schedlowski, M., Falk, A., Rohne, A., Wagner, T.O., Jacobs, R., Tewes, U. and Schmidt, R.E. (1993) Catecholamines induce alterations of distribution and activity of human natural killer (NK) cells. *J. Clin. Immunol.*, **13**, 344–351.

41 Okutsu, M., Ishii, K., Niu, K.J. and Nagatomi, R. (2005) Cortisol-induced CXCR4 augmentation mobilizes T lymphocytes after acute physical stress. *Am. J. Physiol. Regul. Integr. Comp. Physiol.*, **288**, R591–R599.

42 Tarcic, N., Levitan, G., Ben-Yosef, D., Prous, D., Ovadia, H. and Weiss, D.W. (1995) Restraint stress-induced changes in lymphocyte subsets and the expression of adhesion molecules. *Neuroimmunomodulation*, **2**, 249–257.

43 Zola-Morgan, S.M. and Squire, L.R. (1990) The primate hippocampal formation: evidence for a time-limited role in memory storage. *Science*, **250**, 288–290.

44 Jacobson, L. and Sapolsky, R. (1991) The role of the hippocampus in feedback regulation of the hypothalamic-pituitary-adrenocortical axis. *Endocr. Rev.*, **12**, 118–134.

45 van Haarst, A.D., Oitzl, M.S. and de Kloet, E.R. (1997) Facilitation of feedback inhibition through blockade of glucocorticoid receptors in the hippocampus. *Neurochem. Res.*, **22**, 1323–1328.

46 Brewin, C.R., Dalgleish, T. and Joseph, S. (1996) A dual representation theory of posttraumatic stress disorder. *Psychol. Rev.*, **103**, 670–686.

47 Sarac-Hadzihalilovic, A. and Dilberovic, F. (2003) Asymmetry of limbic structure (hippocampal formation and amygdaloidal complex) at ptsd. *Bosn. J. Basic Med. Sci.*, **3**, 17–24.

48 Bossini, L., Tavanti, M., Calossi, S., Lombardelli, A., Polizzotto, N.R., Galli, R., Vatti, G., Pieraccini, F. and Castrogiovanni, P. (2008) Magnetic resonance imaging volumes of the hippocampus in drug-naive patients with post-traumatic stress disorder without comorbidity conditions. *J. Psychiatr. Res.*, **42**, 752–762.

49 Komatsuzaki, Y., Murakami, G., Tsurugizawa, T., Mukai, H., Tanabe, N., Mitsuhashi, K., Kawata, M., Kimoto, T., Ooishi, Y. and Kawato, S. (2005) Rapid spinogenesis of pyramidal neurons induced by activation of glucocorticoid receptors in adult male rat hippocampus. *Biochem. Biophys. Res. Commun.*, **335**, 1002–1007.

50 Eriksson, P.S. and Wallin, L. (2004) Functional consequences of stress-related suppression of adult hippocampal neurogenesis – a novel hypothesis on the neurobiology of burnout. *Acta Neurol. Scand.*, **110**, 275–280.

51 Zhao, H., Xu, H., Xu, X. and Young, D. (2007) Predatory stress induces hippocampal cell death by apoptosis in rats. *Neurosci. Lett.*, **421**, 115–120.

52 Lee, A.L., Ogle, W.O. and Sapolsky, R.M. (2002) Stress and depression: possible links to neuron death in the hippocampus. *Bipolar. Disord.*, **4**, 117–128.

53 Sapolsky, R.M. (2000) Glucocorticoids and hippocampal atrophy in neuropsychiatric disorders. *Arch. Gen. Psychiatry*, **57**, 925–935.

54 Allan, S.M. and Rothwell, N.J. (2001) Cytokines and acute neurodegeneration. *Nat. Rev. Neurosci.*, **2**, 734–744.

55 Palumbo, M.L., Fosser, N.S., Rios, H., Zorrilla Zubilete, M.A., Guelman, L.R., Cremaschi, G.A. and Genaro, A.M. (2007) Loss of hippocampal neuronal nitric oxide synthase contributes to the stress-related deficit in learning and memory. *J. Neurochem.*, **102**, 261–274.

56 Duman, R.S., Malberg, J., Nakagawa, S. and D'Sa, C. (2000) Neuronal plasticity and survival in mood disorders. *Biol. Psychiatry*, **48**, 732–739.

57 Moalem, G., Leibowitz-Amit, R., Yoles, E., Mor, F., Cohen, I.R. and Schwartz, M. (1999) Autoimmune T cells protect neurons from secondary degeneration after central nervous system axotomy. *Nat. Med.*, **5**, 49–55.

58 Hauben, E., Butovsky, O., Nevo, U., Yoles, E., Moalem, G., Agranov, E., Mor, F., Leibowitz-Amit, R., Pevsner, E., Akselrod, S., Neeman, M., Cohen, I.R. and Schwartz, M. (2000) Passive or active immunization with myelin basic protein promotes recovery from spinal cord contusion. *J. Neurosci.*, **20**, 6421–6430.

59 Mizrahi, T., Hauben, E. and Schwartz, M. (2002) The tissue-specific self-pathogen is the protective self-antigen: the case of uveitis. *J. Immunol.*, **169**, 5971–5977.

60 Schori, H., Kipnis, J., Yoles, E., WoldeMussie, E., Ruiz, G., Wheeler, L.A. and Schwartz, M. (2001) Vaccination for protection of retinal ganglion cells against death from glutamate cytotoxicity and ocular hypertension: implications for glaucoma. *Proc. Natl. Acad. Sci. U.S.A.*, **98**, 3398–3403.

61 Kipnis, J., Cohen, H., Cardon, M., Ziv, Y. and Schwartz, M. (2004) T cell deficiency leads to cognitive dysfunction: implications for therapeutic vaccination for schizophrenia and other psychiatric conditions. *Proc. Natl. Acad. Sci. U.S.A.*, **101**, 8180–8185.

62 Kipnis, J., Yoles, E., Schori, H., Hauben, E., Shaked, I. and Schwartz, M. (2001) Neuronal survival after CNS insult is determined by a genetically encoded autoimmune response. *J. Neurosci.*, **21**, 4564–4571.

63 Lewitus, G.M., Kipnis, J., Avidan, H., Ben-Nun, A. and Schwartz, M. (2006) Neuroprotection induced by mucosal tolerance is epitope-dependent: conflicting effects in different strains. *J. Neuroimmunol.*, **175**, 31–38.

64 Brynskikh, A., Warren, T., Zhu, J. and Kipnis, J. (2008) Adaptive immunity affects learning behavior in mice. *Brain Behav. Immun.*, **22**, 861–869.

65 Ziv, Y., Ron, N., Butovsky, O., Landa, G., Sudai, E., Greenberg, N., Cohen, H., Kipnis, J. and Schwartz, M. (2006) Immune cells contribute to the maintenance of neurogenesis and spatial learning abilities in adulthood. *Nat. Neurosci.*, **9**, 268–275.

66 Barouch, R. and Schwartz, M. (2002) Autoreactive T cells induce neurotrophin production by immune and neural cells in injured rat optic nerve: implications for protective autoimmunity. *FASEB J.*, **16**, 1304–1306.

67 Gielen, A., Khademi, M., Muhallab, S., Olsson, T. and Piehl, F. (2003) Increased brain-derived neurotrophic factor expression in white blood cells of relapsing-remitting multiple sclerosis patients. *Scand. J. Immunol.*, **57**, 493–497.

68 Shaked, I., Porat, Z., Gersner, R., Kipnis, J. and Schwartz, M. (2004) Early activation of microglia as antigen-presenting cells correlates with T cell-mediated protection and repair of the injured central nervous system. *J. Neuroimmunol.*, **146**, 84–93.

69 Garmezy, N., Masten, A.S. and Tellegen, A. (1984) The study of stress and competence in children: a building block for developmental psychopathology. *Child Dev.*, **55**, 97–111.

70 Holmes, A.T.J.a.F.B. (1935) Experimental study of the fears of young children. Child Development Monographs 20, pp. 167–296, Teachers College, Columbia University, New York.

71 Ogawa, T., Mikuni, M., Kuroda, Y., Muneoka, K., Mori, K.J. and Takahashi, K. (1994) Periodic maternal deprivation alters stress response in adult offspring: potentiates the negative feedback regulation of restraint stress-induced adrenocortical response and reduces the frequencies of open field-induced behaviors. *Pharmacol. Biochem. Behav.*, **49**, 961–967.

72 Parker, K.J., Buckmaster, C.L., Schatzberg, A.F. and Lyons, D.M. (2004) Prospective investigation of stress inoculation in young monkeys. *Arch. Gen. Psychiatry*, **61**, 933–941.

73 Stacey, B. and Davies, J. (1970) Drinking behavior in childhood and adolescence: an evaluative review. *Br. J. Addict. Alcohol. Other Drugs*, **65**, 203–212.

74 Boyce, W.T. and Chesterman, E. (1990) Life events, social support, and cardiovascular reactivity in adolescence. *J. Dev. Behav. Pediatr.*, **11**, 105–111.

75 Seligman, M.E. and Maier, S.F. (1967) Failure to escape traumatic shock. *J. Exp. Psychol.*, **74**, 1–9.

76 Seligman, M.E., Rosellini, R.A. and Kozak, M.J. (1975) Learned helplessness in the rat: time course, immunization, and reversibility. *J. Comp. Physiol. Psychol.*, **88**, 542–547.

77 Kohut, H. (1984) *How Does Analysis Cure?* University of Chicago Press, Chicago.

78 Siegel, A.M. (1996) *Heinz Kohut and the Psychology of the Self (Makers of Modern Psychotherapy)*, Routledge, New York.

13

Brain Interleukin-1 (IL-1) Mediates Stress-Induced Alterations in HPA Activation, Memory Functioning and Neural Plasticity

Inbal Goshen and Raz Yirmiya

13.1
Introduction

Over the past two decades it became evident that multiple bi-directional communication pathways connect the brain, the endocrine and the immune systems, which may be relevant to stress responses. Initially, most research in this area focused on the impact of stress on immune functions and vulnerability to various disease conditions. However, over the past decade ample research focused on the involvement of immune cells and molecules in stress responses. Immune-like processes were found to underlie not only the stress responses that accompany immune challenges, such as injury and infection, but also the responses to physiological and psychological stressors. In particular, it became evident that pro-inflammatory cytokines, which are produced predominantly by activated cells of the innate immune system such as monocytes, macrophages and brain microglia, play an important role in the neuroendocrine and behavioral responses to various stressors. In the present chapter, we focus on the role of one pro-inflammatory cytokine, interleukin-1 (IL-1), in mediating neuroendocrine and behavioral alterations following exposure to immunological and psychological stressors.

IL-1 was the first cytokine to be associated with modulation of neuroendocrine systems, particularly the hypothalamic–pituitary–adrenal axis (HPA) [1–3], in the 1980s. The studies reviewed below elucidate the role of IL-1 in stress-induced modulation of HPA axis activation, memory functioning and neural plasticity, indicating that this cytokine is a critical mediator of adaptive stress responses as well as stress-associated neurobehavioral pathology.

13.2
The Bi-Directional Interaction Between IL-1 and the HPA Axis

The IL-1 System To elucidate the role of IL-1 in stress responses, it is important to understand the complex regulation of IL-1 signaling. IL-1 is produced by many

Stress – From Molecules to Behavior. Edited by Hermona Soreq, Alon Friedman, and Daniela Kaufer
Copyright © 2010 WILEY-VCH Verlag GmbH & Co. KGaA, Weinheim
ISBN: 978-3-527-32374-6

types of cells, including immune cells in the periphery as well as glia and neurons within the brain [4]. IL-1 signaling is mediated by a complex system, which includes various ligands and receptors: The cytokines IL-1α and IL-1β activate signaling, whereas IL-1 receptor antagonist (IL-1ra) serves to block the effects of IL-1 by binding to the same receptors without triggering signaling. The IL-1 receptor type I (IL-1rI) appears to mediate all of the known biological functions of IL-1, whereas IL-1 receptor type II is a decoy receptor, which antagonizes IL-1 signaling [4].

IL-1 Influences the Brain   In order to induce its neurobehavioral and neurophysiological effects, IL-1 directly influences the brain, as suggested by the presence of its receptors in various brain structures [5, 6] and by the fact that administration of IL-1ra into the brain blocks most of the effects of peripheral IL-1 administration [4, 7]. The source of IL-1 can be either local synthesis by glia cells and neurons, or the passage of peripherally produced IL-1 into the brain through the blood–brain barrier (BBB), either passively in circumventricular organs, or via active transport in other areas [8]. Alternatively, IL-1 can induce the synthesis and secretion of smaller mediators that can easily cross the BBB, such as prostaglandins [8]. In addition, peripheral IL-1 can influence the brain via the activation of vagal afferent fibers. Indeed, vagotomy blocks many centrally mediated effects of peripheral immune activation and attenuates the behavioral effects of peripherally administered IL-1 [8].

IL-1 Activates the HPA Axis   IL-1 is expressed throughout the components of the HPA axis [9, 10] and influences all of its levels. Specifically, IL-1 directly induces the secretion of corticotropin-releasing hormone (CRH) from the paraventricular nucleus (PVN) of the hypothalamus [2, 3] and adrenocorticotropic hormone (ACTH) from the pituitary [1], via IL-1 type I receptors in these regions [9, 10]. Additionally, IL-1 stimulates the secretion of glucocorticoids (GCs) from the adrenal, although no known IL-1 receptors have been found in this gland [9]. The central IL-1-induced activation of the HPA axis is mediated by changes in hypothalamic noradrenergic neurotransmission (e.g., [11]).

Glucocorticoids Inhibit IL-1 Production   The interaction between IL-1 and the HPA axis is bi-directional. On the one hand, IL-1 activates the HPA axis, as detailed above. On the other hand, GCs suppress the production of IL-1 (Figure 13.1a). This suppression is achieved via multiple mechanisms: (i) by decreasing IL-1 mRNA levels, both by inhibiting its transcription and by destabilizing it [12]; (ii) by blockade of post-transcriptional IL-1 synthesis via cAMP [13]; (iii) by inhibition of the release of IL-1β into the extra-cellular fluid [14]. These suppressive effects can be demonstrated by measuring the effect of GCs removal. For example, we found that IL-1 gene expression in astrocyte cultures can be detected only when cortisol is removed from the culture medium [15].

Brain IL-1 is Involved in the Feedback Regulation of the HPA Axis   HPA axis regulation is considered to involve inhibitory actions of GCs via GC receptors.

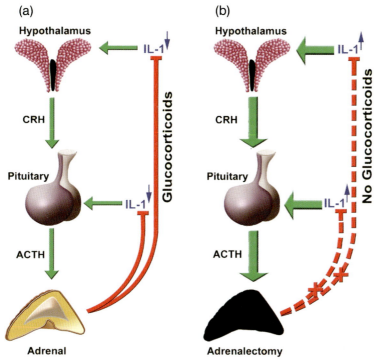

Figure 13.1 The bi-directional interaction between IL-1 and the HPA axis. Under physiological conditions, IL-1 induces the secretion of corticotropin-releasing hormone (CRH) from the hypothalamus and adreno-corticotropic hormone (ACTH) from the pituitary, which result in glucocorticoids secretion from the adrenal gland. These glucocorticoids exert a tonic inhibitory effect on the production of IL-1 (a). Following the removal of endogenous glucocorticoids by adrenalectomy, their feedback inhibition on the hypothalamus and pituitary is removed, but this effect is not sufficient for adrenal-ectomy-associated ACTH hyper-secretion. Rather, an additional excitatory signal, provided by the elevated levels of IL-1 is also necessary (b).

Several studies suggest that the primary target for GC feedback inhibition is not only hypothalamic CRH neurons, but also other hypothalamic and extrahypotha-lamic (e.g., hippocampus, pituitary) systems that regulate CRH expression and release [16]. The action of the feedback system is exemplified by the marked increase in ACTH levels following removal of endogenous GCs by adrenalectomy (ADX). In addition to this ACTH hyper-secretion, ADX is also associated with enhanced expression of the IL-1 gene in the hypothalamus, pituitary and brain stem, as well as increased production of IL-1 protein in the pituitary [15, 17]. These findings suggest that GCs may have a tonic inhibitory effect on brain IL-1. Recently, we demonstrated that following removal of this inhibitory effect by ADX, ACTH levels were markedly elevated in WT controls, but did not increase in mice with IL-1 receptor type I deficiency (IL-1rKO) or mice that overexpress IL-1ra within their nervous system (IL-1raTG) [18]. Even prenatal IL-1 blockade resulted in a significant attenuation of ACTH hyper-secretion [18]. These findings suggest that the removal of a direct inhibitory signal exerted by GCs is not sufficient for ADX-associated ACTH hyper-secretion. Rather, this hyper-secretion is also critically dependent on an excitatory signal, provided by the elevated levels of IL-1, which is a potent HPA axis activator [9, 10] and may be also involved in the GC negative feedback mechanism by altering the levels of brain GC receptors in the hippocampus (Figure 13.1b). Furthermore, during fetal development, IL-1 may have a role in the maturation of brain systems that regulate the HPA axis.

13.3
Stress-Induced HPA Axis Activation is Regulated by Brain IL-1

13.3.1
IL-1 Mediates Immunological Stress-Induced HPA Axis Activation

Immune challenges, such as local inflammation, viral or bacterial infection, were found to induce the production and secretion of brain IL-1, followed by activation of all components of the HPA axis [10]. For example, we demonstrated a marked increase in cortisol levels following administration of the bacterial endotoxin lipopolysaccharide (LPS) in healthy human volunteers [19]. The critical role of IL-1 in illness-induced HPA axis activation is clearly demonstrated by studies on the effects of IL-1 blockade during immune activation, using IL-1ra. For example, when HSV-1 was injected intracerebroventricularly (ICV) together with IL-1ra, the increase in ACTH and CS was blocked [20]. The role of IL-1 in immunological stress was further demonstrated by the finding of a shorter duration of HPA axis activation in IL-1α/β deficient mice [21]. These mice demonstrate a normal increase in corticosterone levels 2 h following peripheral local inflammation, but their level returned to baseline 8 h after the injection, whereas their wild-type controls still continue to evidence high corticosterone levels [21]. The same pattern of impaired HPA axis activation is found in IL-1β deficient but not in IL-1α or IL-1ra deficient mice, suggesting that IL-1β is the primary molecule responsible for inflammation-induced HPA axis activation [21].

13.3.2
Brain IL-1 Mediates Psychological Stress-Induced HPA Axis Activation

The production, secretion and influence of IL-1 have been traditionally considered in the context of illness-associated immune activation. However, over the past decade it became evident that IL-1 also participates in neuromodulation under conditions that do not involve the immune system [22–24]. An important support for this view comes from studies demonstrating that psychological stressors activate the IL-1 system and that endogenous IL-1 activates the HPA axis not only during sickness but also in response to various psychological stressors. Interestingly, the physiological outcomes of both immune activation and exposure to other stressors are very similar: besides sympathetic and HPA axis activation, both immunological and non-immunological stressors produce a very similar behavioral phenotype [25]. Thus, exposure to non-immunological stressors provokes a variety of sickness behavior-like symptoms, including fever, increased sleep and reduced exploratory, social and sexual behavior [26]. Furthermore, stress exposure can even activate components of the peripheral immune system, such as elevation in acute-phase proteins like haptoglobin or α-acid glycoprotein, macrophage priming, increase in white blood cell count and elevation in circulating cytokines [26].

Psychological Stress Increases Brain IL-1 Levels   In animal models, IL-1β (and IL-1ra) gene expression and protein levels in the brain are increased following restraint stress and inescapable shock [27–32]. In the brain regions that are part of the HPA axis (hypothalamus and pituitary), the increase in IL-1 protein levels is immediate, whereas in other brain regions, such as the NTS and the hippocampus, the increase in IL-1 levels is observed only 24h following stress termination [31]. Milder stressors can also activate the IL-1 system when they are administered chronically. Five weeks of chronic mild stress (CMS) induces IL-1 secretion in both the brain and the periphery [33, 34]; and four weeks of social isolation results in elevated hippocampal IL-1 levels [35]. Shorter isolation (1–3h) also resulted in increased hippocampal IL-1 levels [36]. However, other research groups showed no effect of chronic naturalistic stress (predator exposure) or CMS on IL-1 levels [37, 38]; and chronic social stress was even found to cause a reduction in hippocampal IL-1β gene expression [39].

The inhibitory effect of the HPA axis on IL-1 is exemplified by the finding that the increase in IL-1 protein levels in the hypothalamus, hippocampus, cerebellum and NTS following inescapable shock is more distinct in adrenalectomized rats [30, 31]. Specifically, the level of IL-1 in the hypothalamus immediately following shock administration is almost three times higher in ADX rats compared to sham-operated rats [31]. In addition, ADX can unmask undetectable increases in IL-1 following stress, for example, in ADX rats the levels of IL-1 in the hippocampus were significantly increased 2h following exposure to stress, whereas in sham-operated rats no IL-1 increase was evident at this time point [31].

Norepinephrine Triggers Stress-Induced Secretion of IL-1 from Glia Cells   In contrast to the activation of the IL-1 system following immunological stress, the immediate trigger for IL-1 signaling in physical and psychological stress is less clear. Some evidence indicates that secretion of catecholamines, the first neurochemical response to stress, plays an important role. Two research groups have shown that the β-adrenergic blocker propranolol inhibits the increase in IL-1 protein levels following footshock in the hypothalamus and hippocampus [40, 41]; likewise, ablation of noradrenergic projections from the locus coeruleus prevented stress-induced elevation in hippocampal IL-1β [40]. Furthermore, the noradrenergic reuptake inhibitor desipramine facilitated stress-induced hypothalamic IL-1 secretion [41] and systemic administration of the β-adrenergic agonist isoproterenol was sufficient to elevate IL-1 secretion in the serum, as well as in the hypothalamus, pituitary and hippocampus [40]. Thus, the stress-induced stimulation of NE secretion within the brain, acting via β-adrenergic receptors, can trigger a rapid *de novo* production and/or secretion of pre-stored IL-1. The cellular source of psychological stress-induced IL-1 may be activated microglia, as stress increases the hippocampal expression of the microglia activation marker MHC II [42] and minocycline, a selective microglial inhibitor, blocks stress-induced hypothalamic IL-1 secretion [41]. Furthermore, repeated stress exposure induces microglia proliferation, mediated by corticosterone-induced activation of the N-methyl-D-aspartate (NMDA) receptor within the CNS; and exogenous corticosterone administration to non-stressed mice also results in NMDA-dependent microglia proliferation [43].

IL-1 Mediates Psychological Stress-Induced HPA Axis Activation in Animal Models
The role of stress-induced IL-1 in mediating the activation of the HPA axis has been studied by both pharmacological and genetic approaches. Shintani *et al.* [44] demonstrated that the biological activity of IL-1 in the hypothalamus was increased following restraint stress. Furthermore, blockade of IL-1 signaling by administration of IL-1ra into the anterior hypothalamus diminished the increase in ACTH secretion following restraint stress when it was applied 60 min or 5 min before the beginning of restraint, but not 5 min or 30 min afterward [44]. These findings suggest a rapid release of pre-synthesized IL-1, because IL-1 protein synthesis cannot be completed within 5 min.

We further demonstrated the role of IL-1 in mediating stress responses by examining IL-1rKO mice [18], which were previously reported to have a defective response to both IL-1α and IL-1β. IL-1rKO mice displayed impaired HPA axis responses following mild psychological and metabolic stressors. Specifically, when exposed to either auditory stress, or a low dose of 2-deoxyglucose, which induces metabolic stress by causing cytoglucopenia, the wild-type controls displayed a significant increase in corticosterone secretion, whereas corticosterone levels in IL-1rKO mice remained unchanged [18]. However, following exposure to severe stressors, these mice demonstrated a normal HPA axis activation, that is, when exposed to either 60 min of restraint stress or a high dose of 2-deoxyglucose, IL-1rKO mice demonstrated a significant increase in corticosterone secretion, similar to that of their wild-type controls. These findings suggest that IL-1 signaling via its type I receptor is important for HPA axis activation only following mild stress. The response to stronger stressful stimuli probably involves other mediators, which may compensate for the absence of IL-1 signaling. One of these mediators may be the cytokine tumor necrosis factor-α (TNF-α), which often works in synergism with IL-1 [4, 45].

Psychological Stress-Induced IL-1 is Necessary for HPA Axis Activation in Humans
The psychological stress that accompanies academic oral presentation resulted in higher levels of serum IL-1β, as well as cortisol, compared with the control level of the participants [46]. Interview stress and sleep deprivation also caused an increase in plasma IL-1β [47]. Speech stress elevated mitogen-stimulated production of IL-1β from macrophages during the stressor and for 60 min

afterward [48]. Similar increases in salivary cortisol and IL-1 were found following another model of stress: the performance of frustrating cognitive tasks (e.g., subjects were asked to count aloud backwards from 1022 in intervals of 13 and to start over again from the first subtraction after every mistake) [49]. Stressful cognitive tasks also increased IL-1β gene expression in mononuclear cells [50]. We recently demonstrated that the psychological stress associated with surgery (i.e., in patients waiting to enter the operating room, before the beginning of any medical intervention) induces a significant increase in serum IL-1β levels [51]. In another study, IL-1β levels were also elevated in the crevicular fluid following academic stress [52]. A recent meta-analysis reported robust effects for increased levels of circulating IL-1β in humans following acute psychological stress [53].

Elevated serum IL-1β levels have been reported in patients with combat-related post-traumatic stress disorder (PTSD) and these levels correlate with the duration of PTSD symptoms, but not with the severity of PTSD [54]. A recent study [55] replicated the finding of high serum IL-1β levels in patients suffering from PTSD, induced by different traumatic experiences (e.g., sexual or physical abuse, life-threatening events, etc.). In these patients, a reduction in the psychological symptoms following treatment was accompanied by a decrease in IL-1β levels [55].

13.4
Stress-Induced Alterations in Memory and Neural Plasticity are Regulated by Brain IL-1

In addition to its involvement in inflammatory processes per-se, IL-1 mediates a collection of neurobehavioral and neurophysiological symptoms, which are termed together the "sickness behavior syndrome" [56]. It is fairly accepted that this syndrome is adaptive, at least under certain circumstances, helping the organism recuperate from the disease [57]. The symptoms comprising the sickness behavior syndrome include: anorexia, body weight loss, impaired cognitive function, altered sleep patterns, psychomotor retardation, fatigue, reduced exploratory behavior, impaired social behavior, impaired sexual activity, altered pain perception and anhedonia (the diminished capacity to experience pleasure and gratification from activities that previously brought enjoyment) [58, 59]. Mediation of these symptoms by IL-1 is indicated by the findings that: (i) both pharmacological administration (either peripherally, or directly into the brain) and genetically engineered transgenic overproduction of IL-1 produce sickness behavior symptoms; and (ii) IL-1 signaling inhibitors block the behavioral effects of sickness and cytokine elevations [7, 26]. IL-1-induced sickness behavior is accompanied by HPA axis activation and the behavioral symptoms are correlated with ACTH and corticosterone levels [60, 61].

Recently, cytokines were implicated in the regulation of neurobehavioral processes not only during immune activation but in healthy animals as well [22]. Most studies in this area rely on genetic mouse strains lacking normal cytokine signaling. For example, we used several mouse strains with genetic impairments in IL-1 signaling to demonstrate a regulatory role for IL-1 in normal memory consolidation, neural plasticity processes, basal pain sensitivity, morphine-induced analgesia and the development of tolerance following repeated morphine administration [23, 62, 63]. Furthermore, the interactions between stress, IL-1 and the activation of the HPA axis were found to be relevant to several stress-induced behavioral alterations. The following section presents the involvement of stress-induced IL-1 cognitive performance, as well as the neural plasticity and neurogenesis processes that underlie some of these behavioral changes.

13.4.1
Brain IL-1 Underlies Stress-Induced Memory Impairment by Modulating Glucocorticoids Secretion

Memory and Its Underlying Plastic Changes   The ability of all organisms to adapt to new situations and to react properly to previously encountered stimuli depends on their ability to remember. Memories are acquired by the learning process, during which experiences cause lasting changes in the nervous system. Memory can be divided into three stages: encoding, storage (which can last from seconds to years) and retrieval. Failure of any one of these processes will result in impaired memory functioning. Synaptic changes constitute the basis of memory formation as demonstrated by the finding that brief high-frequency stimulation of hippocampal afferents results in persistent augmentation of their synaptic strength, a phenomenon termed long-term potentiation (LTP). Synaptic plasticity was implicated in memory storage, in the hippocampus and in other brain areas [64]. In addition to the biochemical and structural processes of synaptic plasticity, the brain is capable of plasticity in the whole neuronal level, that is, via the formation of new neurons, a process termed neurogenesis, which is induced by memory tasks [65–67].

Stress Changes Memory, for Better or for Worse   One of the most profound modulators of learning and memory is stress exposure. Acute stress can facilitate memory, particularly to stimuli associated with the stressful experience itself, and can be considered as an adaptive cognitive response that enables the organism to cope during the exposure to a stressor and on subsequent encounters with it. Conversely, stress can also impair subsequent learning and even induce profound amnesia [68]. These opposing effects of stress seem to involve GCs and are mainly related to the functioning of the hippocampus [68]. Many research studies examined the effects of endogenous (stress-induced) or exogenous GCs administration on hippocampal-dependent memory and plasticity, reporting that the effects of GCs on hippocampal-dependent memory and LTP can be either inhibitory (particularly at high levels) or facilitatory (particularly at low levels) [68, 69]. Thus, GCs and stress exert an inverted U-shaped influence on memory and plasticity [70, 71].

IL-1 Has a Dual Role in Memory Processes   The picture arising from all the results that accumulated so far regarding the role of IL-1 in memory processes is that physiological levels of IL-1 are needed for memory formation whereas any deviation from the physiological range, either by excess elevation in IL-1 levels or by blockade of IL-1 signaling, results in impaired memory [62, 72]. Many studies have demonstrated the detrimental effect of excess IL-1 levels on memory formation using various doses (either IL-1α or IL-1β), administration routs and species, in several memory testing paradigms (reviewed in [72, 73]). All of these studies found the negative influence of IL-1 to be specific to memory tasks that depend on normal hippocampal functioning, whereas the performance of hippocampus-independent tasks was spared. Other studies also showed that IL-1 mediates the memory impairments caused by various inflammatory agents (reviewed in [72]). In contrast, several studies by our research group as well as others demonstrated the pivotal role of low, "physiological" levels of IL-1 in hippocampus-dependent memory processes, using either pharmacological or genetic approaches [23, 24, 62, 74–76]. The beneficial role of IL-1 in memory processes is further supported by recent reports of learning-induced induction of IL-1 gene expression in the hippocampus [62, 77]. Together, these findings suggest that the influence of IL-1 on memory follows an inverted U-shape pattern, that is, learning-associated increases in IL-1 levels are needed for memory formation; however, any deviation from the physiological range, either by excess elevation in IL-1 levels (induced by immunological, physical, or psychological stressors) or by blockade of IL-1 signaling, results in memory impairment. Similarly to the involvement of IL-1 in memory processes, an inverted U-curve pattern exists for its influence on hippocampal plasticity as well [78]. Whereas some studies report a detrimental effect of elevated IL-1 levels on LTP both *in vivo* and *in vitro* (reviewed in [79]), others report that hippocampal IL-1 expression is induced by high frequency stimulation that also induces LTP [80] and that LTP is impaired when IL-1 signaling is blocked (e.g., [23, 78, 80, 81]).

The findings that the influences of IL-1 and the influence of stress or GCs on memory follow the same inverted U shape and that stress can induce IL-1 production, which in turn activates the HPA axis, may lead to two hypotheses. First, IL-1 may mediate the effects of stress on memory. Second, adrenocortical activation may be involved in the influence of IL-1 on memory formation (both detrimental and beneficial), which is usually assessed in the context of stressful paradigms (e.g., the Morris water maze or fear-conditioning). Both of these hypotheses have been experimentally assessed, as described below.

IL-1 Mediates the Detrimental Effect of Psychological Stress on Memory Maier and Watkins [82] were the first to demonstrate the role of IL-1 in stress-induced modulation of memory functioning, using the learned helplessness procedure, in which rats subjected to a severe inescapable tailshock stress displayed marked impairment in subsequent learning of an active avoidance task (i.e., they learned to be helpless) [83]. ICV administration of IL-1ra (100 μg/rat) before the inescapable shock administration blocked the development of learned helplessness [82]. The same group further showed the involvement of IL-1 in stress-induced memory impairment in the fear-conditioning paradigm, in which animals learned to associate an aversive stimulus (footshock) with either a novel context (hippocampus-dependent memory task) or with an auditory cue (hippocampus-independent memory task). Specifically, when rats were isolated for 5 or 6 h following fear-conditioning, contextual memory was impaired, whereas auditory-cued memory remained intact [36, 84]. Mediation of this effect by IL-1 was evidenced by the increase in hippocampal IL-1β levels following social isolation and by the rescue of the impaired contextual memory by ICV administration of IL-1ra before the isolation stress [36]. The detrimental effect of stress-induced IL-1 on hippocampal memory in this paradigm may be mediated by a reduction in BDNF levels, as isolation reduced BDNF expression in the dentate gyrus and CA3 regions of the hippocampus, but this reduction was blocked by ICV IL-1ra administration before isolation [84]. Recently, we showed the involvement of IL-1 in chronic isolation-induced memory loss [35]. We reported that chronically isolated mice displayed impaired contextual fear-conditioning and reduced spatial memory in the water maze. Moreover, transplantation of neural precursor cells (NPCs), obtained from neonatal mice with transgenic over-expression of IL-1ra (IL-1raTG) under the glial fibrillary acidic protein (GFAP) promoter (which chronically elevated hippocampal IL-1ra levels), completely rescued the stress-induced memory impairments [35].

Glucocorticoids Mediate the Opposing Effects of IL-1 on Memory The possible involvement of GCs in IL-1-induced memory impairment was demonstrated by two studies [85, 86] reporting a concomitant modulation of the effects of IL-1 on memory and corticosterone secretion. These studies found a detrimental effect of ICV administered IL-1β on spatial memory in the water maze as well as on working memory in the radial arm maze, in which animals have to enter previously marked arms in order to get a food reward. Eight weeks of feeding with a diet enriched in the anti-inflammatory omega-3 fatty acid ethyl-eicosapentaenoic acid attenuated the IL-1-induced memory impairment and blocked the IL-1-induced increase in serum corticosterone concentration [85, 86], suggesting that this increase is involved in the effect of IL-1 on memory. This hypothesis was later substantiated by showing that IL-1-induced memory impairment in the radial arm maze was blocked when IL-1 was co-administered with the GC receptor antagonist RU486 [87].

The role of GCs in IL-1-induced memory improvement was directly assessed using GC receptor blockade: Song *et al.* [74] reported improved contextual passive avoidance response concomitantly with increased corticosterone secretion in rats that were ICV injected with IL-1β. However, when IL-1β was co-administered together with RU486, the beneficial effect on memory was eliminated, as was the increase in corticosterone levels [74]. Additionally, IL-1rKO mice, which displayed impaired memory performance [23], also showed diminished corticosterone secretion in response to mild stressors [18], suggesting that impaired HPA axis activation may mediate the poor memory performance of these mice.

To conclude, the findings reported to date show that, on the one hand IL-1 mediates the detrimental effects of stress on memory and, on the other hand HPA axis activation, which affects memory in an inverted U pattern similarly to IL-1, may be involved in both the detrimental and the beneficial effects of IL-1 on memory formation (which were assessed using stressful paradigms; Figure 13.2).

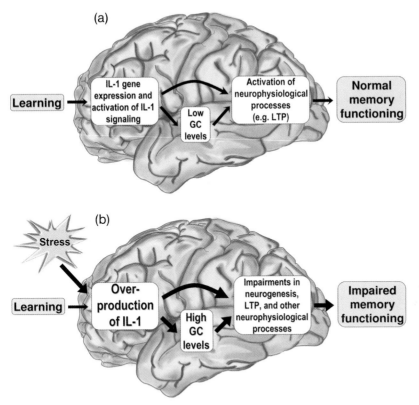

Figure 13.2 IL-1 and glucocorticoids affect memory, for better or for worse. Under low stress conditions (a), learning induces IL-1 gene expression, which activates the HPA axis, resulting in low levels of glucocorticoids (GCs) secretion, which facilitates LTP and memory functioning. Under more stressful conditions (b), high levels of brain IL-1 are produced, resulting in markedly elevated GCs levels, which suppress neurogenesis and impair LTP and memory functioning.

13.4.2
Stress-Induced IL-1 Reduces Hippocampal Neurogenesis: Implications for Memory Impairments

Stress and IL-1 Impair Neurogenesis Impaired hippocampal neurogenesis has been reported following exposure to various acute and chronic stress protocols, as well as corticosterone administration. Adrenalectomy, in contrast, induces a burst of hippocampal cell proliferation in the short term (reviewed in [88]). Similarly, immune activation was recently demonstrated to have a detrimental influence on neurogenesis, suggesting a negative role for pro-inflammatory cytokines on this process. Specifically, microglia, the macrophage-like cells within the brain, were suggested as the mediators of this detrimental effect [89]. Moreover, treatment with IL-1 inducers, including LPS and radiation, resulted in marked suppression of hippocampal neurogenesis [90, 91]. We [33] and others [92] provided direct evidence for the influence of IL-1 on neurogenesis by showing that both acute and chronic IL-1 exposure (4 weeks via osmotic minipumps) results in impaired hippocampal cytogenesis and neurogenesis. Activation of IL-1RI by IL-1β results in a decreased percent of proliferating adult hippocampal progenitors via NfκB signaling pathway activation and can be blocked by IL-1ra [92].

IL-1 Mediates the Anti-Neurogenic Effect of Stress   Based on the similar effects of stress and IL-1 on neurogenesis and the induction of IL-1 by stress, it was hypothesized that IL-1 mediates the anti-neurogenic effect of stress, an hypothesis that was recently confirmed by several studies [33, 35, 92]. A role for IL-1 in mediating the effects of acute stress was recently demonstrated by showing that IL-1ra administration blocks the decrease in neurogenesis induced by two acute stressors (footshock, immobilization) in rats [92]. Similar protection from the anti-neurogenic affect of acute stress was found in IL-1rKO mice as well [92]. We recently showed that IL-1 mediates the detrimental effect of isolation stress on hippocampal neurogenesis and we demonstrated its functional significance for memory processes. As detailed above, subjecting mice to chronic isolation stress elevated hippocampal IL-1 levels and impaired contextual fear-conditioning. Furthermore, chronic isolation stress produced a dramatic decrease in hippocampal neurogenesis [35]. However, intrahippocampal transplantation of IL-1raTG neural precursor cells, which chronically elevated the levels of IL-1ra throughout the stress exposure period, thus blocking IL-1 signaling, completely abolished the detrimental effect of isolation stress on neurogenesis and blocked the cognitive impairment [35].

Based on all the data presented in this section, the following model may be proposed: stressful stimuli induce an increase in brain IL-1 levels, which in turn contribute to the activation of the HPA axis. Subsequently, the secretion of GCs affects memory and plasticity processes in an inverted U-shaped pattern. Finally, brain IL-1 mediates the effects of both chronic and acute stressors on hippocampal neurogenesis, which in turn may be involved in memory impairments (Figure 13.2).

Summary

The pro-inflammatory cytokine interleukin-1 (IL-1), produced following exposure to immunological and psychological challenges, plays an important role in the neuroendocrine and behavioral stress responses. Specifically, brain IL-1 induces the activation of the hypothalamus–pituitary–adrenal (HPA) axis and secretion of glucocorticoids (GCs) in response to both immunological and psychological stressors in humans and animal models. Furthermore, IL-1 mediates the detrimental effects of stress on memory and neural plasticity. Interestingly, both GCs and IL-1 exert beneficial effects on memory at low levels and detrimental effects at high levels; and GCs may be involved in both the detrimental and the beneficial effects of IL-1 on memory formation. Finally, brain IL-1 mediates the effects of both chronic and acute stressors on hippocampal neurogenesis, which may in turn be involved in memory impairments.

Acknowledgment

This research was supported by the Israel Science Foundation.

References

1 Bernton, E.W., Beach, J.E., Holaday, J.W., Smallridge, R.C. and Fein, H.G. (1987) Release of multiple hormones by a direct action of interleukin-1 on pituitary cells. *Science*, **238**, 519–521.

2 Berkenbosch, F., van Oers, J., del Rey, A., Tilders, F. and Besedovsky, H. (1987) Corticotropin-releasing factor-producing neurons in the rat activated by inter-leukin-1. *Science*, **238**, 524–526.

3 Sapolsky, R., Rivier, C., Yamamoto, G., Plotsky, P. and Vale, W. (1987) Interleukin-1 stimulates the secretion of hypothalamic corticotropin-releasing factor. *Science*, **238**, 522–524.

4 Dinarello, C.A. (1996) Biologic basis for interleukin-1 in disease. *Blood*, **87**, 2095–2147.

5 Loddick, S.A., Liu, C., Takao, T., Hashimoto, K. and De Souza, E.B. (1998) Interleukin-1 receptors: cloning studies and role in central nervous system disorders. *Brain Res. Brain Res. Rev.*, **26**, 306–319.

6 Cunningham, E.T. Jr. and De Souza, E.B. (1993) Interleukin 1 receptors in the brain and endocrine tissues. *Immunol. Today*, **14**, 171–176.

7 Rothwell, N.J. and Luheshi, G.N. (2000) Interleukin 1 in the brain: biology, pathology and therapeutic target. *Trends Neurosci.*, **23**, 618–625.

8 Watkins, L.R., Maier, S.F. and Goehler, L.E. (1995) Cytokine-to-brain communication: a review & analysis of alternative mechanisms. *Life Sci.*, **57**, 1011–1026.

9 Besedovsky, H.O. and del Rey, A. (1996) Immune-neuro-endocrine interactions: facts and hypotheses. *Endocr. Rev.*, **17**, 64–102.

10 Turnbull, A.V. and Rivier, C.L. (1999) Regulation of the hypothalamic-pituitary-adrenal axis by cytokines: actions and mechanisms of action. *Physiol. Rev.*, **79**, 1–71.

11 Ericsson, A., Kovacs, K.J. and Sawchenko, P.E. (1994) A functional anatomical analysis of central pathways subserving the effects of interleukin-1 on stress-related neuroendocrine neurons. *J. Neurosci.*, **14**, 897–913.

12 Lee, S.W., Tsou, A.P., Chan, H., Thomas, J., Petrie, K., Eugui, E.M. and Allison, A.C. (1988) Glucocorticoids selectively inhibit the transcription of the interleukin 1 beta gene and decrease the stability of interleukin 1 beta mRNA. *Proc. Natl. Acad. Sci. U. S. A.*, **85**, 1204–1208.

13 Knudsen, P.J., Dinarello, C.A. and Strom, T.B. (1987) Glucocorticoids inhibit transcriptional and post-transcriptional expression of interleukin 1 in U937 cells. *J. Immunol.*, **139**, 4129–4134.

14 Kern, J.A., Lamb, R.J., Reed, J.C., Daniele, R.P. and Nowell, P.C. (1988) Dexamethasone inhibition of interleukin 1 beta production by human monocytes. Posttranscriptional mechanisms. *J. Clin. Invest.*, **81**, 237–244.

15 Ben-Hur, T., Cialic, R., Itzik, A., Barak, O., Yirmiya, R. and Weidenfeld, J. (2001) A novel permissive role for gluco-corticoids in induction of febrile and behavioral signs of experimental herpes simplex virus encephalitis. *Neuroscience*, **108**, 119–127.

16 Jacobson, L. and Sapolsky, R. (1991) The role of the hippocampus in feedback regulation of the hypothalamic-pituitary-adrenocortical axis. *Endocr. Rev.*, **12**, 118–134.

17 Goujon, E., Parnet, P., Laye, S., Combe, C. and Dantzer, R. (1996) Adrenalectomy enhances pro-inflammatory cytokines gene expression, in the spleen, pituitary and brain of mice in response to lipopolysaccharide. *Brain Res. Mol. Brain Res.*, **36**, 53–62.

18 Goshen, I., Yirmiya, R., Iverfeldt, K. and Weidenfeld, J. (2003) The role of endogenous interleukin-1 in stress-induced adrenal activation and adrenalectomy-induced adrenocorticotropic hormone hypersecretion. *Endocrinology*, **144**, 4453–4458.

19 Reichenberg, A., Yirmiya, R., Schuld, A., Kraus, T., Haack, M., Morag, A. and Pollmacher, T. (2001) Cytokine-associated emotional and cognitive disturbances in humans. *Arch. Gen. Psychiatry*, **58**, 445–452.

20 Ben-Hur, T., Cialic, R., Itzik, A., Yirmiya, R. and Weidenfeld, J. (2001) Acute effects of purified and UV-inactivated Herpes simplex virus type 1 on the hypothalamo-pituitary-adrenocortical axis. *Neuroendocrinology*, **74**, 160–166.

21 Horai, R., Asano, M., Sudo, K., Kanuka, H., Suzuki, M., Nishihara, M., Takahashi, M. and Iwakura, Y. (1998) Production of mice deficient in genes for interleukin (IL)-1alpha, IL-1beta, IL-1alpha/beta, and IL-1 receptor antagonist shows that IL-1beta is crucial in turpentine-induced fever development and glucocorticoid secretion. *J. Exp. Med.*, **187**, 1463–1475.

22 Vitkovic, L., Bockaert, J. and Jacque, C. (2000) "Inflammatory" cytokines: neuromodulators in normal brain? *J. Neurochem.*, **74**, 457–471.

23 Avital, A., Goshen, I., Kamsler, A., Segal, M., Iverfeldt, K., Richter-Levin, G. and Yirmiya, R. (2003) Impaired interleukin-1 signaling is associated with deficits in hippocampal memory processes and neural plasticity. *Hippocampus*, **13**, 826–834.

24 Yirmiya, R., Winocur, G. and Goshen, I. (2002) Brain interleukin-1 is involved in spatial memory and passive avoidance conditioning. *Neurobiol. Learn. Mem.*, **78**, 379–389.

25 Maier, S.F. (2003) Bi-directional immune-brain communication: implications for understanding stress, pain, and cognition. *Brain Behav. Immun.*, **17**, 69–85.

26 Maier, S.F. and Watkins, L.R. (1998) Cytokines for psychologists: implications of bidirectional immune-to-brain communication for understanding behavior, mood and cognition. *Psychol. Rev.*, **105**, 83–107.

27 Suzuki, E., Shintani, F., Kanba, S., Asai, M. and Nakaki, T. (1997) Immobilization stress increases mRNA levels of interleukin-1 receptor antagonist in various rat brain regions. *Cell. Mol. Neurobiol.*, **17**, 557–562.

28 Minami, M., Kuraishi, Y., Yamaguchi, T., Nakai, S., Hirai, Y. and Satoh, M. (1991) Immobilization stress induces interleukin-1 beta mRNA in the rat hypothalamus. *Neurosci. Lett.*, **123**, 254–256.

29 O'Connor, K.A., Johnson, J.D., Hansen, M.K., Wieseler Frank, J.L., Maksimova, E., Watkins, L.R. and Maier, S.F. (2003) Peripheral and central proinflammatory cytokine response to a severe acute stressor. *Brain Res.*, **991**, 123–132.

30 Nguyen, K.T., Deak, T., Owens, S.M., Kohno, T., Fleshner, M., Watkins, L.R. and Maier, S.F. (1998) Exposure to acute stress induces brain interleukin-1beta protein in the rat. *J. Neurosci.*, **18**, 2239–2246.

31 Nguyen, K.T., Deak, T., Will, M.J., Hansen, M.K., Hunsaker, B.N., Fleshner, M., Watkins, L.R. and Maier, S.F. (2000) Timecourse and corticosterone sensitivity of the brain, pituitary, and serum interleukin-1beta protein response to acute stress. *Brain Res.*, **859**, 193–201.

32 Kwon, M.S., Seo, Y.J., Lee, J.K., Lee, H.K., Jung, J.S., Jang, J.E., Park, S.H. and Suh, H.W. (2008) The repeated immobilization stress increases IL-1beta immunoreactivities in only neuron, but not astrocyte or microglia in hippocampal CA1 region, striatum and paraventricular nucleus. *Neurosci. Lett.*, **430**, 258–263.

33 Goshen, I., Kreisel, T., Ben-Menachem-Zidon, O., Licht, T., Weidenfeld, J., Ben-Hur, T. and Yirmiya, R. (2007) Brain interleukin-1 mediates chronic stress-induced depression in mice via adrenocortical activation and hippocampal neurogenesis suppression. *Mol. Psychiatry*, **13**, 717–728.

34 Grippo, A.J., Francis, J., Beltz, T.G., Felder, R.B. and Johnson, A.K. (2005) Neuroendocrine and cytokine profile of chronic mild stress-induced anhedonia. *Physiol. Behav.*, **84**, 697–706.

35 Ben Menachem-Zidon, O., Goshen, I., Kreisel, T., Ben Menahem, Y., Reinhartz, E., Ben Hur, T. and Yirmiya, R. (2007) Intrahippocampal transplantation of transgenic neural precursor cells overexpressing interleukin-1 receptor antagonist blocks chronic isolation-induced impairment in memory and neurogenesis. *Neuropsychopharmacology*, **33**, 2251–2262.

36 Pugh, C.R., Nguyen, K.T., Gonyea, J.L., Fleshner, M., Wakins, L.R., Maier, S.F. and Rudy, J.W. (1999) Role of interleukin-1 beta in impairment of contextual fear conditioning caused by social isolation. *Behav. Brain Res.*, **106**, 109–118.

37 Plata-Salaman, C.R., Ilyin, S.E., Turrin, N.P., Gayle, D., Flynn, M.C., Bedard, T., Merali, Z. and Anisman, H. (2000) Neither acute nor chronic exposure to a naturalistic (predator) stressor influences the interleukin-1beta system, tumor necrosis factor-alpha, transforming growth factor-beta1, and neuropeptide mRNAs in specific brain regions. *Brain Res. Bull.*, **51**, 187–193.

38 Mormede, C., Castanon, N., Medina, C., Moze, E., Lestage, J., Neveu, P.J. and Dantzer, R. (2002) Chronic mild stress in mice decreases peripheral cytokine and increases central cytokine expression independently of IL-10 regulation of the cytokine network. *Neuroimmunomodulation*, **10**, 359–366.

39 Bartolomucci, A., Palanza, P., Parmigiani, S., Pederzani, T., Merlot, E., Neveu, P.J. and Dantzer, R. (2003) Chronic psycho-social stress down-regulates central cytokines mRNA. *Brain Res. Bull.*, **62**, 173–178.

40 Johnson, J.D., Campisi, J., Sharkey, C.M., Kennedy, S.L., Nickerson, M., Greenwood, B.N. and Fleshner, M. (2005) Catecholamines mediate stress-induced increases in peripheral and central inflammatory cytokines. *Neuroscience*, **135**, 1295–1307.

41 Blandino, P. Jr., Barnum, C.J. and Deak, T. (2006) The involvement of norepine-phrine and microglia in hypothalamic and splenic IL-1beta responses to stress. *J. Neuroimmunol.*, **173**, 87–95.

42 Frank, M.G., Baratta, M.V., Sprunger, D.B., Watkins, L.R. and Maier, S.F. (2007) Microglia serve as a neuroimmune substrate for stress-induced potentiation of CNS pro-inflammatory cytokine responses. *Brain Behav. Immun.*, **21**, 47–59.

43 Nair, A. and Bonneau, R.H. (2006) Stress-induced elevation of glucocorticoids increases microglia proliferation through NMDA receptor activation. *J. Neuroimmunol.*, **171**, 72–85.

44 Shintani, F., Nakaki, T., Kanba, S., Sato, K., Yagi, G., Shiozawa, M., Aiso, S., Kato, R. and Asai, M. (1995) Involvement of interleukin-1 in immobilization stress-induced increase in plasma adrenocorticotropic hormone and in release of hypothalamic monoamines in the rat. *J. Neurosci.*, **15**, 1961–1970.

45 Yirmiya, R., Weidenfeld, J., Barak, O., Avitsur, R., Pollak, Y., Gallily, R., Wohlman, A., Ovadia, H. and Ben-Hur, T. (1999) The role of brain cytokines in mediating the behavioral and neuroendocrine effects of intracerebral mycoplasma fermentans. *Brain Res.*, **829**, 28–38.

46 Heinz, A., Hermann, D., Smolka, M.N., Rieks, M., Graf, K.J., Pohlau, D., Kuhn, W. and Bauer, M. (2003) Effects of acute psychological stress on adhesion molecules, interleukins and sex hormones: implications for coronary heart disease. *Psychopharmacology (Berl.)*, **165**, 111–117.

47 Altemus, M., Rao, B., Dhabhar, F.S., Ding, W. and Granstein, R.D. (2001) Stress-induced changes in skin barrier function in healthy women. *J. Invest. Dermatol.*, **117**, 309–317.

48 Ackerman, K.D., Martino, M., Heyman, R., Moyna, N.M. and Rabin, B.S. (1998) Stressor-induced alteration of cytokine production in multiple sclerosis patients and controls. *Psychosom. Med.*, **60**, 484–491.

49 Mastrolonardo, M., Alicino, D., Zefferino, R., Pasquini, P. and Picardi, A. (2007)

Effect of psychological stress on salivary interleukin-1beta in psoriasis. *Arch. Med. Res.*, **38**, 206–211.

50 Brydon, L., Edwards, S., Jia, H., Mohamed-Ali, V., Zachary, I., Martin, J.F. and Steptoe, A. (2005) Psychological stress activates interleukin-1beta gene expression in human mononuclear cells. *Brain Behav. Immun.*, **19**, 540–546.

51 Shapira-Lichter, I., Beilin, B., Ofek, K., Bessler, H., Gruberger, M., Shavit, Y., Seror, D., Grinevich, G., Posner, E., Reichenberg, A., Soreq, H. and Yirmiya, R. (2008) Cytokines and cholinergic signals co-modulate surgical stress-induced changes in mood and memory. *Brain Behav. Immun.*, **22**, 388–398.

52 Deinzer, R., Forster, P., Fuck, L., Herforth, A., Stiller-Winkler, R. and Idel, H. (1999) Increase of crevicular interleukin 1beta under academic stress at experimental gingivitis sites and at sites of perfect oral hygiene. *J. Clin. Periodontol.*, **26**, 1–8.

53 Steptoe, A., Hamer, M. and Chida, Y. (2007) The effects of acute psychological stress on circulating inflammatory factors in humans: a review and meta-analysis. *Brain Behav. Immun.*, **21**, 901–912.

54 Spivak, B., Shohat, B., Mester, R., Avraham, S., Gil-Ad, I., Bleich, A., Valevski, A. and Weizman, A. (1997) Elevated levels of serum interleukin-1 beta in combat-related posttraumatic stress disorder. *Biol. Psychiatry*, **42**, 345–348.

55 Tucker, P., Ruwe, W.D., Masters, B., Parker, D.E., Hossain, A., Trautman, R.P. and Wyatt, D.B. (2004) Neuroimmune and cortisol changes in selective serotonin reuptake inhibitor and placebo treatment of chronic posttraumatic stress disorder. *Biol. Psychiatry*, **56**, 121–128.

56 Yirmiya, R., Weidenfeld, J., Pollak, Y., Morag, M., Morag, A., Avitsur, R., Barak, O., Reichenberg, A., Cohen, E., Shavit, Y. and Ovadia, H. (1999) Cytokines, "depression due to a general medical condition," and antidepressant drugs. *Adv. Exp. Med. Biol.*, **461**, 283–316.

57 Hart, B.L. (1988) Biological basis of the behavior of sick animals. *Neurosci. Biobehav. Rev.*, **12**, 123–137.

58 Dantzer, R. (2004) Cytokine-induced sickness behaviour: a neuroimmune response to activation of innate immunity. *Eur. J. Pharmacol.*, **500**, 399–411.

59 Yirmiya, R. (1997) Behavioral and psychological effects of immune activation: implications for "depression due to a general medical condition". *Curr. Opin. Psychiatry*, **10**, 470–476.

60 Brebner, K., Hayley, S., Zacharko, R., Merali, Z. and Anisman, H. (2000) Synergistic effects of interleukin-1beta, interleukin-6, and tumor necrosis factor-alpha: central monoamine, corticosterone, and behavioral variations. *Neuropsychopharmacology*, **22**, 566–580.

61 Wieczorek, M. and Dunn, A.J. (2006) Relationships among the behavioral, noradrenergic, and pituitary-adrenal responses to interleukin-1 and the effects of indomethacin. *Brain Behav. Immun.*, **20**, 477–487.

62 Goshen, I., Kreisel, T., Ounallah-Saad, H., Renbaum, P., Zalzstein, Y., Ben-Hur, T., Levy-Lahad, E. and Yirmiya, R. (2007) A dual role for interleukin-1 in hippo-campal-dependent memory processes. *Psychoneuroendocrinology*, **32**, 1106–1115.

63 Shavit, Y., Wolf, G., Goshen, I., Livshits, D. and Yirmiya, R. (2005) Interleukin-1 antagonizes morphine analgesia and underlies morphine tolerance. *Pain*, **115**, 50–59.

64 Bliss, T.V. and Collingridge, G.L. (1993) A synaptic model of memory: long-term potentiation in the hippocampus. *Nature*, **361**, 31–39.

65 Gould, E. and Tanapat, P. (1999) Stress and hippocampal neurogenesis. *Biol. Psychiatry*, **46**, 1472–1479.

66 Gould, E., Tanapat, P., Rydel, T. and Hastings, N. (2000) Regulation of hippocampal neurogenesis in adulthood. *Biol. Psychiatry*, **48**, 715–720.

67 Kempermann, G., Wiskott, L. and Gage, F.H. (2004) Functional significance of adult neurogenesis. *Curr. Opin. Neurobiol.*, **14**, 186–191.

68 Kim, J.J. and Diamond, D.M. (2002) The stressed hippocampus, synaptic plasticity and lost memories. *Nat. Rev. Neurosci.*, **3**, 453–462.

69 de Kloet, E.R., Joels, M. and Holsboer, F. (2005) Stress and the brain: from adaptation to disease. *Nat. Rev. Neurosci.*, **6**, 463–475.

70 Diamond, D.M., Bennett, M.C., Fleshner, M. and Rose, G.M. (1992) Inverted-U relationship between the level of peripheral corticosterone and the magnitude of hippocampal primed burst potentiation. *Hippocampus*, **2**, 421–430.

71 Conrad, C.D., Lupien, S.J. and McEwen, B.S. (1999) Support for a bimodal role for type II adrenal steroid receptors in spatial memory. *Neurobiol. Learn. Mem.*, **72**, 39–46.

72 Goshen, I. and Yirmiya, R. (2007) The role of pro-inflammatory cytokines in memory processes and neural plasticity, in *Psychoneuroimmunology*, 4th edn (ed. R. Ader), Elsevier Academic Press, Amsterdam, pp. 337–377.

73 Rachal Pugh, C., Fleshner, M., Watkins, L.R., Maier, S.F. and Rudy, J.W. (2001) The immune system and memory consolidation: a role for the cytokine IL-1beta. *Neurosci. Biobehav. Rev.*, **25**, 29–41.

74 Song, C., Phillips, A.G. and Leonard, B. (2003) Interleukin 1 beta enhances conditioned fear memory in rats: possible involvement of glucocorticoids. *Eur. J. Neurosci.*, **18**, 1739–1743.

75 Brennan, F.X., Beck, K.D. and Servatius, R.J. (2003) Low doses of interleukin-1-beta improve the leverpress avoidance performance of Sprague-Dawley rats. *Neurobiol. Learn. Mem.*, **80**, 168–171.

76 Brennan, F.X., Beck, K.D. and Servatius, R.J. (2004) Proinflammatory cytokines differentially affect leverpress avoidance acquisition in rats. *Behav. Brain Res.*, **153**, 351–355.

77 Depino, A.M., Alonso, M., Ferrari, C., del Rey, A., Anthony, D., Besedovsky, H., Medina, J.H. and Pitossi, F. (2004) Learning modulation by endogenous hippocampal IL-1, blockade of endogenous IL-1 facilitates memory formation. *Hippocampus*, **14**, 526–535.

78 Loscher, C.E., Mills, K.H. and Lynch, M.A. (2003) Interleukin-1 receptor antagonist exerts agonist activity in the hippocampus independent of the interleukin-1 type I receptor. *J. Neuroimmunol.*, **137**, 117–124.

79 O'Connor, J.J. and Coogan, A.N. (1999) Actions of the pro-inflammatory cytokine IL-1 beta on central synaptic transmission. *Exp. Physiol.*, **84**, 601–614.

80 Schneider, H., Pitossi, F., Balschun, D., Wagner, A., del Rey, A. and Besedovsky, H.O. (1998) A neuromodulatory role of interleukin-1beta in the hippocampus. *Proc. Natl. Acad. Sci. U.S.A.*, **95**, 7778–7783.

81 Ross, F.M., Allan, S.M., Rothwell, N.J. and Verkhratsky, A. (2003) A dual role for interleukin-1 in LTP in mouse hippocampal slices. *J. Neuroimmunol.*, **144**, 61–67.

82 Maier, S.F. and Watkins, L.R. (1995) Intracerebroventricular interleukin-1 receptor antagonist blocks the enhancement of fear conditioning and interference with escape produced by inescapable shock. *Brain Res.*, **695**, 279–282.

83 Maier, S.F. (1990) Role of fear in mediating shuttle escape learning deficit produced by inescapable shock. *J. Exp. Psychol. Anim. Behav. Process.*, **16**, 137–149.

84 Barrientos, R.M., Sprunger, D.B., Campeau, S., Higgins, E.A., Watkins, L.R., Rudy, J.W. and Maier, S.F. (2003) Brain-derived neurotrophic factor mRNA downregulation produced by social isolation is blocked by intrahippocampal interleukin-1 receptor antagonist. *Neuroscience*, **121**, 847–853.

85 Song, C. and Horrobin, D. (2004) Omega-3 fatty acid ethyl-eicosapenta-enoate, but not soybean oil, attenuates memory impairment induced by central IL-1beta administration. *J. Lipid. Res.*, **45**, 1112–1121.

86 Song, C., Leonard, B.E. and Horrobin, D.F. (2004) Dietary ethyl-eicosapentaenoic acid but not soybean oil reverses central interleukin-1-induced changes in behaviour, corticosterone and immune response in rats. *Stress*, **7**, 43–54.

87 Song, C., Phillips, A.G., Leonard, B.E. and Horrobin, D.F. (2004) Ethyl-eicosapentaenoic acid ingestion prevents corticosterone-mediated memory impairment induced by central administration of interleukin-1beta in rats. *Mol. Psychiatry*, **9**, 630–638.

88 Mirescu, C. and Gould, E. (2006) Stress and adult neurogenesis. *Hippocampus*, **16**, 233–238.

89 Kempermann, G. and Neumann, H. (2003) Neuroscience. Microglia: the enemy within? *Science*, **302**, 1689–1690.

90 Ekdahl, C.T., Claasen, J.H., Bonde, S., Kokaia, Z. and Lindvall, O. (2003) Inflammation is detrimental for neurogenesis in adult brain. *Proc. Natl. Acad. Sci. U.S.A.*, **100**, 13632–13637.

91 Monje, M.L., Toda, H. and Palmer, T.D. (2003) Inflammatory blockade restores adult hippocampal neurogenesis. *Science*, **302**, 1760–1765.

92 Koo, J.W. and Duman, R.S. (2008) IL-1beta is an essential mediator of the antineurogenic and anhedonic effects of stress. *Proc. Natl. Acad. Sci. U.S.A.*, **105**, 751–756.

Part V
Post-Traumatic Stress Disorder

Stress – From Molecules to Behavior. Edited by Hermona Soreq, Alon Friedman, and Daniela Kaufer
Copyright © 2010 WILEY-VCH Verlag GmbH & Co. KGaA, Weinheim
ISBN: 978-3-527-32374-6

14
Post-Traumatic Stress Disorder in Animal Models

Hagit Cohen, Nitzan Kozlovsky, Gal Richter-Levin, and Joseph Zohar

14.1
Introduction

Post-traumatic stress disorder (PTSD) is an incapacitating chronic syndrome reflecting a disorder of cognitive, emotional and physiological processing of and/or recovery from exposure to a potentially traumatic experience (PTE) [1]. In the first hours to days following the experience, the vast majority of individuals exposed to an extreme event demonstrate, to a varying degree, symptoms such as intense fear, helplessness, or horror, followed by anxiety, depression, agitation, shock, or dissociation, and may have trouble functioning in their usual manner for a while [2–5].

Retrospective and prospective epidemiological studies indicate that most individuals affected by a PTE adapt within a period of 1–4 weeks following exposure [2, 6] and only a small proportion develop long-term psychopathology [2, 6]. In the United States, studies report that the rate of lifetime exposure to at least one "serious" traumatic event (excluding grief and mourning) is quite high; a conservative estimate reported 61% among men and 51% among women [7]. Other studies found similar rates [7–12]. The lifetime prevalence of PTSD in the general population reaches about 7% overall [13], suggesting that about 20–30% of individuals exposed to severe stressors develop PTSD [8]. This figure varies depending on the type of trauma studied, where male rape victims suffer very high rates and populations exposed to natural disasters significantly less [14]. The discrepancy between the proportion of the general population exposed to PTE and those who eventually fulfill criteria for the disorder suggests qualitative differences in vulnerability and/or resilience.

After extensive study over the past two decades, PTSD was established as a disorder and listed in the *Diagnostic and Statistic Manual* (DSM). A diagnosis of PTSD is made if the required symptoms are present one month or more after exposure to a triggering event: (i) intrusive re-experiencing of the traumatic event in the form of nightmares and flashbacks, with an exaggerated response to trauma-related reminders/cues; (ii) persistent avoidance of stimuli associated with the

Stress – From Molecules to Behavior. Edited by Hermona Soreq, Alon Friedman, and Daniela Kaufer
Copyright © 2010 WILEY-VCH Verlag GmbH & Co. KGaA, Weinheim
ISBN: 978-3-527-32374-6

trauma and emotional numbing; and (iii) persistent symptoms of exaggerated startle response, increased physiological arousal and sustained preparedness for an instant alarm response [1].

PTSD has severe effects on widespread areas of the individual's functioning, severely compromising quality of life, affecting the work-place, the family and social life. Moreover, PTSD is also often comorbid with other disorders such as depressive and anxiety disorders, drug and alcohol abuse, cognitive and memory impairments and sexual dysfunction [1]. The development of PTSD is often a gradual process and extends over time through a series of stages ranging from relatively contained distress to severe disability [15]. As the disorder evolves over time, pathological changes and debilitating comorbidity may become fixed and irreversible. Unlike processes in which exposure to repeated stimuli induces a process of learning or conditioning, implying increased efficiency in processing of data to produce the required response, the psychopathology underlying PTSD produces a paradoxical vulnerability to negative sequelae upon subsequent stress exposure [15].

The sequelae of exposure to a traumatizing stressor are subject to extensive clinical study. Clinical research often gives rise to important questions or hypotheses as to the pathogenesis, clinical course and outcomes of such events. Among the issues raised are those relating to factors that may confer risk or resilience for the development of more severe stress-induced clinical outcomes, such as PTSD. By their nature, clinical studies raise issues concerning pre-morbid factors largely by means of extrapolating retrospectively. Prospective studies are near impossible to conceive and would most probably be prohibitively expensive to put into practice.

An animal model can give a good approximation of certain aspects of the complex clinical disorder, enabling the study of questions raised in clinical research in a prospective study design and under far more controllable conditions.

Animal models of psychiatric disorders offer a complementary research modality that supports clinical research. In order to achieve a satisfactory degree of validity and reliability, animal models of complex and intricate psychiatric disorders must fulfill certain criteria. For example, the behavioral responses must be observable and measurable and must reliably reflect clinical symptomatology, and pharmacological agents that are known to affect symptoms in human subjects should correct, with equal efficacy, measurable parameters that model symptoms of the disorder.

Developing an animal model for PTSD is not a trivial issue. Diagnosis in human patients relies heavily on personal reports of thoughts, dreams and images, which cannot be studied in rats. Furthermore, several of the typical symptoms of PTSD may be unique to humans and thus not be found in rats. Likewise, an important factor of the trauma in humans is the perception of the life-threatening potential of the situation. It is not clear whether rats can make this judgment or which stressors will be most effective for rats. In addition, there is as yet no clearly effective pharmacological treatment for PTSD. It is thus difficult to test a potential rodent model for its pharmacological predictability in relation to PTSD or other traumatic stress-related disorders.

Nevertheless, using animals to study PTSD holds advantages for several reasons. First, unlike many other mental disorders, the diagnostic criteria for PTSD specify an etiological factor, which is an exposure to a life-threatening, traumatic event [16]. In a model for PTSD, variables such as the quality and intensity of the stressor and the degree of exposure to it can be carefully controlled, and the behavioral and concomitant physiological responses to a (valid) threatening stimulus can be studied. Second, little is known about pre-trauma etiological aspects of the disorder, since, naturally, the studies so far have focused on retrospective assessments of the patients after the onset of PTSD. An animal model enables a prospective follow-up design, in which the disorder is triggered at a specified time and in a uniform manner, in controllable and statistically sound population samples, and enables the assessment of behavioral and gross physiological parameters. Moreover, unlike studies in human subjects, animal model studies enable the assessment of concomitant biomolecular changes in dissected brain areas and experimentation with pharmacological agents with potential therapeutic effects.

This chapter presents and discusses findings from various animal models of PTSD, which differ from one another in the rationale for their development. These models use different paradigms but show a range of behavioral and physiological manifestations seen in PTSD patients. Moreover, we present findings from a series of studies employing a model of individual behavioral response classification. A brief introduction to the standard stress paradigm, the standard behavioral methods and the definition of the cut-off behavioral criteria (CBC) employed for classification precedes this.

14.2
Animal Models of PTSD

14.2.1
Trauma/Stress-Based Models

Stress paradigms in animal studies aim to model criterion A of the DSM diagnostic criteria [1]. They thus use extremely stressful experiences aimed at engendering a sense of threat and helplessness in the animal. Some of these focus more on the intensity of the experience, whereas others combine intensity with an attempt to design an ethologically valid experience, one which an animal might encounter in its natural environment.

Exposure of rodents to predator stimuli (cat, cat odor, fox odor or trimethylthiazoline, a synthetic compound isolated from fox feces) is fear-provoking and stressful and produces long-lasting behavioral and physiological responses. Blanchard *et al.* [17–20], Adamec *et al.* [21–25] and others [26–31] established the validity of this paradigm, in which adult rodents are exposed to feline predators for 5–10 min in a closed environment (i.e., inescapable exposure). The resultant freezing response mode is ethologically adaptive for animals in situations where both "fight" and "flight" options are ineffective. Predator stress has ecological validity in that it mimics brief, intense, threatening experiences with lasting affective consequences [23, 24]. The predator stress paradigm has proven to be effective in inducing the expected range of behavioral and physiological responses [16, 21–25]. These include freezing, avoidance, increased secretion of stress hormones and changes in transmission from the hippocampus via the ventral angular bundle to the basolateral amygdala and from the central amygdala to the lateral column of the periaqueductal gray [17, 21, 22, 26–30, 32–41]. These pathways are of interest because neuroplastic changes within them are associated with aversive learning. The potency of predator stimuli is comparable to that of a variety of paradigms in which the threat is more tangible and immediate, such as paradigms based on inescapable pain or electric shock, swimming and near-drowning, a small raised platform and even direct proximity to a kitten or a cat (separated by a mesh divide or a solid divide with an opening large enough for the rodent to slip through).

Richter-Levin [42] developed an interesting stress model, the underwater trauma. Although rats naturally swim well and are able to dive and to cope with exposure to water, brief (30–45 s), uncontrollable restraint under water establishes an ethologically relevant traumatic experience. Exposure of rats to underwater trauma results in long-lasting, heightened anxiety and context-specific spatial memory deficits [42–44]; underwater trauma in a different (out-of-context) water container has no effect on the ability of rats to perform a spatial memory task in the water maze [42]. These results may explain the lack of effect of inescapable tail-shock procedure on spatial performance reported by others [45], because in their study, the stressor was not associated with the context of the maze. Moreover, underwater trauma results in both behavioral and electrophysiological aversive effects. At 20 min after the trauma, the traumatized rats performed poorly in the spatial memory task in the water maze, and 40 min after the tetanic stimulation (100 min after the underwater trauma) they showed a reduced level of LTP. Thus, the underwater trauma induces electrophysiological alterations, which resemble those observed in other models of stress [46]. In addition, the impaired performance in the water maze was significantly correlated with the reduced ability to induce LTP. These findings of a strong correlation between LTP and spatial learning suggest that these two phenomena are related. However, it is possible that the trauma impairs performance not by affecting memory but by affecting memory-related processes such as attention. It was suggested that the underwater trauma could provide an important and potentially powerful model for understanding the mechanisms underlying the relationships between stress, cognition and learning.

14.2.2
Mechanism-Based Models

Another approach in developing animal models of PTSD is to consider potential brain mechanisms that may underlie the disorder and to develop behavioral protocols that would mimic the activation of such mechanisms.

14.2.2.1   Enhanced Fear Conditioning

The persistence of psychological and biological fear responses cannot be satisfactorily explained by the stress theory, leading some to suggest that fear conditioning might underlie the phenomenon [47]. In certain respects, fear conditioning resembles PTSD [48]. During Pavlovian fear conditioning, a neutral conditioned stimulus (CS, usually a tone or light) is repeatedly paired with an unconditioned stressful stimulus (US, usually a footshock). Once the CS-US association has been formed, the CS produces a conditioned fear response (CR, such as freezing (or movement arrest), enhancement of musculature (startle) reflexes, autonomic changes, analgesia and behavioral response suppression) in anticipation of the US [49, 50]. A CR is also evoked when the animal is placed in the environment in which the experiment took place. Translating to PTSD, the traumatic event (US) triggers an unconditioned response (UR), which is characterized by strong arousal and intense fear. This UR becomes associated with cues (CSs), such as smells, voices, or sights that were present during the traumatic event. As a result of this pairing, these cues can trigger similar responses (CRs) even in the absence of the US [47, 51]. Thus, given the association between traumatic recall and seemingly unrelated stimuli and the ensuing fearful response, the mechanism of enhanced fear conditioning has often been suggested as a model for the re-experiencing phenomena in PTSD [52–55].

14.2.2.2   Impaired Extinction

Conditioned fear responses can be extinguished by repeatedly presenting the CS without the US [48]. Pavlov, in his classic investigation of appetitive conditioning in dogs, observed that extinguished responses spontaneously recovered with the passage of time [48]. This suggested that extinction suppresses, rather than erases, the original CS-US association. Thus, extinction is an important behavioral phenomenon that allows the organism to adapt its behavior to a changing environment [56]. Moreover, experimental extinction is a behavioral technique leading to suppression of the acquired fear, that is, a decrease in the amplitude and frequency of a CR as a function of non-reinforced CS presentations [57]. More recently, impaired extinction learning has been proposed as an alternative mechanism for the formation of PTSD symptoms [58, 59].

It has been established that fear extinction: (i) is dependent on NMDA receptors and L-type voltage-gated calcium channels [60]; (ii) is sensitive to modulation of second messenger systems, including kinase and phosphatase activity [61]; and (iii) may require protein synthesis [62–64]. Recently, a prominent role for medial prefrontal cortex (mPFC)–amygdala–hippocampus circuits was suggested in the contextual modulation of the extinction of fear memory. The current neurocircuitry model for PTSD hypothesizes hyper-responsivity within the amygdala to threat-related stimuli, with inadequate top-down governance over the amygdala by ventral/medial prefrontal cortex (mvPFC) (encompassing the rostral anterior cingulated cortex, subcallosal cortex, anterior cingulated cortex), orbitofrontal cortex and the hippocampus [65]. The decreased mPFC inhibition of the amygdala prevents retention of extinction learning, thus allowing reinstatement of the conditioned fear response. Interestingly, neuroimaging data support the current neurocircuitry model of PTSD and provide evidence for heightened responsivity of the amygdala, diminished responsivity of the mPFC, diminished hippocampal volumes and integrity, as well as impaired hippocampal function in PTSD [65–69]. However, PTSD is a complex disorder that involves far more than a fear response and cannot be explained by a simple conditioning model.

14.2.3
Individual Differences in Response to an Exposure to a Traumatic Experience

Researchers who work with animals have long been aware that individual study subjects tend to display a varying range of responses to stimuli, certainly where stress paradigms are concerned. This heterogeneity in responses was accepted for many years and regarded as unavoidable. Since humans clearly do not respond homogeneously to PTE, the heterogeneity in animal responses might be regarded as confirming the validity of animal studies, rather than as a problem. It stands to reason that a model of diagnostic criteria for psychiatric disorders could be applied to animal responses to augment the validity of study data, as long as the criteria for classification are clearly defined, reliably reproducible and yield results that conform to findings in human subjects. Of course, different study paradigms may give rise to different sets of criteria.

14.2.4
Behavioral Assessments

A variety of mazes and open environments have been employed to assess changes in exploratory behavior resulting from stress exposure. These test environments assess behaviors whose disruption indicates anxiety-like fearful behaviors and behaviors reflecting avoidance. Various learning and memory tasks are employed in which both exploration and learned task performance can be assessed. Some studies have investigated social behavior in home cages and in challenge situations. The startle response, which characterizes many PTSD patients, has been employed as one of the more definitively measurable parameters for the hyper-vigilant/hyper-alert component of the behavioral responses [30]. In the studies presented below exploratory behavior on the elevated plus maze (EPM) serves as the main platform for the assessment of overall behavior, and the acoustic startle response (ASR) paradigm provides a precise quantification of hyper-alertness, in terms of magnitude of response and habituation to the stimulus. For details regarding these tests, see Cohen *et al.* [26–28, 44, 70–72].

As to the timing of behavioral assessments, a large number of studies performed in a range of research centers indicate quite clearly that behavioral changes that are observed in rodents at day 7 after stress exposure are unlikely to change significantly over the next 30 days [72]. The average life expectancy for the domestic rat is between 2.5–3.0 years. Hence, behavioral patterns observed at day 7 can reliably be taken to represent PTSD-like responses (i.e., "translating" a week for a rat to a month for a human).

14.2.5
Classification According to Cut-Off Behavioral Criteria

Data from a large series of studies had previously shown that 7 days after a single 10-min predator scent exposure, the overall exposed population displayed significantly decreased time spent in the open arms and increased time in the closed arms of the EPM (which is translated to "avoidant" and "anxiety-like behavior") and higher mean startle responses as compared to control rats (Figure 14.1). It is important to note that the rats' behavior was not uniformly disturbed, but rather demonstrated a broad range of variation in response severity. The pooled data were re-examined for definable behavioral criteria and revealed a group of animals whose behavioral response patterns clearly demonstrated no significant difference from unexposed control animals, and a second group whose responses to both test paradigms were equally significantly at the extreme end on all measures. Each of these groups was significantly distinct from animals whose behavior lay between the extremes.

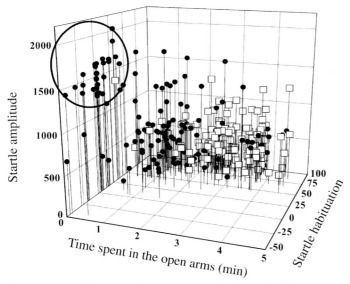

Figure 14.1 The effect of single predator scent stress (PSS) exposure versus unexposed control on rat anxiety-like behavior and acoustic startle response and habituation. The graphic representation of the data from both paradigms (EPM and ASR) reveals two obvious and rather distinct features. Firstly, it is clear that PSS exposure alters the response of the majority of individuals to at least some degree. Secondly the cluster of individuals that forms in the upper left hand corner of the graph (i.e., more extreme responses to exposure) is quite distinct from the majority of individuals.

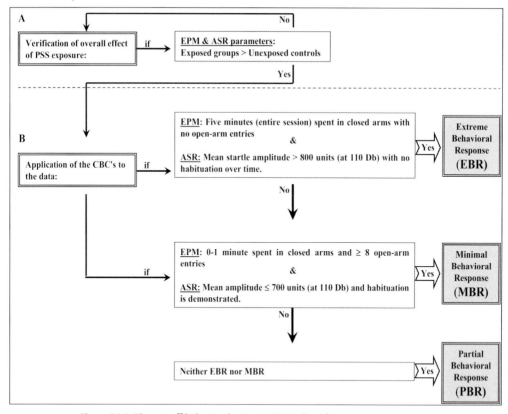

Figure 14.2 The cut-off behavioral criteria (CBC) algorithm.

The behavioral measures for each of these groups on the EPM and ASR tests were employed to define the basic CBCs. Since clinical diagnostic criteria require a sufficient number of symptoms from three symptom clusters in order to achieve satisfactory diagnostic specificity, the CBC response classification process requires that a given rat fulfill all criteria on *both* tests, performed in series. The standard algorithm for the CBC classification model also requires that prior to classification, a significant overall effect be demonstrated (Figure 14.2).

The CBCs enable us to clearly classify a given rat as displaying extreme behavioral response (EBR) or minimal behavioral response (MBR; i.e., extreme responses on *both* EPM and ASR tests lead to classification as EBR, whereas minimal responses are defined as MBR) – both of which have been validated in a large series of studies. The remaining rats display clearly disrupted behavior patterns compared to controls, but the extent of the disruption does not cross the threshold for EBR. These are labeled partial behavioral responders (PBR) and have as yet not been further subclassified [44, 70, 72].

The pooled behavioral data for entire predator scent stress (PSS)-exposed populations were re-examined according to the CBCs, revealing that the overall preva-

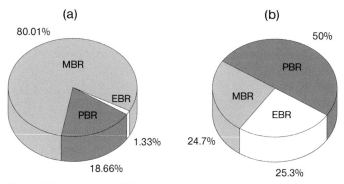

Figure 14.3 Re-analysis of data applying CBC;
(a) unexposed populations, (b) PSS-exposed populations.

lence rate for EBR rats was approximately 25% (Figure 14.3a) as compared to 1.3% in unexposed control populations (Figure 14.3b). The prevalence of MBR rats in the PSS-exposed groups was 24.7% (Figure 14.3a) as compared to 80.0% in the control groups (Figure 14.3b).

The implication of this initial finding was that all prior study analyses must have included a significant proportion of animals whose behavior had not been affected by the stressor (MBR) and many animals whose response was of uncertain significance (PBR), alongside those whose response was unequivocally one of severely disrupted behavioral patterns (EBR). Hence, the method offered a feasible means for classifying animal response patterns to trauma, thereby increasing the conceptual accuracy of the data.

It is of interest to note that the proportion of the entire exposed population fulfilling criteria for extreme responses (EBR) was compatible with epidemiological data for PTSD among trauma-exposed human populations [8], which report that between 15–35% fulfill criteria for PTSD and that approximately 20–30% display partial or subsymptomatic clinical pictures [7, 8, 11]. This compatibility further supports the concept of criterion-based classification in terms of face validity.

14.3
Selected CBC-Based Studies

14.3.1
Behavioral Response Patterns Versus Time

Time is an integral factor in traumatic stress-induced disorders. The prevalence rates of EBR rats were assessed among PSS-exposed rats on days 1, 3, 5, 7, 30 and 90 after exposure (Figure 14.4). Initially (day 1), almost all animals displayed extreme disruptions of behavior (EBR = 90%). The proportion of EBR animals

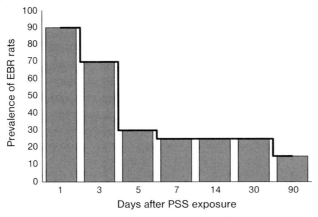

Figure 14.4 Prevalence of EBR rats after single PSS exposure as a function of time.

dropped rapidly over days 1 and 3 and between days 3 and 5 to about 25% at day 7. This proportion remained stable till day 30, dropping to about 15% by day 90. The resulting time curve of EBR prevalence rates parallels the rates of stress-related symptoms in humans, culminating in acute and chronic traumatic stress disorders [72].

14.3.2
Physiological Correlates

Physiological data were correlated with behavioral classification in a series of studies, including the HPA axis (circulating corticosterone, dehydroepiandros-terone (DHEA) and its sulfate derivative DHEA-S levels), autonomic nervous system (heart rate and heart rate variability) [34, 70] and immune system [73]. Although the gross population data had shown that the parameters in each study displayed significant responses to the stressor, CBC classification revealed that animals whose behavior conformed to EBR criteria were characterized by significantly more disturbances on all measures, whereas MBR rats displayed almost none.

14.3.3
Strain/Genetic Studies

The CBC classification model was applied to genetically manipulated rodent strains in order to examine two aspects of PTSD. One study assessed the HPA axis response in rat strains inbred to have either deficient or excessive HPA axis responsiveness, compared to outbred rats. The other examined the heritability of vulnerability versus resilience factors using inbred (near-isogenic) mouse strains exposed to PSS and classified according to the CBC method.

14.3.3.1 HPA Axis Response in Lewis and Fischer Rats

PTSD has been associated with disordered levels of circulating cortisol, an integral component of the stress response (increased levels in some studies, decreased in others) [74–80]. As mentioned above, clinical data show that preventive administration of cortisol reduces the incidence of cases of PTSD following septic shock [81].

The study examined whether low basal cortisol levels represent a consequence of traumatic exposure (i.e., possible neurotoxic effects of trauma) or a predisposing trait for pathological stress reactions in populations of inbred Lewis and Fischer rats compared to outbred Sprague–Dawley rats (controls). Lewis rats exhibit a reduced synthesis and secretion of corticotropin releasing factor (CRF), leading to reduced plasma ACTH and reduced CORT release from the adrenal cortex, whereas Fischer rats possess a hyper-responsive HPA axis. After PSS exposure, prevalence rates of EBR individuals were significantly higher in Lewis (50%) than in Fischer rats (10%) or in controls (25%) [82]. However, exogenous administration of cortisol to Lewis rats before applying the stressor decreased the prevalence of EBR significantly (to 8%) [82]. These results suggest that a blunted HPA axis response to stress may play a role in the susceptibility to experimentally-induced PTSD-like behavioral changes, especially as these effects were reversed by pre-exposure administration of corticosterone.

14.3.3.2 Stress-Induced Behavioral Responses in Inbred Mouse Strains

Twin and family studies of PTSD patients raise questions as to a possible genetic predisposition to PTSD, although the relative contributions of genotype and environment to endophenotypic expression are unclear.

Six inbred strains of mice frequently employed in transgenic research were assessed at baseline and 7 days after PSS exposure [83]. Inbred strains are expected to demonstrate ~97.5% homozygosity of loci, as the result of at least 20 generations of sibling matings. The results, however, revealed an unexpectedly high degree of within-strain individual heterogeneity at baseline and a high degree of heterogeneity in response to stress. This within-strain phenotypic heterogeneity might imply that environmental factors play a significant role in characterizing individual responses, in spite of the significant strain-related (i.e., genetic) underpinnings. This pilot study suggests that heritable factors may be involved only in part of the endophenotypes associated with the PTSD-like behavioral phenotype and may be influenced indirectly by interactions with environmental variables [84, 85].

14.3.4
Molecular Neurobiological Correlates

Selected brain areas, especially the hippocampal substructures and frontal cortex, of rats classified according to the CBC procedure have been studied in correlation to both their behavioral and physiological response patterns. The studies have examined the expression of genes and gene products for key intracellular and intercellular biomolecules associated with neuromodulation, synaptic plasticity and receptor systems. In some studies, these data have also been correlated to individual performance on memory-related tasks.

The development of an extreme behavioral response has been shown to be associated with a distinct pattern of long-term and persistent downregulation of mRNA for brain-derived neurotrophic factor (BDNF) and synaptophysin, and an upregulation of glucocorticoid receptor (GR) protein levels and tyrosine kinase receptor (TrkB) mRNA in the CA1 subregion of the hippocampus, compared to PBR and MBR animals and to unexposed controls [37]. The persistently higher levels of glucocorticoids are associated with attenuation of BDNF, synaptophysin and expression of zif/268 and immediate early genes such as activity-regulated cytoskeletal-associated protein (Arc) in the EBR individuals; this suggests that glucocorticoids reflect or mediate the characteristic changes in neural plasticity and synaptic functioning underlying chronic, stress-induced behavioral disruption. Neurotrophins, and particularly BDNF, are known to modulate many aspects of neuronal plasticity [86] and the selection of functional neuronal connections in the CNS [75, 87, 88]. The decreased expression of BDNF (mRNA and protein) levels in EBR individuals may result in a decrease in synaptic plasticity and impair the stabilization of synaptic connectivity, leading to vulnerability to psychopathology. Decreased hippocampal expression of synaptophysin, a major integral protein on synaptic vesicles, might also be associated with hippocampal damage that may occur during stress exposure [89]. Taken together, decreased hippocampal expression of these genes may have physiological consequences, for example, inducing damage to hippocampal neurons.

SSRIs and other antidepressant drugs have been found to reinforce synaptic strength in mood-related brain regions in a manner akin to that achieved in experimental models of synaptic plasticity [90, 91]. Pei *et al.* [92] reported that repeated administration of the monoamine reuptake inhibitors paroxetine, venlafaxine or desipramine induces region-specific increases in Arc mRNA (in the frontal and parietal cortex and hippocampal CA1 area). Thus, whereas downregulation of Arc mRNA has been shown to promote stress-induced psychopathology in synaptic networks, long-term administration of SSRIs might prevent or reverse this effect. Taken together, SSRIs might, therefore, be able to protect and/or to rescue the functional integrity of neuronal circuitry from the effects of stress [37].

14.3.5
Drug Studies

Acute-phase pharmacotherapeutic interventions that effectively alleviate symptoms and possess potential preventive effects on the development of PTSD and that are founded on large-scale, double-blind, controlled, prospective clinical trials are lacking. The CBC classification model affords distinct advantages for the prospective study of the therapeutic and preventive potential of medications. The model enables the prospective study of associations between the behavioral efficacy of the drug in question, in a quantifiable manner over specific periods of time, and the biomolecular and physiological correlates of these behavioral effects.

The CBC model has been applied to the study of a number of drugs – a selective serotonin reuptake inhibitor (sertraline), corticosteroids and a benzodiazepine (alprazolam).

14.3.5.1 Early Intervention with an SSRI (Sertraline)

Based on the rationale that the acute phase in rodents is represented by the first 7 days following stress exposure (see Section 14.2), rats were randomly allocated to 7 days of treatment, either immediately following exposure or as of day 7, and compared to saline treatment. Behavioral and biomolecular assessments performed at day 7 (or day 14) demonstrated the following. Brief immediate post-exposure intervention with sertraline had an observable short-term effect on stress-induced behavioral changes compared to the later treatment regimen, and compared to the saline-treated control group [93]. Seven days of treatment with sertraline immediately after PSS exposure elicited a statistically significant reduction (14%) in prevalence rates of EBR and an increase of 5% in prevalence rates of minimal response (MBR) compared to the placebo control group. These finding suggest that SSRI drugs represent potential agents for secondary intervention in the acute aftermath of traumatic stress exposure and are thus worthy of further investigation.

14.3.5.2 Early Intervention with Corticosterone

Since corticosteroid treatment is clinically indicated only in cases in which there is significant physical illness or poly-trauma, recurrent clinical reports of a significant preventive effect in terms of the incidence of concomitant PTSD are difficult to interpret, despite their relative frequency and impressive results [81]. The CBC model was employed to examine the effect of a single high-dose intervention with the adrenocorticoid stress hormone CORT given immediately after exposure. This regimen was compared to lower doses, later treatment and saline. Stress-induced behavioral responses were assessed at day 30 and trauma cue-triggered freezing response was assessed on day 31.

The results clearly showed that a single 25 mg/kg dose of CORT administered immediately after exposure to the scent of predator urine resulted in a statistically significant reduction of 13.2% in the prevalence rates of EBR individuals at 30 days, with a concomitant increase of 12.4% in the prevalence of MBR individuals, as compared to saline controls – that is, a significant shift toward less extreme behavioral disruption ensuing from traumatic stress [94]. Rats in the high-dose CORT group responded markedly less extremely to exposure to the trauma cue (24% of time freezing) than the saline-control group (80% of time freezing). This pattern of response suggests that the single high-dose CORT treatment confers some degree of resilience to future trauma-related stress exposure [94]. Lower doses of CORT (0.1–5.0 mg/kg) were ineffective in attenuating behavioral disruptions, significantly increased the prevalence of EBR at day 30 and significantly increased vulnerability to the trauma cue as compared to placebo.

The marked attenuation of the response of treated individuals to the trauma cue 31 days after exposure is of significance. The time frame in which CORT was administered (1 h after stress exposure) conforms to the time frame within which the memory consolidation process takes place at the cellular level (3–6 h after initiation of data acquisition). The time at which the effect was assessed was sufficiently distant from the initial exposure as to suggest that the effect was mediated by memory-related processes. Furthermore, the same pattern was observed in another study, where the protein synthesis inhibitor anisomycin was effective when administered within 1 h after exposure, but not when administered later (after reactivation of the trauma by a trauma cue) [95]. This may suggest that the single high-dose CORT treatment interfered similarly, by disrupting consolidation of short-term memory to long-term memory [94].

14.3.5.3 Early Intervention with Benzodiazepine (Alprazolam)

Benzodiazepines are commonly used to relieve distress. Since it has been claimed that they impede adequate processing of acute grief, their effects on the processing of acute stress were examined.

The CBC model was employed to examine the short-term efficacy and the long-term potential of brief, early post-exposure administration of a commonly-prescribed benzodiazepine (alprazolam) for the prevention of subsequent PTSD-like behavioral changes and to examine its effects on subsequent vulnerability to stress, compared to later treatment and placebo. As expected, the results demonstrated short-term efficacy, but no preventive potential. The finding that caused particular concern was that rats treated immediately after the initial exposure were rendered significantly more vulnerable to the trauma cue and by far more vulnerable to re-exposure to PSS than control groups. Treatment initiated after one week did not affect vulnerability [96]. It will be important to establish whether this finding is replicable and whether it is related to specific benzodiazepines and/or a certain time frame, both in animal and in clinical studies. One possible mechanism might be related to the effect of alprazolam on cortisol secretion. The marked suppression of corticosterone activity during alprazolam treatment and the sharp rebound after its cessation may well be key factors in the pathogenesis of the different behavioral responses observed in the study subjects when treatment was initiated immediately. Since corticosterone plays a major role in the regulation of responses to stress, alterations in timing and polarity of plasma corticosterone levels may be of great pathogenetic significance, especially in the earliest phases.

Summary

Animal models may complement clinical research and enable modalities that are difficult to attain in clinical studies. The animal model presented here, which is a combination of exposure to a predator and a focus on setting apart the affected based on behavioral cut-off criteria, demonstrates high face validity, construct validity and predictive validity. The cumulative results of our studies indicate that the contribution of animal models can be further enhanced by classifying individual animal study subjects according to their response patterns. This approach enables researchers to test interventions that might be impossible (e.g., anisomycin) or difficult (e.g., benzodiazepine, SSRI, cortisol) to do in a clinical setting without any proper preclinical basis. The animal model also enables the researcher to go one step further and correlate specific anatomic biomolecular and physiological parameters with the degree and pattern of individual behavioral response.

References

1 American Psychiatric Association (1994) *Diagnostic and Statistical Manual of Mental Disorders*, 4th edn, American Psychiatric Association, Washington, DC.

2 Bryant, R.A. (2006) Recovery after the tsunami: timeline for rehabilitation. *J. Clin. Psychiatry*, **67**, 50–55.

3 Connor, K.M., Foa, E.B. and Davidson, J.R. (2006) Practical assessment and evaluation of mental health problems following a mass disaster. *J. Clin. Psychiatry*, **67** (Suppl. 2), 26–33.

4 Davidson, J.R. (2006) Pharmacologic treatment of acute and chronic stress following trauma: 2006. *J. Clin. Psychiatry*, **67** (Suppl. 2), 34–39.

5 Shalev, A.Y. (2002) Acute stress reactions in adults. *Biol. Psychiatry*, **51**, 532–543.

6 Foa, E.B., Stein, D.J. and McFarlane, A.C. (2006) Symptomatology and psychopathology of mental health problems after disaster. *J. Clin. Psychiatry*, **67**, 15–25.

7 Breslau, N., Kessler, R.C., Chilcoat, H.D., Schultz, L.R., Davis, G.C. and Andreski, P. (1998) Trauma and posttraumatic stress disorder in the community: the 1996 Detroit Area Survey of Trauma. *Arch. Gen. Psychiatry*, **55**, 626–632.

8 Breslau, N., Davis, G.C., Andreski, P. and Peterson, E. (1991) Traumatic events and posttraumatic stress disorder in an urban population of young adults. *Arch. Gen. Psychiatry*, **48**, 216–222.

9 Helzer, J.E., Robins, L.N. and McEvoy, L. (1987) Post-traumatic stress disorder in the general population. Findings of the epidemiologic catchment area survey. *N. Engl. J. Med.*, **317**, 1630–1634.

10 Perkonigg, A., Kessler, R.C., Storz, S. and Wittchen, H.U. (2000) Traumatic events and post-traumatic stress disorder in the community: prevalence, risk factors and comorbidity. *Acta Psychiatr. Scand.*, **101**, 46–59.

11 Resnick, H., Yehuda, R., Pitman, R. and Foy, D. (1995) Effect of previous trauma on acute plasma cortisol level following rape. *Am. J. Psychiatry*, **152**, 1675–1677.

12 Shore, J.H., Vollmer, W.M. and Tatum, E.L. (1989) Community patterns of posttraumatic stress disorders. *J. Nerv. Ment. Dis.*, **177**, 681–685.

13 Fairbank, J.A., Schlenger, W.E., Saigh, P.A. and Davidson, J.R.T. (1995) An epidemiologic profile of post-traumatic stress disorder: prevalence, comorbidity, and risk factors, in *Neurobiological and Clinical Consequences of Stress: From Normal Adaptation to PTSD* (eds M.J. Friedman, D.S. Charney and A.V. Deutsch), Lippincott-Raven, Philadelphia, pp. 415–427.

14 Peri, T., Ben-Shakhar, G., Orr, S.P. and Shalev, A.Y. (2000) Psychophysiologic assessment of aversive conditioning in posttraumatic stress disorder. *Biol. Psychiatry*, **47**, 512–519.

15 Solomon, Z., Shklar, R. and Mikulincer, M. (2005) Frontline treatment of combat stress reaction: a 20-year longitudinal evaluation study. *Am. J. Psychiatry*, **162**, 2309–2314.

16 Nutt, D. and Davidson, J. (2000) *Post-Traumatic Stress Disorder Diagnosis, Management and Treatment*, Taylor & Francis, London.

17 Blanchard, D.C., Griebel, G. and Blanchard, R.J. (2003) Conditioning and residual emotionality effects of predator stimuli: some reflections on stress and emotion. *Prog. Neuropsychopharmacol. Biol. Psychiatry*, **27**, 1177–1185.

18 Blanchard, R.J. and Blanchard, D.C. (1990) Anti-predator defense as models of fear and anxiety, in *Brain* (eds R.J.

Blanchard and S. Parmigiani), Harwood Academic Publishers, London.

19 Blanchard, R.J., Griebel, G., Henrie, J.A. and Blanchard, D.C. (1997) Differentiation of anxiolytic and panicolytic drugs by effects on rat and mouse defense test batteries. *Neurosci. Biobehav. Rev.*, **21**, 783–789.

20 Blanchard, R.J., Nikulina, J.N., Sakai, R.R., McKittrick, C., McEwen, B. and Blanchard, D.C. (1998) Behavioral and endocrine change following chronic predatory stress. *Physiol. Behav.*, **63**, 561–569.

21 Adamec, R., Head, D., Blundell, J., Burton, P. and Berton, O. (2006) Lasting anxiogenic effects of feline predator stress in mice: sex differences in vulnerability to stress and predicting severity of anxiogenic response from the stress experience. *Physiol. Behav.*, **88**, 12–29. Epub 2006 Apr 2019.

22 Adamec, R., Muir, C., Grimes, M. and Pearcey, K. (2007) Involvement of noradrenergic and corticoid receptors in the consolidation of the lasting anxiogenic effects of predator stress. *Behav. Brain Res.*, **179**, 192–207. Epub 2007 Feb 2006.

23 Adamec, R., Strasser, K., Blundell, J., Burton, P. and McKay, D.W. (2006) Protein synthesis and the mechanisms of lasting change in anxiety induced by severe stress. *Behav. Brain Res.*, **167**, 270–286.

24 Adamec, R.E., Blundell, J. and Burton, P. (2006) Relationship of the predatory attack experience to neural plasticity, pCREB expression and neuroendocrine response. *Neurosci. Biobehav. Rev.*, **30**, 356–375. Epub 2005 Aug 2022.

25 Adamec, R.E. and Shallow, T. (1993) Lasting effects on rodent anxiety of a single exposure to a cat. *Physiol. Behav.*, **54**, 101–109.

26 Cohen, H., Benjamin, J., Kaplan, Z. and Kotler, M. (2000) Administration of high-dose ketoconazole, an inhibitor of steroid synthesis, prevents posttraumatic anxiety in an animal model. *Eur. Neuropsychopharmacol.*, **10**, 429–435.

27 Cohen, H., Friedberg, S., Michael, M., Kotler, M. and Zeev, K. (1996) Interaction of CCK-4 induced anxiety and post-cat

exposure anxiety in rats. *Depress. Anxiety*, **4**, 144–145.

28 Cohen, H., Kaplan, Z., Kotler, M. and C.C. (1999) K-antagonists in a rat exposed to acute stress: implication for anxiety associated with post-traumatic stress disorder. *Depress. Anxiety*, **10**, 8–17.

29 Diamond, D.M., Campbell, A.M., Park, C.R., Woodson, J.C., Conrad, C.D., Bachstetter, A.D. and Mervis, R.F. (2006) Influence of predator stress on the consolidation versus retrieval of long-term spatial memory and hippocampal spinogenesis. *Hippocampus*, **16**, 571–576.

30 File, S.E., Zangrossi, H. Jr., Sanders, F.L. and Mabbutt, P.S. (1993) Dissociation between behavioral and corticosterone responses on repeated exposures to cat odor. *Physiol. Behav.*, **54**, 1109–1111.

31 Griebel, G., Blanchard, D.C., Jung, A., Lee, J.C., Masuda, C.K. and Blanchard, R.J. (1995) Further evidence that the mouse defense test battery is useful for screening anxiolytic and panicolytic drugs: effects of acute and chronic treatment with alprazolam. *Neuropharmacology*, **34**, 1625–1633.

32 Apfelbach, R., Blanchard, C.D., Blanchard, R.J., Hayes, R.A. and McGregor, I.S. (2005) The effects of predator odors in mammalian prey species: a review of field and laboratory studies. *Neurosci. Biobehav. Rev.*, **29**, 1123–1144. Epub 2005 Aug 1128.

33 Blundell, J., Adamec, R. and Burton, P. (2005) Role of NMDA receptors in the syndrome of behavioral changes produced by predator stress. *Physiol. Behav.*, **86**, 233–243.

34 Cohen, H., Maayan, R., Touati-Werner, D., Kaplan, Z., Matar, M., Loewenthal, U., Kozlovsky, N. and Weizman, R. (2007) Decreased circulatory levels of neuroactive steroids in behaviorally more extremely affected rats subsequent to exposure to a potentially traumatic experience. *Int. J. Neuropsychopharmacol.*, **10**, 203–209.

35 Endres, T., Apfelbach, R. and Fendt, M. (2005) Behavioral changes induced in rats by exposure to trimethylthiazoline, a component of fox odor. *Behav. Neurosci.*, **119**, 1004–1010.

36 Kozlovsky, N., Matar, M.A., Kaplan, Z., Kotler, M., Zohar, J. and Cohen, H. (2007) Long-term down-regulation of BDNF mRNA in rat hippocampal CA1 subregion correlates with PTSD-like behavioural stress response. *Int. J. Neuropsychopharmacol.*, 1–18.

37 Kozlovsky, N., Matar, M.A., Kaplan, Z., Kotler, M., Zohar, J. and Cohen, H. (2007) The immediate early gene Arc is associated with behavioral resilience to stress exposure in an animal model of posttraumatic stress disorder. *Eur. Neuropsychopharmacol.*, **2**, 2.

38 Mazor, A., Matar, M., Kozlovsky, N., Zohar, J., Kaplan, Z. and Cohen, H. (2007) Gender-related qualitative differences in baseline and post stress anxiety responses are not reflected in the incidence of criterion-based PTSD-like behavior patterns. *World J. Biol. Psychiatry*, **13**, 1–14.

39 Roseboom, P.H., Nanda, S.A., Bakshi, V.P., Trentani, A., Newman, S.M. and Kalin, N.H. (2007) Predator threat induces behavioral inhibition, pituitary-adrenal activation and changes in amygdala CRF-binding protein gene expression. *Psychoneuroendocrinology*, **32**, 44–55. Epub 2006 Nov 2020.

40 Sullivan, M. and Gratton, A. (1998) Relationships between stress-induced increases in medial prefrontal cortical dopamine and plasma corticosterone levels in rats: role of cerebral laterality. *Neuroscience*, **83**, 81–91.

41 Takahashi, L.K., Nakashima, B.R., Hong, H. and Watanabe, K. (2005) The smell of danger: a behavioral and neural analysis of predator odor-induced fear. *Neurosci. Biobehav. Rev.*, **29**, 1157–1167. Epub 2005 Aug 1110.

42 Richter-Levin, G. (1998) Acute and long-term behavioral correlates of underwater trauma – potential relevance to stress and post-stress syndromes. *Psychiatry Res.*, **79**, 73–83.

43 Wang, J., Akirav, I. and Richter-Levin, G. (2000) Short-term behavioral and electrophysiological consequences of underwater trauma. *Physiol. Behav.*, **70**, 327–332.

44 Cohen, H. and Zohar, J. (2004) Animal models of post traumatic stress disorder: the use of cut off behavioral criteria. *Ann. N. Y. Acad. Sci.*, **1032**, 167–178.

45 Warren, D.A., Castro, C.A., Rudy, J.W. and Maier, S.F. (1991) No spatial learning impairment following exposure to inescapable shock. *Psychobiology*, **19**, 127–134.

46 Diamond, D.M. and Rose, G.M. (1994) Stress impairs LTP and hippocampal-dependent memory. *Ann. N. Y. Acad. Sci.*, **746**, 411–414.

47 Yehuda, R. and LeDoux, J. (2007) Response variation following trauma: a translational neuroscience approach to understanding PTSD. *Neuron*, **56**, 19–32.

48 Milad, M.R., Rauch, S.L., Pitman, R.K. and Quirk, G.J. (2006) Fear extinction in rats: implications for human brain imaging and anxiety disorders. *Biol. Psychol.*, **73**, 61–71. Epub 2006 Feb 2013.

49 Dunsmoor, J.E., Bandettini, P.A. and Knight, D.C. (2007) Neural correlates of unconditioned response diminution during Pavlovian conditioning. *Neuroimage*, **1**, 811–817.

50 LeDoux, J. (1996) Emotional networks and motor control: a fearful view. *Prog. Brain Res.*, **107**, 437–446.

51 Blechert, J., Michael, T., Vriends, N., Margraf, J. and Wilhelm, F.H. (2007) Fear conditioning in posttraumatic stress disorder: evidence for delayed extinction of autonomic, experiential, and behavioural responses. *Behav. Res. Ther.*, **45**, 2019–2033.

52 Foa, E.B. and Kozak, M.J. (1986) Emotional processing of fear: exposure to corrective information. *Psychol. Bull.*, **99**, 20–35.

53 Kolb, L.C. (1987) A neuropsychological hypothesis explaining posttraumatic stress disorders. *Am. J. Psychiatry*, **144**, 989–995.

54 LeDoux, J.E. (2000) Emotion circuits in the brain. *Annu. Rev. Neurosci.*, **23**, 155–184.

55 Maren, S. (2001) Neurobiology of Pavlovian fear conditioning. *Annu. Rev. Neurosci.*, **24**, 897–931.

56 Bouton, M.E. (2004) Context and behavioral processes in extinction. *Learn. Mem.*, **11**, 485–494.

57 Akirav, I. and Maroun, M. (2007) The role of the medial prefrontal cortex-amygdala circuit in stress effects on the extinction of fear. *Neura. Plast.*, **2007**, 30873.

58 Maren, S. and Chang, C.H. (2006) Recent fear is resistant to extinction. *Proc. Natl. Acad. Sci. U. S. A.*, **103**, 18020–18025. Epub 12006 Nov 18027.

59 Myers, K.M. and Davis, M. (2002) Behavioral and neural analysis of extinction. *Neuron*, **36**, 567–584.

60 Chan, C.S., Weeber, E.J., Kurup, S., Sweatt, J.D. and Davis, R.L. (2003) Integrin requirement for hippocampal synaptic plasticity and spatial memory. *J. Neurosci.*, **23**, 7107–7116.

61 Szapiro, G., Vianna, M.R., McGaugh, J.L., Medina, J.H. and Izquierdo, I. (2003) The role of NMDA glutamate receptors, PKA, MAPK, and CAMKII in the hippocampus in extinction of conditioned fear. *Hippocampus*, **13**, 53–58.

62 Davis, M., Myers, K.M., Chhatwal, J. and Ressler, K.J. (2006) Pharmacological treatments that facilitate extinction of fear: relevance to psychotherapy. *NeuroRx*, **3**, 82–96.

63 Lattal, K.M. and Abel, T. (2001) Different requirements for protein synthesis in acquisition and extinction of spatial preferences and context-evoked fear. *J. Neurosci.*, **21**, 5773–5780.

64 Santini, E., Ge, H., Ren, K., Pena de Ortiz, S. and Quirk, G.J. (2004) Consolidation of fear extinction requires protein synthesis in the medial prefrontal cortex. *J. Neurosci.*, **24**, 5704–5710.

65 Rauch, S.L., Shin, L.M. and Phelps, E.A. (2006) Neurocircuitry models of posttraumatic stress disorder and extinction: human neuroimaging research—past, present, and future. *Biol. Psychiatry*, **60**, 376–382.

66 De Bellis, M.D., Keshavan, M.S., Shifflett, H., Iyengar, S., Beers, S.R., Hall, J. and Moritz, G. (2002) Brain structures in pediatric maltreatment-related posttraumatic stress disorder: a sociodemographically matched study. *Biol. Psychiatry*, **52**, 1066–1078.

67 Etkin, A. and Wager, T.D. (2007) Functional neuroimaging of anxiety: a meta-analysis of emotional processing in PTSD, social anxiety disorder, and specific phobia. *Am. J. Psychiatry*, **164**, 1476–1488.

68 Liberzon, I., Britton, J.C. and Phan, K.L. (2003) Neural correlates of traumatic recall in posttraumatic stress disorder. *Stress*, **6**, 151–156.

69 Liberzon, I. and Sripada, C.S. (2008) The functional neuroanatomy of PTSD: a critical review. *Prog. Brain Res.*, **167**, 151–169.

70 Cohen, H., Joseph, Z. and Matar, M. (2003) The relevance of differential response to trauma in an animal model of post-traumatic stress disorder. *Biol. Psychiatry*, **53**, 463–473.

71 Cohen, H., Matar, M.A., Richter-Levin, G. and Zohar, J. (2006) The contribution of an animal model toward uncovering biological risk factors for PTSD. *Ann. N. Y. Acad. Sci.*, **1071**, 335–350.

72 Cohen, H., Zohar, J., Matar, M.A., Zeev, K., Loewenthal, U. and Richter-Levin, G. (2004) Setting apart the affected: the use of behavioral criteria in animal models of post traumatic stress disorder. *Neuropsychopharmacology*, **29**, 1962–1970.

73 Cohen, H., Ziv, Y., Cardon, M., Kaplan, Z., Matar, M.A., Gidron, Y., Schwartz, M. and Kipnis, J. (2006) Maladaptation to mental stress mitigated by the adaptive immune system via depletion of naturally occurring regulatory CD4+CD25+ cells. *J. Neurobiol.*, **66**, 552–563.

74 Bremner, J.D., Licinio, J., Darnell, A., Krystal, J.H., Owens, M.J., Southwick, S.M., Nemeroff, C.B., Charney, D.S. and Elevated, C.S. (1997) F corticotropin-releasing factor concentrations in posttraumatic stress disorder. *Am. J. Psychiatry*, **154**, 624–629.

75 Delahanty, D., Raimonde, A. and Spoonster, E. (2000) Initial posttraumatic urinary cortisol levels predict subsequent PTSD symptoms in motor vehicle accident victims. *Biol. Psychiatry*, **48**, 940–947.

76 Mason, J.W., Giller, E.L., Kosten, T.R., Ostroff, R.B. and Podd, L. (1986) Urinary

free-cortisol levels in posttraumatic stress disorder patients. *J. Nerv. Ment. Dis.*, **174**, 145–149.

77 Pitman, R. and Orr, S. (1990) Twenty-four hour urinary cortisol and catecholamine excretion in combat-related posttraumatic stress disorder. *Biol. Psychiatry*, **27**, 245–247.

78 Rasmusson, A.M., Lipschitz, D.S., Wang, S., Hu, S., Vojvoda, D., Bremner, J.D., Southwick, S.M. and Charney, D.S. (2001) Increased pituitary and adrenal reactivity in premenopausal women with posttraumatic stress disorder. *Biol. Psychiatry*, **50**, 965–977.

79 Yehuda, R. (2005) Neuroendocrine aspects of PTSD. *Handb. Exp. Pharmacol.*, **169**, 371–403.

80 Yehuda, R., Morris, A., Labinsky, E., Zemelman, S. and Schmeidler, J. (2007) Ten-year follow-up study of cortisol levels in aging holocaust survivors with and without PTSD. *J. Trauma. Stress*, **20**, 757–761.

81 Schelling, G., Kilger, E., Roozendaal, B., de Quervain, D.J., Briegel, J., Dagge, A., Rothenhausler, H.B., Krauseneck, T., Nollert, G. and Kapfhammer, H.P. (2004) Stress doses of hydrocortisone, traumatic memories, and symptoms of posttraumatic stress disorder in patients after cardiac surgery: a randomized study. *Biol. Psychiatry*, **55**, 627–633.

82 Cohen, H., Zohar, J., Gidron, Y., Matar, M.A., Belkind, D., Loewenthal, U., Kozlovsky, N. and Kaplan, Z. (2006) Blunted HPA axis response to stress influences susceptibility to posttraumatic stress response in rats. *Biol. Psychiatry*, **59**, 1208–1218.

83 Cohen, H., Zohar, J., Matar, M., Loewenthal, U. and Kaplan, Z. (2007) The impact of environment factors in determining post-exposure responses in isogenic strains of mice: can genetic predisposition explain phenotypic vulnerability? *Int. J. Neuropsychopharmacol.*, **11**, 331–349.

84 Caspi, A. and Moffitt, T. (2006) Gene–environment interactions in psychiatry: joining forces with neuroscience. *Nat. Rev. Neurosci.*, **7**, 583–590.

85 Moffitt, T., Caspi, A. and Rutter, M. (2006) Measured gene–environment interactions in psychopathology. *Perspect. Psychol. Sci.*, **1**, 5–27.

86 Thoenen, H. (1995) Neurotrophins and neuronal plasticity. *Science*, **270**, 593–598.

87 Wang, M.J., Huang, H.M., Chen, H.L., Kuo, J.S. and Jeng, K.C. (2001) Dehydroepiandrosterone inhibits lipopolysaccharide-induced nitric oxide production in BV-2 microglia. *J. Neurochem.*, **77**, 830–838.

88 Mamounas, L.A., Altar, C.A., Blue, M.E., Kaplan, D.R., Tessarollo, L. and Lyons, W.E. (2000) BDNF promotes the regenerative sprouting, but not survival, of injured serotonergic axons in the adult rat brain. *J. Neurosci.*, **20**, 771–782.

89 Thome, J., Pesold, B., Baader, M., Hu, M., Gewirtz, J.C., Duman, R.S. and Henn, F.A. (2001) Stress differentially regulates synaptophysin and synaptotagmin expression in hippocampus. *Biol. Psychiatry*, **50**, 809–812.

90 Duman, R.S. (2002) Synaptic plasticity and mood disorders. *Mol. Psychiatry*, **7**, 29–34.

91 Manji, H.K., Drevets, W.C. and Charney, D.S. (2001) The cellular neurobiology of depression. *Nat. Med.*, **7**, 541–547.

92 Pei, Q., Zetterstrom, T.S., Sprakes, M., Tordera, R. and Sharp, T. (2003) Antidepressant drug treatment induces Arc gene expression in the rat brain. *Neuroscience*, **121**, 975–982.

93 Matar, M.A., Cohen, H., Kaplan, Z. and Zohar, J. (2006) The effect of early poststressor intervention with sertraline on behavioral responses in an animal model of post-traumatic stress disorder. *Neuropsychopharmacology*, **31**, 2610–2618.

94 Cohen, H., Matar, M., Buskila, D., Kaplan, Z. and Joseph, Z. (2008) Early post-stressor intervention with high dose corticosterone attenuates post traumatic stress response in an animal model of PTSD. *Biol. Psychiatry*, **15**, 708–717.

95 Cohen, H., Kaplan, Z., Matar, M., Loewenthal, U., Kozlovsky, N. and Zohar, J. (2006) Anisomycin, a protein synthesis

inhibitor, disrupts traumatic memory consolidation and attenuates post traumatic stress response in rats. *Biol. Psychiatry*, **60**, 767–776.

96 Matar, M., Zohar, J., Kaplan, Z. and Cohen, H. (2009) Alprazolam treatment immediately after stress exposure interferes with the normal HPA-stress response and increases vulnerability to subsequent stress in an animal model of PTSD. *Eur. Neuropsychopharm.*, **19**, 283–295.

15
The Cholinergic Model for PTSD: from Acute Stress to PTSD, from Neuron to Network and Behavior

Alon Friedman and Lev Pavlovsky

15.1
From Acute Stress to PTSD

Emotional stress is often followed by long-term changes in brain functions. In humans, such changes can be clinically diagnosed as post-traumatic stress disorder (PTSD), a mental condition characterized by re-experiencing of the event, avoidance, numbing of responsiveness and hyper-arousal. While the neurobiological mechanisms of PTSD are not known, accumulating clinical observations suggest that in most cases: (i) the acute traumatic event has a significant emotional impact on the individual and is perceived as "life threatening"; (ii) the emotional experience associated with the stressful event is often re-experienced by the individual; and (iii) both emotional and cognitive brain functions are disturbed. Thus, in the search for a biological mechanism underlying PTSD, it is reasonable to assume that: (i) it is associated with brain circuits activated by emotional stimuli; (ii) it is activated fast (during or immediately after the stimuli) but remains altered for long period of time; and (iii) it is associated with the proper function of both "emotional" and "cognitive" related brain circuits. We present the *brain cholinergic system* as an appropriate candidate system to underlie PTSD, as it has been shown to be rapidly responding to stressful stimulus by long-term changes and it has a key role in both emotional and cognitive high brain functions.

15.2
Stress and Brain Cholinergic Pathways

Understanding the basis for post-stress modifications in central nervous system (CNS) functioning requires an understanding of the anatomical and physiological principles determining high brain functions. In this chapter we focus on the cholinergic system as a well described transmitter system involved in both emotional and cognitive aspects. The central cholinergic system is, to a large extent, considered *modulatory* in nature: it is composed of relatively few networks with a

Stress – From Molecules to Behavior. Edited by Hermona Soreq, Alon Friedman, and Daniela Kaufer
Copyright © 2010 WILEY-VCH Verlag GmbH & Co. KGaA, Weinheim
ISBN: 978-3-527-32374-6

small number of neurons which give widespread innervations to most cells within the limbic system and cerebral cortex, modulating their activity. Parallel anatomical and physiological studies are consistent with the notion that cholinergic transmission is involved in determining the response of sensory-motor neurons, thus affecting both the perception of events and the performed action by the individual.

15.3
Anatomical Considerations

It is now established that certain brain structures and pathways are crucial for the mediation and experience of emotion and that these are part of our evolutionary inheritance. The key system crucial for the understanding of emotional and stress-related behavior is the limbic system, which includes groups of complex neuronal networks, including the hippocampus and amygdala as the most important. The amygdala is located in the anterior part of the temporal lobes and is central to the brain's regulation of emotions. The massive sensory input which it receives from the neocortex makes the amygdala the main provider of affective valence to sensory representations. The reciprocal connections between the amygdala and neocortical sensory systems allow the direct influence of emotional tone on the perception and interpretation of sensory information. The hippocampus is also situated in the temporal lobes. It has reciprocal connections with the entorhinal cortex, which receives polysensory information and sends a widespread output (via the fornix) into the mamillary bodies of the hypothalamus, making it an excellent link structure between the neocortex and autonomic circuits regulating behavior. The hippocampus is considered a key structure for visuospatial processing as well as memory consolidation and retrieval.

The hippocampus and parahippocampal region receive a considerable cholinergic input, which originates primarily from the basal forebrain through the fimbria-fornix. This projection was shown by anatomical studies using staining for the acetylcholine (ACh) hydrolyzing enzyme acetylcholinesterase (AChE), and the ACh synthesizing enzyme choline acyltransferase (ChAT) to be derived primarily from the medial septum and diagonal band of Broca [1–3]. Indeed, stimulation of this pathway leads to increased ACh levels in the hippocampus. Little is known about the detailed synaptic connectivity of cholinergic inputs to the hippocampus and parahippocampal region. There is an abundance of cholinergic fibers in the stratum oriens of CA1 and dentate hilus. However, the detailed anatomical targets of these synapses are not known. Interestingly, in the neocortex, cholinergic boutons are frequently observed in close apposition to asymmetrical (presumably excitatory) synapses on dendritic spines and shafts without forming recognizable synaptic specializations. Immunohistochemical localization of muscarinic receptors, the predominant cholinergic receptors in the adult brain, confirms the anatomical data on cholinergic innervation; muscarinic receptors are found on postsynaptic elements opposed to symmetrical synapses, predominantly on dendritic shafts. In addition, muscarinic receptors are found in pre- and post-synaptic elements of non-cholinergic, asymmetrical synapses, including those on dendritic spines. Thus, anatomical data suggest that ACh affects synaptic integration both through activation of classic cholinergic synapses and through direct modulation of non-cholinergic synapses.

In cortical cells, there are currently five known muscarinic receptor subtypes expressed [4]. Immunocytochemical studies regarding the localization of different muscarinic receptor subtypes show complex patterns and region specificity. For example, Rouse and colleagues [5] showed that M1–M4 receptors are differentially expressed in laminar patterns in the molecular layer of the dentate gyrus. Moreover, while M1 receptors were found in dendrites and spines, M2–M4 were most frequently found on pre-synaptic terminals. These differences might be functionally important as, for example, shown in physiological experiments where muscarinic receptor agonists inhibit synaptic transmission at the medial but not lateral perforant path [6]. These physiological studies, as well as lesion studies, suggest that different muscarinic receptor subtypes are differentially expressed in different sites of the CNS synapse and have different functions.

15.4
Neurophysiological Considerations

While anatomical details are necessary to elucidate to possible candidates for cholinergic action in cortical regions, physiological experiments allow an understanding of the detailed effects of receptor activation. Neurophysiological experiments present a complex and sometimes conflicting results of how ACh affects neuronal networks. Some of these differences are related to experimental conditions, region of study and species. We focus on some of these findings which help to illuminate wider aspects of cholinergic functions, specifically consistent experimental results which clarify how stress-induced cholinergic dysfunctions alter high brain functions and behavior.

As suggested by the anatomical studies described above, it should be noted that the release of ACh may affect the function of the network at multiple levels, as described below. Using different methodologies, researchers described significant cholinergic influences on both *intrinsic properties* of neurons as well as on *synaptic transmission* – both critically determine *synaptic integration* and finally *network functions* and *behavior*. Here we briefly summarize the effects observed in response to the activation of cholinergic receptor on brain physiology at different levels.

Intrinsic Properties The electrical activity of a single neuron and its output is determined in a complex manner by the properties of its intrinsic membrane properties together with the synaptic input to this neuron in any given time. Intrinsic membrane properties include resting membrane potential, membrane resistance to current flow, time constant and electrotonic length. These properties determine the way the individual cell responds to incoming synaptic information. Intrinsic properties also determine the "active" response of the cell, or its "*excitability*", for example, its threshold to evoke an action potential, the frequency and pattern of firing. Indeed, a significant advancement of our understanding of cholinergic actions in the brain comes from electrophysiological recordings from single neurons. Using sharp microelectrodes and the current-clamp technique, Cole and Nicoll [7, 8] showed muscarinic-dependent slow depolarization and decreased frequency adaptation in CA1 pyramidal neurons. They and others [9] concluded that muscarinic receptor activation suppresses a slow K$^+$ conductance, leading to an increase in excitability. At least two different potassium conductances (I_m and a slower I_{AHP}) are inactivated by muscarinic activation. Although the distribution of muscarinic-inactivating K$^+$ channels in pyramidal neurons is not known, the abundance of receptors along the dendritic tree may suggest significant increase in dendritic membrane resistance and decreased electrotonic distance following release of ACh. Such changes in sub-threshold range membrane potential properties may increase the efficacy of distance synapses in the input–output relationship as well as boosting the action potential back propagation along the dendrites and the effect on synaptic properties [10]. Accumulating evidence suggests that the excitability of neurons within the cerebral cortex may be modulated by several muscarinic-modulated conductances besides potassium. For example, two mixed cation currents, I_h and I_{cat}, are potentiated by muscarinic activation [11–13]. Carbamylcholine (CCh) was also reported to increase intracellular calcium (and hence modulate calcium-dependent processes) by activation of voltage-dependent calcium channels [14–16].

Synaptic Transmission In addition to changes in intrinsic membrane properties, activation of ACh receptors (both muscarinic and nicotinic) modulates synaptic transmission. Muscarinic autoreceptors on pre-synaptic sites in the hippocampus mediate the release of ACh from cholinergic nerve terminals. In a study performed by Zhang and colleagues [17] M2–M4 double-knockout mice were used, and it was found that the non-selective muscarinic agonist, oxotremorine (0.1–10.0 μM), inhibited ACh release up to 80% in wild-type mice, whereas no such inhibition was recorded in knockout mice. In addition, studies with single-knockout mice indicated that the M2 receptor was principally responsible for inhibition in the hippocampus [17]. However, a pharmacological study by Vannucchi and colleagues [18] found that the M1 and M4 receptor subtypes were most likely responsible for regulation of ACh release. A recent study focusing on cholinergic modulation of neocortical neurons convincingly demonstrated that muscarine inhibits GABA release from fast-spiking neurons in somatosensory cortex [19]. Thus, cholinergic activation significantly decreased feedforward inhibitory response, resulting in an overall increase in thalamic excitation of the cortex. ACh was also shown to affect glutamatergic transmission, although there are apparently conflicting results described in the literature. As paired-pulse fascilitation enhancement during CCh-induced depression of field potentials may serve as an indirect evidence for decreased excitatory release, a more direct measure of spontaneous excitatory post-synaptic currents (EPSCs) showed that activation of muscarinic receptors increased the frequency of synaptic events, even in the presence of a Na$^+$ channel blocker (TTX), suggesting direct cholinergic effect of ACh on release mechanisms [20]. Others suggested ACh-mediated direct enhancement of the current through NMDA receptors [21].

ACh Affects Network Activity, Plasticity and Rhythmogenesis The above-described cellular and synaptic mechanisms of ACh actions on cortical neurons suggest overall increased excitability of cortical neurons and enhanced firing to incoming sensory input. In order to study directly ACh actions on neuronal networks, electrophysiological recordings are made from small groups of functionally connected neurons ("neuronal networks"). Many of these experiments have been done using *in vitro* slice preparations, which allows hours of stable recordings from defined brain regions under specific pharmacological conditions. However, it should be remembered that brain slices are in most cases deafferented from the remotely cholinergic nuclei and thus lack a normal cholinergic input. An interesting network property in several brain regions is a long-term change in the network response to an identical stimulus. The most intensively studied is *long-term potentiation* (LTP). Bliss and Lomo [22] originally described LTP of excitatory postsynaptic potentials (EPSPs) in hippocampal neurons. Since that time, LTP has been studied extensively as a candidate electrophysiological model for memory. ACh enhances LTP in many areas, including the hippocampus [23, 24], the entorhinal cortex [25] and the pyriform cortex [26]. In region CA1 of the hippocampus, induction of LTP has been described to depend on the phase relative to oscillatory activity [23]. Stimulation of the medial septum enhances LTP induction *in vivo* [27] and the muscarinic antagonist, scopolamine, blocks LTP enhancement associated with medial septum stimulation [28]. In the entorhinal cortex, it is possible to induce LTP in the presence of atropine, but it is significantly suppressed [25]. Interestingly, a low dose of CCh (<1 μM) causes a transient, reversible depression of field potentials, followed by a long-lasting potentiation [29], sometimes referred to as muscarinic LTP (mLTP) [30]. mLTP in the rat hippocampus is similar to LTP produced by tetanic stimulation in that it shows increased activity to afferent stimulation which starts slowly and persists over time. However, its effects are mediated by a post-synaptic M2 receptor instead of an NMDA receptor, although it appears that the same post-receptor mechanism is used for both mLTP and tetanic, NMDA-dependent LTP. Notably, very low doses of CCh (0.1 μM) facilitated induction of tetanic LTP without producing an effect on their own. mLTP could be a pre-synaptic effect [20], a post-synaptic effect [29, 31] or both. Overall, it seems that while the induction of LTP in cortical structures does not necessitate cholinergic input, increasing ACh levels do play a significant role in facilitation of synaptic plasticity.

An important effect of cholinergic activation on brain excitability is its robust effect on rhythmic network activity. Excitation of cholinergic neurons in the basal forebrain is followed by a shift in the electroencephalogram from low frequency, high amplitude to high frequency, low amplitude fluctuations. Muscarinic activation of the septal nucleus is known to induce a prominent rhythmical slow activity at the theta range (4–12 Hz) [32] in both the hippocampus and entorhinal cortex [33–36]. Theta rhythm oscillations have been shown to facilitate synaptic plasticity in CA1 of the hippocampus [37] and, as such, to have an important role in memory and its cellular representative (i.e., LTP). Another rhythm frequently studied is the gamma rhythm (~30–80 Hz) [38], which is reported to be important for information processing [39]. Muscarinic receptor activation (apparently mediated by the M1 receptor subtype) induces gamma network oscillations in the hippocampus [40]. In mice, knockout of the M1 but not the M2–M5 receptor results in inability to induce gamma oscillations in the CA3 region [41]. Cholinergic input is also crucial for the initiation of gamma activity within the cerebral cortex, activity which is associated with alertness during wake and absence during sleep [42, 43]. In single unit recordings, basal forebrain neurons were shown to fire at their highest rate during cortical gamma activity and at their lowest rate during slow delta activity of sleep. Cholinergic input has also been suggested to modulate rhythmic theta activity in the cerebral cortex, including the medial prefrontal and posterior cingulated [43]. Thus, it is evident that cholinergic modulation of cortical network is crucial to normal brain function and has a key role in adapting the network response to incoming stimuli under arousal and stress. Furthermore, common brain diseases (e.g., epilepsy, Alzheimer's disease) are often characterized by abnormal rhythmogenesis [44–46]. Although not thoroughly studied, some reports do suggest abnormal brain rhythms in PTSD patients, specifically increased theta over central regions and increased beta, mainly over frontal regions [47]. Recent data further suggest that abnormal cholinergic-related rhythmogenesis within the entorhinal cortex, as found in animal models of epilepsy may be associated with the pathogenesis of the disease or related cognitive disturbances [48].

15.4.1
The Role of ACh in Behavior, Learning and Memory

The role of ACh in behavior and high cognitive functions has been known for decades since the early use of toxins to block AChE (e.g., [49]), followed by controlled pharmacological studies in experimental animals and human subjects. Generally, these pharmacological studies conclusively demonstrated that blockade of the muscarinic cholinergic receptors by drugs such as scopolamine impairs the encoding of new memories (for a review, see [50]). Localized infusions of scopolamine in parahippocampal structures revealed a role for cholinergic receptors for the encoding of information for subsequent recognition for rats and monkeys [51]. Similar experiments in the hippocampus demonstrated impaired spatial encoding [52], pointing to the general role of ACh in encoding information while anatomical specificity remains crucial to the understanding of its complete effect.

15.5
Stress Induces Cholinergic Dysfunction

The data presented above indicate ACh as a modulatory neurotransmitter which determines the response of brain sensory networks to incoming stimulus, or in other words sensory perception. This property together with its role in consolidation and retrieval of memories makes the cholinergic system a candidate associated with the perception of stressful events by the individual, the storage of the stressful events and its retrieval (such as in PTSD). As mentioned, a model for PTSD should involve network modifications which are initiated by the acute stress episode and – at least in affected individuals – a dysfunction will persist for long period of time. Thus, in order for the brain cholinergic pathways to be a candidate system underlying PTSD: (i) it should be affected by a single stressful event; (ii) changes in its function should persist for a long period of time (at least in affected individuals); and (iii) its dysfunction should be associated with emotional and cognitive disturbances. Here, we briefly summarize recent data suggesting that the cholinergic system fulfills these conditions.

15.5.1
Stress Event Affects Cholinergic Functions

It is well established that different stress protocols induce that upregulation of early immediate genes, including c-FOS. The presence of c-FOS and STRE stress element binding sites in the promoters of key cholinergic elements, such as the ACh hydrolyzing enzyme, acetylcholinesterase (AChE), the acetylcholine synthesizing enzyme choline acetyltransferase (ChAT) and the vesicular acetylcholine transporter (VAChT), suggests these are likely to be regulated by the stress-induced elevated c-FOS, leading to changes in the levels of proteins mediating brain cholinergic neurotransmission. Indeed it has been shown that, following c-FOS upregulation, marked increase in mRNA levels of AChE are measured together with downregulation of ChAT and VAChT mRNAs [53]. The combined effects of these changes should reduce the bioavailability of ACh through suppressed synthesis/packaging and enhanced hydrolysis. It has been shown that the changes observed in transcription are associated with parallel changes in protein levels. Microdialysis measurements showing a transient increase and delayed reduction in ACh levels during stress support these molecular studies [54]. Interestingly, similar results were obtained following exposure of animals or brain slices to AChE inhibitors [53, 55–57]. These results indicate that: (i) ACh released during stress or exposure to AChE inhibitors may underlie the consequent transcriptional regulation; and (ii) the response is a local brain response and, while corticosterone enhances this response [58], it is not essential. The observation of similar transcriptional changes *in vitro* allows electrophysiological experiments to confirm that, in the hippocampus, AChE inhibitors (thus increasing ACh levels) cause initial increase following decrease in glutamatergic-mediated synaptic transmission, supporting a functional consequence for the molecular changes [53].

15.5.2
Cholinergic Dysfunction Persists After Stress

Importantly for considering a role for cholinergic transmission in brain dysfunction observed in PTSD patients, is the long-term nature of the stress effect. Indeed, in animal studies it has been shown that stress is associated with a long-term (weeks) shift for the normal synaptic form of AChE (AChE-S) to the normally rare alternative transcript AChE-R. Interestingly, the shift of transcripts is associated with the appearance of AChE-R transcripts within the dendritic tree of hippocampal neurons. What could be the consequence of alternative splicing of the AChE protein for cholinergic functions? Functionally speaking, while *in vitro* AChE-R hydrolyzes ACh similarly to the synaptic form, its structure predicts that this water-soluble form (in contrast to the membrane-bound AChE-S), diffuses in the extracellular solution, suggesting reduced AChE within the synaptic cleft, allowing increased ACh actions on the receptors. Supporting this possibility are electrophysiological experiments showing enhanced network excitability in response to AChE inhibitors one month after stress [58, 59]. Similar enhanced cholinergic effects were observed in chronic epileptic animals, also showing a replacement of the synaptic membrane-bound tetrameric enzyme with the soluble monomeric one [48]. The hypersensitivity observed in response to AChE inhibitors (endogenous ACh release) or in response to applied muscarinic agonists was mediated by muscarinic receptors. While other changes may contribute to the observed hypersensitivity (e.g., altered receptor levels), there is currently no experimental evidence to allow such conclusion.

15.5.3
Cholinergic Dysfunction May Underlie Short- and Long-Term Clinical Symptoms

There is little direct evidence to correlate cholinergic dysfunction and clinical symptoms. Acute inhibition of AChE during intoxication with organophosphates (OPs), causes (depending on the dose) confusion, decreased levels of consciousness, epileptic seizures and coma. Accumulating evidence suggests that chronic exposure to OPs, even at low doses, results in decreased cognitive performances and altered brain activity recorded using EEG [60–62]. A recent study in individuals chronically exposed to low levels of OPs [63] showed significantly reduced serum AChE activity confirming exposure and altered cholinergic homeostasis. Brain activity recorded using electroencephalography (EEG) showed decreased theta activity in the hippocampus, parahippocampal regions and the cingulate cortex, together with increased beta activity in the prefrontal cortex of exposed individuals – areas known to play a role in cholinergic-associated cognitive functions (increased beta was also described in [64]). These EEG changes observed in rest were associated with deficits in visual recall, supporting their functional significance. It is interesting to note that emotional disturbances similar to those observed in PTSD were described following exposure to low levels of OP [65] provide further support for a cholinergic role in the disorder.

15.6
The Cholinergic Basis for PTSD: the Model

The presented data suggests a sequence of events and mechanisms leading from acute stress into long-term changes in specific brain functions, as observed in

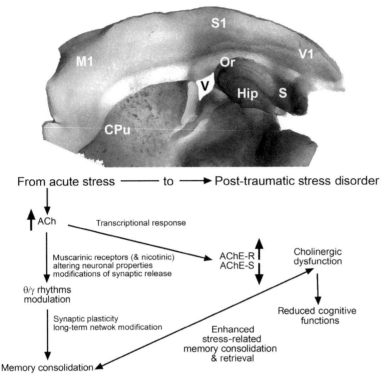

Figure 15.1 The cholinergic model for PTSD: A sagital section through the mice brain. Gray intensity represents AChE activity (Karnovsky staining). Note the high cholinergic activity in sub-cortical regions, specifically the hippocampus (Hip), but also in the primary visual and sensory cortices. CPu, caudate putamen (striatum); Hip, hippocampus; M1, primary motor cortex; Or, oriens layer; S, subiculum; S1, primary sensory cortex; V, ventricle; V1, primary visual cortex. Below – sequence of events in the central cholinergic system leading for acute stress to PTSD (see text for details).

PTSD (Figure 15.1). While the presented "cholinergic hypothesis" is no doubt a simplification of the complex changes found in brain functions after stress, it nevertheless may significantly contribute to disease generation after stress.

The "cholinergic model" (Figure 15.1) suggests that a rapid increase in ACh within specific brain regions during stress leads to behavioral and cognitive changes as well as the altered perception required for the survival of the organism. These include ACh-mediated cellular activity allowing the animal to memorize the stressful event. When changes in cholinergic gene expression are extreme – or occur in sensitive individuals (see below) – they lead to a shut-down of the cholinergic system in the short term, which behaviorally is expressed as certain dissociative phenomena observed following the acute stress episode [66]. In the long run, hypersensitivity of specific cortical networks to cholinergic activation predicts spontaneous retrieval of memories associated with the stressful episode, together with reduced cognitive performance, learning disability and abnormal sleep pattern. Interestingly, recent studies suggest that abnormal cholinergic feedback response may arise from mutations in the AChE gene promoter or PON1 enzyme, associated with increased sensitivity to OP exposure or increased anxiety levels[63, 67]. Future studies are needed to elucidate how critical is the role of cholinergic changes in human stress-related brain changes, and to what extent cholinergic dysfunction – on the molecular or functional level – may become a useful diagnostic tool, or a target for therapeutic intervention.

Summary

Stress is frequently followed by long-term changes in brain functions. The most often mentioned diagnosis is post-traumatic stress disorder (PTSD), which affects up to 40% of individuals exposed to a single stressful episode. While recent studies suggest that individual differences in response to stressful events stem from complex components contributing to stress responses, including genetic vulnerability, it is generally accepted that perception of the acute stressful event is an important factor in the individual vulnerability to long-term squeal. In this chapter we present a neurobiological approach, modeling potential mechanisms underlying the transfer of a short emotional input into a long-term permanent change in specific neuronal networks within the brain. We propose the cholinergic system as a model for a neuromodulatory transmitter system showing molecular and physiological changes following stress. We emphasize that the principles presented in the cholinergic model can probably be generalized to other transmitter system, leading to the complex changes in variability observed in patients with PTSD.

Acknowledgments

This study was supported by the DFG-SFB/TR3 (C8) and the Israel Science Foundation (566/07).

References

1 Lewis, P.R., Shute, C.C. and Silver, A. (1967) Confirmation from choline acetylase analyses of a massive cholinergic innervation to the rat hippocampus. *J. Physiol.*, **191**, 215–224.

2 Milner, T.A., Loy, R. and Amaral, D.G. (1983) An anatomical study of the development of the septo-hippocampal projection in the rat. *Brain Res.*, **284**, 343–371.

3 Mellgren, S.I. and Srebro, B. (1973) Changes in acetylcholinesterase and distribution of degenerating fibres in the hippocampal region after septal lesions in the rat. *Brain Res.*, **52**, 19–36.

4 Levey, A.I., Edmunds, S.M., Koliatsos, V., Wiley, R.G. and Heilman, C.J. (1995) Expression of m1-m4 muscarinic acetylcholine receptor proteins in rat hippocampus and regulation by

cholinergic innervation. *J. Neurosci.*, **15**, 4077–4092.

5 Rouse, S.T., Marino, M.J., Potter, L.T., Conn, P.J. and Levey, A.I. (1999) Muscarinic receptor subtypes involved in hippocampal circuits. *Life Sci.*, **64**, 501–509.

6 Kahle, J.S. and Cotman, C.W. (1989) Carbachol depresses synaptic responses in the medial but not the lateral perforant path. *Brain Res.*, **482**, 159–163.

7 Cole, A.E. and Nicoll, R.A. (1983) Acetylcholine mediates a slow synaptic potential in hippocampal pyramidal cells. *Science*, **221**, 1299–1301.

8 Cole, A.E. and Nicoll, R.A. (1984) The pharmacology of cholinergic excitatory responses in hippocampal pyramidal cells. *Brain Res.*, **305**, 283–290.

9 Madison, D.V., Lancaster, B. and Nicoll, R.A. (1987) Voltage clamp analysis of cholinergic action in the hippocampus. *J. Neurosci.*, **7**, 733–741.

10 Tsubokawa, H. and Ross, W.N. (1997) Muscarinic modulation of spike backpropagation in the apical dendrites of hippocampal CA1 pyramidal neurons. *J. Neurosci.*, **17**, 5782–5791.

11 Brown, D.A. and Adams, P.R. (1980) Muscarinic suppression of a novel voltage-sensitive K+ current in a vertebrate neurone. *Nature*, **283**, 673–676.

12 Halliwell, J.V. and Adams, P.R. (1982) Voltage-clamp analysis of muscarinic excitation in hippocampal neurons. *Brain Res.*, **250**, 71–92.

13 Colino, A. and Halliwell, J.V. (1993) Carbachol potentiates Q current and activates a calcium-dependent non-specific conductance in rat hippocampus in vitro. *Eur. J. Neurosci.*, **5**, 1198–1209.

14 Muller, W. and Connor, J.A. (1991) Cholinergic input uncouples Ca2+ changes from K+ conductance activation and amplifies intradendritic Ca2+ changes in hippocampal neurons. *Neuron*, **6**, 901–905.

15 Egorov, A.V., Gloveli, T. and Muller, W. (1999) Muscarinic control of dendritic excitability and Ca(2+) signaling in CA1 pyramidal neurons in rat hippocampal slice. *J. Neurophysiol.*, **82**, 1909–1915.

16 Gloveli, T., Egorov, A.V., Schmitz, D., Heinemann, U. and Muller, W. (1999) Carbachol-induced changes in excitability and [Ca2+]i signalling in projection cells of medial entorhinal cortex layers II and III. *Eur. J. Neurosci.*, **11**, 3626–3636.

17 Zhang, W., Basile, A.S., Gomeza, J., Volpicelli, L.A., Levey, A.I. and Wess, J. (2002) Characterization of central inhibitory muscarinic autoreceptors by the use of muscarinic acetylcholine receptor knock-out mice. *J. Neurosci.*, **22**, 1709–1717.

18 Vannucchi, M.G. and Pepeu, G. (1995) Muscarinic receptor modulation of acetylcholine release from rat cerebral cortex and hippocampus. *Neurosci. Lett.*, **190**, 53–56.

19 Kruglikov, I. and Rudy, B. (2008) Perisomatic GABA release and thalamocortical integration onto neocortical excitatory cells are regulated by neuromodulators. *Neuron*, **58**, 911–924.

20 Pavlovsky, L., Browne, R.O. and Friedman, A. (2003) Pyridostigmine enhances glutamatergic transmission in hippocampal CA1 neurons. *Exp. Neurol.*, **179**, 181–187.

21 Markram, H. and Segal, M. (1990) Acetylcholine potentiates responses to N-methyl-D-aspartate in the rat hippocampus. *Neurosci. Lett.*, **113**, 62–65.

22 Bliss, T.V. and Lomo, T. (1973) Long-lasting potentiation of synaptic transmission in the dentate area of the anaesthetized rabbit following stimulation of the perforant path. *J. Physiol.*, **232**, 331–356.

23 Huerta, P.T. and Lisman, J.E. (1995) Bidirectional synaptic plasticity induced by a single burst during cholinergic theta oscillation in CA1 in vitro. *Neuron*, **15**, 1053–1063.

24 Adams, S.V., Winterer, J. and Muller, W. (2004) Muscarinic signaling is required for spike-pairing induction of long-term potentiation at rat Schaffer collateral-CA1 synapses. *Hippocampus*, **14**, 413–416.

25 Cheong, M.Y., Yun, S.H., Mook-Jung, I., Joo, I., Huh, K. and Jung, M.W. (2001) Cholinergic modulation of synaptic physiology in deep layer entorhinal cortex of the rat. *J. Neurosci. Res.*, **66**, 117–121.

26 Patil, M.M., Linster, C., Lubenov, E. and Hasselmo, M.E. (1998) Cholinergic agonist carbachol enables associative long-term potentiation in piriform cortex slices. *J. Neurophysiol.*, **80**, 2467–2474.

27 Ovsepian, S.V., Anwyl, R. and Rowan, M.J. (2004) Endogenous acetylcholine lowers the threshold for long-term potentiation induction in the CA1 area through muscarinic receptor activation: *in vivo* study. *Eur. J. Neurosci.*, **20**, 1267–1275.

28 Leung, L.S., Shen, B., Rajakumar, N. and Ma, J. (2003) Cholinergic activity enhances hippocampal long-term potentiation in CA1 during walking in rats. *J. Neurosci.*, **23**, 9297–9304.

29 Yun, S.H., Cheong, M.Y., Mook-Jung, I., Huh, K., Lee, C. and Jung, M.W. (2000) Cholinergic modulation of synaptic transmission and plasticity in entorhinal cortex and hippocampus of the rat. *Neuroscience*, **97**, 671–676.

30 Segal, M. and Auerbach, J.M. (1997) Muscarinic receptors involved in hippocampal plasticity. *Life Sci.*, **60**, 1085–1091.

31 Auerbach, J.M. and Segal, M. (1996) Muscarinic receptors mediating depression and long-term potentiation in rat hippocampus. *J. Physiol.*, **492** (Pt 2), 479–493.

32 Bland, B.H. (1986) The physiology and pharmacology of hippocampal formation theta rhythms. *Prog. Neurobiol.*, **26**, 1–54.

33 Dickson, C.T., Kirk, I.J., Oddie, S.D. and Bland, B.H. (1995) Classification of theta-related cells in the entorhinal cortex: cell discharges are controlled by the ascending brainstem synchronizing pathway in parallel with hippocampal theta-related cells. *Hippocampus*, **5**, 306–319.

34 Klink, R. and Alonso, A. (1997) Ionic mechanisms of muscarinic depolarization in entorhinal cortex layer II neurons. *J. Neurophysiol.*, **77**, 1829–1843.

35 Klink, R. and Alonso, A. (1997) Muscarinic modulation of the oscillatory and repetitive firing properties of entorhinal cortex layer II neurons. *J. Neurophysiol.*, **77**, 1813–1828.

36 Olpe, H.R., Klebs, K., Kung, E., Campiche, P., Glatt, A., Ortmann, R., D'Amato, F., Pozza, M.F. and Mondadori, C. (1987) Cholinomimetics induce theta rhythm and reduce hippocampal pyramidal cell excitability. *Eur. J. Pharmacol.*, **142**, 275–283.

37 Huerta, P.T. and Lisman, J.E. (1993) Heightened synaptic plasticity of hippocampal CA1 neurons during a cholinergically induced rhythmic state. *Nature*, **364**, 723–725.

38 Chrobak, J.J. and Buzsaki, G. (1998) Gamma oscillations in the entorhinal cortex of the freely behaving rat. *J. Neurosci.*, **18**, 388–398.

39 Behrens, C.J., van den Boom, L.P., de Hoz, L., Friedman, A. and Heinemann, U. (2005) Induction of sharp wave-ripple complexes *in vitro* and reorganization of hippocampal networks. *Nat. Neurosci.*, **8**, 1560–1567.

40 Fisahn, A., Pike, F.G., Buhl, E.H. and Paulsen, O. (1998) Cholinergic induction of network oscillations at 40 Hz in the hippocampus *in vitro*. *Nature*, **394**, 186–189.

41 Fisahn, A., Yamada, M., Duttaroy, A., Gan, J.W., Deng, C.X., McBain, C.J. and Wess, J. (2002) Muscarinic induction of hippocampal gamma oscillations requires coupling of the M1 receptor to two mixed cation currents. *Neuron*, **33**, 615–624.

42 Jones, B.E. (2004) Activity, modulation and role of basal forebrain cholinergic neurons innervating the cerebral cortex. *Prog. Brain Res.*, **145**, 157–169.

43 Jones, B.E. (2008) Modulation of cortical activation and behavioral arousal by cholinergic and orexinergic systems. *Ann. N. Y. Acad. Sci.*, **1129**, 26–34.

44 Rutecki, P.A. and Yang, Y. (1998) Ictal epileptiform activity in the CA3 region of hippocampal slices produced by pilocarpine. *J. Neurophysiol.*, **79**, 3019–3029.

45 Lothman, E.W., Bertram, E.H. III and Stringer, J.L. (1991) Functional anatomy of hippocampal seizures. *Prog. Neurobiol.*, **37**, 1–82.

46 Bragin, A., Csicsvari, J., Penttonen, M. and Buzsaki, G. (1997) Epileptic afterdischarge in the hippocampal-

entorhinal system: current source density and unit studies. *Neuroscience*, **76**, 1187–1203.

47 Begic, D., Hotujac, L. and Jokic-Begic, N. (2001) Electroencephalographic comparison of veterans with combat-related post-traumatic stress disorder and healthy subjects. *Int. J. Psychophysiol.*, **40**, 167–172.

48 Zimmerman, G., Njunting, M., Ivens, S., Tolner, E., Behrens, C.J., Gross, M., Soreq, H., Heinemann, U. and Friedman, A. (2008) Acetylcholine-induced seizure-like activity and modified cholinergic gene expression in chronically epileptic rats. *Eur. J. Neurosci.*, **27**, 965–975.

49 Levin, H.S. and Rodnitzky, R.L. (1976) Behavioral effects of organophosphate in man. *Clin. Toxicol.*, **9**, 391–403.

50 Hasselmo, M.E. (2006) The role of acetylcholine in learning and memory. *Curr. Opin. Neurobiol.*, **16**, 710–715.

51 Tang, Y. and Aigner, T.G. (1996) Release of cerebral acetylcholine increases during visually mediated behavior in monkeys. *Neuroreport*, **7**, 2231–2235.

52 Blokland, A., Honig, W. and Raaijmakers, W.G. (1992) Effects of intra-hippocampal scopolamine injections in a repeated spatial acquisition task in the rat. *Psychopharmacology (Berl.)*, **109**, 373–376.

53 Kaufer, D., Friedman, A., Seidman, S. and Soreq, H. (1998) Acute stress facilitates long-lasting changes in cholinergic gene expression. *Nature*, **393**, 373–377.

54 Imperato, A., Puglisi-Allegra, S., Casolini, P. and Angelucci, L. (1991) Changes in brain dopamine and acetylcholine release during and following stress are independent of the pituitary-adreno-cortical axis. *Brain Res.*, **538**, 111–117.

55 Friedman, A., Kaufer, D., Shemer, J., Hendler, I., Soreq, H. and Tur-Kaspa, I. (1996) Pyridostigmine brain penetration under stress enhances neuronal excitability and induces early immediate transcriptional response. *Nat. Med.*, **2**, 1382–1385.

56 Kaufer, D., Friedman, A., Seidman, S. and Soreq, H. (1999) Anticholinesterases induce multigenic transcriptional

feedback response suppressing cholinergic neurotransmission. *Chem. Biol. Interact.*, **119–120**, 349–360.

57 Kaufer, D., Friedman, A. and Soreq, H. (1999) The vicious circle of stress and anticholinesterase responses. *Neuroscientist*, **5**, 173–183.

58 Meshorer, E., Erb, C., Gazit, R., Pavlovsky, L., Kaufer, D., Friedman, A., Glick, D., Ben Arie, N. and Soreq, H. (2002) Alternative splicing and neuritic mRNA translocation under long-term neuronal hypersensitivity. *Science*, **295**, 508–512.

59 Meshorer, E. and Soreq, H. (2006) Virtues and woes of AChE alternative splicing in stress-related neuropathologies. *Trends Neurosci.*, **29**, 216–224.

60 Ray, D.E. and Richards, P.G. (2001) The potential for toxic effects of chronic, low-dose exposure to organophosphates. *Toxicol. Lett.*, **120**, 343–351.

61 Brown, M.A. and Brix, K.A. (1998) Review of health consequences from high-, intermediate- and low-level exposure to organophosphorus nerve agents. *J. Appl. Toxicol.*, **18**, 393–408.

62 Fiedler, N., Kipen, H., Kelly-McNeil, K. and Fenske, R. (1997) Long-term use of organophosphates and neuropsychological performance. *Am. J. Ind. Med.*, **32**, 487–496.

63 Browne, R.O., Moyal-Segal, L.B., Zumsteg, D., David, Y., Kofman, O., Berger, A., Soreq, H. and Friedman, A. (2006) Coding region paraoxonase polymorphisms dictate accentuated neuronal reactions in chronic, sub-threshold pesticide exposure. *FASEB J.*, **20**, 1733–1735.

64 Burchfiel, J.L. and Duffy, F.H. (1982) Organophosphate neurotoxicity: chronic effects of sarin on the electroencephalogram of monkey and man. *Neurobehav. Toxicol. Teratol.*, **4**, 767–778.

65 Levin, H.S., Rodnitzky, R.L. and Mick, D.L. (1976) Anxiety associated with exposure to organophosphate compounds. *Arch. Gen. Psychiatry*, **33**, 225–228.

66 Gershuny, B.S. and Thayer, J.F. (1999) Relations among psychological trauma,

dissociative phenomena, and trauma-related distress: a review and integration. *Clin. Psychol. Rev.*, **19**, 631–657.

67 Sklan, E.H., Lowenthal, A., Korner, M., Ritov, Y., Landers, D.M., Rankinen, T., Bouchard, C., Leon, A.S., Rice, T., Rao, D.C. *et al.* (2004) Acetylcholinesterase/paraoxonase genotype and expression predict anxiety scores in Health, Risk Factors, Exercise Training, and Genetics study. *Proc. Natl. Acad. Sci. U. S A.*, **101**, 5512–5517.

Part VI
Vulnerability to Disease

16
Stress and Neurodegeneration: Adding Insult to Injury?

Amit Berson, Mor Hanan, and Hermona Soreq

16.1
Abstract

A widely accepted notion states that mammalian stress reactions act to protect the organism in the short time range and that the price paid for that is manifested as delayed damages. Intriguingly, glucocorticoid induction, neuronal impairments and hippocampal atrophy are all involved in both stress reactions and Alzheimer's disease (AD), two situations in which changes in acetylcholinesterase (*ACHE*) gene expression have been demonstrated. Here, we describe common features of neuronal responses to psychological stress and AD-induced toxicity, review current data suggesting that stress may exacerbate AD pathology, and propose a molecular model in which alternative splicing-mediated changes in the cholinergic system inversely exert stress-related neuroprotection in the context of neurodegenerative diseases.

16.2
Alzheimer's Disease Shares Phenotypic Biomarkers with Stress Reactions

Alzheimer's disease (AD), a progressive neurodegenerative disorder, is the most common cause of dementia in aged humans today. A century ago, Alois Alzheimer diagnosed several neuropathologic hallmarks in the AD brain [1]. These include extracellular proteinaceous depositions named neuritic plaques and neurofibrillary tangles (NFT), intraneuronal pairs of filaments wound into helices. At the behavioral level, AD involves short-term memory and learning impairments, anxiety and often depression, all of which also accompany stress [2].

The modern era of AD research began with the identification of amyloid-β (Aβ) as the major neuritic plaque component and with the cloning of the amyloid precursor protein (APP) gene [3]. APP is a single-transmembrane protein, which is ubiquitously expressed in both neuronal and non-neuronal cells. Aβ, an approximately 4-kDa peptide is cleaved off the APP protein by a series of enzymatic steps.

Stress – From Molecules to Behavior. Edited by Hermona Soreq, Alon Friedman, and Daniela Kaufer
Copyright © 2010 WILEY-VCH Verlag GmbH & Co. KGaA, Weinheim
ISBN: 978-3-527-32374-6

The large ectodomain is first shedded off by either one of two membrane-anchored proteases. The α-secretase complex cleaves APP within the Aβ domain and therefore does not result in Aβ release. This pathway is hence termed "non-amyloidogenic". A second protease, β-secretase (BACE) cleaves APP and defines the N-terminus of Aβ. Cleavage of the APP intermembrane stub by the γ-secretase complex generates Aβ, a peptide 38, 40 or 42 amino acids long [4].

The γ-secretase is an aspartyl protease composed of presenilin1 (PS1), or presenilin2 (PS2), nicastrin, anterior pharynx defective 1 (APH1) and PS enhancer 2 (PEN2), all of which are necessary for the reconstitution of γ-secretase activity. The exact point at which γ-secretase cleaves the γ site has profound implications on the amyloidogenesis of Aβ, since Aβ1-42 has the strongest propensity to oligomerize *in vitro* and *in vivo*. It is therefore not surprising that point mutations in the APP or the presenilins, which alter APP processing, are characteristic of familial AD patients [4].

AD is notably characterized by cognitive impairments. However, the relationship between the pathological hallmarks of AD and the cognitive status is not simple. Increasing evidence suggests that the amount of soluble Aβ is a much better correlate to AD cognitive decline than the simple plaque burden [5]. More direct proof for the role of specific soluble Aβ oligomers has recently been published [6], demonstrating that a 56-kDa Aβ assembly is one of the major AD neurotoxic agents responsible for memory loss in transgenic mice. These findings raise the question of why do Aβ oligomers impair memory functions. Several observations demonstrate that Aβ can impair long-term potentiation (LTP) both in hippocampal slices and *in vivo* [7, 8], a characteristic electrophysiological impairment known under stress. Interestingly, Aβ has been shown to induce endocytosis of N-methyl-D-aspartate (NMDA) receptors through initial binding to α7 nicotinic receptors [9], but the molecular mechanisms underlying Aβ-induced toxicity are largely unknown [4]. Several neurotoxic roles of Aβ have been proposed, including the generation of reactive oxygen species, calcium dishomeostasis, inflammatory reactions and mitochondrial dysfunction [10], but a unifying theory for AD etiology is still required.

16.3
Molecular Underpinnings of Stress-Induced Cognitive Impairments

Stress is a commonly used term to describe any change that induces imbalance and disturbs the homeostasis of a certain system. However, since the pioneering work of Hans Selye [11], the behavioral and molecular knowledge that has been gained allows a more precise definition. Kim and Diamond [12] define stress as "a condition in which an individual is aroused by an aversive situation, and the magnitude of the stress and its physiological consequences are influenced greatly by the individual's perception of its ability to control the presence or intensity of the stimulation". Thus, stress is a subjective experience, and both the stimulus and the current physical/psychological status of the individual perceiving it must be considered.

A large body of evidence suggests that acute stress induces adaptive responses which allow an individual to cope with the aversive situation, whereas chronic stress induces a harmful cascade of events which may lead to abnormal neuro-endocrine signaling, neuronal demise and finally cognitive and behavioral impairments. The main dogma has been extensively reviewed [2, 12, 13] and hence is only briefly summarized here.

Upon experiencing a stressful situation, the hypothalamic–pituitary–adrenal (HPA) axis is activated by limbic and ascending brain stem pathways projecting to the hypothalamic paraventricular nucleus (PVN). These inputs induce the release of corticotrophin-releasing hormone (CRH) which activates the synthesis of pro-opiomelanocortin in the anterior pituitary, from which corticotropin (ACTH), opioid and melanocortin are produced. ACTH in turn stimulates the adrenal cortex to secrete cortisol (in humans) and corticosterone (in humans and rodents). To exert their effects on transcription, glucocorticoids (GCs) bind to two classes of nuclear receptors: the glucocorticoid receptor (GR) and the mineralocorticoid receptor (MR). Functioning as either homo- or heterodimers, these receptors bind to glucocorticoid response elements (GRE) where they recruit either co-activators or repressors in order to regulate gene expression. Therefore, cortisol can up- and/or downregulate the expression of different genes, depending on the cellular context [2]. GR and MR like other members of the nuclear receptors superfamily, reside in the cytoplasm as inactive complexes, bound to the heat-shock protein chaperons HSP70 and HSP90 in addition to co-chaperons, which impede their translocation into the nucleus [14]. Once cortisol is bound, the complex is activated, and following conformational changes and alteration of its components, the complex enters the nucleus where it dissociates and the GR is converted to its DNA-binding form.

A bottom-up hypothesis regarding the harmful effect on cognition of prolonged exposure to stress or to GCs can start with the concentration of GR in the relevant brain region [15], then the effect on gene expression, followed by structural neuronal impairments and altered neuronal plasticity. Reduced neurogenesis (either due to impaired stem cells proliferation or because of mitigated survival of newly created neurons) and consequent increases of neuronal loss lead the way to gross morphological changes such as hippocampal atrophy and contribute to the final outcome of cognitive impairments. Strikingly, a large part of these changes in brain structure and function have similar characteristics to those observed in neurodegenerative diseases, most notably in AD. These observations have several implications. First, similar gene expression patterns might be shared between the two syndromes. Second, chronic stress or extreme acute stress may be risk factors for AD, most likely in genetically predisposed individuals. Third, coping with AD imposes harsh and chronic stress on both the patients and their caregivers. Therefore, the stress component of the disease itself might exacerbate AD pathology. Nevertheless, stress reactions may alternatively induce a feedback response attempting to attenuate neuronal damage. Together, this implies that distinct aspects of the cellular response to stress might serve as neuroprotective strategies in AD and should be examined as possible kernels for the development of new therapeutics. In light of the above, we considered the parallels between stress reactions and Alzheimer's disease.

Neuronal Structure Chronic stress has been shown to result in apical dendritic atrophy of hippocampal CA3 neurons of tree shrews [16] and rats [17] and induces shrinkage and reduced number of CA3 neurons in primates [18]. Importantly, stress-induced synaptic loss in rats can be reversed following new learning tasks [19], emphasizing the importance of neuronal plasticity in the process. Similarly, AD is notably characterized by dendritic and axonal degeneration and synapse loss (for a review, see [20, 21]). Among the major hallmarks of AD, including amyloid plaques and neurofibrillary tangles, loss of synapses is the best correlate to cognitive decline [22]. Further, synapse loss is induced by soluble Aβ and precedes the formation of plaques in transgenic animals [23]. Fibrillar Aβ is also neurotoxic, and dendrites near fibrillar amyloid deposits undergo spine loss and atrophy; also, nearby axons develop large varicosities, which eventually leads to additional neurite breakage and disruption of neuronal connections [24].

Neuronal Plasticity Impairments of hippocampal long-term potentiation (LTP) following stress have been documented for over 30 years [12]. Importantly, controllability has a pronounced effect over the modulation of LTP, as indicated by reports that rats that can escape an aversive shock display a more robust LTP than rats that cannot escape [25]. Given that the degree of control over a stressor is a strong modulator of the cognitive and behavioral outcome [26], the LTP changes seem to be one of the best correlates of this phenomenon. Conversely to LTP, long-term depression (LTD) is enhanced following stress, and both synaptic changes can be blocked by NMDA antagonists [27]. LTP disruption is also an important hallmark in AD mouse models. Naturally secreted Aβ oligomers, but not fibrils or monomers, inhibit LTP in rats *in vivo* [8]. Moreover, organotypic hippocampal slices from APP_{SWE} transgenic mice exhibit markedly reduced potentiation time compared to controls [28]. Given that LTP is fundamental to neuronal plasticity and learning and memory processes, cognitive impairments characterizing both situations may result from malfunctioning LTP and LTD.

Adult Neurogenesis Both stress [29] and GCs [30] interefere with adult neurogenesis in the hippocampus. The reversal of stress-induced suppression of neurogenesis by CRH and vasopressin receptor antagonists further support this model [31]. Given the importance of newly generated neurons for specific aspects of learning and memory [32], it is likely that stress-induced cognitive impairments are at least partially caused by hampered neurogenesis. However, the changes in adult neurogenesis in AD are still controversial. Several studies report increased neurogenesis in AD [33] and AD mouse models [34], whereas others show reduced proliferation or no change in proliferation but dramatic reduction in the survival of newborn neurons (see [35] for discussion). The individual studies include the use of human postmortem tissue or animal models, a large array of transgenic lines, genetic background strains and different staining procedures, all contributing to the complicated picture. Therefore, a thorough examination of neurogenesis in AD should be performed before definitive conclusions can be drawn regarding the similarities between AD and stress reactions in this aspect.

Glutamate Toxicity   Hyper-excitation of glutamatergic neurons can induce ionic imbalance, leading to massive apoptotic and necrotic cell death, described as glutamate toxicity. Glutamate neuronal toxicity may be triggered by either excessive glutamate release or shortage in glutamate reuptake. Increased levels of excitatory amino acids have been demonstrated both following stress and in neurodegenerative diseases. For example, glutamate outflow induced by restraint stress, which was attenuated by adrenalectomy, was abolished following glucocorticoid replacement [36]. Glutamate elevates intracellular calcium concentrations by direct binding to α-amino-3-hydroxy-5-methylisoxazole-4-propionate acid (AMPA) and NMDA receptors and by indirect activation of voltage-dependent calcium channels (VDCC). Further, metabotropic glutamate receptors coupled to the GTP-binding protein stimulate the release of inositol triphosphate (IP_3), which activates calcium channels in the endoplasmic reticulum (ER) [37]. Calcium overload activates calpains and caspases which degrade key cellular components. It also triggers apoptotic cascades and induces oxidative stress. Calcium dishomeostasis is prominent in AD [38] and the enhanced calcium influx following stress-induced glutamate release may predispose neurons to different insults [2]. Stress reactions and AD may thus be causally associated.

16.4
Direct Evidence Linking Stress Reactions and AD

Depression [39], anxiety [40] and proneness to psychological distress [41] are all risk factors for AD. Many studies show hypersecretion of GC in early and late AD [42, 43], and cortisol suppression after dexamethasone challenge is also impaired [44]. Furthermore, plasma cortisol levels associate with cognitive decline in AD patients [45]. Direct evidence for the harmful effects of GCs in AD comes from a clinical trial of prednisone, a GC which was used for its anti-inflammatory properties. Strikingly, prednisone-treated AD patients exhibited worse cognitive performance than the placebo group [46]. A genetic association study further implicates GCs involvement in AD. de Quervain *et al.* show that a rare haplotype in the 5′ regulatory region of the gene encoding 11beta-hydroxysteroid dehydrogenase type 1 (HSD11B1) is associated with a sixfold increased risk for sporadic AD and that the rare risk-associated haplotype alters HSD11B1 transcription [47].

Novel studies with animal models suggest molecular mechanisms underlying stress exacerbation of AD. *In vitro*, GCs increase Aβ formation specifically dependent on glucocorticoid but not mineralocorticoid receptors [48]. Further, GCs enhance both mRNA and protein levels of steady state APP and the protein levels of the immediate Aβ precursor, C99. *In vivo*, GC treatment increases $A\beta_{40}$ and $A\beta_{42}$ levels through APP and BACE upregulation. In addition, GCs also increase tau protein levels without affecting its phosphorylation status [48]. However, various stressors do induce tau phosphorylation [49–51], suggesting that GCs do not mediate this effect of stress. This conclusion is further supported by reports showing that restraint stress-induced tau phosphorylation is not affected by preventing the concomitant rise of GCs [52]. However, disruption of signaling via CRF G protein coupled receptor 1 (CRFR1) does block stress-induced phosphorylation of tau while blocking CRFR2 exaggerates this response. Therefore, CRFR receptors differentially regulate stress-induced tau phosphorylation.

Interestingly, plasma corticosterone levels in triple (APP/PS1/Tau) transgenic mice are elevated in a relatively late phase after considerable Aβ build-up, suggesting that AD like pathology induces the HPA response and not vice versa [48]. Intra-cerebro-ventricular infusion of Aβ increased C99, BACE1 and nicastrin levels. These changes were even more robust in previously stressed animals, suggesting that stress history is an important modulator of APP processing [53]. In mouse studies, stress, GCs administration and Aβ infusion impair cognitive performance to similar extents, leading to similar failures in the Morris water maze. The combination of all three worsened this phenotype, suggesting additive effects. Kang *et al.* [54] show that chronic isolation stress increases $Aβ_{40}$ and $Aβ_{42}$ levels in the interstitial fluid (ISF) by enhancing Aβ processing, without changing the levels of full length APP. Importantly, acute stress increases Aβ levels in the ISF. In contrast to the chronic administration of GC [48], acute GC treatment does not change Aβ levels, suggesting that drastic changes in GC levels do not mediate the acute effect of stress to the Aβ pathway [54]. However, acute CRF infusion enhanced Aβ in the ISF, and administration of a CRF receptor antagonist blocked the increase of Aβ following acute stress. In contrast to acute stress, chronic stress did not enhance CRF levels, suggesting that acute and chronic stress enhance Aβ through distinct mechanisms [54]. Furthermore, blockade of neuronal activity (by tetrodotoxin [TTX] administration) attenuated the stress-induced increases in Aβ, compatible with studies showing that neuronal activity regulates Aβ release. Interestingly, another study reports that, *in vitro*, CRF induces protection against amyloid, glutamate and oxidative insults [55]. Therefore, the big picture of stress involvement in AD seems complicated, and compensatory reactions promoting neuro-protection are likely to participate. A relevant example involves the complex roles of acetylcholinesterase (AChE) variants in stress and AD.

16.5
The Cholinergic System Connects Stress Reactions and AD

The well established role of AChE is to terminate cholinergic neurotransmission by hydrolyzing synaptic acetylcholine (ACh) in the nervous system. However, AChE is not one, but a combinatorial series of proteins having variant N- and C-termini due to alternate promoter usage and 3′ alternative splicing. Neuronal AChE variants show indistinguishable enzymatic activity yet differ in their expression, multimeric assembly, membrane association patterns and non-hydrolytic properties. The canonical tetrameric AChE-S ("S" for synaptic) isoform (also named AChE-T; "T" for tailed [56]) is replaced under stress through 3′ alternative

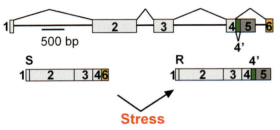

Figure 16.1 AChE-S and AChE-R are produced from alternatively spliced transcripts of the single *ACHE* gene. Exons 2, 3 and 4 (light gray) encode the common core of these variants. AChE-R mRNA includes pseudo-intron 4 (4') and exon 5, whereas AChE-S mRNA includes exon 6. Their open reading frames (ORF) and protein products deviate in their C-terminal domains. Stress reactions induce a splice shift toward the R variant.

splicing with the "read-through" variant of AChE, designated AChE-R [57, 58] (Figure 16.1). AChE-R has a distinct, shorter C-terminus termed ARP [59] which lacks the cysteine residue present in the C-terminus of AChE-S. This enables the multimerization of AChE as well as its synaptic adherence. Therefore, AChE-R remains monomeric and soluble [60], unlike the dimeric AChE-E (bound to the plasma membrane of red blood cells) and tetrameric AChE-S units. Tetrameric AChE-S can further interact with the collagen Q (ColQ) protein, enabling anchorage to neuromuscular junctions (NMJ) [56]. Alternatively, it adheres to brain synapses through the proline-rich membrane anchor protein PRiMA [61], or to peripheral synapses through laminin [62]. The synaptic docking of AChE-S may thus occur in three different ways at least.

Alzheimer's disease (AD) has been identified as a cholinergic deficit in the early eighties [63], which led to the development of cholinesterase inhibitors as the leading currently used AD drugs [64]. The cognitive symptoms in AD involve learning and memory impairments accompanied by neuropsychiatric disturbances, considered to be partially induced by the loss of cholinergic neurons [65]. These neurons notably express AChE in considerably higher levels than other neuron types [66]. That the brain's ACh hydrolyzing activity decreases relatively late in the AD process, raised doubts regarding the causal involvement of AChE in AD [67]. However, identification of AChE in the amyloid plaques characteristic of AD brains [68], the non-catalytic enhancement of amyloid-β (Aβ) fibrils formation by the major "synaptic" AChE-S variant [69] and its suppression by the highly homologous enzyme butyrylcholinesterase (BChE; [70]) renewed the interest in both ACh and cholinesterases as potential modulators of the AD neuropathology. Supporting this notion, muscarinic agonists were found to attenuate both Aβ and tau pathologies and delay the development of cognitive impairments in transgenic mice [71]. Importantly, alternative splicing and alternate promoter usage generate various forms of AChE, differing in their N- and C- termini [59] but all capable of hydrolyzing ACh. For comparison, splice variants of other genes often show inverse properties (e.g., pro-apoptotic, or anti-apoptotic features of splice variants of Bcl2 [72]). This raised the possibility that distinct AChE variants exert different effects on the AD process. AChE-S is by far the prominent variant in the brain; the minor AChE-R accumulates following swim stress [57], head injury, immobilization stress or exposure to cholinesterase inhibitors [59]. This upregulation is dependent on an interplay between two antagonistic splicing factors, SC35 and ASF/SF2 [73]. Under or following stress, neuronal SC35 levels are increased and this promotes an alternative splicing shift from AChE-S to AChE-R. Further evidence linking AChE-R and stress reactions comes from its involvement in fear memory. Selective downregulation of AChE-R by antisense oligonucleotides abolished stress-associated increase of AChE-R, elevation of contextual fear and LTP enhancement in the hippocampal CA1 region. Reciprocally, intrahippocampally injected synthetic ARP, the C-terminal peptide of AChE-R enhanced contextual fear [74]. In healthy humans AChE-R blood levels increase during normal aging [75]; however transgenic overexpression of AChE-R attenuates aging and stress-associated neuropathologies, unlike AChE-S which enhances these features [76]. Given all of the above, we recently addressed the specific role of AChE-R in AD progression [77]: *in vitro* we found that highly purified AChE-R suppresses the formation of insoluble Aβ oligomers and fibrils. AChE-R also abolishes Aβ toxicity to cultured cells, unlike the prevalent AChE-S protein which facilitates these processes. *In vivo*, double transgenic APPsw/AChE-R mice show lower plaque burden, fewer reactive astrocytes and less dendritic damage than single APPsw mice, inverse to the acceleration of these features in double APPsw/AChE-S mice [78]. In hippocampi from AD patients, dentate gyrus neurons show significantly elevated AChE-R mRNA and reduced AChE-S mRNA. As AChE-S was shown to enhance Aβ fibrils formation *in vitro* [69], enhance their neurotoxic effects *ex vivo* [79] and facilitate AD plaque formation *in vivo* [78], this evidence suggests that stress-associated alternative splicing shifts the normally prevalent AChE-S to the stress-induced neuroprotective AChE-R. However, the full effect of such neuroprotection is prevented in AD, so that the mRNA changes are not reflected at the level of the corresponding protein products. Rather, immunoblot analyses reveal drastic reductions in the levels of intact AChE-R protein, and increased amounts of degradation products, suggesting that AChE-R loss in the AD brain exacerbates Aβ-induced damages.

Given the many parallels between AD and stress reactions described here (Figure 16.2), it is not surprising that a similar pattern of alternative splicing of AChE was detected in both cases. It seems however, that activation of proteases in AD [80] does not permit the accumulation of the neuro-protective AChE-R.

The alternative splicing of AChE in stress and AD suggests that other transcripts may be similarly alternatively spliced. Further, epigenetic mechanisms such as chromatin remodeling have been demonstrated to be involved [81]. The extensively packaged nucleosome, represses transcription by blocking the access to the transcription start site. However, the nuclear GC receptor (GR) is exceptional in its ability to access the hormone-responsive elements (GRE), thereby releasing the tight structure of the nucleosome and allowing access of molecules involved in chromatin remodelling and gene transcription. [82]. The contribution of these mechanisms to stress reactions and neurodegeneration is summarized below.

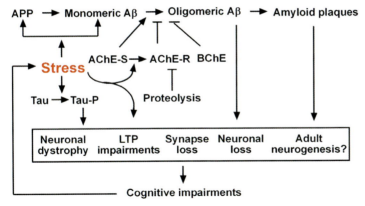

Figure 16.2 Stress reactions and AD share strikingly similar molecular mechanisms which eventually lead to neuronal toxicity and cognitive impairments. These in turn re-propagate stress reactions, resulting in a vicious cycle. In both cases, alternative splicing of AChE is shifted toward a neuroprotective variant, AChE-R, suggesting an attempt to cope with these insults. In AD, though, AChE-R is selectively degraded, exacerbating amyloid pathology. In contrast, AChE-S promotes neurodegeneration and its downregulation in stress and AD might be beneficial (see text for details).

16.6
Chronic Stress Affects Behavior Thorough Chromatin Remodeling and Alternative Splicing

The close link between stress and depression has long been established, but the molecular mechanisms binding these two conditions and the mechanism of action of antidepressant drugs are poorly understood. Recent evidence suggests that the brain-derived neurotrophic factor (BDNF) may play a role in the relevant therapeutic mechanism, as BDNF infusion to the hippocampus generates antidepressant-like effects in an animal model of depression, whereas BDNF deficiency produces impaired antidepressant responses [83].

The BDNF gene undergoes alternative splicing that adds five different non-coding exons to an identical coding one. This creates five different transcripts, all yielding an identical BDNF protein. Each of these non-coding exons associates with a unique promoter region with a typical chromatin architecture, regulating BDNF expression.

Chronic stress and chronic antidepressant treatment have both been shown to change transcription of the different BDNF variants, modifying both its total expression level and in its splicing pattern. This change is associated with long-lasting changes in chromatin structure at the BDNF promoter regions. In a mouse model of chronic defeat stress, this change can be reversed in response to a chronic antidepressant treatment with imipramine [84], a tertiary amine that is thought to affect transmitter uptake in monoaminergic neurons.

Interestingly, BDNF polymorphism is associated in humans with verbal episodic memory, hippocampal physiological activity, neuronal integrity and synaptic abundance, due to altered BDNF secretion. *In vitro*, a single nucleotide polymorphism affects the distribution, packaging and release of BDNF, possibly leading to impairments in hippocampal functioning *in vivo* [85].

In the mouse hippocampus, chronic stress selectively induces a decrease in the transcription of BDNF variants III and IV (controlled by promoters P3 and P4, respectively), leading to a threefold decrease in total transcription. This is accompanied by increase in H3-K27 dimethylation, a repressive histone modification at the BDNF P3 and P4 promoters, detectable even one month after the termination of stress [84]. A significant decrease of BDNF serum concentration has been established in AD patients at the early stages of the disease in comparison to healthy age-matched controls [86]. In an animal model a significant increase of BDNF serum level is observed 15 months after treatment with the AChE inhibitor donepezil [87]. The molecular mechanism yielding BDNF increase following AChE inhibitor administration may involve upregulation of BDNF which contributes to the neuroprotective effects of AChE inhibitors. It has been demonstrated that BDNF promotes the expression of somatostatin, which increases neprilysin activity, a degrading protein of Aβ, in primary cortical neurons [88, 89]. In addition BDNF is capable of inactivating glycogen synthase kinase 3 beta (GSK-3beta), which is involved in hyperphosphorylation of the tau protein [90].

Epigenetic mechanisms and chromatin remodeling may initiate molecular processes involved in the pathogenesis of several psychiatric conditions including stress. Hence, proteins underlaying these mechanisms should be considered as possible therapeutic targets. An example is HDAC5, capable of repressing gene expression by generating a less accessible chromatin structure by removing an acetyl group from the histone proteins that wrap the DNA in a tight nucleosome structure. HDAC5 is a member of the class II HDAC proteins, known to be activity-dependent regulators in neurons [91]. Chronic stress reduces HDAC5 expression in the nucleus accumbens (NA) [81], an important region for stress and depression-like behavior [92]. This reduction propagates the sensitivity to stress expressed as social avoidance. Chronic treatment with imipramine reverses the downregulation of HDAC5 mRNA levels, and restores the stress-induced behavioral outcome [84].

Importantly, mice lacking HDAC5 showed dysregulation of histone acetylation and gene expression, accompanied by hypersensitivity to chronic but not acute stress.

Together, these findings suggest that HDAC5 acts as a regulator of the adaptive stress response, which likely participates in the complicated molecular machinery involved in the developing long-lasting pathological state [81].

Importantly, HDAC5's function in other brain regions is very different. HDAC5 acts as an antidepressant agent in the NA and demonstrates opposite prodepressant properties when expressed in the hippocampus [47], similar to other signaling proteins with different effects in these two regions [93, 94]. Chronic imipramine administration to the hippocampus of mice in a state of chronic stress causes antidepressant-like effect, in spite of the HDAC5 downregulation. In addition, HDAC overexpression blocked this antidepressant effect caused by imipramine

[84]. Moreover, sodium butyrate, a nonspecific HDAC inhibitor, also causes an antidepressant-like effect after stress. Sodium butyrate treatment also reinstated learning abilities in mice exhibiting severe neurodegeneration, and impaired access to long term memory and spatial memory [95]. These results suggest a model by which chronic stress induces repression of BDNF gene expression, mediated mainly by histone methylation and HDAC expression.

16.7
GCs Regulate Neuronal Excitability by Modifying Alternative Splicing

A well established example of the role of alternative splicing in stress reactions is that of the large conductance calcium-activated potassium (BK) channels, important for neuronal excitability and action potential firing properties [96].

BK channels undergo alternative pre-mRNA splicing that generates two major splice variants from the single SLO gene, with clearly distinct properties. The alternative stress-regulated exon (STREX) encodes 58 amino acids which change calcium sensitivity and channel regulation compared to the insertless (ZERO) variant. Therefore, by modulating splicing decisions it is possible to adjust and increase the ability of chromaffin cells to secrete epinephrine when provoked [97].

The expression of BK channels may be activated or inhibited depending on GCs, suggesting that BK channels are important targets for glucocorticoid action.

Hypophysectomy reduces the inclusion of the STREX exon and this decline can be prevented by ACTH injection [98]. Higher depolarization activation of the cells and reduced firing patterns by BK channels in GC-treated animals yield lower epinephrine secretions.

The stress-hyposensitive period (SHRP) is a period of reduced responsiveness of the pituitary–adrenal axis. Through this period, prenatal in humans and postnatal in rats, there is a reduction in GCs concentration in the serum. This decline is correlated with a drop of STREX exon inclusion of BK channels in anterior pituitary cells. Likewise, in cultured cells the drop in STREX inclusion can be overturned by the addition of GCs to the culture medium [99]. It would be interesting to test whether STREX-containing exons are alternatively spliced in AD as well.

In addition to the long-term effect of GCs caused by modified alternative splicing, GCs induce the activation of phosphatase 2A-like protein (PPase) that blocks the activation of BK channels by PKA. This effect can provide a short-term switch of cellular excitability [98]. Altogether, this data supports the hypothesis that GCs have an important role in shaping the excitability of neuroendocrine cells during development and adaptation to stress, both short and long term. It also adds another dimension to stress-related plasticity in the adrenal tissue.

The afferent synaptic input into the adrenomedullary chromaffin cells triggers rapid release of adrenaline by activating Na^+ and Ca^{2+} channel-dependent action potentials.

Importantly, the stress activated β-adrenergic receptor (β2-AR) increases Aβ production *in vitro* and amyloid plaque formation *in vivo* by enhancing γ-secretase activity, mostly by promoting γ-secretase localization to the endosomal and lysosomal systems [100]. Compatible with the reported localizing of APP and Aβ in these systems, the activation of β2-AR, controlled by adrenaline secretion can mediate the environmental stress effects by promoting Aβ aggregation and inducing AD pathology.

Summary

In summary, an increasingly growing body of evidence suggests that stress reactions and AD exert bidirectionsl effects, creating a vicious cycle. Similar neurotoxic characteristics are shared by both syndromes and stress appears to be a prominent risk factor. Neuroprotective mechanisms employed by brain neurons are therefore highly important for attenuating neurodegeneration. Indeed, an alternative splicing shift from AChE-S to the neuroprotective AChE-R occurs in both situations. In AD, however, AChE-R is selectively degraded and therefore cannot interfere with the disease progress, exacerbating amyloid pathology. Mimicking the increase of AChE-R may therefore be therapeutically beneficial in AD and perhaps also in stress conditions, especially in individuals in which the capacity to overproduce AChE-R has been exhausted.

Acknowledgments

The Israel Science Foundation (Grant No. 399/07) and the ROSETREES Foundation have supported this study.

References

1 Alzheimer, A., Stelzmann, R.A., Schnitzlein, H.N. and Murtagh, F.R. (1995) An English translation of Alzheimer's 1907 paper, "Uber eine eigenartige Erkankung der Hirnrinde". *Clin. Anat.*, **8**, 429–431.

2 de Kloet, E.R., Joels, M. and Holsboer, F. (2005) Stress and the brain: from adaptation to disease. *Nat. Rev. Neurosci.*, **6**, 463–475.

3 Kang, J., Lemaire, H.G., Unterbeck, A., Salbaum, J.M., Masters, C.L., Grzeschik, K.H., Multhaup, G., Beyreuther, K. and Muller-Hill, B. (1987) The precursor of Alzheimer's disease amyloid A4 protein resembles a cell-surface receptor. *Nature*, **325**, 733–736.

4 Haass, C. and Selkoe, D.J. (2007) Soluble protein oligomers in neurodegeneration: lessons from the Alzheimer's amyloid beta-peptide. *Nat. Rev. Mol. Cell. Biol.*, **8**, 101–112.

5 Naslund, J., Haroutunian, V., Mohs, R., Davis, K.L., Davies, P., Greengard, P. and Buxbaum, J.D. (2000) Correlation between elevated levels of amyloid beta-peptide in the brain and cognitive decline. *JAMA*, **283**, 1571–1577.

6 Lesne, S., Koh, M.T., Kotilinek, L., Kayed, R., Glabe, C.G., Yang, A., Gallagher, M. and Ashe, K.H. (2006) A specific amyloid-beta protein assembly in the brain impairs memory. *Nature*, **440**, 352–357.

7 Townsend, M., Shankar, G.M., Mehta, T., Walsh, D.M. and Selkoe, D.J. (2006) Effects of secreted oligomers of amyloid beta-protein on hippocampal synaptic plasticity: a potent role for trimers. *J. Physiol.*, **572**, 477–492.

8 Walsh, D.M., Klyubin, I., Fadeeva, J.V., Cullen, W.K., Anwyl, R., Wolfe, M.S., Rowan, M.J. and Selkoe, D.J. (2002) Naturally secreted oligomers of amyloid beta protein potently inhibit hippo-campal long-term potentiation *in vivo*. *Nature*, **416**, 535–539.

9 Snyder, E.M., Nong, Y., Almeida, C.G., Paul, S., Moran, T., Choi, E.Y., Nairn, A.C., Salter, M.W., Lombroso, P.J., Gouras, G.K. and Greengard, P. (2005) Regulation of NMDA receptor trafficking by amyloid-beta. *Nat. Neurosci.*, **8**, 1051–1058.

10 Mattson, M.P. (2004) Pathways towards and away from Alzheimer's disease. *Nature*, **430**, 631–639.

11 Selye, H. (1998) A syndrome produced by diverse nocuous agents. 1936. *J. Neuropsychiatry Clin. Neurosci.*, **10**, 230–231.

12 Kim, J.J. and Diamond, D.M. (2002) The stressed hippocampus, synaptic plasticity and lost memories. *Nat. Rev. Neurosci.*, **3**, 453–462.

13 McEwen, B.S. (1999) Stress and hippocampal plasticity. *Annu. Rev. Neurosci.*, **22**, 105–122.

14 Pratt, W.B. and Toft, D.O. (1997) Steroid receptor interactions with heat shock protein and immunophilin chaperones. *Endocr. Rev.*, **18**, 306–360.

15 Reul, J.M. and de Kloet, E.R. (1985) Two receptor systems for corticosterone in rat brain: microdistribution and differential occupation. *Endocrinology*, **117**, 2505–2511.

16 Magarinos, A.M., McEwen, B.S., Flugge, G. and Fuchs, E. (1996) Chronic psychosocial stress causes apical dendritic atrophy of hippocampal CA3 pyramidal neurons in subordinate tree shrews. *J. Neurosci.*, **16**, 3534–3540.

17 Magarinos, A.M., Verdugo, J.M. and McEwen, B.S. (1997) Chronic stress alters synaptic terminal structure in hippocampus. *Proc. Natl. Acad. Sci. U.S.A.*, **94**, 14002–14008.

18 Uno, H., Tarara, R., Else, J.G., Suleman, M.A. and Sapolsky, R.M. (1989) Hippocampal damage associated with prolonged and fatal stress in primates. *J. Neurosci.*, **9**, 1705–1711.

19 Sandi, C., Davies, H.A., Cordero, M.I., Rodriguez, J.J., Popov, V.I. and Stewart, M.G. (2003) Rapid reversal of stress induced loss of synapses in CA3 of rat hippocampus following water maze training. *Eur. J. Neurosci.*, **17**, 2447–2456.

20 Conforti, L., Adalbert, R. and Coleman, M.P. (2007) Neuronal death: where does the end begin? *Trends Neurosci.*, **30**, 159–166.

21 Selkoe, D.J. (2002) Alzheimer's disease is a synaptic failure. *Science*, **298**, 789–791.

22 Terry, R.D., Masliah, E., Salmon, D.P., Butters, N., DeTeresa, R., Hill, R., Hansen, L.A. and Katzman, R. (1991) Physical basis of cognitive alterations in Alzheimer's disease: synapse loss is the major correlate of cognitive impairment. *Ann. Neurol.*, **30**, 572–580.

23 Hsia, A.Y., Masliah, E., McConlogue, L., Yu, G.Q., Tatsuno, G., Hu, K., Kholodenko, D., Malenka, R.C., Nicoll, R.A. and Mucke, L. (1999) Plaque-independent disruption of neural circuits in Alzheimer's disease mouse models. *Proc. Natl. Acad. Sci. U.S.A.*, **96**, 3228–3233.

24 Tsai, J., Grutzendler, J., Duff, K. and Gan, W.B. (2004) Fibrillar amyloid deposition leads to local synaptic abnormalities and breakage of neuronal branches. *Nat. Neurosci.*, **7**, 1181–1183.

25 Shors, T.J., Seib, T.B., Levine, S. and Thompson, R.F. (1989) Inescapable versus escapable shock modulates long-term potentiation in the rat hippocampus. *Science*, **244**, 224–226.

26 Maier, S.F. and Watkins, L.R. (2005) Stressor controllability and learned helplessness: the roles of the dorsal raphe nucleus, serotonin, and corticotropin-releasing factor. *Neurosci. Biobehav. Rev.*, **29**, 829–841.

27 Kim, J.J., Foy, M.R. and Thompson, R.F. (1996) Behavioral stress modifies

hippocampal plasticity through
N-methyl-D-aspartate receptor activation.
Proc. Natl. Acad. Sci. U.S.A., **93**,
4750–4753.

28 Kamenetz, F., Tomita, T., Hsieh, H.,
Seabrook, G., Borchelt, D., Iwatsubo, T.,
Sisodia, S. and Malinow, R. (2003) APP
processing and synaptic function.
Neuron, **37**, 925–937.

29 Heine, V.M., Maslam, S., Zareno, J.,
Joels, M. and Lucassen, P.J. (2004)
Suppressed proliferation and apoptotic
changes in the rat dentate gyrus
after acute and chronic stress are
reversible. *Eur. J. Neurosci.*, **19**,
131–144.

30 Wong, E.Y. and Herbert, J. (2004) The
corticoid environment: a determining
factor for neural progenitors' survival in
the adult hippocampus. *Eur. J. Neurosci.*,
20, 2491–2498.

31 Alonso, R., Griebel, G., Pavone, G.,
Stemmelin, J., Le Fur, G. and Soubrie,
P. (2004) Blockade of CRF(1) or V(1b)
receptors reverses stress-induced
suppression of neurogenesis in a mouse
model of depression. *Mol. Psychiatry*, **9**,
278–286, 224.

32 Zhang, C.L., Zou, Y., He, W., Gage, F.H.
and Evans, R.M. (2008) A role for adult
TLX-positive neural stem cells in
learning and behaviour. *Nature*, **451**,
1004–1007.

33 Jin, K., Peel, A.L., Mao, X.O., Xie, L.,
Cottrell, B.A., Henshall, D.C. and
Greenberg, D.A. (2004) Increased
hippocampal neurogenesis in
Alzheimer's disease. *Proc. Natl. Acad.
Sci. U.S.A.*, **101**, 343–347.

34 Jin, K., Galvan, V., Xie, L., Mao, X.O.,
Gorostiza, O.F., Bredesen, D.E. and
Greenberg, D.A. (2004) Enhanced
neurogenesis in Alzheimer's disease
transgenic (PDGF-APPSw,Ind) mice.
Proc. Natl. Acad. Sci. U.S.A., **101**,
13363–13367.

35 Verret, L., Jankowsky, J.L., Xu, G.M.,
Borchelt, D.R. and Rampon, C. (2007)
Alzheimer's-type amyloidosis in
transgenic mice impairs survival of
newborn neurons derived from adult
hippocampal neurogenesis. *J. Neurosci.*,
27, 6771–6780.

36 Moghaddam, B., Bolinao, M.L.,
Stein-Behrens, B. and Sapolsky, R.
(1994) Glucocorticoids mediate the
stress-induced extracellular accumulation
of glutamate. *Brain Res.*, **655**, 251–254.

37 Mattson, M.P. (2007) Calcium and
neurodegeneration. *Aging Cell*, **6**,
337–350.

38 LaFerla, F.M. (2002) Calcium
dyshomeostasis and intracellular
signalling in Alzheimer's disease. *Nat.
Rev. Neurosci.*, **3**, 862–872.

39 Ownby, R.L., Harwood, D.G., Barker,
W.W. and Duara, R. (2000) Predictors of
anxiety in patients with Alzheimer's
disease. *Depress. Anxiety*, **11**, 38–42.

40 Ownby, R.L., Crocco, E., Acevedo, A.,
John, V. and Loewenstein, D. (2006)
Depression and risk for Alzheimer
disease: systematic review, meta-analysis,
and metaregression analysis. *Arch. Gen.
Psychiatry*, **63**, 530–538.

41 Wilson, R.S., Barnes, L.L., Bennett, D.A.,
Li, Y., Bienias, J.L., Mendes de Leon,
C.F. and Evans, D.A. (2005) Proneness
to psychological distress and risk of
Alzheimer disease in a biracial
community. *Neurology*, **64**, 380–382.

42 Masugi, F., Ogihara, T., Sakaguchi, K.,
Otsuka, A., Tsuchiya, Y., Morimoto, S.,
Kumahara, Y., Saeki, S. and Nishide, M.
(1989) High plasma levels of cortisol in
patients with senile dementia of the
Alzheimer's type. *Methods Find. Exp.
Clin. Pharmacol.*, **11**, 707–710.

43 Rasmuson, S., Andrew, R., Nasman, B.,
Seckl, J.R., Walker, B.R. and Olsson, T.
(2001) Increased glucocorticoid
production and altered cortisol
metabolism in women with mild to
moderate Alzheimer's disease. *Biol.
Psychiatry*, **49**, 547–552.

44 Elgh, E., Lindqvist Astot, A.,
Fagerlund, M., Eriksson, S., Olsson, T.
and Nasman, B. (2006) Cognitive
dysfunction, hippocampal atrophy
and glucocorticoid feedback in
Alzheimer's disease. *Biol. Psychiatry*,
59, 155–161.

45 Csernansky, J.G., Dong, H., Fagan,
A.M., Wang, L., Xiong, C., Holtzman,
D.M. and Morris, J.C. (2006) Plasma
cortisol and progression of dementia in

subjects with Alzheimer-type dementia. *Am. J. Psychiatry*, **163**, 2164–2169.

46 Aisen, P.S., Davis, K.L., Berg, J.D., Schafer, K., Campbell, K., Thomas, R.G., Weiner, M.F., Farlow, M.R., Sano, M., Grundman, M. and Thal, L.J. (2000) A randomized controlled trial of prednisone in Alzheimer's disease. Alzheimer's Disease Cooperative Study. *Neurology*, **54**, 588–593.

47 de Quervain, D.J., Poirier, R., Wollmer, M.A., Grimaldi, L.M., Tsolaki, M., Streffer, J.R., Hock, C., Nitsch, R.M., Mohajeri, M.H. and Papassotiropoulos, A. (2004) Glucocorticoid-related genetic susceptibility for Alzheimer's disease. *Hum. Mol. Genet.*, **13**, 47–52.

48 Green, K.N., Billings, L.M., Roozendaal, B., McGaugh, J.L. and LaFerla, F.M. (2006) Glucocorticoids increase amyloid-beta and tau pathology in a mouse model of Alzheimer's disease. *J. Neurosci.*, **26**, 9047–9056.

49 Feng, Q., Cheng, B., Yang, R., Sun, F.Y. and Zhu, C.Q. (2005) Dynamic changes of phosphorylated tau in mouse hippocampus after cold water stress. *Neurosci. Lett.*, **388**, 13–16.

50 Korneyev, A., Binder, L. and Bernardis, J. (1995) Rapid reversible phosphorylation of rat brain tau proteins in response to cold water stress. *Neurosci. Lett.*, **191**, 19–22.

51 Yanagisawa, M., Planel, E., Ishiguro, K. and Fujita, S.C. (1999) Starvation induces tau hyperphosphorylation in mouse brain: implications for Alzheimer's disease. *FEBS Lett.*, **461**, 329–333.

52 Rissman, R.A., Lee, K.F., Vale, W. and Sawchenko, P.E. (2007) Corticotropin-releasing factor receptors differentially regulate stress-induced tau phosphorylation. *J. Neurosci.*, **27**, 6552–6562.

53 Catania, C., Sotiropoulos, I., Silva, R., Onofri, C., Breen, K.C., Sousa, N. and Almeida, O.F. (2009) The amyloidogenic potential and behavioral correlates of stress. *Mol. Psychiatry*, **14**, 95–105.

54 Kang, J.E., Cirrito, J.R., Dong, H., Csernansky, J.G. and Holtzman, D.M. (2007) Acute stress increases interstitial fluid amyloid-beta via corticotropin-releasing factor and neuronal activity. *Proc. Natl. Acad. Sci. U.S.A.*, **104**, 10673–10678.

55 Pedersen, W.A., McCullers, D., Culmsee, C., Haughey, N.J., Herman, J.P. and Mattson, M.P. (2001) Corticotropin-releasing hormone protects neurons against insults relevant to the pathogenesis of Alzheimer's disease. *Neurobiol. Dis.*, **8**, 492–503.

56 Massoulie, J. (2002) The origin of the molecular diversity and functional anchoring of cholinesterases. *Neurosignals*, **11**, 130–143.

57 Kaufer, D., Friedman, A., Seidman, S. and Soreq, H. (1998) Acute stress facilitates long-lasting changes in cholinergic gene expression. *Nature*, **393**, 373–377.

58 Meshorer, E., Erb, C., Gazit, R., Pavlovsky, L., Kaufer, D., Friedman, A., Glick, D., Ben-Arie, N. and Soreq, H. (2002) Alternative splicing and neuritic mRNA translocation under long-term neuronal hypersensitivity. *Science*, **295**, 508–512.

59 Meshorer, E. and Soreq, H. (2006) Virtues and woes of AChE alternative splicing in stress-related neuro-pathologies. *Trends Neurosci.*, **29**, 216–224.

60 Soreq, H. and Seidman, S. (2001) Acetylcholinesterase–new roles for an old actor. *Nat. Rev. Neurosci.*, **2**, 294–302.

61 Perrier, A.L., Massoulie, J. and Krejci, E. (2002) PRiMA: the membrane anchor of acetylcholinesterase in the brain. *Neuron*, **33**, 275–285.

62 Johnson, G. and Moore, S.W. (2003) Human acetylcholinesterase binds to mouse laminin-1 and human collagen IV by an electrostatic mechanism at the peripheral anionic site. *Neurosci. Lett.*, **337**, 37–40.

63 Coyle, J.T., Price, D.L. and DeLong, M.R. (1983) Alzheimer's disease: a disorder of cortical cholinergic innervation. *Science*, **219**, 1184–1190.

64 Lleo, A., Greenberg, S.M. and Growdon, J.H. (2006) Current pharmacotherapy for Alzheimer's disease. *Annu. Rev. Med.*, **57**, 513–533.

65 Cummings, J.L. and Back, C. (1998) The cholinergic hypothesis of neuropsychiatric symptoms in Alzheimer's disease. *Am. J. Geriatr. Psychiatry*, **6**, S64–78.

66 Landwehrmeyer, B., Probst, A., Palacios, J.M. and Mengod, G. (1993) Expression of acetylcholinesterase messenger RNA in human brain: an in situ hybridization study. *Neuroscience*, **57**, 615–634.

67 Davis, K.L., Mohs, R.C., Marin, D., Purohit, D.P., Perl, D.P., Lantz, M., Austin, G. and Haroutunian, V. (1999) Cholinergic markers in elderly patients with early signs of Alzheimer disease. *JAMA*, **281**, 1401–1406.

68 Carson, K.A., Geula, C. and Mesulam, M.M. (1991) Electron microscopic localization of cholinesterase activity in Alzheimer brain tissue. *Brain Res.*, **540**, 204–208.

69 Inestrosa, N.C., Alvarez, A., Perez, C.A., Moreno, R.D., Vicente, M., Linker, C., Casanueva, O.I., Soto, C. and Garrido, J. (1996) Acetylcholinesterase accelerates assembly of amyloid-beta-peptides into Alzheimer's fibrils: possible role of the peripheral site of the enzyme. *Neuron*, **16**, 881–891.

70 Diamant, S., Podoly, E., Friedler, A., Ligumsky, H., Livnah, O. and Soreq, H. (2006) Butyrylcholinesterase attenuates amyloid fibril formation *in vitro*. *Proc. Natl. Acad. Sci. U.S.A.*, **103**, 8628–8633.

71 Caccamo, A., Oddo, S., Billings, L.M., Green, K.N., Martinez-Coria, H., Fisher, A. and LaFerla, F.M. (2006) M1 receptors play a central role in modulating AD-like pathology in transgenic mice. *Neuron*, **49**, 671–682.

72 Shin, C. and Manley, J.L. (2004) Cell signalling and the control of pre-mRNA splicing. *Nat. Rev. Mol. Cell Biol.*, **5**, 727–738.

73 Meshorer, E., Bryk, B., Toiber, D., Cohen, J., Podoly, E., Dori, A. and Soreq, H. (2005) SC35 promotes sustainable stress-induced alternative splicing of neuronal acetylcholinesterase mRNA. *Mol. Psychiatry*, **10**, 985–997.

74 Nijholt, I., Farchi, N., Kye, M., Sklan, E.H., Shoham, S., Verbeure, B., Owen, D., Hochner, B., Spiess, J., Soreq, H. and Blank, T. (2004) Stress-induced alternative splicing of acetylcholinesterase results in enhanced fear memory and long-term potentiation. *Mol. Psychiatry*, **9**, 174–183.

75 Sklan, E.H., Lowenthal, A., Korner, M., Ritov, Y., Landers, D.M., Rankinen, T., Bouchard, C., Leon, A.S., Rice, T., Rao, D.C., Wilmore, J.H., Skinner, J.S. and Soreq, H., (2004) Acetylcholinesterase/paraoxonase genotype and expression predict anxiety scores in Health, Risk Factors, Exercise Training, and Genetics study. *Proc. Natl. Acad. Sci. U. S. A.*, **101**, 5512–5517.

76 Sternfeld, M., Shoham, S., Klein, O., Flores-Flores, C., Evron, T., Idelson, G.H., Kitsberg, D., Patrick, J.W. and Soreq, H. (2000) Excess "read-through" acetylcholinesterase attenuates but the "synaptic" variant intensifies neuro-deterioration correlates. *Proc. Natl. Acad. Sci. U. S. A.*, **97**, 8647–8652.

77 Berson, A., Knobloch, M., Hanan, M., Diamant, S., Sharoni, M., Schuppli, D., Geyer, B.C., Ravid, R., Mor, T.S., Nitsch, R.M. and Soreq, H. (2008) Changes in readthrough acetylcholinesterase expression modulate amyloid-beta pathology. *Brain*, **131**, 109–119.

78 Rees, T.M., Berson, A., Sklan, E.H., Younkin, L., Younkin, S., Brimijoin, S. and Soreq, H. (2005) Memory deficits correlating with acetylcholinesterase splice shift and amyloid burden in doubly transgenic mice. *Curr. Alzheimer Res.*, **2**, 291–300.

79 Alvarez, A., Alarcon, R., Opazo, C., Campos, E.O., Munoz, F.J., Calderon, F.H., Dajas, F., Gentry, M.K., Doctor, B.P., De Mello, F.G. and Inestrosa, N.C. (1998) Stable complexes involving acetylcholinesterase and amyloid-beta peptide change the biochemical properties of the enzyme and increase the neurotoxicity of Alzheimer's fibrils. *J. Neurosci.*, **18**, 3213–3223.

80 Saito, K., Elce, J.S., Hamos, J.E. and Nixon, R.A. (1993) Widespread activation of calcium-activated neutral proteinase (calpain) in the brain in Alzheimer disease: a potential molecular basis for

neuronal degeneration. *Proc. Natl. Acad. Sci. U. S. A.*, **90**, 2628–2632.

81 Renthal, W., Maze, I., Krishnan, V., Covington, H.E. 3rd, Xiao, G., Kumar, A., Russo, S.J., Graham, A., Tsankova, N., Kippin, T.E., Kerstetter, K.A., Neve, R.L., Haggarty, S.J., McKinsey, T.A., Bassel-Duby, R., Olson, E.N. and Nestler, E.J. (2007) Histone deacetylase 5 epigenetically controls behavioral adaptations to chronic emotional stimuli. *Neuron*, **56**, 517–529.

82 Hebbar, P.B. and Archer, T.K. (2003) Chromatin remodeling by nuclear receptors. *Chromosoma*, **111**, 495–504.

83 Monteggia, L.M., Barrot, M., Powell, C.M., Berton, O., Galanis, V., Gemelli, T., Meuth, S., Nagy, A., Greene, R.W. and Nestler, E.J. (2004) Essential role of brain-derived neurotrophic factor in adult hippocampal function. *Proc. Natl. Acad. Sci. U. S. A.*, **101**, 10827–10832.

84 Tsankova, N.M., Berton, O., Renthal, W., Kumar, A., Neve, R.L. and Nestler, E.J. (2006) Sustained hippocampal chromatin regulation in a mouse model of depression and antidepressant action. *Nat. Neurosci.*, **9**, 519–525.

85 Egan, M.F., Kojima, M., Callicott, J.H., Goldberg, T.E., Kolachana, B.S., Bertolino, A., Zaitsev, E., Gold, B., Goldman, D., Dean, M., Lu, B. and Weinberger, D.R. (2003) The BDNF val66met polymorphism affects activity-dependent secretion of BDNF and human memory and hippocampal function. *Cell*, **112**, 257–269.

86 Laske, C., Stransky, E., Leyhe, T., Eschweiler, G.W., Maetzler, W., Wittorf, A., Soekadar, S., Richartz, E., Koehler, N., Bartels, M., Buchkremer, G. and Schott, K. (2007) BDNF serum and CSF concentrations in Alzheimer's disease, normal pressure hydrocephalus and healthy controls. *J. Psychiatr. Res.*, **41**, 387–394.

87 Leyhe, T., Stransky, E., Eschweiler, G.W., Buchkremer, G. and Laske, C. (2008) Increase of BDNF serum concentration during donepezil treatment of patients with early Alzheimer's disease. *Eur. Arch. Psychiatry Clin. Neurosci.*, **258**, 124–128.

88 Marmigere, F., Choby, C., Rage, F., Richard, S. and Tapia-Arancibia, L. (2001) Rapid stimulatory effects of brain-derived neurotrophic factor and neurotrophin-3 on somatostatin release and intracellular calcium rise in primary hypothalamic cell cultures. *Neuroendocrinology*, **74**, 43–54.

89 Saito, T., Iwata, N., Tsubuki, S., Takaki, Y., Takano, J., Huang, S.M., Suemoto, T., Higuchi, M. and Saido, T.C. (2005) Somatostatin regulates brain amyloid beta peptide Abeta42 through modulation of proteolytic degradation. *Nat. Med.*, **11**, 434–439.

90 Brion, J.P., Anderton, B.H., Authelet, M., Dayanandan, R., Leroy, K., Lovestone, S., Octave, J.N., Pradier, L., Touchet, N. and Tremp, G. (2001) Neurofibrillary tangles and tau phosphorylation. *Biochem. Soc. Symp.*, **67**, 81–88.

91 Chawla, S., Vanhoutte, P., Arnold, F.J., Huang, C.L. and Bading, H. (2003) Neuronal activity-dependent nucleo-cytoplasmic shuttling of HDAC4 and HDAC5. *J. Neurochem.*, **85**, 151–159.

92 Nestler, E.J. and Carlezon, W.A. Jr. (2006) The mesolimbic dopamine reward circuit in depression. *Biol. Psychiatry*, **59**, 1151–1159.

93 Berton, O., McClung, C.A., Dileone, R.J., Krishnan, V., Renthal, W., Russo, S.J., Graham, D., Tsankova, N.M., Bolanos, C.A., Rios, M., Monteggia, L.M., Self, D.W. and Nestler, E.J. (2006) Essential role of BDNF in the mesolimbic dopamine pathway in social defeat stress. *Science*, **311**, 864–868.

94 Eisch, A.J., Bolanos, C.A., de Wit, J., Simonak, R.D., Pudiak, C.M., Barrot, M., Verhaagen, J. and Nestler, E.J. (2003) Brain-derived neurotrophic factor in the ventral midbrain-nucleus accumbens pathway: a role in depression. *Biol. Psychiatry*, **54**, 994–1005.

95 Fischer, A., Sananbenesi, F., Wang, X., Dobbin, M. and Tsai, L.H. (2007) Recovery of learning and memory is associated with chromatin remodelling. *Nature*, **447**, 178–182.

96 Xie, J. and Black, D.L. (2001) A CaMK IV responsive RNA element mediates

depolarization-induced alternative splicing of ion channels. *Nature*, **410**, 936–939.

97 Lai, G.J. and McCobb, D.P. (2006) Regulation of alternative splicing of Slo K+ channels in adrenal and pituitary during the stress-hyporesponsive period of rat development. *Endocrinology*, **147**, 3961–3967.

98 Xie, J. and McCobb, D.P. (1998) Control of alternative splicing of potassium channels by stress hormones. *Science*, **280**, 443–446.

99 Tian, L., Hammond, M.S., Florance, H., Antoni, F.A. and Shipston, M.J. (2001) Alternative splicing determines sensitivity of murine calcium-activated potassium channels to glucocorticoids. *J. Physiol.*, **537**, 57–68.

100 Ni, Y., Zhao, X., Bao, G., Zou, L., Teng, L., Wang, Z., Song, M., Xiong, J., Bai, Y. and Pei, G. (2006) Activation of beta2-adrenergic receptor stimulates gamma-secretase activity and accelerates amyloid plaque formation. *Nat. Med.*, **12**, 1390–1396.

17
Stress and Neurotransmission:
Clinical Evidence and Therapeutic Implications

Hadar Shalev and Jonathan Cohen

17.1
Introduction

Stress is abundant in daily life but for the most part does not cardinally impair functioning and daily living. However the stress response is a continuum of physiological and emotional reactions, leading in severe cases to significant distress. It is these cases from which we wish to generalize to the neurotransmitter physiology of the human stress response.

Emotional stress may be the result of universally distressing life events such as loss of a beloved, unemployment, life-threatening events, or illness [1]; for others, stress may be caused by specific situations that to others pose no threat such as public speaking, heights, or closed chambers. For some a stress response arises with no apparent external stressor. Similarly, the scope of reactions to an event perceived as being stressful might range from an improvement in performance and motivation (as described by the Yerkes–Dodson law [2]), through physiological discomfort, hampered performance and at the extreme a significant impairment in function and intense distress. A stressor might be short-lived or prolonged as may be the reaction to it; at times the stress reaction might actually considerably outlive the stressor.

In psychiatry the obvious cluster of conditions related to stress are the anxiety disorders, though additional disorders such as mood disorders, personality disorders, somatoform and impulse control disorders may be triggered or aggravated by stress. Anxiety disorders are considered to be the most common of psychiatric disorders in the general population with an estimated prevalence of up to 30% [3]. Interestingly, a significant comorbidity exists for different anxiety disorders [4] implying a common biological and genetic basis for the anxiety disorders. Anxiety disorders include specific phobia, social phobia, panic disorder, obsessive–compulsive disorder (OCD), generalized anxiety disorder (GAD) and post-traumatic stress disorder (PTSD). The anxiety disorders share a combination of emotional and physiological symptoms. Physiological symptoms are mostly related to sympathetic autonomic activation and include palpitations, sweating and

Stress – From Molecules to Behavior. Edited by Hermona Soreq, Alon Friedman, and Daniela Kaufer
Copyright © 2010 WILEY-VCH Verlag GmbH & Co. KGaA, Weinheim
ISBN: 978-3-527-32374-6

tremor, to mention a few. The common emotional symptoms are tension and fear accompanied by feelings of impending death or intense tension and often a specific cognitive pattern of catastrophization and pessimism.

Post-traumatic stress disorder is the only anxiety disorder that by definition has an obvious stressful trigger that in most cases shortly precedes the development of the disorder (though this can evolve at any time following a traumatic event). Following a traumatic event, many individuals display an acute stress reaction and some an acute stress disorder; the majority of whom experience resolution of symptoms by one month from the trauma. PTSD is a chronic stress condition and thus the DSM-IV requires that the symptoms occur at least one month following the traumatic event. PTSD is not only triggered by a stressful event but can be intensified in situations that are related to or provoke recall of the traumatic event. GAD is also a chronic stress disorder in which patients are preoccupied with distressing thoughts and tension. In panic disorder, patients suffer sudden attacks of intense fear and a concurrent burst of sympathetic physiological symptoms; these in turn are followed by less intense but chronic fear which is the consequence of anticipation of pending attacks (i.e., anticipation anxiety).

Depression is closely related to stress both biologically and in its frequent comorbidity with the various types of anxiety disorders. In the National Comorbidity Survey Replication, Kessler *et al.* [5] reported a lifetime prevalence of 16.2% for major depression and notably also a significant 60% comorbidity of depression and anxiety disorders, emphasizing that comorbidity of these disorders is the rule rather than the exception. Both disorder spectrums may be the consequence of chronic or acute stress and either can be the primary or secondary condition. Depression has been causally related to stressful life events [6] even after controlling for genetic and other predisposing factors. Even though some parameters are different, the same structures, pathways and substances are thought to be modified during stress exposure, anxiety disorders and depressive episodes, as attested to by a common genetic predisposition [7].

Despite the wide spectrum of symptoms and differing clinical courses between these disorders there is a limited and similar group of drugs prescribed for them.

In the following sections we discuss from a clinical point of view the three major neurotransmitter systems targeted by drugs prescribed for stress-related disorders, that is, serotonin, GABA and noradrenalin. The discussion of additional neurotransmitter systems such as neuropeptide Y, neuropeptide P, glutamate, CRF and others emerging as potentially relevant therapeutic targets not in clinical use is left to chapters dealing with more basic science [8]. This chapter provides clinical data to complement other chapters in this book presenting animal data and basic science aspects of the stress response; therefore this chapter presents mainly human data. For many therapies employed by medicine, therapeutic use has been empirically based and preceded knowledge of pathophyisiological mechanisms; this is the case for many of the drugs we discuss. We believe that from the clinical realm knowledge should be returned to basic science laboratories guiding further research.

17.2
Serotonin

Drugs targeting the serotonergic system the prototype of which are the serotonin-specific reuptake inhibitors (SSRI) are the most widely employed class of agents for stress related psycho-pathologies. These drugs are prescribed for acute stress as well as chronic stress disorders and the prolonged consequences of stress exposure. Several other drug classes target the serotonergic system with varying degrees of specificity and are mentioned, including: tricyclic antidepressants (TCAs), monoamine oxidase inhibitors (MAO-I) and serotonin neurepinephrine reuptake inhibitors (SNRIs).

Serotonin (5-HT) is an amine neurotransmitter synthesized from tryptophan through the actions of tryptophan hydroxylase and L-amino acid decarboxylase. 5-HT is synthesized mainly in the raphe nucleus, from which ascending projections lead to cortical, hippocampal, limbic and other regions; and descending projections lead to the spinal cord. 5-HT is released to the synaptic cleft where it acts on multiple subtypes of postsynaptic and pre-synaptic auto- and hetero-receptors. 5-HT removal from the synaptic cleft is accomplished by 5-HT transporter reuptake; this is followed by recycling into pre-synaptic vesicles or degradation by monoamine oxidase (mainly the MAO-A enzyme) [9].

The 5-HT transporter is the common site of action of all SSRIs. The immediate consequence of SSRI administration is an increase in serotonin within the synaptic cleft; this in turn is thought to augment serotoninergic transmission though it conversely has been suggested to decrease such transmission due to a preferential negative feedback effect on pre-synaptic auto-receptors [10]. Thus short-term negative effects of SSRI treatment such as increased anxiety could result from either an increase or decrease in available 5-HT. Specific drugs within the SSRI family have additional direct actions on 5-HT receptor subtypes (e.g., fluoxetine binds 5-HT$_{2A}$ and $_{2C}$ receptors) affecting mainly their side effect profile.

Clinically SSRIs do not provide immediate benefit and in certain conditions may actually cause short-term exacerbations. Therefore, in order to understand the source of SSRI benefit, it is important to understand the long-term alterations caused by these agents. Here again, both increased and decreased serotoninergic transmission have been advocated [11, 12]. SSRI treatment has been demonstrated in animal models to result in delayed prolonged downregulation of 5-HT receptors. However, results vary with regard to the dominantly affected 5-HT receptor population, that is, pre- or post-synaptic.

In an attempt to resolve the 5-HT increase/decrease debate regarding SSRI mode of action, several tryptophan depletion (TD) studies have been conducted in humans. As mentioned, tryptophan is the precursor of 5-HT, the brain levels of which depend on both plasma levels and active transport across the blood brain barrier. Intake of a tryptophan depleted diet reduces its plasma levels; this followed by ingestion of a tryptophan lacking–amino acid-rich drink results in active transport competition further lowering brain tryptophan levels and consequently 5-HT [13]. In healthy controls, TD enhanced the response to fearful faces, similar to the response to a single SSRI dose and opposite to the effect of long-term SSRI treatment. The opposing effects of TD and long-term SSRI treatment suggest that such treatment produces its effect through enhancement of 5-HT transmission (reviewed in [9]). TD has been shown to induce temporary depressive symptoms in depression patients successfully treated with SSRIs. Similarly, in patients effectively treated with SSRIs for panic disorders, TD but not a tryptophan-rich diet produced sensitivity to the panic provoking flumazenil challenge. These studies indicate that the effect of SSRIs is dependent on the availability of 5-HT. Conversely, in obsessive compulsive disorder, TD is not associated with symptom relapse following successful SSRI treatment. Thus, while 5-HT levels seem crucial for the beneficial effects of SSRIs in some mood and anxiety disorders, it may not be so for others (reviewed in [14]).

The serotonin transporter (5-HTT) plays an important role in serotonergic neurotransmission by facilitating reuptake of 5-HT from the synaptic cleft. A repeat of 20–23 base pairs has been observed as a motif within a polymorphic region of the serotonin transporter gene, and it occurs as two prevalent alleles: one consisting of 14 repeats (the short allele variant) and another of 16 repeats (the long allele variant), though subdivision of these alleles to ten variants has been described. This functional polymorphism in the 5′ regulatory promoter region, termed 5-HTTLPR (5′ HTT gene-linked polymorphic region), alters transcription of the serotonin transporter gene [15]. Specifically, the short allele of the 5-HTTLPR polymorphism is thought to reduce transcription efficiency for the gene, resulting in decreased expression of the serotonin transporter. The short allele has been shown to be related to a decrease in availability of the serotonin transporter in human imaging studies as well as decreased serotonin uptake in lymphoblast cell lines [16].

During the past decade a growing number of studies have examined the association between allelic variations in 5-HTTLPR and personality traits related to depression and anxiety. The results of these studies have been mixed, with some studies demonstrating an association between allelic variation and negative traits such as anxiety, hostility, depression, tension and harm avoidance [17, 18] and others reporting no association (e.g., [19, 20]). One noted reason for varied findings concerns the stress-diathesis model. Caspi *et al.* illustrated this point with findings indicating that allelic variation in 5-HTTLPR interacts with stressful life events to predict the occurrence of depression [21]. Specifically, carriers of the less transcriptionally efficient 5-HTTLPR short (s) allele had increased rates of major depression as a function of increased numbers of varied past life stressors. Others have replicated this interaction by examination of environmental stressors such as unemployment [22] and low socio-economic status [23]. Otte *et al.* [24] found that among patients with chronic illness, carriers of the s allele of 5-HTTLPR are more vulnerable to depression, perceived stress, and high norepinephrine secretion. Kilpatrick *et al.* [25] recruited 589 adults aged 18 or older, 6–9 months after the 2004 hurricanes; subjects were interviewed regarding hurricane exposure, social support and post-hurricane PTSD and major depression. DNA was collected from a subset of participants. The results showed that the low-expression variant of 5-HTTLPR increased risk of post-hurricane PTSD and major depression under conditions of high hurricane exposure and low social support after adjustment for sex, ancestry and age. Similar effects were found for major depression. High-risk individuals (high hurricane exposure, low-expression 5-HTTLPR variant, low social support) were at 4.5× the risk of developing PTSD and major depression compared with low-risk individuals. Hariri *et al.* [26] reported that individuals with at least one copy of the 5HTTLPR **s** variant exhibit greater amygdala responsivity, as assessed by fMRI, in response to fearful stimuli compared with individuals homozygous for the high-activity l allele. This result confirms that genetically driven variations of serotonergic function contribute to the response of brain regions underlying human emotional behavior and indicate that differential excitability of the amygdala to emotional stimuli may contribute to increased fear and anxiety-related responses.

Genetic variation in 5HTT function may lead to both increased susceptibility to anxious or depressive features as well as altered antidepressant responses in patients already affected by these disorders. An association between SSRI response and 5-HTTLPR polymorphism was first reported by Smeraldi *et al.*, in 1998 [27], in major depression patients with improved response to fluvoxamine demonstrated for the l allele carriers (l/l and l/s) as opposed to homozygotes for the s variant (s/s). Stein and colleagues [28] examined whether variations in 5HTTLPR influence the efficacy of selective serotonin reuptake inhibitors (paroxetine or fluvoxamne) in 32 patients suffering from generalized social anxiety disorder (GSAD). This study reported a trend toward a linear association between 5HTTLPR genotype and likelihood of response to SSRI: l/l (88%), l/s (67%), s/s (33%), $P = 0.051$. Yu *et al.* [29] reported that depressed patients with the l/l genotype had a significantly better response to SSRIs (fluoxetine) when compared with s allele carriers, as evaluated on the basis of percentage change in total ($P = 0.013$), core ($P = 0.011$), psychic anxiety ($P = 0.005$) and somatic anxiety ($P = 0.002$) on the Hamilton depression rating scale.

To date, SSRIs are approved by the FDA for the treatment of major depressive disorder, panic disorder, post-traumatic stress disorder, generalized anxiety disorder, and obssessive–compulsive disorder; and the family members of these drugs are considered to be the drug of choice for these disorders. Thus, clinical evidence supports a fundamental role for 5-HT in both acute and chronic stress response.

17.3
Adrenergic Antagonists

There is an increasing interest in adrenergic antagonists for the treatment and prevention of maladaptive stress response. Beta blockers are prescribed for performance anxiety and are advocated at times for other anxiety disorders as is clonidine, a central acting α2 agonist. Recent work and ongoing trials suggest that such agents may also have a significant role in prevention and treatment of PTSD.

Noradrenalin (NA) is a catecholamine neurotransmitter in the central nervous system. The precursor for NA synthesis is the amino acid tyrosine. Through the rate-limiting action of tyrosine hydroxylase and following decarboxylation, tyrosine is transformed to dopamine; dopamine is then converted to NA through the action of dopamine β-hydroxylase. Noradrenergic neurons are located in the locus coeruleus from where they send projections to cortical and limbic structures including the amygdala and hippocampus. Secreted NA acts on several types of G coupled receptors: α2 pre-synaptic inhibitory auto-receptors and the post-synaptic α1, β1, 2 and 3.

The classic "fight or flight" stress response involves sympathetic activation with an increase in both CNS and peripheral NA and adrenaline. Somatic consequences of sympathetic activation include among others an increase in heart rate, tremor and sweating, all characteristic of anxiety and panic attacks. In the CNS, NA is considered a modulatory neurotransmitter with significant contribution to memory consolidation and retrieval. In healthy controls, guanfacine (a specific α2A agonist) has been shown in some but not all studies to improve memory, planning and associative learning. Augmenting NA release via clonidine improves memory recall and working memory measures also in several disease populations (e.g., Parkinson's disease, Alzheimer's disease, Korsakoff's amnesia). However, animal research suggests that excess NA may also adversely affects cognitive function probably through α1 activation. Receptor specific data also suggests that β adrenergic stimulation improves memory consolidation with β antagonists conversely impairing it (reviewed in [30]). In clinical practice the somatic manifestations are dominant in acute stress while in delayed and chronic stress the CNS manifestation take precedence. Thus beta blockers are prescribed for performance anxiety successfully reducing short term somatic debilitating symptoms.

Following acute stress, increased heart rate a measure of adrenergic activity, is a predictor of PTSD development [31], though some suggest this is not true for resting heart rate [32]. In acutely ill patients, long-term traumatic memories of hospitalization are correlated with cumulative norepinephrine administered during the acute phase and inversely correlated with beta blocker administration [33]. Thus, an exaggerated adrenergic spontaneous or administered response may contribute to consolidation of stress-related memories and through this to the pathophysiology of PTSD (reviewed in [34]). A large body of evidence indicates that patients suffering from PTSD have increased basal peripheral and CNS adrenergic activation; as well as an exaggerated adrenergic response to stressful stimuli. This has been shown through biochemical measures such as urine catecholamines and metabolites as well as by a somatic reflection of peripheral adrenergic tone, that is, heart rate; CNS adrenergic tone has been shown to be elevated via CSF norepinephrine concentrations (reviewed in [35]).

Researchers posited that an exaggerated β adrenergic response to traumatic stress may play a role in pathological "over-consolidation" of traumatic memories. Hence, it has been suggested that treatment of trauma victims with β antagonists could prevent the development of PTSD. In support of this theory, animal and human studies demonstrated that administering propranolol interferes with memory consolidation, re-consolidation and recall (reviewed in [35]). Brunet *et al.* [36] reported that administering propranolol to chronic PTSD patients while they recall the trauma; results in a reduction in physiological hyper-arousal in a subsequent traumatic imagery session. Immediately following stress exposure, propranolol was shown in a small preliminary study to reduce likelihood of physiological hyper-arousal and PTSD symptoms as measured with the clinician-administered PTSD scale (CAPS) [37]. These data provided the impetus for a flurry of ongoing clinical trials aimed at preventing PTSD through early pro-pranolol administration (e.g., trials NCT00648375, NCT00158262) or at extinction of pathological memories (e.g., NCT 00465608, NCT00611871, registered on www.clinicaltrials.gov).

Adrenergic alpha antagonists may have a role in treatment of existing PTSD symptoms [35]. Two recent randomized clinical trials examined the effect of Pra-zosin on PTSD symptoms in combat-related [38] and civilian PTSD [39]. In both studies, the central acting α1 antagonist improved sleep quality and reduced nightmares. The authors suggest that the benefit of this agent stems from an increase in REM sleep over light sleep in which nightmares in PTSD are frequent. However the true mechanism for α1 antagonist effects in PTSD is still unclear.

Clinical evidence thus points to an important role for adrenergic response in both susceptibility and maintenance of pathological response to stress.

17.4
γ-Aminobutyric Acid

γ-Aminobutyric acid (GABA) is the major inhibitory neurotransmitter in the brain, present in more than one-third of the brain synapses. GABA is synthesized from glutamate by the rate-limiting enzyme glutamic acid decarboxylase (GAD), requir-ing pyridoxine (vitamin B_6) as a cofactor. There are three GABA receptor subtypes: A, B and C. GABA-A and C receptor subtypes generate a rapid effect via ligand-gated chloride channels, while GABA-B receptors have a slower effect via G pro-teins coupled to calcium and potassium channels. In the hippocampus and other limbic regions, GABAergic projections modulate the activity of noradrenergic and serotonergic neurons an effect implicated in their anxiolytic properties.

The GABA-A receptor is also known as the benzodiazepine receptor, containing a glycoprotein binding site for the benzodiazepine group members. Benzodi-azepines (BZD) bind to the benzodiazepine site on the receptor and while they do not directly activate the channel they enhance the influx of chloride through it generating hyper-polarization of the neuron and resultant neuronal inhibition. GABA-BZD receptors are widely distributed in the thalamus, cerebellum, subcorti-cal areas (e.g., hippocampus, amygdala) as well as in the brainstem and spinal cord [40]. The amygdala, hippocampus and medial prefrontal cortex in particular are important in anxiety signaling, and hence, BZD receptors in these regions could play a major role in anxiety and stress responses. Indeed, in healthy controls, administration of lorazepam during viewing of emotional faces was found to attenuate the fMRI BOLD signal in the amygdala and insula [41]. In the amygdala GABAergic circuits modulate glutamate and corticotropin releasing factor (CRF) signaling, most likely responsible for the immediate anxiolytic effects of BZD [42].

Morgan *et al.* suggested that the intrusiveness, hyperarousal and disinhibited emotional behavior of PTSD patients are due to deficient GABAergic function [43]. Several experiments have been carried out to determine whether there is indeed a difference in GABAergic activity between anxious and non-anxious individuals. Roy-Byrne *et al.* demonstrated a reduced effect of diazepam on eye movement velocity and accuracy (a measure of sedation) in panic disorder patients [44] and several studies found reduced GABA-A benzodiazepine binding in PTSD and panic disorder [45–47]. In a prospective study [48], plasma GABA levels were measured in 108 road traffic accident victims on arrival at a traumatology department; six weeks later these patients were assessed for PTSD. GABA concentration in those who went on to develop PTSD was significantly lower compared with members of the trauma-exposed group who did not develop PTSD. Thus evidence points to altered GABA response in anxiety disorders.

In line with the above evidence for altered GABA-BZD physiology in anxiety, human and animal studies demonstrate that positive modulators of GABA receptors generally possess anxiolytic activity, while negative modulators such as flumazenil or benzodiazepine receptor inverse agonists produce anxiogenic effects [49–51]. In addition to the BZD binding site, GABA-A receptors contain binding sites for additional positive modulators of which ethanol is of special clinical interest. Maes *et al.* examined the association between alcohol consumption and risk for PTSD, reporting that alcohol consumption and intoxication prior to traumatic events diminishes the relative risk of developing PTSD in response to that trauma [52]; while the mechanism for this effect is not clear it may involve attenuation of traumatic memory consolidation. In a nationally representative sample of 43 093 United States civilians, Robison *et al.* [53] examined self-medicating behavior within four anxiety disorder categories (panic, social phobia, specific phobia, generalized anxiety disorder). Specifically, participants were asked if at any time in the past they had used alcohol or other substances for the purpose of reducing their fear, anxiety, or avoidance (of a feared object or situation). Highest rates of self-medication using alcohol alone were found among those diagnosed with generalized anxiety disorder (18.3%), followed by social phobia (16.9%) and panic disorder with agoraphobia (15.0%). These findings remained significant after adjusting for other substance abuse suggesting a specific benefit from ethanol in alleviating anxiety symptoms.

The clinical use of benzodiazepines in the treatment of anxiety disorders is rather complicated and is currently indicated on an "as needed" basis and not as core treatment. In cases of chronic stress such as GAD and PTSD, results of BZD treatment are inconsistent. In a double-blind placebo-controlled trial the efficacy of antidepressants (imipramine or trazodone) and benzodiazepines (diazepam) were compared in patients with GAD. Diazepam showed rapid and superior efficacy at week 2 but was inferior to imipramine at week 8 [54]. Similar results were found in a study comparing imipramine, paroxetine and 2′chlordesmethyldiazepam [55]. In a systematic review and meta-analysis [56] of clinical trials however, the authors failed to find convincing evidence for short-term effectiveness of benzodiazepines in the treatment of GAD but did find evidence in favor of BZD regarding their efficacy. Efficacy in this regard suggests the intervention is potentially beneficial if it were to be properly implemented under controlled conditions, whereas the lack of effectiveness implies that under actual clinical conditions there is no clinical benefit.

BZD have been used commonly for the treatment of PTSD both for ameliorating arousal symptoms and in the past for the prevention of PTSD after acute stress reactions. Experimental data however, do not support the use of BZD for these indications. Prospective studies concluded that contrary to expectations, early administration of benzodiazepines to trauma survivors with high levels of initial distress did not have a significant beneficial effect on the course of the illness [57, 58]. A review of pharmacological treatments following trauma concluded that benzodiazepines cannot be recommended for management of acute stress disorder or early PTSD because they have either no effect or a harmful effect on patients [59]. Even for the induction of sleep, a common indication in healthy individuals, in PTSD patients the use of BZD is controversial; there have been reports showing no advantage for clonazepam in sleep induction [60], though others reported moderate effectiveness for this indication [60, 61], again suggesting altered GABA transmission in this disorder.

When attempting to summarize clinical evidence regarding the role of GABA in the stress response a confusing picture arises. While BZD appear to provide some immediate relief of anxiety as doe's ethanol, these agents appear to be inferior to other agents in treating chronic stress and at times detrimental to the long-term outcome. It is reasonable to conclude that GABA transmission is important in the stress response however this may be the case more in the immediate stress reaction with a less clear role in the long term consequences of stress.

Summary

Hitherto we have discussed the clinical evidence for involvement of distinct neurotransmitter systems (i.e., serotoninergic, adrenergic, GABAergic) in the stress response. However it stands to reason that these systems do not act independently of one another but rather interact across brain regions to produce the behavioral and physiological phenotype. In clinical practice, drugs targeting the various systems are often prescribed in combination in an attempt to control various symptoms or as a temporary measure to bridge over the period until a slow-acting agent takes effect [62]. Surprisingly little research has been carried out however on the effects of combination therapy. One study carried out in older patients found a high rate of benzodiazepine use in patients also taking SSRIs or serotonin/norepinephrine reuptake inhibitors (SNRI, discussed below) [63] suggesting an additive role for these agents; though BZD dependence is a factor that could account at least in part for continued BZD use. In panic disorder, combined

SSRI-BZD continued therapy produced no long-term benefit over SSRI treatment with only a short initial course of BZD [64]. Similarly, in a naturalistic follow-up study of panic disorder patients, BZD was administered for 12 weeks in conjunction with an SSRI followed by SSRI mono-therapy; again there was no long-term benefit for the combined therapy over SSRI mono-therapy alone, though there was a benefit at 4 and 8 weeks from therapy initiation [65]. These studies suggest that the GABAergic involvement in stress response is important in the acute phase but not in the chronic one. Alternatively, it is possible that whatever the role of GABA in the stress response, this is overridden by activity of the 5-HT system.

Several additional classes of psychoactive drugs are employed as second line therapy in anxiety and depression, SNRIs (e.g., Venlafaxine, Duloxetine, Milnaciprane) inhibit reuptake of both serotonin and norepinephrine from the synaptic cleft. Most tricyclic antidepressants (TCA) also share this characteristic dual 5-HT- norepinephrine reuptake inhibition (e.g., imipramine, desipramine), though their clinical use has been in the decline due to their side effect profile. Finally, a third class of drugs – the monoamine oxidase inhibitors (MAO-I) – inhibit the metabolism of 5-HT, NA and dopamine; however this group also holds a poor side effect and toxic profile and is accordingly currently sparsely prescribed. A recent meta-analysis [66] concluded that SNRIs are significantly albeit modestly superior to SSRIs in the treatment of major depression. Venlafaxine is approved for the treatment of several anxiety disorders (e.g., GAD, panic disorder, social anxiety disorder) and has been shown to be superior to placebo in double-blind controlled trials [67]. Few studies directly compared SNRIs with SSRIs for the treatment of anxiety disorders. A recent randomized trial compared Venlafaxine, Paroxetine and placebo for the treatment of panic disorder; both treatment groups were superior to placebo, however at low and medium Venlafaxine doses (75–150 mg/day) there was no advantage over Paroxetine (40 mg/day) treatment [68]. Venlafaxine at a high dose of 225 mg/day produced greater improvements in the panic disorder severity scale and a higher proportion of patients free of full-symptom panic attacks as compared with the Paroxetine treatment [68]. The authors of this study suggest that only at this high dose the additional norepinephrine effect takes place indicating a simultaneous role for both 5-HT and norepinephrine. To date direct clinical evidence for the interaction of the serotoninergic and noradrenergic systems in the stress response is sparse, however the continued use of combined agents such as SNRIs and TCAs is likely to provide further insight in this regard. Emerging interest in pharmacogenomics is likely to further point to interactions between neurotransmitter systems. Many studies have recently investigated the interaction between genetic polymorphisms in various neurotransmitter systems and antidepressant drug response (reviewed in [69]); as these studies accumulate a better understanding should emerge of the involved interactions.

The clinical evidence we have presented obviously portrays only a small part of the accumulated knowledge regarding neurotransmitter involvement in the stress reaction. Much data supporting the clinical evidence and exploring mechanisms involved has emerged from animal and other models of stress, anxiety and depression. Nevertheless we believe the clinical perspective is an important one, keeping in mind that much of the stress research is ultimately aimed at returning to and treating human suffering.

References

1 Holmes, T.H. (1978) Life situations, emotions, and disease. *Psychosomatics*, **19** (12), 747–754.

2 Yerkes, R.M. and Dodson, J.D. (1908) The relation of strength of stimulus to rapidity of habit-formation. *J. Comp. Neurol. Psychol.*, **18**, 459–482.

3 Kessler, R.C. *et al.* (2007) Lifetime prevalence and age-of-onset distributions of mental disorders in the World Health Organization's World Mental Health Survey Initiative. *World Psychiatry*, **6** (3), 168–176.

4 Kessler, R.C. *et al.* (2005) Prevalence, severity, and comorbidity of 12-month DSM-IV disorders in the National Comorbidity Survey Replication. *Arch. Gen. Psychiatry*, **62** (6), 617–627.

5 Kessler, R.C. *et al.* (2003) The epidemiology of major depressive disorder: results from the National Comorbidity Survey Replication (NCS-R). *JAMA*, **289** (23), 3095–3105.

6 Kendler, K.S., Karkowski, L.M. and Prescott, C.A. (1999) Causal relationship between stressful life events and the onset of major depression. *Am. J. Psychiatry*, **156** (6), 837–841.

7 Hettema, J.M. (2008) What is the genetic relationship between anxiety and depression? *Am. J. Med. Genet. C Semin. Med. Genet.*, **148** (2), 140–146.

8 Mathew, S.J., Price, R.B. and Charney, D.S. (2008) Recent advances in the neurobiology of anxiety disorders: implications for novel therapeutics. *Am. J. Med. Genet. C Semin. Med. Genet.*, **148** (2), 89–98.

9 Cools, R., Roberts, A.C. and Robbins, T.W. (2008) Serotoninergic regulation of emotional and behavioural control processes. *Trends Cogn. Sci.*, **12** (1), 31–40.

10 Hjorth, S. and Auerbach, S.B. (1996) 5-HT1A autoreceptors and the mode of action of selective serotonin reuptake inhibitors (SSRI). *Behav. Brain Res.*, **73** (1–2), 281–283.

11 Stahl, S.M. (1998) Mechanism of action of serotonin selective reuptake inhibitors. Serotonin receptors and pathways mediate therapeutic effects and side effects. *J. Affect. Disord.*, **51** (3), 215–235.

12 Stahl, S. (1994) 5HT1A receptors and pharmacotherapy. Is serotonin receptor down-regulation linked to the mechanism of action of antidepressant drugs? *Psychopharmacol. Bull.*, **30** (1), 39–43.

13 Nishizawa, S. *et al.* (1997) Differences between males and females in rates of serotonin synthesis in human brain. *Proc. Natl. Acad. Sci. U.S.A.*, **94** (10), 5308–5313.

14 Nutt, D.J. *et al.* (1999) Mechanisms of action of selective serotonin reuptake inhibitors in the treatment of psychiatric disorders. *Eur. Neuropsychopharmacol.*, **9** (Suppl. 3), S81–S86.

15 Nakamura, M. *et al.* (2000) The human serotonin transporter gene linked polymorphism (5-HTTLPR) shows ten novel allelic variants. *Mol. Psychiatry*, **5** (1), 32–38.

16 Bradley, S.L. *et al.* (2005) Relationship of serotonin transporter gene polymorphisms and haplotypes to mRNA transcription. *Am. J. Med. Genet. B Neuropsychiatr. Genet.*, **136B** (1), 58–61.

17 Greenberg, B.D. *et al.* (2000) Association between the serotonin transporter promoter polymorphism and personality traits in a primarily female population sample. *Am. J. Med. Genet.*, **96** (2), 202–216.

18 Lesch, K.P. *et al.* (1996) Association of anxiety-related traits with a polymorphism in the serotonin transporter gene regulatory region. *Science*, **274** (5292), 1527–1531.

19 Gustavsson, J.P. *et al.* (1999) No association between serotonin transporter gene polymorphisms and personality traits. *Am. J. Med. Genet.*, **88** (4), 430–436.

20 Jorm, A.F. *et al.* (1998) An association study of a functional polymorphism of the serotonin transporter gene with personality and psychiatric symptoms. *Mol. Psychiatry*, **3** (5), 449–451.

21 Caspi, A. *et al.* (2003) Influence of life stress on depression: moderation by a polymorphism in the 5-HTT gene. *Science*, **301** (5631), 386–389.

22 Grabe, H.J. *et al.* (2005) Mental and physical distress is modulated by a polymorphism in the 5-HT transporter gene interacting with social stressors and chronic disease burden. *Mol. Psychiatry*, **10** (2), 220–224.

23 Manuck, S.B. *et al.* (2004) Socio-economic status covaries with central nervous system serotonergic responsivity as a function of allelic variation in the serotonin transporter gene-linked polymorphic region. *Psychoneuro-endocrinology*, **29** (5), 651–668.

24 Otte, C. *et al.* (2007) Association of a serotonin transporter polymorphism (5-HTTLPR) with depression, perceived stress, and norepinephrine in patients with coronary disease: the heart and soul study. *Am. J. Psychiatry*, **164** (9), 1379–1384.

25 Kilpatrick, D.G. *et al.* (2007) The serotonin transporter genotype and social support and moderation of posttraumatic stress disorder and depression in hurricane-exposed adults. *Am. J. Psychiatry*, **164** (11), 1693–1699.

26 Hariri, A.R. *et al.* (2002) Serotonin transporter genetic variation and the response of the human amygdala. *Science*, **297** (5580), 400–403.

27 Smeraldi, E. *et al.* (1998) Polymorphism within the promoter of the serotonin transporter gene and antidepressant efficacy of fluvoxamine. *Mol. Psychiatry*, **3** (6), 508–511.

28 Stein, M.B., Seedat, S. and Gelernter, J. (2006) Serotonin transporter gene promoter polymorphism predicts SSRI response in generalized social anxiety disorder. *Psychopharmacology (Berl.)*, **87** (1), 68–72.

29 Yu, Y.W. *et al.* (2002) Association study of the serotonin transporter promoter polymorphism and symptomatology and antidepressant response in major depressive disorders. *Mol. Psychiatry*, **7** (10), 1115–1119.

30 Ramos, B.P. and Arnsten, A.F. (2007) Adrenergic pharmacology and cognition: focus on the prefrontal cortex. *Pharmacol. Ther.*, **113** (3), 523–536.

31 Shalev, A.Y. *et al.* (1998) A prospective study of heart rate response following trauma and the subsequent development of posttraumatic stress disorder. *Arch. Gen. Psychiatry*, **55** (6), 553–559.

32 O'Donnell, M.L. *et al.* (2007) Tonic and phasic heart rate as predictors of posttraumatic stress disorder. *Psychosom. Med.*, **69** (3), 256–261.

33 Schelling, G. *et al.* (2003) Exposure to high stress in the intensive care unit may have negative effects on health-related quality-of-life outcomes after cardiac surgery. *Crit. Care Med.*, **31** (7), 1971–1980.

34 Schelling, G. (2008) Post-traumatic stress disorder in somatic disease: lessons from critically ill patients. *Prog. Brain Res.*, **167**, 229–237.

35 Strawn, J.R. and Geracioti, T.D. Jr. (2008) Noradrenergic dysfunction and the psychopharmacology of posttraumatic stress disorder. *Depress. Anxiety*, **25** (3), 260–271.

36 Brunet, A. *et al.* (2008) Effect of post-retrieval propranolol on psychophysiologic responding during subsequent script-driven traumatic imagery in post-traumatic stress disorder. *J. Psychiatr. Res.*, **42** (6), 503–506.

37 Pitman, R.K. *et al.* (2002) Pilot study of secondary prevention of posttraumatic stress disorder with propranolol. *Biol. Psychiatry*, **51** (2), 189–192.

38 Raskind, M.A. *et al.* (2007) A parallel group placebo controlled study of prazosin for trauma nightmares and sleep disturbance in combat veterans with post-traumatic stress disorder. *Biol. Psychiatry*, **61** (8), 928–934.

39 Taylor, F.B. *et al.* (2008) Prazosin effects on objective sleep measures and clinical symptoms in civilian trauma posttraumatic stress disorder: a placebo-controlled study. *Biol. Psychiatry*, **63** (6), 629–632.

40 Roy-Byrne, P.P. (2005) The GABA-benzodiazepine receptor complex: structure, function, and role in anxiety. *J. Clin. Psychiatry*, **66** (Suppl. 2), 14–20.

41 Paulus, M.P. *et al.* (2005) Dose-dependent decrease of activation in bilateral amygdala and insula by lorazepam during emotion processing. *Arch. Gen. Psychiatry*, **62** (3), 282–288.

42 Kaufmann, W.A. *et al.* (2003) Compartmentation of alpha 1 and alpha 2 GABA(A) receptor subunits within rat extended amygdala: implications for benzodiazepine action. *Brain Res.*, **964** (1), 91–99.

43 Morgan, C.A. 3rd, Krystal, J.H. and Southwick, S.M. (2003) Toward early pharmacological posttraumatic stress intervention. *Biol. Psychiatry*, **53** (9), 834–843.

44 Roy-Byrne, P. *et al.* (1996) Reduced benzodiazepine sensitivity in patients with panic disorder: comparison with patients with obsessive-compulsive disorder and normal subjects. *Am. J. Psychiatry*, **153** (11), 1444–1449.

45 Cameron, O.G. *et al.* (2007) Reduced gamma-aminobutyric acid(A)-benzodiazepine binding sites in insular cortex of individuals with panic disorder. *Arch. Gen. Psychiatry*, **64** (7), 793–800.

46 Geuze, E. *et al.* (2008) Reduced GABAA benzodiazepine receptor binding in veterans with post-traumatic stress disorder. *Mol. Psychiatry*, **13** (1), 74–83.

47 Kaschka, W., Feistel, H. and Ebert, D. (1995) Reduced benzodiazepine receptor binding in panic disorders measured by iomazenil SPECT. *J. Psychiatr. Res.*, **29** (5), 427–434.

48 Vaiva, G. *et al.* (2004) Low posttrauma GABA plasma levels as a predictive factor in the development of acute posttraumatic stress disorder. *Biol. Psychiatry*, **55** (3), 250–254.

49 Kalueff, A.V. and Nutt, D.J. (2007) Role of GABA in anxiety and depression. *Depress. Anxiety*, **24** (7), 495–517.

50 Bueno, C.H., Zangrossi, H. Jr. and Viana Mde B. (2007) GABA/benzodiazepine receptors in the ventromedial hypothalamic nucleus regulate both anxiety and panic-related defensive responses in the elevated T-maze. *Brain Res. Bull.*, **74** (1–3), 134–141.

51 Liu, G.X. *et al.* (2007) Reduced anxiety and depression-like behaviors in mice lacking GABA transporter subtype 1. *Neuropsychopharmacology*, **32** (7), 1531–1539.

52 Maes, M. *et al.* (2001) Risk and preventive factors of post-traumatic stress disorder (PTSD), alcohol consumption and intoxication prior to a traumatic event diminishes the relative risk to develop PTSD in response to that trauma. *J. Affect. Disord.*, **63** (1–3), 113–121.

53 Robinson, J. *et al.* (2008) Self-medication of anxiety disorders with alcohol and drugs: results from a nationally representative sample. *J. Anxiety Disord.*, **23**, 38–45.

54 Rickels, K. *et al.* (1993) Antidepressants for the treatment of generalized anxiety disorder. A placebo-controlled comparison of imipramine, trazodone, and diazepam. *Arch. Gen. Psychiatry*, **50** (11), 884–895.

55 Rocca, P. *et al.* (1997) Paroxetine efficacy in the treatment of generalized anxiety disorder. *Acta Psychiatr. Scand.*, **95** (5), 444–450.

56 Martin, J.L. *et al.* (2007) Benzodiazepines in generalized anxiety disorder: heterogeneity of outcomes based on a systematic review and meta-analysis of clinical trials. *J. Psychopharmacol.*, **21** (7), 774–782.

57 Braun, P. *et al.* (1990) Core symptoms of posttraumatic stress disorder unimproved by alprazolam treatment. *J. Clin. Psychiatry*, **51** (6), 236–238.

58 Gelpin, E. *et al.* (1996) Treatment of recent trauma survivors with benzo-diazepines: a prospective study. *J. Clin. Psychiatry*, **57** (9), 390–394.

59 Davidson, J.R. (2006) Pharmacologic treatment of acute and chronic stress following trauma: 2006. *J. Clin. Psychiatry*, **67** (Suppl. 2), 34–39.

60 Cates, M.E. *et al.* (2004) Clonazepam for treatment of sleep disturbances associated with combat-related posttraumatic stress disorder. *Ann. Pharmacother.*, **38** (9), 1395–1399.

61 Mellman, T.A., Byers, P.M. and Augenstein, J.S. (1998) Pilot evaluation of hypnotic medication during acute traumatic stress response. *J. Trauma. Stress*, **11** (3), 563–569.

62 Dunlop, B.W. and Davis, P.G. (2008) Combination treatment with benzo-diazepines and SSRIs for comorbid anxiety and depression: a review. *Prim. Care Companion J. Clin. Psychiatry*, **10** (3), 222–228.

63 Benitez, C.I. *et al.* (2008) Use of benzodiazepines and selective serotonin reuptake inhibitors in middle-aged and older adults with anxiety disorders: a longitudinal and prospective study. *Am. J. Geriatr. Psychiatry*, **16** (1), 5–13.

64 Pollack, M.H. *et al.* (2003) Combined paroxetine and clonazepam treatment strategies compared to paroxetine monotherapy for panic disorder. *J. Psychopharmacol.*, **17** (3), 276–282.

65 Dannon, P.N. *et al.* (2005) Clonazepam augmentation of paroxetine in the treatment of panic disorder: a one year naturalistic follow-up study. *Int. J. Ment. Health*, **2** (2).

66 Papakostas, G.I. *et al.* (2007) Are antidepressant drugs that combine serotonergic and noradrenergic mechanisms of action more effective than the selective serotonin reuptake inhibitors in treating major depressive disorder? A meta-analysis of studies of newer agents. *Biol. Psychiatry*, **62** (11), 1217–1227.

67 Thase, M.E. (2006) Treatment of anxiety disorders with venlafaxine XR. *Expert Rev. Neurother.*, **6** (3), 269–282.

68 Pollack, M. *et al.* (2007) A randomized controlled trial of venlafaxine ER and paroxetine in the treatment of outpatients with panic disorder. *Psychopharmacology (Berl.)*, **194** (2), 233–242.

69 Binder, E.B. and Holsboer, F. (2006) Pharmacogenomics and antidepressant drugs. *Ann. Med.*, **38** (2), 82–94.

18
Metabolic Components of Neuroendocrine Allostatic Responses: Implications in Lifestyle-Related Diseases

Ronan M.G. Berg and Bente Klarlund Pedersen

18.1
Introduction

Chronic non-communicable diseases (CNCDs) which include cardiovascular disease (CVD), some cancers, chronic respiratory conditions and type 2 diabetes (T2D), affect people of all ages, nationalities and classes and are reaching epidemic proportions worldwide [1]. CNCDs cause the greatest global share of death and disability, accounting for around 60% of all deaths worldwide. Approximately 80% of chronic-disease deaths occur in low- and middle-income countries and account for 44% of premature deaths worldwide. It is estimated that the number of deaths from these diseases is double the number of deaths that result from a combination of infectious diseases (including HIV/AIDS, tuberculosis and malaria), maternal and perinatal conditions and nutritional deficiencies [1].

CVD and T2D are lifestyle-related diseases caused by obesity and a physical inactive lifestyle in genetically predisposed individuals. Psychosocial factors, such as stress and job control, are widely held to be important determinants of physical health [2]. Without doubt, there is a strong association between psychological stress and psychosocial factors on one hand and chronic diseases such as T2D and CVD on the other [2, 3].

The foremost risk factors for CVD and T2D are encompassed in the so-called metabolic syndrome [4], which comprises glucose intolerance, visceral obesity, dyslipidemia and hypertension. In addition, both the occurrence and the impact of the metabolic syndrome are related to various psychosocial factors, that is, psychological and/or emotional stress [5]; thus, indices of psychological stress in modern Western societies have increased in concert with the increasing incidence of lifestyle-related diseases.

Life exists through maintenance of homeostasis, which is constantly challenged by intrinsic and extrinsic stressors [6–8]. An intricate repertoire of physiological and behavioural responses is mobilized in stressful situations forming the adaptive stress response that aims to re-establish the challenged body equilibrium

Stress – From Molecules to Behavior. Edited by Hermona Soreq, Alon Friedman, and Daniela Kaufer
Copyright © 2010 WILEY-VCH Verlag GmbH & Co. KGaA, Weinheim
ISBN: 978-3-527-32374-6

[6, 7]. This ability to preserve stability through change, termed *allostasis* [6, 8], is manifested as part of a generalized effort to utilize every available energy resource against the perceived stressor. This catabolic shift reverses upon retraction of the threat and soon after the previous metabolic state is re-established. A variety of allostatic responses exist; nonetheless, they all involve the neuroendocrine systems that encompass the hypothalamic–pituitary–adrenocortical (HPA) axis and the sympatho-adrenomedullary (SAM) unit, in which the main effectors are glucocorticoids (GCs) and catecholamines, respectively [6, 7].

As allostatic responses are mainly designed to manage "fight or flight" reactions they may be rendered inappropriate in the context of chronic and/or psychological stress, both of which are endorsed by the profound changes in lifestyle that modern society has brought with it. Hence, it is possible that components of the allostatic systems contribute to the metabolic derangements that signify lifestyle-related diseases. This notion is supported by a number of animal and human studies that demonstrate subtle alterations of HPA axis activity in obesity, the metabolic syndrome, T2D and CVD [9]. In addition, the same patient groups show indices of altered SAM unit activity [10] and increased levels of circulating catecholamines [11, 12].

The scope of the present chapter is to review the effects of GCs and catecholamines on fat and glucose metabolism in the context of physiological stress responses, in order to elucidate the implications of neuroendocrine function in the pathophysiology and pathogenesis of the metabolic syndrome and the associated lifestyle-related diseases.

18.2
Neuroendocrine Allostatic Responses and Fat Metabolism

18.2.1
Background

The impact of GCs on fat metabolism is imminently evident in conditions with GC excess such as Cushing's syndrome and adrenal adenoma, which cause obesity, specifically with fat accumulation in visceral regions [4, 9]. In general, visceral obesity is frequently accompanied by insulin resistance and dyslipidemia [13] and is strongly associated with T2D and an adverse outcome [4]. From an evolutionary point of view, visceral fat accumulation may be beneficial as energy substrates and gluconeogenetic precursors are readily available for the liver, hence triacylglycerols (TGCs) are hydrolyzed and released as free fatty acids (FFAs), which are consumed in the liver in the context of an allostatic response. Furthermore FFAs provide energy for skeletal muscle in the "fight or flight" reaction. This presupposes that the allostatic response is accompanied by exercise, which is usually not the case in emotional stress within the framework of a sedentary lifestyle.

Obesity is well recognized as a condition associated with increased lipolysis [14, 15]. On thermodynamic grounds, the increased lipolysis of obesity appears paradoxical. It implies an excess mobilization of fat which should mitigate against obesity, because even if FFA are re-esterified, such ineffectual cycling back to TGC should consume energy and favor weight loss. However, the energy cost of TGC-FFA cycling is low, and rather than being consumed by the liver and skeletal muscle, the lipolytic products contribute to dyslipidemia with high levels of very-low-density lipoprotein (VLDL), low-density lipoprotein (LDL) and TGC, along with low levels of high-density lipoprotein (HDL), which entails a substantial cardiovascular risk [4, 13].

The present section reviews the effects of GCs and catecholamines on fat metabolism and their implications in obesity and dyslipidemia.

18.2.2
Effects of GCs on Fat Metabolism and Distribution

Exogenous GCs stimulates fat depletion over their entire dose range by inducing lipolysis in adipose tissue. This involves the hormone-sensitive lipase (HSL) which is the main lipolytic enzyme responsible for the hydrolysis of TGC into glycerol and non-esterified FFAs [16]. These are released to the blood stream and taken up by the liver and oxidized, esterified, converted into ketone bodies or utilized for energy in skeletal muscle. Early *in vivo* studies showed that both endogenous and exogenous GC excess inhibits lipolysis, whereas others found no effect. However, these contradicting findings are likely explained by compensatory hyperinsulinemia, as insulin antagonizes the effects of GCs on HSL [16].

The circulating levels of FFAs rise when the lipolytic rate exceeds the demand for FFAs as fuels. So, increases in circulating FFAs have consistently been reported in both humans and animals during GC treatment [17]. FFAs have important pathophysiological implications as they act as feed-forward activators for phosphatidate phosphohydrolase [17], a regulatory enzyme in that it particularly enables the liver to synthesize TGCs. GCs further contribute to this by increasing the synthesis of phosphodiate phosphohydrolase [17], which may cause a fatty liver in some pathological states. The alternative to TGC accumulation in the liver is to secrete TGCs in the form of VLDL, a process that is also stimulated by GCs [18].

The hydrolysis of TGCs in VLDL by peripheral tissues is controlled by lipoprotein lipase (LPL), which is located on the capillary endothelium [19]. Its activity in adipose tissue is increased mainly by insulin [19]. The presence of GCs further enhances the insulin-dependent activity increase. GCs also elicit independent responses [20], conceivably by increasing LPL expression.

LDL, which is formed after the degradation of VLDL by LPL, is removed from the circulation by the liver via receptor-mediated endocytosis [18]. This step in lipid metabolism is regulated by the HPA axis, as GCs attenuate the binding and degradation of LDL by rat hepatocytes [18] and further decrease the expression of LDL receptors [18].

From the above it is clear that GCs cause dyslipidemia and fatty liver by simultaneously increasing circulating FFAs, VLDL and LDL levels as well as hepatic GC synthesis through effects on numerous steps in lipid metabolism.

Apart from affecting lipolysis, GCs have detrimental effects on body fat distribution. When present in excess, GCs induce fat accumulation in visceral regions [21]. Experimental data unequivocally support this notion; GC treatment stimulate the accumulation of both body fat and visceral fat and when measured relative to body weight, visceral fat contents were found to be approximately 50% greater in GC-treated mice than in controls [22]. These GC effects probably act in synergy with insulin as GCs synergizes with insulin to promote pre-adipocyte differentiation and increase lipid accumulation in visceral fat depots [23]. The reason for the topical predilection of GC-induced fat accumulation in visceral regions may be that adipocyte GC receptor density depends on the specific fat depot [24] and that adipocyte depots from different regions differ in hormonal sensivity [25]. Accordingly, the highest density of GC receptors is found in internal fat regions [24]. This may contribute to increases in abdominal fat mass in relation to stress-related weight gain.

18.2.3
Effects of Catecholaminergic Signaling on Lipolysis and Dyslipidemia

Catecholamines influence several aspects of lipid metabolism and are the major hormones to stimulate lipolysis during fasting, mainly through activation of β-adrenergic receptors [26]. β-Adrenergic stimulation enhances HSL activity, resulting in FFA release, and inhibits triglyceride secretion from the liver [26], effects that are similar to those of GCs. However, catecholaminergic signaling impairs LPL activity [27], thus potentially promoting dyslipidemia by specifically increasing VLDL levels.

The implications of catecholamines in dyslipidemia are supported by a number of experimental and clinical studies. Administration of adrenaline in doses that yield plasma levels similar to those seen during emotional stress results in significant increases of plasma cholesterol [28]. Plasma adrenaline correlates with non-HDL cholesterol in untreated hypertensive men [29], and treatment with α-adrenergic antagonists amends whereas treatment with β-adrenergic antagonists aggravates dyslipidemia in humans [29, 30]. In addition, β-adrenergic agonists have been shown to increase HDL levels [31].

Together these data indicate that the SAM unit is involved in the pathogenesis of dyslipidemia through β-adrenergic effects on HSL and α-adrenergic effects on LPL. Of the two, the latter appears to be essential in pathological states, when considering the beneficial effects of α-adrenergic antagonists in patients with dyslipidemia. In theory, this should impair the ability to utilize FFAs as energy substrates; curiously, such a defect has been described in skeletal muscle of patients with T2D [32].

18.3
Neuroendocrine Allostatic Responses and Glucose Metabolism

18.3.1
Background

As indicated by the term "glucocorticoid", GCs have conspicuous effects on glucose homeostasis. GCs mainly exert their effects on glucose metabolism by increasing hepatic glycogen synthesis and gluconeogenesis and rapidly inducing peripheral insulin resistance [33, 34]. Catecholamines facilitate and utilize these effects; the liver is richly innervated by the splanchnic sympathetic nerves, and activation of these nerves enhances glycogenolysis through the actions of catecholamines [35]. Hence, in the context of a stress response, this adds to the maintenance of normal concentrations of glucose in blood in order to conserve fuels for neuronal tissues and exercising skeletal muscle [36]. An increases in liver glycogen by GCs has preparative actions, as it ensures available fuel for forthcoming stress responses, which is then readily mobilized, for example, by catecholamines as well as other humoral mediators.

Disturbances in glucose homeostasis are hallmarks in the pathophysiology of the metabolic syndrome and T2D [4, 37, 38]. These include peripheral insulin resistance, hepatic glycogen depletion [37] and increased endogenous glucose production [38], of which the latter is considered the major cause of fasting hyperglycemia in patients with T2D [39, 40]. This mainly involves augmented gluconeogenesis, as demonstrated in both clinical and experimental studies [39, 40].

There are three rate-controlling enzymes in the process of hepatic gluconeogenesis [41] , all of which are implicated in obesity and T2D: (i) phosphoenolpyruvate carboxykinase (PEPCK); (ii) fructose 1,6-bisphosphatase (Fru-1,6-P2ase); and (iii) glucose-6-phosphatase (Glu-6-Pase). Studies show that mice overexpressing the PEPCK gene develop a diabetic phenotype [42]. With regards to Fru-1,6-P2ase, experimental studies show that obese and diabetic mice display an impaired regulation of this enzyme [43], and Fru-1,6-P2ase inhibition improves glycaemic control in obese and diabetic rats [44]. Likewise, Glu-6-Pase activity is increased in patients with T2D [40].

Considering the aforementioned findings of increased HPA axis and SAM unit activity in the metabolic syndrome, obesity and T2D, it is an intriguing hypothesis that GC and catecholaminergic signaling contribute to the abnormalities in glucose homeostasis during these pathological states. This section reviews the effects of GC and catecholaminergic signaling on glucose metabolism.

18.3.2
The Effects of GCs on Gluconeogenesis and Glycogen Synthesis

GCs increase the expression of several key enzymes in gluconeogenesis as well as the availability of amino acids essential for the process [41]. Thus, the gluconeogenetic pathway results in the synthesis of glucose from other non-hexose organic molecules, such as pyruvate, lactate, glycerol and amino acids [41].

PEPCK is the rate-limiting enzyme in gluconeogenesis. The major determinant of its activity is expression of the gene coding for its synthesis [41]. GCs increase PEPCK by stimulating transcription of the gene and by stabilizing PEPCK mRNA [45]. In this regard, it is noteworthy that GCs are thought to increase the responsiveness of the gluconeogenetic pathway to cyclic AMP (cAMP) [46], which may exert a permissive function to the effects of various other metabolic hormones on PEPCK activity, including those of catecholamines, glucagon and insulin.

The activity of Fru-1,6-P_2ase is modulated by various metabolic hormones and dietary status [47]. This enzyme produces fructose-1,6-biphosphate (Fru-l,6-P_2) for the generation of glucose in the gluconeogenetic pathway, and the glycolytic enzyme 6-phosphofructokinase antagonises these effects [47]. These two opposing enzymes are regulated by the intracellular levels of fructose-2,6-bisphosphate (Fru-2,6-P_2), which inhibit Fru-1,6-P_2ase and enhance 6-phosphofructokinase, thus counteracting gluconeogenesis [47]. Fru-2,6-P_2 levels are regulated by a unique bifunctional enzyme, the 6-phosphofructo-2-kinase/fructose-2,6-bisphosphatase (6-PF-2-K/Fru-2,6-P_2) complex [47]. Phosphorylation of the 6-PF-2-K/Fru-2,6-P_2 complex by cAMP-dependent protein kinase decreases Fru-2,6-P_2 levels and thus facilitates gluconeogenesis, whereas dephosphorylation results in opposite changes [47].

Although GCs and insulin are usually metabolic antagononists, they both increase Fru-1,6-P_2ase levels through effects on the expression of the 6-PF-2-K/Fru-2,6-P_2 complex. The amount of 6-PF-2-K/Fru-2,6-P_2ase and its cognate mRNA are reduced in the liver of adrenalectomized rats [47], and administration of exogenous GCs to adrenalectomized rats increases mRNA levels of the bifunctional enzyme; and a stimulatory effect of GCs has also been observed *in vitro* [47]. Hence, rather than directly affecting the activity of the bifunctional enzyme, GCs increase the levels of the bifunctional enzyme along with insulin, rendering the gluconeogenetic pathway more susceptible to regulation by other metabolic hormones through cAMP-signaling.

GCs increase hepatic Glu-6-Pase activity levels in adrenalectomised rats, although Glu-6-Pase mRNA levels are unaffected [48]. In primary hepatocyte cultures the stimulatory effect is seen only when GCs and cAMP are added together [48]; thus, the effects of GCs on Glu-6-Pase appear to involve direct effects on enzyme activity, possibly through cAMP-dependent pathways.

Apart from effects on the rate-controlling gluconeogenetic enzymes, GCs have a profound impact on skeletal muscle proteolysis. Hence, alanine and glutamine are released from skeletal muscle and serve as gluconeogenetic precursors in the liver during systemic stress [49]. Several studies have thus demonstrated physiologic as well as supraphysiological elevations of circulating GCs increases hepatic gluconeogenesis *in vivo* through enhanced substrate delivery [50, 51]. Moreover, as increases in lipolysis are known to enhance gluconeogenesis through similar mechanisms, the aforementioned effects of GCs on fat metabolism may further promote hepatic glucose output.

Stimulation of liver glycogen synthesis is yet another facet of the metabolic responses to stress initiated by GCs. Coherent with this notion, chronic elevation of circulating GCs into the stress range in dogs causes profound increases in hepatic glycogenesis [50]. Deficiency in liver glycogen deposition following glucose administration in adrenalectomized rats has long been appreciated [52]. GCs activate glycogen synthase by dephosphorylation through GC-induced proteins and increases in hepatic glycogen synthase transcription [53]. Hence, these effects of GCs facilitate, and may in fact depend on, the actions of insulin signaling on glycogenesis. Rather than being a component in the immediate allostatic response, the glycogenetic effects of GCs are preparative, in the sense that they enable the organism to store fuel available for maintaining homeostasis in the context of future stressors.

18.3.3
Effects of Catecholaminergic Signaling on Hepatic Glucose Metabolism

SAM unit activation causes rapid increases in the hepatic glycogen phosphorylase and Glu-6-Pase, with a concomitant decrease in liver glycogen content [35]. Although circulating noradrenaline levels increase markedly in response to stress, circulating noradrenaline contributes less to the rapid increases in glycogenolysis than does synaptic noradrenaline [54]. Hence, synaptic noradrenaline enhances net hepatic gluconeogenetic efficiency and the net hepatic fractional extraction of lactate by the liver [54, 55]. Furthermore, combined increases in gluconeogenetic precursors are released from peripheral tissues and contribute to this increase [55]. Under these circumstances, the actions of the hepatic sympathetic nerves are modulated by hormones, including glucagon, insulin and adrenaline. Among these factors, adrenaline is by far the strongest effector [35] and appears to be the principal insulin antagonist on hepatic glucose production [56]. Consequently, circulating adrenaline is known to play an important role in augmenting hepatic glucose production through glycogenolysis during acute stress, mainly through increases in hepatic glycogenolysis [57]. This may contribute to increased hepatic glucose production and glycogen depletion in T2D [37].

18.3.4
Effects of GCs on Peripheral Insulin Sensitivity

The first hints of GC-induced insulin resistance in humans came to light in 1966 when Perly and Kipnis demonstrated that GC treatment causes a mild increase in plasma glucose along with a considerable increase in insulin [58]. This was preceded by experimental studies that had shown similar effects in rodents [58, 59]; animal studies subsequently showed that adrenalectomy produces a decrease in plasma insulin [60]. It was, however, unclear whether the GC-induced rise in plasma insulin was due to direct a GC effect on insulin secretion or if it was secondary to the metabolic actions of GCs. It turned out that GCs in fact have adverse effects on pancreatic β-cells and attenuate insulin release [61]. An igniting amount of studies has subsequently demonstrated that GCs impair insulin-mediated glucose uptake, oxidation and storage in extrahepatic tissues [34, 62, 63]. Thus, hyperinsulinemia after GC treatment in undoubtedly a secondary effect.

The mechanisms by which GCs induce peripheral insulin resistance include translocation of glucose transporters, namely the glucose transporter 4 (GLUT4), from the cell membrane [34, 64], which is the rate-limiting step in skeletal muscle glucose utilization. Several additional mechanisms contribute through GLUT4 independent mechanisms. In a recent study GC treatment resulted in 310% greater intramyocellular lipid levels compared to controls in a rodent model of obesity [22]; intramyocellular lipid accumulation has been associated with insulin resistance in both humans and rats [22]. Furthermore, GC-induced increases in FFA concentrations may inhibit glucose oxidation and cause hepatic insulin resistance by preventing binding of insulin to its receptor, hence aggravating hyperinsulinemia and hyperglycemia [13].

18.3.5
Effects of Catecholaminergic Signaling on Peripheral Glucose Transport

Skeletal muscle is considered the principal site of insulin-stimulated glucose uptake and a major tissue for blood glucose utilization. However, skeletal muscle glucose uptake is not entirely insulin-dependent; activation of the sympathetic nervous system through central neurochemical or electrical stimulation causes increased glucose uptake in skeletal muscle without an increase in the insulin concentration in plasma [65]. Furthermore, β-adrenergic receptor agonists increase glucose transport into myocytes independent of insulin, without affecting the GLUT4 content in the plasma membrane [66], in this respect, noradrenaline rather than adrenaline seems to be of relevance, as increases in skeletal muscle glucose uptake in response to sympathetic activation are prevented by pretreatment with a noradrenergic antagonist, but not by adrenal demedullation [65].

Adrenaline does affect glucose uptake, but rather through insulin-dependent pathways, as it causes translocation of GLUT4 to the myocyte cell surface [67, 68]. Intriguingly, it simultaneously *lowers* glucose transport across the plasma membrane [67], possibly by attenuating the activity of the surface GLUT4 transporters. Accordingly, low-rate infusions of adrenalin cause glucose intolerance in humans, even at physiologic concentrations [69].

From the above it is clear that the effects of the SAM unit on peripheral glucose transport are dichotomous; glucose uptake in skeletal muscle is enhanced through insulin-independent signaling, whereas insulin-dependent glucose uptake is ameliorated. The opposing actions of adrenalin and noradrenalin are probably mediated through α- and β-adrenergic pathways, respectively, as α-adrenergic antagonists ameliorate, whereas β-adrenergic antagonists aggravate insulin resistance and glucose tolerance in hypertensive subjects [70, 71]. In addition, long-term treatment with β-adrenergic receptor agonists in obese rats increases glucose uptake in skeletal muscle and improves insulin responsiveness and glucose tolerance [72].

18.4
Discussion

From the present review it is evident that neuroendocrine allostatic responses have profound effects on glucose and fat metabolism; hence the HPA axis induces glucose intolerance with enhanced gluconeogenesis and peripheral insulin resistance and contributes to visceral obesity and dyslipidemia through GC effects. Some components of SAM unit function counteract these effects, whereas others aggravate them in synergy with GCs; specifically catecholamines increase lipolysis and contribute to dyslipidemia, and they may further cause hyperglycemia through enhanced hepatic glycogenolysis. These metabolic aberrations are hallmarks of the lifestyle-related diseases, specifically the metabolic syndrome, obesity and T2D, which have been demonstrated to exhibit enhanced HPA axis and SAM unit function in both human and animal studies. Thus, GCs and catecholamines have profound pathogenetic and pathophysiological implications, and this may well explain the clustering between psychological stress and lifestyle-related diseases and add to the impact of CNCDs on all-cause mortality.

In support of GCs having a pathogenetic role in lifestyle-related diseases, both steroidal and non-steroidal GC receptor antagonists have been reported to improve hyperglycemia and dyslipidemia in rodent models of the metabolic syndrome and T2D [73, 74]. Furthermore, inhibition of 11 β-hydroxysteroid dehydrogenase, the enzyme that converts inactive endogenous GCs into active GCs in cells, has a therapeutic effect in mouse models of the metabolic syndrome and T2D [75]. Hence, it lowers body weight and improves insulin resistance, hepatic glucose output and dyslipidemia [75], and it furthermore slows plaque progression in experimental atherosclerosis [75]. Similarly, catecholamine antagonists influence the metabolic derangements that accompany obesity, T2D and CVD, specifically reflected in the copious human studies that demonstrate the effects of α- and β-adrenergic antagonists on dyslipidemia and hyperglycemia, as reviewed above.

From a historical perspective, inflammation is considered the natural host response to tissue injury, namely an infectious episode; through the past decades it has however become increasingly clear that inflammation is a key player in the pathogenesis of a number of CNCDs. Thus, obesity, physical inactivity, T2D and CVD are now widely recognized as states of so-called chronic low-grade inflammation, the latter being defined by two- to threefold increases in circulating proinflammatory cytokines [76, 77]. Accordingly, there is an increasing appreciation of the role of inflammation both in the pathogenesis of atherosclerosis [77, 78] and as a key factor in insulin resistance [76].

The circulating levels of proinflammatory cytokines are predictive of risk of disease and all-cause mortality in various populations [79, 80]. The origin of these cytokines is disputed; they may derive from the immune system *per se*, as well as from skeletal muscle, adipose tissue, or from other tissues [81, 82]. Low-grade inflammation has imperative allusions to lifestyle-related diseases, as proinflammatory cytokines have considerable effects on various metabolic parameters. Hence, tumor necrosis factor (TNF)-α induces insulin resistance *in vitro* and *in vivo* both in animals and humans [81, 83], and both TNF-α and interleukin (IL)-6 enhance lipolysis in humans [84, 85]. Moreover, cytokine expression in adipose tissue and skeletal muscle is related to obesity and glucose intolerance in human subjects [82].

Intriguingly, inflammatory mediators exert direct effects on the neuroendocrine allostatic systems [7, 86]. Hence, IL-6 is a potent activator of the HPA axis both in humans and in animals, and similar effects have been reported for TNF-α and other cytokines such as IL-1β [86]; the effects of cytokines on the HPA axis act both independently and in synergy [86]. Furthermore, inflammatory reactions provoke increases in circulating catecholamines, e.g. through IL-6 [84]. This opens up to complex feedback circuits, due to immune-modulatory effects of the neuro-endocrine allostatic systems. The immunosuppressive and anti-inflammatory actions of GCs have been recognized for decades [7] as this leads back to the initial observations by Selye in 1936 [87]. The most general effect of GCs is to inhibit synthesis and release of cytokines and other mediators that promote inflammatory reactions, both in cell culture systems and in whole organisms, and to suppress activation and trafficking of leukocytes and immune accessory cells [7]. These effects depend on alterations of the transcription rates of GC-responsive genes or changes in the stability of messenger RNA of several inflammatory proteins [88]. It is generally believed that the homeostatic value of GC suppression of inflammatory responses is to prevent excessive immune activation and the potentially damaging effects of cytokines [7].

The effects of catecholamines on proinflammatory cytokines are equivocal; adrenaline increases IL-6 production and release in adipose tissue [89] and skeletal muscle [90]. In contrast, β-adrenergic pathways inhibit TNF-α production and release upon adrenaline stimulation, whereas data regarding the effects of noradrenaline are less clear [91].

Neuroendocrine–immune interactions are highly pertinent when considering the pathophysiology of psychological stress. Animal studies have revealed that inflammatory responses with increases in temperature [92] and circulating cytokines [92] are triggered by psychological stress, e.g. placement of rats in open-field settings or conditioned aversion stress. Hence, numerous studies have consistently demonstrated increases in circulating IL-6 levels [92–94], a response that occurs even before there is a rise in GC concentrations [95–97]. Accordingly, a more pronounced IL-6 response has been demonstrated in subjects of low socio-economic status in response to psychological stress [98].

The fever response, leukocytosis and increases in IL-6 after psychological stress are all attenuated by treatment with β-adrenergic antagonists [92–94, 99]. Thus, the inflammatory response may be regarded as both a ubiquitous stressor and an allostatic response that mediates bidirectional signals between the HPA axis and the SAM unit. Given that the lifestyle-related diseases are characterized by chronic low-grade inflammation and their impact and occurrence are affected by psycho-social factors, it is conceivable that neuroendocrine–immune interactions comprise a missing link that explains the peculiar association between psychological stress and lifestyle-related diseases.

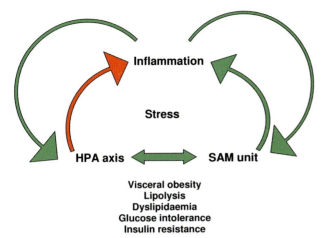

Figure 18.1 An allostatic ouroboros (consisting of inflammation, HPA axis and SAM unit activity) may form a pathogenetic core in lifestyle-related diseases. Chronic low-grade inflammation enhances lipolysis and induces insulin resistance while continually activating the HPA axis and the SAM unit. The HPA axis exacerbates the dysmetabolic state by causing visceral obesity and promoting dyslipidemia, insulin resistance and gluconeogenesis, while inefficiently counteracting inflammation. The SAM unit contributes to dyslipidemia and glucose intolerance, while promoting inflammation. It can thus be hypothesized that, when extrinsic or intrinsic stressors affect any one of the three allostatic systems, they push the other two toward an aggravated dysmetabolic state, hence endorsing allostatic load and disease.

The interrelations between chronic low-grade inflammation, HPA axis and SAM unit activity may be regarded as an allostatic ourobouros that constitutes a pathogenetic core in lifestyle-related diseases (see Figure 18.1). Thus, low-grade inflammation continually activates the HPA axis and the SAM unit, of which the HPA unit aggravates the dysmetabolic state with visceral obesity, dyslipidemia, insulin resistance and gluconeogenesis, while inefficiently counteracting inflammation. Also, the SAM unit may contribute to dyslipidemia and glycogen deficiency, while promoting inflammation. In addition, the HPA axis and SAM unit directly activate each other [6]. If any of the three pathophysiological pillars are promoted through extrinsic or intrinsic stressors, allostatic mechanisms push the other two toward an aggravated dysmetabolic state; accordingly, infections are known to severely worsen the clinical course and glycemic control in T2D and may in fact cause the metabolic syndrome to progress into T2D. Likewise, periods of psychological and/or emotional stress are known to aggravate the clinical and metabolic state in these patients [17, 100]. The clinical entities that accompany low-grade inflammation may thus be regarded as consequences of so-called allostatic load [8], this being defined as the wear and tear on the body that results from chronic imbalance of the allostatic systems.

Summary

The foremost risk factors for lifestyle-related diseases are encompassed in the so-called metabolic syndrome, which may be related to neuroendocrine allostatic responses, thus involving the hypothalamic–pituitary–adrenocortical (HPA) axis and the sympatho-adrenomedullary (SAM) unit. Accordingly, a variety of lifestyle-related diseases exhibit enhanced HPA axis and SAM unit function. The HPA axis induces glucose intolerance with enhanced gluconeogenesis and peripheral insulin resistance and contributes to visceral obesity and dyslipidaemia. The SAM unit increases lipolysis and contributes to dyslipidaemia and may further cause hyperglycaemia through enhanced hepatic glycogenolysis. Intriguingly, inflammatory mediators, which are notably involved in the metabolic derangements that accompany lifestyle-related diseases through so-called low-grade inflammation, exert direct effects on the neuroendocrine allostatic systems. This opens up to complex feedback circuits, due to immune-modulatory effects of the neuroendocrine allostatic systems. In this context, inflammation may be regarded as both a ubiquitous stressor and an allostatic response that mediates bidirectional signals between the neuroendocrine allostatic responses. Thus, imbalances between the neuroendocrine allostatic systems and inflammation may contribute to lifestyle-related diseases by facilitating an aggravated dysmetabolic state.

References

1 Daar, A.S. *et al.* (2007) Grand challenges in chronic non-communicable diseases. *Nature*, **450**, 494–496.

2 Hemingway, H. and Marmot, M. (1999) Evidence based cardiology: psychosocial factors in the aetiology and prognosis of coronary heart disease. Systematic review of prospective cohort studies. *BMJ*, **318**, 1460–1467.

3 Brotman, D.J., Golden, S.H. and Wittstein, I.S. (2007) The cardiovascular toll of stress. *Lancet*, **370**, 1089–1100.

4 Alberti, K.G., Zimmet, P. and Shaw, J. (2005) The metabolic syndrome – a new worldwide definition. *Lancet*, **366**, 1059–1062.

5 Stewart-Knox, B.J. (2005) Psychological underpinnings of metabolic syndrome. *Proc. Nutr. Soc.*, **64**, 363–369.

6 Chrousos, G.P. and Gold, P.W. (1992) The concepts of stress and stress system disorders. Overview of physical and behavioral homeostasis. *JAMA*, **267**, 1244–1252.

7 Sapolsky, R., Romero, L.M. and Munck, A.U. (2000) How do glucocorticoids influence stress responses? Integrating permissive, suppressive, stimulatory, and preparative actions. *Endocr. Rev.*, **21**, 55–89.

8 McEwen, B. (1998) Protective and damaging effects of stress mediators. *N. Engl. J. Med.*, **338**, 171–179.

9 Chrousos, G.P. (2000) The role of stress and the hypothalamic-pituitary-adrenal axis in the pathogenesis of the metabolic syndrome: neuro-endocrine and target tissue-related causes. *Int. J. Obes. Relat. Metab. Disord.*, **24** (Suppl. 2), S50–S55.

10 Peterson, H.R. *et al.* (1988) Body fat and the activity of the autonomic nervous system. *N. Engl. J. Med.*, **318**, 1077–1083.

11 Reaven, G., Lithell, H. and Landsberg, L. (1996) Hypertension and associated metabolic abnormalities – the role of insulin resistance and the sympatho-adrenal system. *N. Engl. J. Med.*, **334**, 374–381.

12 Brunner, E.J. *et al.* (2002) Adrenocortical, autonomic, and inflammatory causes of the metabolic syndrome: nested case-control study. *Circulation*, **106**, 2659–2665.

13 Björntorp, P. (1990) Classification of obese patients and complications related to the distribution of surplus fat. *Nutrition*, **6**, 131–137.

14 Jensen, M.D., Haymond, M.W., Rizza, R., Cryer, P.E. and Miles, J.M. (1989) Influence of body fat distribution on free fatty acid metabolism in obesity. *J. Clin. Invest.*, **83**, 1168–1173.

15 Martin, M.L. and Jensen, M.D. (1991) Effects of body fat distribution on regional lipolysis in obesity. *J. Clin. Invest.*, **88**, 609–613.

16 Divertie, G.D., Jensen, M.D. and Miles, J.M. (1991) Stimulation of lipolysis in humans by physiological hypercortisolemia. *Diabetes*, **40**, 1228–1232.

17 Brindley, D.N. and Rolland, Y. (1989) Possible connections between stress, diabetes, obesity, hypertension and altered lipoprotein metabolism that may result in atherosclerosis. *Clin. Sci. (Lond.)*, **77**, 453–461.

18 Brindley, D.N. and Salter, A.M. (1991) Hormonal regulation of the hepatic low density lipoprotein receptor and the catabolism of low density lipoproteins: relationship with the secretion of very low density lipoproteins. *Prog. Lipid Res.*, **30**, 349–360.

19 Ong, J.M., Simsolo, R.B., Saffari, B. and Kern, P.A. (1992) The regulation of lipoprotein lipase gene expression by dexamethasone in isolated rat adipocytes. *Endocrinology*, **130**, 2310–2316.

20 Ashby, P. and Robinson, D.S. (1980) Effects of insulin, glucocorticoids and adrenaline on the activity of rat adipose-tissue lipoprotein lipids. *Biochem. J.*, **188**, 185–192.

21 Ross, E.J. and Linch, D.C. (1982) Cushing's syndrome–killing disease: discriminatory value of signs and symptoms aiding early diagnosis. *Lancet*, 2, 646–649.

22 Gounarides, J.S. *et al.* (2007) Effect of dexamethasone on glucose tolerance and fat metabolism in a diet-induced obesity mouse model. *Endocrinology*, **149**, 758–766.

23 Björntorp, P. (1991) Adipose tissue distribution and function. *Int. J. Obes.*, **15** (Suppl. 2), 67–81.

24 Rebuffe-Scrivé, M. *et al.* (1990) Steroid hormone receptors in human adipose tissues. *J. Clin. Endocrinol. Metab.*, **71**, 1215–1219.

25 Gregoire, F., Genart, C., Hauser, N. and Remacle, C. (1991) Glucocorticoids induce a drastic inhibition of proliferation and stimulate differentiation of adult rat fat cell precursors. *Exp. Cell Res.*, **196**, 270–278.

26 Dall'Aglio, E., Chang, H. and Reaven, G. (1983) Disparate effects of prazosin and propranolol on lipid metabolism in a rat model. *Metabolism*, **32**, 510–513.

27 Ong, J.M., Saffari, B., Simsolo, R.B. and Kern, P.A. (1992) Epinephrine inhibits lipoprotein lipase gene expression in rat adipocytes through multiple steps in posttranscriptional processing. *Mol. Endocrinol.*, **6**, 61–69.

28 Dimsdale, J.E., Herd, J.A. and Hartley, L.H. (1983) Epinephrine mediated increases in plasma cholesterol. *Psychosom. Med.*, **45**, 227–232.

29 Kjeldsen, S.E. *et al.* (1983) Increased beta-thromboglobulin in essential hypertension: interactions between arterial plasma adrenaline, platelet function and blood lipids. *Acta Med. Scand.*, **213**, 369–373.

30 Leren, P. *et al.* (1980) Effect of propranolol and prazosin on blood lipids. The Oslo study. *Lancet*, **2**, 4–6.

31 Hooper, P.L., Woo, W., Visconti, L. and Pathak, D.R. (1981) Terbutaline raises high-density-lipoprotein-cholesterol levels. *N. Engl. J. Med.*, **305**, 1455–1457.

32 Kelley, D.E. and Simoneau, J.A. (1994) Impaired free fatty acid utilization by skeletal muscle in non-insulin-dependent diabetes mellitus. *J. Clin. Invest.*, **94**, 2349–2356.

33 Caro, J.F. and Amatruda, J.M. (1982) Glucocorticoid-induced insulin resistance: the importance of post-binding events in the regulation of insulin binding, action, and degradation

in freshly isolated and primary cultures of rat hepatocytes. *J. Clin. Invest.*, **69**, 866–875.

34 Weinstein, S.P., Paquin, T., Pritsker, A. and Haber, R.S. (1995) Glucocorticoid-induced insulin resistance: dexamethasone inhibits the activation of glucose transport in rat skeletal muscle by both insulin- and non-insulin-related stimuli. *Diabetes*, **44**, 441–445.

35 Shimazu, T. (1996) Innervation of the liver and glucoregulation: roles of the hypothalamus and autonomic nerves. *Nutrition*, **12**, 65–66.

36 McMahon, M., Gerich, J. and Rizza, R. (1988) Effects of glucocorticoids on carbohydrate metabolism. *Diabetes Metab. Rev.*, **4**, 17–30.

37 Basu, R., Chandramouli, V., Dicke, B., Landau, B. and Rizza, R. (2005) Obesity and type 2 diabetes impair insulin-induced suppression of glycogenolysis as well as gluconeogenesis. *Diabetes*, **54**, 1942–1948.

38 Bogardus, C., Lillioja, S., Howard, B.V., Reaven, G. and Mott, D. (1984) Relationships between insulin secretion, insulin action, and fasting plasma glucose concentration in nondiabetic and noninsulin-dependent diabetic subjects. *J. Clin. Invest.*, **74**, 1238–1246.

39 Perriello, G. *et al.* (1997) Evidence of increased systemic glucose production and gluconeogenesis in an early stage of NIDDM. *Diabetes*, **46**, 1010–1016.

40 Mevorach, M. *et al.* (1998) Regulation of endogenous glucose production by glucose per se is impaired in type 2 diabetes mellitus. *J. Clin. Invest.*, **102**, 744–753.

41 Pilkis, S.J. and Granner, D.K. (1992) Molecular physiology of the regulation of hepatic gluconeogenesis and glycolysis. *Annu. Rev. Physiol.*, **54**, 885–909.

42 Valera, A., Pujol, A., Pelegrin, M. and Bosch, F. (1994) Transgenic mice overexpressing phosphoenolpyruvate carboxykinase develop non-insulin-dependent diabetes mellitus. *Proc. Natl. Acad. Sci. U.S.A.*, **91**, 9151–9154.

43 Wu, C., Okar, D.A., Newgard, C.B. and Lange, A.J. (2001) Overexpression of

6-phosphofructo-2-kinase/fructose-2, 6-bisphosphatase in mouse liver lowers blood glucose by suppressing hepatic glucose production. *J. Clin. Invest.*, **107**, 91–98.

44 van Poelje, P.D. *et al.* (2006) Inhibition of fructose 1,6-bisphosphatase reduces excessive endogenous glucose production and attenuates hyperglycemia in Zucker diabetic fatty rats. *Diabetes*, **55**, 1747–1754.

45 Petersen, D.D., Koch, S.R. and Granner, D.K. (1989) 3' noncoding region of phosphoenolpyruvate carboxykinase mRNA contains a glucocorticoid-responsive mRNA-stabilizing element. *Proc. Natl. Acad. Sci. U.S.A.*, **86**, 7800–7804.

46 Exton, J.H. (1987) Mechanisms of hormonal regulation of hepatic glucose metabolism. *Diabetes Metab. Rev.*, **3**, 163–183.

47 Pilkis, S.J., Claus, T.H., Kurland, I.J. and Lange, A.J. (1995) 6-Phosphofructo-2-kinase/fructose-2,6-bisphosphatase: a metabolic signaling enzyme. *Annu. Rev. Biochem.*, **64**, 799–835.

48 Argaud, D. *et al.* (1996) Regulation of rat liver glucose-6-phosphatase gene expression in different nutritional and hormonal states: gene structure and 5'-flanking sequence. *Diabetes*, **45**, 1563–1571.

49 Darmaun, D., Matthews, D.E. and Bier, D.M. (1988) Physiological hypercortisolemia increases proteolysis, glutamine, and alanine production. *Am. J. Physiol.*, **255**, E366–E373.

50 Goldstein, R.E. *et al.* (1993) Effects of chronic elevation in plasma cortisol on hepatic carbohydrate metabolism. *Am. J. Physiol.*, **264**, E119–E127.

51 McGuinness, O.P. *et al.* (1993) Impact of chronic stress hormone infusion on hepatic carbohydrate metabolism in the conscious dog. *Am. J. Physiol.*, **265**, E314–E322.

52 Bishayi, B. and Ghosh, S. (2003) Metabolic and immunological responses associated with *in vivo* glucocorticoid depletion by adrenalectomy in mature Swiss albino rats. *Life Sci.*, **73**, 3159–3174.

53 Bollen, M. and Stalmans, W. (1992) The structure, role, and regulation of type 1 protein phosphatases. *Crit. Rev. Biochem. Mol. Biol.*, **27**, 227–281.

54 McGuinness, O.P. *et al.* (1997) Role of epinephrine and norepinephrine in the metabolic response to stress hormone infusion in the conscious dog. *Am. J. Physiol.*, **273**, E674–E681.

55 Chu, C.A. *et al.* (1998) Effect of a selective rise in sinusoidal norepinephrine on HGP is due to an increase in glycogenolysis. *Am. J. Physiol.*, **274**, E162–E171.

56 Stevenson, R.W. *et al.* (1991) Dose-related effects of epinephrine on glucose production in conscious dogs. *Am. J. Physiol.*, **260**, E363–E370.

57 Connolly, C.C. *et al.* (1991) Regulation of glucose metabolism by norepinephrine in conscious dogs. *Am. J. Physiol.*, **261**, E764–E772.

58 Perley, M. and Kipnis, D.M. (1966) Effect of glucocorticoids on plasma insulin. *N. Engl. J. Med.*, **274**, 1237–1241.

59 Ashmore, J. (1964) Hormonal modifiers of insulin action. New York *diabetes* symposium. II. The effects of glucocorticoids on insulin action. *Diabetes*, **13**, 349–354.

60 Kawai, A. and Kuzuya, N. (1977) On the role of glucocorticoid in glucose-induced insulin secretion. *Horm. Metab. Res.*, **9**, 361–365.

61 Delaunay, F. *et al.* (1997) Pancreatic beta cells are important targets for the diabetogenic effects of glucocorticoids. *J. Clin. Invest.*, **100**, 2094–2098.

62 Eigler, N., Sacca, L. and Sherwin, R.S. (1979) Synergistic interactions of physiologic increments of glucagon, epinephrine, and cortisol in the dog: a model for stress-induced hyperglycemia. *J. Clin. Invest.*, **63**, 114–123.

63 Haber, R.S. and Weinstein, S.P. (1992) Role of glucose transporters in glucocorticoid-induced insulin resistance. GLUT4 isoform in rat skeletal muscle is not decreased by dexamethasone. *Diabetes*, **41**, 728–735.

64 Dimitriadis, G. *et al.* (1997) Effects of glucocorticoid excess on the sensitivity of glucose transport and metabolism to insulin in rat skeletal muscle. *Biochem. J.*, **321** (Pt 3), 707–712.

65 Minokoshi, Y., Okano, Y. and Shimazu, T. (1994) Regulatory mechanism of the ventromedial hypothalamus in enhancing glucose uptake in skeletal muscles. *Brain Res.*, **649**, 343–347.

66 Tanishita, T., Shimizu, Y., Minokoshi, Y. and Shimazu, T. (1997) The beta3-adrenergic agonist BRL37344 increases glucose transport into L6 myocytes through a mechanism different from that of insulin. *J. Biochem.*, **122**, 90–95.

67 Han, X.X. and Bonen, A. (1998) Epinephrine translocates GLUT-4 but inhibits insulin-stimulated glucose transport in rat muscle. *Am. J. Physiol.*, **274**, E700–E707.

68 Bonen, A., Megeney, L.A., McCarthy, S.C., McDermott, J.C. and Tan, M.H. (1992) Epinephrine administration stimulates GLUT4 translocation but reduces glucose transport in muscle. *Biochem. Biophys. Res. Commun.*, **187**, 685–691.

69 Hamburg, S., Hendler, R. and Sherwin, R.S. (1980) Influence of small increments of epinephrine on glucose tolerance in normal humans. *Ann. Intern. Med.*, **93**, 566–568.

70 Pollare, T. *et al.* (1989) Metabolic effects of diltiazem and atenolol: results from a randomized, double-blind study with parallel groups. *J. Hypertens.*, **7**, 551–559.

71 Pollare, T., Lithell, H., Selinus, I. and Berne, C. (1988) Application of prazosin is associated with an increase of insulin sensitivity in obese patients with hypertension. *Diabetologia*, **31**, 415–420.

72 Liu, X., Perusse, F. and Bukowiecki, L.J. (1998) Mechanisms of the antidiabetic effects of the beta 3-adrenergic agonist CL-316243 in obese Zucker-ZDF rats. *Am. J. Physiol.*, **274**, R1212–R1219.

73 Morgan, B.P. *et al.* (2002) Discovery of potent, nonsteroidal, and highly selective glucocorticoid receptor antagonists. *J. Med. Chem.*, **45**, 2417–2424.

74 Watts, L.M. *et al.* (2005) Reduction of hepatic and adipose tissue glucocorticoid receptor expression with antisense oligonucleotides improves hyperglycemia and hyperlipidemia in diabetic rodents

without causing systemic glucocorticoid antagonism. *Diabetes*, **54**, 1846–1853.

75 Hermanowski-Vosatka, A. *et al.* (2005) 11beta-HSD1 inhibition ameliorates metabolic syndrome and prevents progression of atherosclerosis in mice. *J. Exp. Med.*, **202**, 517–527.

76 Pradhan, A.D., Manson, J.E., Rifai, N., Buring, J.E. and Ridker, P.M. (2001) C-reactive protein, interleukin 6, and risk of developing type 2 diabetes mellitus. *JAMA*, **286**, 327–334.

77 Ross, R. (1999) Atherosclerosis – an inflammatory disease. *N. Engl. J. Med.*, **340**, 115–126.

78 Libby, P. (2002) Inflammation in atherosclerosis. *Nature*, **420**, 868–874.

79 Bruunsgaard, H., Andersen-Ranberg, K., Hjelmborg, J.B., Pedersen, B.K. and Jeune, B. (2003) Elevated levels of tumor necrosis factor alpha and mortality in centenarians. *Am. J. Med.*, **115**, 278–283.

80 Pai, J.K. *et al.* (2004) Inflammatory markers and the risk of coronary heart disease in men and women. *N. Engl. J. Med.*, **351**, 2599–2610.

81 Krogh-Madsen, R., Plomgaard, P., Møller, K., Mittendorfer, B. and Pedersen, B.K. (2006) Influence of TNF-alpha and IL-6 infusions on insulin sensitivity and expression of IL-18 in humans. *Am. J. Physiol. Endocrinol. Metab.*, **291**, E108–E114.

82 Plomgaard, P. *et al.* (2007) Associations between insulin resistance and TNF-alpha in plasma, skeletal muscle and adipose tissue in humans with and without type 2 diabetes. *Diabetologia*, **50**, 2562–2571.

83 Plomgaard, P. *et al.* (2005) Tumor necrosis factor-alpha induces skeletal muscle insulin resistance in healthy human subjects via inhibition of Akt substrate 160 phosphorylation. *Diabetes*, **54**, 2939–2945.

84 van Hall, G. *et al.* (2003) Interleukin-6 stimulates lipolysis and fat oxidation in humans. *J. Clin. Endocrinol. Metab.*, **88**, 3005–3010.

85 Plomgaard, P., Fischer, C.P., Ibfelt, T., Pedersen, B.K. and van Hall, G. (2007) TNF-alpha modulates human *in vivo* lipolysis. *J. Clin. Endocrinol. Metab.*, **93**, 543–549.

86 Turnbull, A.V. and Rivier, C. (1999) Regulation of the hypothalamic-pituitary-adrenal axis by cytokines: actions and mechanisms of action. *Physiol. Rev.*, **79**, 1–71.

87 Selye, H. (1936) A syndrome produced by diverse nocuous agents. *Nature*, **138**, 32.

88 Lee, S.W. *et al.* (1988) Glucocorticoids selectively inhibit the transcription of the interleukin 1 beta gene and decrease the stability of interleukin 1 beta mRNA. *Proc. Natl. Acad. Sci. U.S.A.*, **85**, 1204–1208.

89 Keller, P., Keller, C., Robinson, L.E. and Pedersen, B.K. (2004) Epinephrine infusion increases adipose interleukin-6 gene expression and systemic levels in humans. *J. Appl. Physiol.*, **97**, 1309–1312.

90 Frost, R.A., Nystrom, G.J. and Lang, C.H. (2004) Epinephrine stimulates IL-6 expression in skeletal muscle and C2C12 myoblasts: role of c-Jun NH2-terminal kinase and histone deacetylase activity. *Am. J. Physiol. Endocrinol. Metab.*, **286**, E809–E817.

91 Guirao, X. *et al.* (1997) Catecholamines increase monocyte TNF receptors and inhibit TNF through beta 2-adreno-receptor activation. *Am. J. Physiol.*, **273**, E1203–E1208.

92 Soszynski, D., Kozak, W., Conn, C.A., Rudolph, K. and Kluger, M.J. (1996) Beta-adrenoceptor antagonists suppress elevation in body temperature and increase in plasma IL-6 in rats exposed to open field. *Neuroendocrinology*, **63**, 459–467.

93 van Gool, J., van Vugt, H., Helle, M. and Aarden, L.A. (1990) The relation among stress, adrenalin, interleukin 6 and acute phase proteins in the rat. *Clin. Immunol. Immunopathol.*, **57**, 200–210.

94 Johnson, J.D. *et al.* (2005) Catecholamines mediate stress-induced increases in peripheral and central inflammatory cytokines. *Neuroscience*, **135**, 1295–1307.

95 Morrow, L.E., McClellan, J.L., Conn, C.A. and Kluger, M.J. (1993) Glucocorticoids alter fever and IL-6

responses to psychological stress and to lipopolysaccharide. *Am. J. Physiol.*, **264**, R1010–R1016.

96 LeMay, L.G., Vander, A.J. and Kluger, M.J. (1990) The effects of psychological stress on plasma interleukin-6 activity in rats. *Physiol. Behav.*, **47**, 957–961.

97 Zhou, D., Kusnecov, A.W., Shurin, M.R., DePaoli, M. and Rabin, B.S. (1993) Exposure to physical and psychological stressors elevates plasma interleukin 6: relationship to the activation of hypothalamic-pituitary-adrenal axis. *Endocrinology*, **133**, 2523–2530.

98 Brydon, L., Edwards, S., Mohamed-Ali, V. and Steptoe, A. (2004) Socioeconomic status and stress-induced increases in interleukin-6. *Brain Behav. Immun.*, **18**, 281–290.

99 Engler, H. *et al.* (2004) Effects of social stress on blood leukocyte distribution: the role of alpha- and beta-adrenergic mechanisms. *J. Neuroimmunol.*, **156**, 153–162.

100 Surwit, R.S. *et al.* (2002) Stress management improves long-term glycemic control in type 2 diabetes. *Diabetes Care*, **25**, 30–34.

19
Environmental Stress is Not Always Vicious:
a Lesson from Heat Acclimation-Mediated Neuroprotection
After Traumatic Brain Injury

Michal Horowitz and Esther Shohami

19.1
Introduction

"Stress is any unexpected, large and sudden perturbation of the cellular network, to which the network (i) does not have a prepared adaptive response or (ii) does not have enough time to mobilize the adaptive response" [1]. The cellular response to stress involves the activation of a number of specific signaling pathways that may lead to either damage or repair. Under this broad definition of stress, proposed by Selye [2] half a century ago, many pathological events may be viewed as "stress".

This chapter focuses on traumatic brain injury (TBI)-induced cellular stress responses and their modulation by "preconditioning" achieved via heat acclimation (HA), a unique model of long-term environmental stress. Remodeling of the: (i) antioxidative cascade; (ii) hypoxia inducible factor-1-erythropoietin receptor signaling; and (iii) anti-inflammatory capacity and neurotrophic factors by HA are highlightened.

19.2
TBI Damage – Detrimental Signaling and Possible Protective Pathways

19.2.1
Short Background

Central nervous system (CNS) injury, namely, traumatic brain injury (TBI) and spinal cord injury (SCI) are major health problems in developed countries, with annual incidence rates estimated around 2×10^6 for TBI and 11 000 for SCI in the United States and persisting neurological impairment in many of these victims [3–6]. CNS injury involves a primary mechanical impact that usually causes skull fracture and abruptly disrupts the brain parenchyma with shearing and tearing of blood vessels and brain tissue [7, 8]. This in turn, leads to disruption of the blood–

Stress – From Molecules to Behavior. Edited by Hermona Soreq, Alon Friedman, and Daniela Kaufer
Copyright © 2010 WILEY-VCH Verlag GmbH & Co. KGaA, Weinheim
ISBN: 978-3-527-32374-6

brain barrier, development of vasogenic and cytotoxic edema and impairment of energy metabolism and ionic homeostasis.

Moreover, the primary impact triggers a cascade of events characterized by activation of molecular and cellular responses that lead to a secondary injury. The evolvement of secondary damage is an active process in which many biochemical pathways are involved (for a review, see [9]). Most of these pathways are detrimental, yet there is large body of literature indicating the concomitant activation of endogenous, self-protective mechanisms.

The outcome of CNS injury depends not only on the extent of the primary injury but also on the amount of secondary degeneration that subsequently develops [10, 11]. Neurons that are directly damaged will inevitably die, regardless of the location (white or gray matter) of the lesion. The relatively rapid process of primary death is followed by the secondary degeneration of adjacent neurons that escaped the initial insult. The primary death of brain cells causes accumulation of harmful physiological substances such as glutamate, reactive oxygen and nitrogen species and pro-inflammatory cytokines, creating a toxic environment for the neighboring neurons [12–16]. These events have been shown to be the cause of the majority of nerve cell loss and disruption of neuronal circuitry, resulting in functional deficits. Understanding of the cellular/molecular signaling leading to these events provides a link to possible neuroprotective, therapeutic tools.

19.2.2
TBI-Induced Tissue and Cellular Stress Response

19.2.2.1 Glutamate
The robust increase of extracellular glutamate levels following TBI and stroke set the ground for the notion that activation of the NMDA receptor is toxic. The term excitotoxicity was coined and numerous drugs which act as NMDA antagonists were developed in nearly two decades. Some of the drugs that were most efficacious in animal studies were tested in the clinic, yet all of them failed [17]. Despite the fact that animal models clearly showed loss of efficacy within 1 h after injury [18], NMDAR antagonists drugs were administered in clinical trials for several days rather than once after the injury. Furthermore, some of these trials had to be stopped prematurely because of increases in mortality and morbidity in the drug arm of the stroke trials [17, 19, 20], suggesting that prolonged blockade of NMDAR may actually be harmful in the post-traumatic or post-ischemic patient. The study of Biegon et al. [21] demonstrated that the NMDAR are downregulated from hours to seven days post-injury. Their findings strongly suggest that administration of NMDAR blockers to brain injured subjects beyond the acute post-injury period is not only useless (as has been shown by both animal and clinical studies) but actually counterproductive (as suggested by the results of clinical trials). Furthermore, brain-injured subjects may stand to benefit from stimulation of NMDAR in the sub-acute post-injury phase, as was also shown recently by Yaka et al. [22].

19.2.2.2 Reactive Oxygen Species

Reactive oxygen species (ROS) are widely implicated in the pathogenesis of secondary neuronal damage after TBI [23, 24]. Despite methodological difficulties in measuring *in vivo* generation of ROS, oxidative modification of essential cellular macromolecules and their subsequent functional derangement are often used as indirect markers of tissue oxidative stress [24–26]. These biochemical and molecular processes are involved in post-traumatic apoptotic and/or necrotic cell death [23, 27]. In addition, it is suggested that ROS also regulate signal transduction pathways such as NF-kappaB and AP-1 [28], which in turn govern both harmful and defensive cellular responses to primary injurious events [29, 30], including brain trauma [31]. Sustained activation of NF-kappaB after TBI was shown to be inhibited by melatonin, an endogenous molecule known for its antioxidant properties, and this was associated with enhanced recovery [32]. Continuous exposure to various types of oxidative stress from a variety of sources has led the cell and the entire organism to develop specific defense mechanisms and molecules. One such group of molecules are the low molecular weight antioxidants (LMWA) which are capable of preventing oxidative damage by interaction with ROS. These molecules share a similar chemical trait that allows them to scavenge oxygen radical and prevent it from attacking the biological targets. The cell can regulate the concentrations of these scavenger molecules, for example, by being regenerated within the cell (for a review, see [33]).

19.2.2.3 Neuroinflammation

Different cell types in the CNS (neurons, glia, endothelial cells) produce and respond to the pro-inflammatory cytokines, such as tumor necrosis factor (TNFα) and interleukin-1 (IL-1β), prostaglandins, free radicals and complement. The presence of inflammatory cells around the traumatic region may increase cellular damage. Polymorphonuclear cells may cause tissue damage by the effects of their toxic enzymes such as myeloperoxidase, which further damage cell membranes. These cytokines may induce an inflammatory reaction and also act as chemoattractants to leukocytes. In many experimental TBI models and clinical studies, a rapid, robust increase of the inflammatory mediators was reported [34, 35], as well as the induction of chemokines and adhesion molecules. These, in turn activate immune and glial cells. The local inflammatory response is now believed to play an important role in mediating and exacerbating secondary tissue damage after CNS injury. In recent years, the concept of a "dual role" evolved with regard to concomitant beneficial and adverse effects of pro-inflammatory mediators, depending on the kinetic of their expression and post-traumatic regulation in the injured brain [14, 36, 37]. Thus, inhibition of TNFα within this time frame led to an improvement in functional outcome and reduced edema formation, BBB disruption and hippocampal cell loss [38, 39]. Also, blockage of IL-1 signaling in mice with central overexpression of IL-1ra was shown to be neuroprotective after TBI, resulting in delayed cytokine induction and improved neurological recovery [40]. Although during the early post-injury period the pro-inflammatory cytokines are harmful, they may play a pivotal beneficial role in later, long-term, recovery processes [14, 36, 41–43].

19.2.3
Treatments

The quest for effective treatments for TBI remains an urgent priority. Despite advances in basic and clinical research as well as improved neurointensive care in recent years, no specific pharmacological therapy for TBI is available. To date, the majority of the clinical trials in TBI have failed, yet all proved effective in the preclinical studies. One of the major discrepancies between drug efficacy at bench and the clinic lies in the different time frame for treatment. Whereas in most experimental studies treatment was begun within 1–4 h, in the clinical setting this interval is too short, and by the time the patient is ready for treatment, the therapeutic window has been missed [17, 20]. Further studies into the multifactorial mechanisms of the progressive damage following CNS injury and identifying endogenous mechanisms that are activated in order to offer neuroprotection are needed in order to develop novel therapeutic targets for TBI and SCI.

19.3
Preconditioning: Inherent Protective Mechanisms Are Induced by Preceding Sub-Lethal (Multiple) Stressors

Cells have many mechanisms for adapting and surviving stress. Some of these coping mechanisms are constitutive; others are inducible and therefore require time for their expression. An example of an inducible response to stress occurs where protection is conveyed to limit undesirable recurrent stress effects by prior repetitive short episodes of sub-lethal stress [1, 2, 44, 45]. This preconditioning (PC) effect is considered to be part of a rapid adaptive mechanism induced by many physiological and pharmacological stressors and is shared by many cells and organs. PC was initially discovered following intermittent ischemia in the heart [46] but less than 10 years later brain PC (e.g., via thermal/ischemic stress) was widely recognized as a powerful cytoprotective mechanism, thereby providing an innovative approach for the discovery of novel cerebroprotective strategies [47]. The recognition that preconditioning is achieved via diverse stressors and provides protection against novel stressors including against TBI [48] introduced the term *cross-tolerance*, a term encompassing wide forms of PC manipulations [49–51]. The vast number of studies on PC focus on short/limited time boundaries, in which transient cross-tolerance is conferred following a recovery period from an initial, sub-lethal short duration stress. This short-term protective process comprises the first window of protection, which lasts for approximately 1 h and involves the activation of immediate salvage signaling pathways. These are later replaced by transcriptional activation and enhanced expression of cytoprotective components, thereby affording protection for a few days (the second window of protection, [52, 53]). There are universal pathways shared by all types of PC (e.g., adenosine for the first window, HSP72 or NO for the second window of protection); however, gene chip analyses revealed that there are substantial subsets of the differentially expressed genes, unique to each preconditioning stimulus, thus suggesting that different PC manipulations may determine a specific neuroprotective phenotypes [54].

In contrast to the classic PC effect, we have demonstrated that long-term acclimation to adverse thermal environment (heat acclimation) or exercise training confers long-lasting protection for approximately 2–3 weeks against a diversity of novel stressors, comprising heat stress, ischemia, TBI, hyperoxia and ionized irradiation in both the heart and the brain [51]. Because heat acclimation cross-tolerance (HACT) is the outcome of constitutive changes that have developed while shaping the acclimatory homeostasis [51, 55], it differs, conceptually, from classic PC [52] and relies on enhanced, predisposed protective molecules or pathways. HACT occurs in a wide range of taxonomic groups, including invertebrate worms (*Caenorhabditis elegans* [56]), insects (*Folsomia candida* [57]) and ectothermic vertebrates, such as fish (the tidepool sculpin [50]) and rodents (rats and mice [51, 58], for a review, see [59]), implying that this feature is evolutionarily conserved. The HACT is committed to memory and reappears (even following two months of de-acclimation) shortly (two days) after a return to the acclimating conditions (instead of the one month required for the initial acclimation [60]).

19.4
Heat Acclimation and Neuroprotection

19.4.1
Heat Acclimation

Heat acclimation (HA) is a conserved phenotypic adaptive response to a prolonged transfer to higher ambient temperatures that not only confers protection against acute heat stress, but also delays thermal injury [51, 55]. Conceptually, the process of heat acclimation represents a transition to very "efficient" cellular performance [51, 55, 61]. In terms of integrative physiological mechanisms, HA is characterized by decreased heat production, core temperature (Tc), Tc threshold for heat dissipation and heart rate. The capacities of the vascular and the evaporative cooling systems for heat dissipation are augmented [55, 61]. Plasticity of the autonomic nervous system and remodeling of cytoprotective networks, respectively, are responsible for these changes [62, 102]. Currently, both the neuronal electrophysiological activity under situations of chronic stress and peripheral changes in the effector organs are believed to depend on structural flexibility and molecular changes [63]. Therefore, new concepts of the core processes leading to acclimation plasticity have developed [51] and promoted the study on molecular and cellular mechanisms underlying the integrative responses. A hallmark of the acclimation process is the enhancement of cytoprotective networks – including the heat shock proteins, anti-oxidative and anti-apoptotic networks [64–68] and the stabilization of HIF-1α, the master regulator of oxygen homeostasis [56, 69, 70]. The HIF-1-mediated hypoxic response appeared early in metazoan evolution to regulate metabolic responses and is highly conserved [71, 72]. We hypothesize that HIF-1 could be exploited by a variety of physiological adaptive mechanisms requiring metabolic changes, as in the case of heat acclimation, which shows enhancement of the metabolic machinery to elevate energy potential upon insults [73]. An important HIF-1 target observed in our HA rats and mice is the epo-epoR pathway [69, 70].

Using HSP72 as a prototype model we demonstrated that HA: (i) hastens the transcriptional response upon stress [74, 75]; and (ii) increases the cellular reserves of the inducible species of HSP 72, implying that in the acclimated phenotype cytoprotection can be accomplished without *de novo* HSP synthesis. The buildup of HSP cellular reserves is also fascinating; due to the fact that heat acclimation in mammals does not induce extreme hyperthermia, mediators other than severely elevated body temperature must be able to increase *hsp* gene transcription.

The induction of the acclimated phenotype requires reprogramming of gene expression. Gene chips analyses using RNA from rat hypothalamic and heart tissues sampled during the course of heat acclimation indicated a two-tier genomic expression program of genes linked with physiological integrative functions to match the stress loads at each acclimation phase. The immediate transient acclimatory response is associated with: (i) maintenance of DNA and cellular integrity; and (ii) a transient upregulation or downregulation of genes encoding voltage-gated ion channels, ion pumps and transporters, as well as hormone/transmitter receptors and cellular messengers. Collectively, this reprogrammed profile points to enhanced membrane depolarization, and in turn, a release of neurotransmitters, which agrees with the enhanced neuronal excitability during the strain developed at the onset of acclimation. The sustained response, in contrast, correlates with continuous, long-lasting, adaptive cytoprotective signaling networks. Genes associated with a variety of homeostatic functions returned to pre-acclimation expression levels, except that genes encoding hormones and neuropeptides linked with metabolic rate and food intake were downregulated in both short and long acclimation. Such a temporal genomic profile is consistent with our previously established integrative acclimation model, demonstrating enhanced autonomic excitability during STHA and its resumption upon LTHA [51, 55, 76]. Changes in the level of the encoded proteins, response rate of various pathways as well as post-translational modifications in the heat-acclimated phenotype underlie the cross-tolerance response.

19.4.2
Heat Acclimation Affords Neuroprotection After TBI

19.4.2.1 Physiological Evidence
The effects of HA on the outcome of TBI have been examined in rodents, using a model closed head injury (CHI) [77, 78] that reproduces the post-traumatic

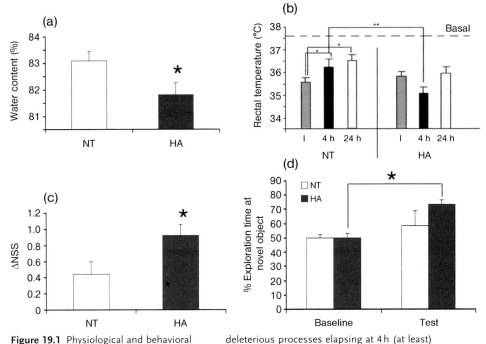

Figure 19.1 Physiological and behavioral evidence of heat acclimation-mediated neuroprotection against traumatic brain injury. Heat-acclimated (HA) and normo-thermic (NT) mice were subjected to the closed head injuries of similar severity. Neuroprotection was evident by: (a) reduced edema formation, expressed as percent water content in the brain 24 h post-injury [80]; (b) hypothermia, leading to attenuation of deleterious processes elapsing at 4 h (at least) post-injury in HA mice vs a return to normal body temperature in NT mice [81].
(c) Recovery of motor function, expressed by the change in neurological severity scores (ΔNSS) between 1 and 24 h is higher in HA vs NT mice [70]. (d) HA mice perform better in object recognition test 3 days after injury, indicating their better ability to memorize familiar objects [70].

sequence of events observed in the injured human. This model consistently recreates several pathophysiological features such as edema formation, blood–brain barrier (BBB) disruption as well as cognitive and motor dysfunction [79]. Superiority of the HA groups was unequivocally assessed when multiple neurobehavioral functions were compared in non-acclimated and HA rats and mice (Figure 19.1a–d). Heat-acclimated animals demonstrated significantly smaller degree of dysfunction following injury as compared with normothermic controls. Likewise, HA mice performed better in a memory test when examined by the object recognition test [70]. Additionally, brain edema and BBB disruption were reduced in the HA group. TBI induces temporal hypothermia [81]. Noteworthy is the finding that HA mice display an ability to maintain sustained hypothermia for a longer period of time as compared to non-acclimated mice. Among the integrative physiological features studied by our team, maintenance of integrated BBB and hypothermia are the likely features providing a potential physiological link to neuroprotection.

19.4.3
Underlying Signaling Pathways

Taken together our findings of improved neurobehavioral functional recovery of the HA animals and our previous knowledge of enhanced cytoprotection in the acclimated phenotype, we tested the assumption that enriched cytoprotective pathways play an important role in shaping the neuroprotective features of the HA phenotype.

We focus here on three central functional categories: (i) antioxidation; (ii) hypoxia inducible factor-1-erythropoietin receptor signaling (both stress and metabolic effectors); and (iii) anti-inflammatory capacity and the expression of beneficial neurotrophic factors. A large body of publications on the important role played by these cascades in neuroprotection has been recently published.

19.4.3.1 Oxidative Stress

Oxidative stress is an early cellular and tissue response to injury, and reactive oxygen species (ROS) are widely implicated in the pathogenesis of secondary neuronal damage after traumatic and ischemic brain injuries [9, 23]. ROS bring about oxidative modification of essential cellular macromolecules and their subsequent functional derangement involved in post-traumatic apoptotic and/or necrotic cell death and also regulates signal transduction pathways such as NF-kappaB and AP-1, which in turn govern both harmful and defensive cellular responses.

We have examined the total reducing capacity, as determined by the levels of the endogenous antioxidants, in the brains of normothermic and HA rats, before and after TBI. A cyclic voltammetry study performed on the water-soluble fraction of low molecular weight antioxidants (LMWA) showed that the chemical nature of LMWA is not altered by HA. Interestingly, the concentrations of these reducing equivalents were significantly lower in the HA group, that is, the basal LMWA levels were found to be lower in HA rats [66]. However, while the relative changes in anodic currents during the post-TBI period revealed a decrease from basal (sham) levels in the non-acclimated animals, this did not occur in the HA group which sustained levels higher than those measured in sham controls, for up to a week following trauma.

19.4.3.2 HIF-1-Erythropoietin Pathway

Hypoxia-inducible factor 1 (HIF-1), a transcriptional activator, the master regulator of oxygen homeostasis has been shown to be essential for the development of HA [56]. Cai *et al.*, 2003 [82] have demonstrated that erythropoietin (Epo) a-HIF-1 targeted pathway induces cardio-protection against ischemia–reperfusion injury in mice [82]. Preceding this finding, Buemi *et al.* [83], reported that Epo and its receptor (EpoR) are expressed not only in the kidneys, where the hematopoietic system is activated but in other tissues, including the nervous system [84, 85]. This finding yielded a large number of studies on the protective role of Epo, including its role in neuroprotective effect in our model of CHI [86] and in HA cross-tolerance-mediated pathways both in the heart and the brain [69, 70]. Except for its "supportive" importance in energy metabolism Epo has been implicated as neuroprotective against neurotoxicity [87], attenuating apoptotic cascades [88]. Our data demonstrate that HA alone led to an upregulation of HIF-1α within the cell nucleus and of EpoR, collectively leading to an increased EpoR/Epo ratio, namely increased signaling capacity. Following CHI (4 h), HIF-1α expression and in turn EpoR were further elevated in HA but not in control mice. The changes observed after acclimation were far more pronounced than those induced by the injury. This profile is similar to that described for HSPs after HA in the heart [76] and brain [89] as well as to several other proteins (see Section 19.4.3.3). We therefore suggest that the improved outcome of CHI in the acclimated phenotype is by large protein reserves, because of the establishment of larger cytoprotective protein reserves during the acclimation process. This conclusion is further confirmed by our finding that treatment with anti-Epo antibody given to non-acclimated mice increased edema formation, whereas rhEpo induced no beneficial effect in HA mice [90], suggesting that HA-induced neuroprotection is shaped by pre-existing mediators but cannot be modified by post-injury treatment aimed at increasing the levels of neuroprotective agents.

Akt phosphorylation (EpoR-mediated Akt activation) was evident in HA, but not in normothermic mice. This elevation, however, was only evident following CHI, suggesting that this effect is an altered dynamic response induced by HA. Selective inhibition of Akt-P (by Tricitibine) abolished the functional benefits (namely, improved functional recovery and reduced edema formation) induced by HA [80]. A large increase in Bad (pro-apoptotic member of the Bcl 2 family) mRNA, occurring in response to CHI in normothermic mice, was abolished in the HA mice, thus further supporting predominance of an anti-apoptotic balance in the CHI HA mice. Similar findings were recorded in heat-stressed and ischemic/reperfused hearts of HA rats [64].

19.4.3.3 Acute Inflammatory Response

Acute inflammatory response is initiated by TBI and has been shown to involve the upregulation of pro-inflammatory cytokines in the brain, including tumor necrosis factor alpha (TNF-α) and interleukin-1β (IL-1b) [34, 39]. Investigating the levels of pro- and anti-inflammatory cytokines in HA and normothermic mice revealed that HA mice do indeed display changes in pro- and anti-inflammatory capacity as part of the cross-tolerance response. HA was found to induce multiple changes in the inflammation-related factors. These included higher pre-injury levels of the anti-inflammatory cytokine interleukin 10 (IL-10) and interleukin 4 (IL-4) and reduced post-injury expression of the pro-inflamatory cytokine TNFα mRNA [81].

The brain-derived neurotrophic factor (BDNF), discovered in the early 1950s as a member of the neurotrophin family, has been shown to regulate survival, proliferation and maintenance of function in various neuronal cell populations [91–93]. Several studies demonstrated beneficial effects of BDNF administration in SCI and TBI models [94–98]. Glial cells have been shown to express BDNF [99]

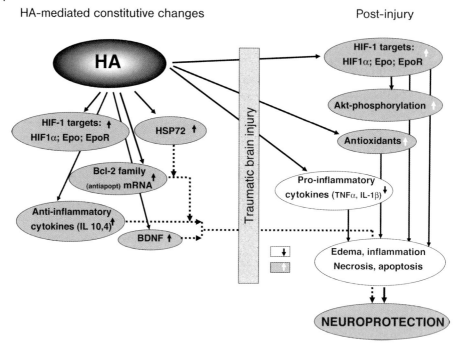

Figure 19.2 Conceptual model of the mechanisms leading to heat acclimation-mediated neuroprotection: HA provides protection both by the induction of constitutive elevation in cytoprotective components observed following the acclimation period (left panel) and by dynamic processes which are only apparent post-injury (right panel). Hence, neuro-protection is achieved by preconditioning events providing protection without the need for rapid *de novo* synthesis of cyto-protective proteins and by post-injury events, mostly post-translational modification such as Akt-phosphorylation and antioxidants formation. Collectively, these embroidered events bring about lesser extent of edema, necrosis and apoptosis as well as reduced inflammatory response. These changes are accompanied by improved functional outcome. Post-injury, the black down arrow indicates decrease, the white up arrows indicate increases. From left to right, the dashed lines depict preconditioning constitutive events which give no further changes post-injury. Data were compiled from [13, 59, 66, 70, 80, 81, 89].

and beneficial roles of glial cells, and microglia in particular, may be associated with their ability to supply trophic factors [100], such as BDNF. Interestingly, HA mice were found to express higher levels of BDNF at the end of the 30-day acclimation period, before injury was induced. During the post-injury period, HA induced an increase in the amount of BDNF-positive ramified microglia, but a global increase of this factor was not noted [103]. Our findings suggest that the BDNF-positive ramified subpopulation of microglia may indeed play a part in HA-induced neuroprotection.

Summary

The data provided in this chapter establish a neuroprotective potential in the heat acclimation procedure via potentiation of endogenous cytoprotection as well as attenuation of detrimental injury processes. Some of the changes are constitutive and are observed following the acclimation period, while others are dynamic and only apparent post-injury. This scenario suggests the establishment of an alerted system "on call" that can be further modulated by a dynamic components collectively conferring neuroprotection. The heat-acclimation neuroprotective pathways studied so far are summarized in Figure 19.2.

References

1 Szalaya, M.S., Kovácsa, I.A., Korcsmárosa, T., Bödeb, C. and Csermely, P. (2007) Stress-induced rearrangements of cellular networks: consequences for protection and drug design. *FEBS Lett.*, **581**, 3675–3680.

2 Selye, H. (1955) Stress and disease. *Science*, **122**, 625–631.

3 Marshall, L.F. (2000) Head injury: recent past, present, and future. *Neurosurgery*, **47**, 546–561.

4 Bayir, H., Clark, R.S. and Kochanek, P.M. (2003) Promising strategies to minimize secondary brain injury after head trauma. *Crit. Care Med.*, **31**, S112–S117.

5 Nortje, J. and Menon, D.K. (2004) Traumatic brain injury: physiology, mechanisms, and outcome. *Curr. Opin. Neurol.*, **17**, 711–718.

6 Nolan, S. (2005) Traumatic brain injury: a review. *Crit. Care Nurs. Q.*, **28**, 188–194.

7 Gentleman, S.M., Roberts, G.W., Gennarelli, T.A., Maxwell, W.L., Adams, J.H., Kerr, S. and Graham, D.I. (1995) Axonal injury: a universal consequence of fatal closed head injury? *Acta Neuropathol. Berl.*, **89**, 537–543.

8 Povlishock, J.T. and Christman, C.W. (1995) The pathobiology of traumatically induced axonal injury in animals and humans: a review of current thoughts. *J. Neurotrauma*, **12**, 555–564.

9 Bramlett, H.M. and Dietrich, W.D. (2007) Progressive damage after brain and spinal cord injury: pathomechanisms and treatment strategies. *Prog. Brain Res.*, **161**, 125–141 (Review).

10 Yoles, E. and Schwartz, M. (1998) Degeneration of spared axons following partial white matter lesion: implications for optic nerve neuropathies. *Exp. Neurol.*, **153**, 1–7.

11 Harrop, J.S., Sharan, A.D., Vaccaro, A.R. and Przybylski, G.J. (2001) The cause of neurologic deterioration after acute cervical spinal cord injury. *Spine*, **26**, 340–346.

12 Tymianski, M. and Tator, C.H. (1996) Normal and abnormal calcium homeostasis in neurons: a basis for the pathophysiology of traumatic and ischemic central nervous system injury. *Neurosurgery*, **38**, 1176–1195.

13 Shohami, E., Beit-Yannai, E., Horowitz, M., Kohen, R., Beit-Yannai, E., Horowitz, M. and Kohen, R. (1997) Oxidative stress in closed-head injury: brain antioxidant capacity as an indicator of functional outcome. *J. Cereb. Blood Flow Metab.*, **17**, 1007–1019.

14 Shohami, E., Ginis, I. and Hallenbeck, J.M. (1999) Dual role of tumor necrosis factor alpha in brain injury. *Cytokine Growth Factor Rev.*, **10**, 119–130.

15 Beattie, M.S., Farooqui, A.A. and Bresnahan, J.C. (2000) Review of current evidence for apoptosis after spinal cord injury. *J. Neurotrauma*, **17**, 915–925.

16 Schwartz, M. and Yoles, E. (2000) Self-destructive and self-protective processes in the damaged optic nerve:

implications for glaucoma [comment]. *Invest. Ophthalmol. Vis. Sci.*, **41**, 349–351.

17 Narayan, R.K., Michel, M.E., Ansell, B., Baethmann, A., Biegon, A., Bracken, M.B. *et al.* (2002) Clinical trials in head injury. *J. Neurotrauma*, **19**, 503–557.

18 Kroppenstedt, S.N., Schneider, G.H., Thomale, U.W. and Unterberg, A.W.J. (1998) Protective effects of aptiganel HCl (Cerestat) following controlled cortical impact injury in the rat. *Neurotrauma*, **15**, 191–197.

19 Fisher, M. (1998) The travails of neuroprotective drug development for acute ischemic stroke. *Eur. Neurol.*, **40**, 65–66.

20 Maas, A.I., Steyerberg, E.W., Murray, G.D., Bullock, R., Baethmann, A., Marshall, L.F. and Teasdale, G.M. (1999) Why have recent trials of neuroprotective agents in head injury failed to show convincing efficacy? A pragmatic analysis and theoretical considerations. *Neurosurgery*, **44**, 1286–1298.

21 Biegon, A., Fry, P.A., Paden, C.M., Alexandrovich, A., Tsenter, J. and Shohami, E. (2004) Delayed activation, rather than inhibition, of glutamate NMDA receptors improves neurological outcome after closed head injury in mice. *Proc. Natl. Acad. Sci. U.S.A.*, **101**, 5117–5122.

22 Yaka, R., Biegon, A., Grigoriadis, N., Simeonidou, C., Grigoriadis, S., Alexandrovich, A.G., Matzner, H. *et al.* (2007) D-cycloserine improves functional recovery and reinstates long-term potentiation (LTP) in a mouse model of closed head injury. *FASEB J.*, **21**, 2033–2041.

23 Chan, P.H. (2001) Reactive oxygen radicals in signaling and damage in the ischemic brain. *J. Cereb. Blood Flow Metab.*, **21**, 2–14.

24 Lewen, A. and Hillered, L. (1998) Involvement of reactive oxygen species in membrane phospholipid breakdown and energy perturbation after traumatic brain injury in the rat. *J. Neurotrauma*, **15**, 521–530.

25 McDonald, R.P., Horsburgh, K.J., Graham, D.I. and Nicoll, J.A. (1999) Mitochondrial DNA deletions in acute brain injury. *Neuroreport*, **10**, 1875–1878.

26 Vagnozzi, R., Marmarou, A., Tavazzi, B., Signoretti, S., Di Pierro, D., del Bolgia, F., Amorini, A.M. *et al.* (1999) Changes in cerebral energy metabolism and lipid peroxidation in rats leading to mitochondrial disfuncion after diffuse brain injury. *J. Neurotrauma*, **16**, 903–913.

27 Fiskum, G. (2000) Mitochondrial participation in ischemic and traumatic neural cell death. *J. Neurotrauma*, **17**, 843–855.

28 Das, D.K. and Maulik, N. (2003) Preconditioning potentiates redox signaling and converts death signal into survival signal. *Arch. Biochem. Biophys.*, **420**, 305–311.

29 Martindale, J.L. and Holbrook, N.J. (2002) Cellular response to oxidative stress: signaling for suicide and survival. *J. Cell Physiol.*, **192**, 1–15.

30 Muller, J.M., Rupec, R.A. and Baeuerle, P.A. (1997) Study of gene regulation by NF-kappa B and AP-1 in response to reactive oxygen intermediates. *Methods*, **11**, 301–312.

31 Nonaka, M., Chen, X.H., Pierce, J.E., Leoni, M.J., McIntosh, T.K., Wolf, J.A. and Smith, D.H. (1999) Prolonged activation of NF-kappaB following traumatic brain injury in rats. *J. Neurotrauma*, **16**, 1023–1034.

32 Beni, S.M., Kohen, R., Reiter, R.J., Tan, D.-T. and Shohami, E. (2004) Melatonin-induced neuroprotection after closed head injury is associated with increased brain antioxidants and attenuated late-phase activation of NF-kappaB and AP-1. *FASEB J.*, **18**, 149–151.

33 Kohen, R. and Nyska, A. (2002) Oxidation of biological systems: oxidative stress phenomena, antioxidants, redox reactions, and methods for their quantification. *Toxicol. Pathol.*, **30**, 620–650.

34 Shohami, E., Novikov, M., Bass, R., Yamin, A. and Gallily, R. (1994) Closed head injury triggers early production of TNFα and IL-6 by brain tissue. *J. Cereb. Blood Flow Metab.*, **14**, 615–619.

35 Hutchinson, P.J., O'Connell, M.T., Rothwell, N.J., Hopkins, S.J., Nortje, J., Carpenter, K.L., Timofeev, I. *et al.* (2007) Inflammation in human brain injury: intracerebral concentrations of IL-1alpha, IL-1beta, and their endogenous inhibitor IL-1ra. *J. Neurotrauma*, **24**, 1545–1557.

36 Morganti-Kossmann, M.C., Rancan, M., Stahel, P.F. and Kossmann, T. (2002) Inflammatory response in acute traumatic brain injury: a double-edged sword. *Curr. Opin. Crit. Care*, **8**, 101–105 (Review).

37 Lenzlinger, P.M., Morganti-Kossmann, M.C., Laurer, H.L. and McIntosh, T.K. (2001) The duality of the inflammatory response to traumatic brain injury. *Mol. Neurobiol.*, **24**, 169–181 (Review).

38 Shohami, E., Bass, R., Wallach, D., Yamin, A. and Gallily, R. (1996) Inhibition of tumor necrosis factorα activity in rat brain is associated with cerebroprotection after closed head injury. *J. Cereb. Blood Flow Metabol.*, **16**, 378–384.

39 Shohami, E., Gallily, R., Mechoulam, R., Bass, R. and Ben-Hur, T. (1997) Cytokine production in the brain following closed head injury: dexanabinol (HU-211) is a novel TNFα inhibitor and an effective neuroprotectant. *J. Neuroimmunol.*, **72**, 169–177.

40 Tehranian, R., Andell-Jonsson, S., Beni, S.M., Yatsiv, I., Shohami, E., Bartfai, T., Lundkvist, J., Iverfeldt, K., Recovery, I. and Cytokine, D. (2002) Induction after closed head injury in mice with central overexpression of the secreted isoform of the interleukin-1 receptor antagonist. *J. Neurotrauma.*, **19**, 939–951.

41 Bethea, J.R. and Dietrich, W.D. (2002) Targeting the host inflammatory response in traumatic spinal cord injury. *Curr. Opin. Neurol.*, **15**, 355–360.

42 Morganti-Kossman, M.C., Lenzlinger, P.M., Hans, V., Stahel, P., Csuka, E., Ammann, E., Stocker, R. *et al.* (1997) Production of cytokines following brain injury: beneficial and deleterious for the damaged tissue. *Mol. Psychiatry*, **2**, 133–136.

43 Scherbel, U., Raghupathi, R., Nakamura, M., Saatman, K.E., Trojanowski, J.Q., Neugebauer, E., Marino, M.W. and McIntosh, T.K. (1999) Differential acute and chronic responses of tumor necrosis factor-deficient mice to experimental brain injury. *Proc. Natl. Acad. Sci. U.S.A.*, **96**, 8721–8726.

44 Sommerschild, H.T. and Kirkeboen, K.A. (2002) Preconditioning endogenous defense mechanisms of the heart. *Acta Anaesthesiol. Scand.*, **46**, 123–137.

45 Xi, L., Tekin, D., Bhargava, P. and Kukreja, R.C. (2001) Whole body hyperthermia and preconditioning of the heart: basic concepts, complexity, and potential mechanisms. *Int. J. Hyperthermia*, **17**, 439–455.

46 Murry, C.E., Jennings, R.B. and Reimer, K.A. (1986) Preconditioning with ischemia: a delay of lethal cell injury in ischemic myocardium. *Circulation*, **74**, 1124–1136.

47 Obrenovitch, T.P. (2008) Molecular physiology of preconditioning-induced brain tolerance to ischemia. *Physiol. Rev.*, **88**, 211–247.

48 Perez Pinzon, M.A., Alonso, O., Kraydieh, S. and Dietrich, W.D. (1999) Induction of tolerance against traumatic brain injury by ischemic preconditioning. *Neuroreport*, **10**, 2951–2954.

49 Kirino, T. (2002) Ischemic tolerance. *J. Cereb. Blood Flow Metab.*, **22**, 1283–1296.

50 Todgham, A.E., Schulte, P.M. and Iwana, G.K. (2005) Cross-tolerance in the tidepool sculpin: the role of heat shock proteins. *Physiol. Biochem. Zool.*, **144**, 133–144.

51 Horowitz, M. (2007) Heat acclimation and cross-tolerance against novel stressors: genomic-physiological linkage. *Prog. Brain Res.*, **162**, 373–392.

52 Bolli, R. (2007) Preconditioning: a paradigm shift in the biology of myocardial ischemia. *Am. J. Physiol. Heart Circ. Physiol.*, **292**, H19–H27.

53 Hausenloy, D.J. and Yellon, D.M. (2007) Preconditioning and postconditioning: united at reperfusion. *Pharmacol. Therapeut.*, **116**, 173–191.

54 Stenzel-Poore, M.P., Stevens, S.L., King, J.S. and Simon, R.P. (2007) Preconditioning reprograms the response to ischemic injury and primes the emergence of unique endogenous neuroprotective phenotypes: a speculative synthesis. *Stroke*, **38**, 680–685.

55 Horowitz, M. (2002) Molecular and cellular to integrative heat defense during exposure to chronic heat. *Comp. Biochem. Physiol. A Mol. Integr. Physiol.*, **131**, 475–483 (Review).

56 Treinin, M., Shliar, J., Jiang, H., Powell-Coffman, J.A., Bromberg, Z. and Horowitz, M. (2003) HIF-1 is required for heat acclimation in the nematode caenorhabditis elegans. *Physiol. Genomics*, **14**, 17–24.

57 Bayley, M., Petersen, S.O., Knigge, T., Köhler, H. and Holmstrup, M. (2001) Drought acclimation confers cold tolerance in the soil collembolan folsomia candida. *J. Insect Physiol.*, **47**, 1197–1204.

58 Bromberg, Z. and Horowitz, M. (2004) Mus Musculus: a model studying the dynamics of heat acclimation and cross tolerance between heat acclimation and oxygen deprivation. Proceedings of the 1st Integrated Meeting on Physiol. and Pharmacol. of Thermoregulation (Rhodes, Greece PPTR 2004), p. 198 (Abstract).

59 Shein, N.A., Horowitz, M. and Shohami, E. (2007) Heat acclimation: a unique model of physiologically mediated global preconditioning against traumatic brain injury. *Prog. Brain Res.*, **161**, 353–363.

60 Tetievsky, A., Cohen, O., Eli-Berchoer, L., Gerstenblith, G., Stern, M.D., Wapinski, I., Friedman, N. and Horowitz, M. (2008) Physiological and molecular evidence of heat acclimation memory: a lesson from thermal responses and ischemic cross-tolerance in the heart. *Physiol. Genomics*, **12**, 78–87.

61 Horowitz, M. and Cellular, D. (1998) Heat acclimation responses modulate central thermoregulatory activity? *News Physiol. Sci.*, **13**, 218–225.

62 Horowitz, M. and Meiri, U. (1993) Central and peripheral contributions to control of heart rate during heat acclimation. *Pflugers Arch.*, **422**, 386–392.

63 Armstrong, L. and Stopani, J. (2002) Central nervous system control of heat acclimation adaptations: an emerging paradigm. *Rev. Neurosci.*, **13**, 271–285.

64 Horowitz, M., Eli-Berchoer, L., Wapinski, I., Friedman, N. and Kodesh, E. (2004) Stress related genomic responses during the course of heat acclimation and its association with ischemic/reperfusion cross-tolerance. *J. Appl. Physiol.*, **97**, 1496–1507.

65 Maloyan, A. and Horowitz, M. (2002) Beta-adrenergic signaling and thyroid hormones affect HSP72 expression during heat acclimation. *J. Appl. Physiol.*, **93**, 107–115.

66 Beit-Yannai, E., Kohen, R., Horowitz, M., Trembovler, V. and Shohami, E. (1997) Changes of biological reducing activity in rat brain following closed head injury: a cyclic voltammetry study in normal and heat-acclimated rats. *J. Cereb. Blood Flow Metab.*, **17**, 273–279.

67 Schwimmer, H., Eli-Berchoer, L. and Horowitz, M. (2006) Acclimatory-phase specificity of gene expression during the course of heat acclimation and superimposed hypohydration in the rat hypothalamus. *J. Appl. Physiol.*, **100**, 1992–2003.

68 Assayag, M., Gerstenblith, G. and Horowitz, M. (2007) Mitochondria mediated anti-apoptosis cascade is enhanced in heat acclimated rats. 15th Euroconference on Apoptosis and the 4th training course on "Concepts and Methods in Programmed Cell Death" Portoroz, Slovenia (Abst).

69 Maloyan, A., Eli-Berchoer, L., Semenza, G.L., Gerstenblith, G., Stern, M.D. and Horowitz, M. (2005) HIF-1alpha-targeted pathways are activated by heat acclimation and contribute to acclimation-ischemic cross-tolerance in the heart. *Physiol. Genomics*, **23**, 79–88.

70 Shein, N.A., Horowitz, M., Alexandrovich, A., Tsenter, J., Shohami, E. and Acclimation, H. (2005) Increases hypoxia inducible factor 1α and erythropoietin receptor expression: implication for neuroprotection after closed head injury in mice. *J. Cereb. Blood Flow Metabol.*, **25**, 1456–1465.

71 Semenza, G.L. (2004) Hydroxylation of HIF-1, oxygen sensing at the molecular level. *Physiology*, **19**, 176–182.

72 Zelzer, E., Levy, Y., Kahana, C., Shilo, B., Rubinstein, M. and Cohen, B. (1998) Insulin induces transcription of target genes through the hypoxia-inducible factor HIF-1/ARNT. *EMBO J.*, **17**, 5085–5094.

73 Wright, G.L., Hanlon, P., Amin, K., Steenbergen, C., Murphy, E. and Arcasoy, M.O. (2004) Erythropoietin receptor expression in adult rat cardiomyocytes is associated with an acute cardioprotective effect for recombinant erythropoietin during ischemia-reperfusion injury. *FASEB J.*, **18**, 1031–1033.

74 Horowitz, M., Maloyan, A. and Shlaier, J. (1997) HSP 70 kDa dynamics in animals undergoing heat stress superimposed on heat acclimation. *Ann. N.Y. Acad. Sci.*, **813**, 617–619.

75 Maloyan, A., Palmon, A. and Horowitz, M. (1999) Heat acclimation increases the basal HSP72 level and alters its production dynamics during heat stress. *Am. J. Physiol.*, **276**, R1506–R1515.

76 Horowitz, M. and Meiri, U. (1985) Altered responsiveness to parasympathetic activation of submaxillary salivary gland in the heat-acclimated rat. *Comp. Biochem. Physiol. A*, **80**, 57–60.

77 Shapira, Y., Shohami, E., Sidi, A., Soffer, D., Freeman, S. and Cotev, S. (1988) Experimental closed head injury in rats: mechanical, pathophysiologic, and neurologic properties. *Crit. Care Med.*, **16**, 258–265.

78 Chen, Y., Constantini, S., Trembovler, V., Weinstock, M. and Shohami, E. (1996) An experimental model of closed head injury in mice: pathophysiology, histopathology and cognitive deficits. *J. Neurotrauma*, **13**, 557–568.

79 Tsenter, J., Beni-Adani, L., Assaf, Y., Alexandrovich, A.G., Trembovler, V. and Shohami, E. (2008) Dynamic changes in the recovery after traumatic brain injury in mice: effect of injury severity on T2- weighted MRI abnormalities, motor and cognitive function. *J. Neurotrauma*, **25**, 324–333.

80 Shein, N.A., Tsenter, J., Alexandrovich, A.G., Horowitz, M. and Shohami, E. (2007) Akt phosphorylation is required for heat acclimation-induced neuroprotection. *J. Neurochem.*, **103**, 1523–1529.

81 Shein, N.A., Doron, H., Horowitz, M., Trembovler, V., Alexandrovich, A.G., Shohami, E. (2007) Altered cytokine expression and sustained hypothermia following traumatic brain injury in heat acclimated mice. *Brain Res.*, **1185**, 313–320.

82 Cai, Z., Manalo, D.J., Wei, G., Rodriguez, E.R., Fox-Talbot, K., Lu, H., Zweier, J.L. and Semenza, G.L. (2003) Hearts from rodents exposed to intermittent hypoxia or erythropoietin are protected against ischemia-reperfusion injury. *Circulation*, **108**, 79–85.

83 Buemi, M., Cavallaro, E., Floccari, F., Sturiale, A., Aloisi, C., Trimarchi, M., Corica, F. and Frisina, N. (2003) The pleiotropic effects of erythropoietin in the central nervous system. *J. Neuropathol. Exp. Neurol.*, **62**, 228–236 (Review).

84 Bernaudin, M., Marti, H.H., Roussel, S., Divoux, D., Nouvelot, A., MacKenzie, E.T. and Petit, E. (1999) A potential role for erythropoietin in focal permanent cerebral ischemia in mice. *J. Cereb. Blood Flow Metab.*, **19**, 643–651.

85 Bernaudin, M., Bellail, A., Marti, H.H., Yvon, A., Vivien, D., Duchatelle, I., Mackenzie, E.T. and Petit, E. (2000) Neurons and astrocytes express EPO mRNA: oxygen-sensing mechanisms that involve the redox-state of the brain. *Glia*, **30**, 271–278.

86 Yatsiv, I., Grigoriadis, N., Simeonidou, C., Stahel, P.F., Schmidt, O.I., Alexandrovitch, A.G., Tsenter, J. and Shohami, E. (2005) Erythropoietin is neuroprotective, improves functional recovery, and reduces neuronal apoptosis and inflammation in a rodent model of experimental closed head injury. *FASEB J.*, **19**, 1701–1703.

87 Ehrenreich, H., Hasselblatt, M., Dembowski, C., Cepek, L., Lewczuk, P., Stiefel, M., Rustenbeck, H.H. *et al.* (2002) Erythropoietin therapy for acute stroke is both safe and beneficial. *Mol. Med.*, **8**, 495–505.

88 Sirén, A.L., Fratelli, M., Brines, M., Goemans, C., Casagrande, S., Lewczuk, P., Keenan, S. *et al.* (2001) Erythropoietin prevents neuronal apoptosis after cerebral ischemia and metabolic stress. *Proc. Natl. Acad. Sci. U. S. A.*, **98**, 4044–4049.

89 Openheim, A., Beit-Yannai, E., Horowitz, M. and Shohami, E. (1996) Production of heat shock protein-72 in rat brain after closed head injury: study in heat acclimated and non acclimated. *Isr. J. Med. Sci.*, **32**, S38.

90 Shein, N.A., Grigoriadis, N., Alexandrovich, A.G., Simeonidou, C., Spandou, E., Tsenter, J., Yatsiv, I. *et al.* (2008) Differential neuroprotective properties of endogenous and exogenous erythropoietin in a mouse model of traumatic brain injury. *J. Neurotrauma*, **25**, 112–123.

91 Bibel, M. and Barde, Y.A. (2000) Neurotrophins: key regulators of cell fate and cell shape in the vertebrate nervous system. *Genes Dev.*, **14**, 2919–2937.

92 Binder, D.K. and Scharfman, H.E. (2004) Brain-derived neurotrophic factor. *Growth Factors*, **22**, 123–131.

93 Sharma, H.S. (2006) Post-traumatic application of brain-derived neurotrophic factor and glia-derived neurotrophic factor on the rat spinal cord enhances neuroprotection and improves motor function. *Acta Neurochir. Suppl.*, **96**, 329–334.

94 Blesch, A. and Tuszynski, M.H. (2007) Transient growth factor delivery sustains regenerated axons after spinal cord injury. *J. Neurosci.*, **27**, 10535–10545.

95 Faden, A.I., Knoblach, S.M., Movsesyan, V.A. and Cernak, I. (2004) Novel small peptides with neuroprotective and nootropic properties. *J. Alzheimers Dis.*, **6**, S93–S97.

96 Mahmood, A., Lu, D., Wang, L. and Chopp, M. (2002) Intracerebral transplantation of marrow stromal cells cultured with neurotrophic factors promotes functional recovery in adult rats subjected to traumatic brain injury. *J. Neurotrauma*, **19**, 1609–1617.

97 Nakajima, H., Uchida, K., Kobayashi, S., Inukai, T., Horiuchi, Y., Yayama, T., Sato, R. and Baba, H. (2007) Rescue of rat anterior horn neurons after spinal cord injury by retrograde transfection of adenovirus vector carrying brain-derived neurotrophic factor gene. *J. Neurotrauma*, **24**, 703–712.

98 Vavrek, R., Girgis, J., Tetzlaff, W., Hiebert, G.W. and Fouad, K. (2006) BDNF promotes connections of corticospinal neurons onto spared descending interneurons in spinal cord injured rats. *Brain*, **129**, 1534–1545.

99 Conner, J.M., Lauterborn, J.C., Yan, Q., Gall, C.M. and Varon, S. (1997) Distribution of brain-derived neurotrophic factor (BDNF) protein and mRNA in the normal adult rat CNS: evidence for anterograde axonal transport. *J. Neurosci.*, **17**, 2295–2313.

100 Kempermann, G. and Neumann, H. (2003) Neuroscience. Microglia: the enemy within? *Science*, **302**, 1689–1690.

101 Reiter, R.J., Tan, D.-X., Manchester, L.C. and Qi, W. (2001) Biochemical reactivity of melatonin with reactive oxygen and nitrogen species: a review of the evidence. *Cell Biochem. Biophys.*, **34**, 237–256.

102 Schwimmer, H., Gerstberger, R. and Horowitz, M. (2004) Heat acclimation affects the neuromodulatory role of AngII and nitric oxide during combined heat and hypohydration stress. *Brain Res. Mol. Brain Res.*, **130**, 95–108.

103 Shein, N.A., Grigoriadis, N., Horowitz, M., Umschwief, G., Alexandrovich, A.G., Simeonidou, C., Grigoriadis, S., Touloumi, O. and Shohami, E. (2008) Microglial Involvement in Neuroprotection Following Experimental Traumatic Brain Injury in Heat Acclimated Mice. *Brain Res.*, **1244**, 132–141.

Index

Stress – From Molecules to Behavior. Edited by Hermona Soreq, Alon Friedman, and Daniela Kaufer
Copyright © 2010 WILEY-VCH Verlag GmbH & Co. KGaA, Weinheim
ISBN: 978-3-527-32374-6